Don Hovda

Principles of Secure Communication Systems

The Artech House Communication and Electronic Defense Library

Principles of Secure Communication Systems by Don J. Torrieri
Introduction to Electronic Warfare by D. Curtis Schleher
Electronic Intelligence: The Analysis of Radar Signals by Richard G. Wiley
Electronic Intelligence: The Interception of Radar Signals by Richard G. Wiley
Signal Theory and Random Processes by Harry Urkowitz
Signal Theory and Processing by Frederic de Coulon
Digital Signal Processing by Murat Kunt
Analysis and Synthesis of Logic Systems by Daniel Mange
Mathematical Methods of Information Transmission by Kurt Arbenz and Jean-Claude Martin
Advanced Mathematics for Practicing Engineers by Kurt Arbenz and Alfred Wohlhauser
Codes for Error-Control and Synchronization by Djimitri Wiggert
Machine Cryptography and Modern Cryptanalysis by Cipher A. Deavours and Louis Kruh
Microcomputer Tools for Communication Engineering by S.T. Li, J.C. Logan, J.W. Rockway, and D.W.S. Tam

Principles of Secure Communication Systems

Don J. Torrieri

ARTECH HOUSE, INC.
685 Canton Street
Norwood, MA 02062

International Standard Book Number: 0-89006-139-4
Library of Congress Catalog Card Number: 85-47815

10 9 8 7 6 5 4 3 2

To Nancy

Contents

Appendices

Preface

Secure communication systems resist unintentional interference, jamming, detection by an opponent, and the unauthorized extraction of information from a transmitted waveform. They are epitomized by military communication systems, which must be designed with the presumption that they will operate in hostile environments. Special techniques that might be irrelevant or even harmful in ordinary communication systems are needed for systems that are to be viable under duress.

This book presents an analytical study of those parts of communication theory and other disciplines that have special relevance to secure communication systems. The book concentrates on principles, concepts, and systems-level analyses. Specific implementations and hardware, which are more susceptible to obsolescence due to the relentless progress of technology, are given much less emphasis.

This new edition is a major expansion and revision of the first edition (*Principles of Military Communication Systems*, Artech House, 1981). Much more material is presented, and the original material has been clarified and streamlined. The word "secure" has replaced the word "military" in the title to reflect and encourage the broadening of the potential applications of the subject. Among the additions are new or enhanced treatments of digital frequency modulation, error-correcting codes, direct-sequence and frequency-hopping acquisition and tracking systems, the Adcock array, the Howells-Applebaum algorithm, the combination of adaptive antenna systems and frequency hopping, the Data Encryption Standard, and analog cryptography.

This book is intended for graduate students and practicing engineers with a solid background in traditional communication theory. However, any prerequisites can be found in the cited textbooks. Signal representations that are used in Chapters 1 and 4 are derived in Appendix A. The Cramer-Rao bound and associated results, which are used in Chapter 4, are derived in Appendix B. Matrix analysis, which is extensively used in Chapter 5, is briefly summarized in Appendix C. Although an adequate mathematical background is necessary for a full appreciation of the issues, the reader can often skip the mathematical detail and concentrate on the final results without a great loss of understanding.

Chapters 1 to 4 are much more closely linked to the traditional concerns of communication engineers than the remaining two chapters. Chapters 4 to 6 are almost completely independent of Chapters 1 to 3 and of each other. With little or no modification, the results of Chapters 4 and 5 can be applied to radar signals and systems.

The prevailing terminology of various fields has been adopted when it does not conflict with standard English. Acronyms are used sparingly. The notation for the huge number of distinct mathematical entities reflects the delicate balancing of the demands of current usage, internal consistency, simplicity, and clarity. In choosing the references, preference was given to readily accessible textbooks and articles.

In undertaking this new edition, I was impelled by the explosion of publications and the growth of interest in secure communications. Much of the material was drawn from papers, reports, and notes prepared in the course of my work for the United States Army.

I am thankful to many friends for their direct and indirect assistance in the effort, particularly Djimitri Wiggert, who carefully reviewed the entire manuscript. I am grateful to my wife, Nancy, who provided me with unwavering support and endured all the inconveniences that accompanied the writing of this book. My daughter, Marisa, and son, Peter, provided the additional inspiration needed to bring the book to fruition.

Chapter 1
Modulation and Coding

1.1 MILITARY COMMUNICATIONS

Military communications pit potential communicators against hostile personnel who seek to intercept or disrupt their communications. Because the development of covert communication techniques has severely curtailed the possibility of intercepting and interpreting communications, it seems inevitable that military communications in the battlefield will be forced to operate in a jamming environment. However, to conserve power, the potential jammer usually must first detect the communications and perhaps locate the receivers.

To reduce the susceptibility of communications to interception and jamming, various general measures may be adopted. If transmission frequencies exceed 30 MHz, ionospheric reflections are small, thereby reducing long-range interception and jamming possibilities. Highly directional laser beams or millimeter waves are difficult to intercept. Their receivers are difficult to jam because of the narrow radiation patterns of the receiving antennas. Although atmospheric attenuation may hinder interception and jamming, attenuation and acquisition problems may limit the use of optical or millimeter-wave transmissions through the air to special applications such as satellite-to-satellite communications.

Cable communications by metallic wires, coaxial cables, and waveguides and by optical fibers provide advantages similar to those of directional beams, but without the acquisition problem. The main problems with cables are their impracticality in mobile, rapidly changing tactical environments and their susceptibility to damage.

Because *optical fibers* do not emit a significant amount of electromagnetic energy, they are very effective in preventing the interception of communications by an opponent. Tapping is more difficult than it is for a metallic cable. Because ambient electromagnetic energy does not interfere significantly with the propagation of optical waves in fibers, communication by means of optical fibers is nearly invulnerable to jamming. Other advantages of optical fibers are their light weight, resistance to fire, lack of crosstalk among fibers, and freedom from short circuits. Although it may not be necessary in many military communication systems, optical fibers can carry a much higher message density than metallic cables of comparable dimensions.

The use of *relays* increases the jamming resistance of a communication network. However, network management problems and equipment costs limit the number of relays in a network. If the communicators store, compress, and rapidly transmit all messages, effective jamming becomes less likely. Another tactic is to relocate one or more elements of a disrupted communication network to establish line-of-sight paths between the transmitters and the receivers and, if the jammer's location is known, to mask the receivers from the jammer by means of terrain obstacles.

If the communication frequencies can be accurately estimated, narrowband *spot jamming* can be used against the communicators. Spot jamming not only allows the economical use of power, but also helps the jammer to avoid jamming friendly communications. If frequency-estimation equipment is not used, or if accurate frequency estimation is impossible because of rapid frequency changes or other unfavorable conditions, the jammer may use *barrage jamming.* In this case, the jamming energy is spread over large spectral regions to increase the probability that some of the energy interferes with the enemy communications.

1.2 POWER AND PROPAGATION

The *interference-to-signal ratio* at the receiver is a basic quantity that is needed in the assessment of communication system performance. To relate this power ratio to other quantities, we examine the power relations at the receiver. The deployment of interest is depicted in Figure 1.1. The interference may be jamming or unintentional interference.

When two communicators are separated by a distance D_T, the average signal power at the input of a receiver is [1]

$$R_s = \frac{P_T G_{TR} G_{RT} \lambda^2}{(4\pi)^2 D_T^2 L_{TR}} \tag{1.2.1}$$

where P_T is the average transmitted power, G_{TR} is the gain of the transmitter antenna in the direction of the receiver, G_{RT} is the gain of the receiver antenna in the direction of the transmitter, and λ is the wavelength. The loss factor L_{TR} accounts for the deviation of the propagation from ideal free-space propagation. Other system losses are incorporated into L_{TR}, G_{RT}, or P_T. Unless the communicators are airborne and the atmospheric attenuation is negligible, L_{TR} is a function of D_T.

If one or both of the communication elements are on the ground, an accurate calculation of R_s must allow for the curvature of the spherical earth, the

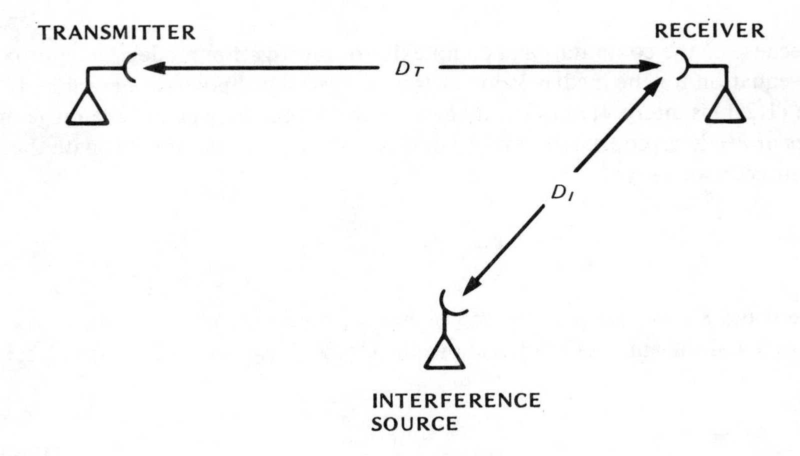

Figure 1.1 Deployment of communicators and interference source.

dielectric constant, the antenna heights, the presence of obstacles, multipath, and other effects. Many modern propagation models are based upon the *Longley-Rice model* [2], which gives computerized predictions of the transmission loss, $L = L_{TR}D_T^2$, over irregular terrain for frequencies between 20 MHz and 10 GHz.

Measurements of L or R_s for different paths of the same length often show great variation due to path-to-path differences in terrain profiles. Values recorded for a single path over a long period of time sometimes exhibit comparable ranges. The variability of L is strongly influenced by trees or buildings near the path terminals.

The Longley-Rice model can be used to calculate the median and the standard deviation of L in a geographical region. Experimental results suggest that L usually has a distribution that is approximately lognormal; thus, if L is measured in decibels, it has an approximately normal distribution. The standard deviation of L increases with frequency and terrain irregularity, but is relatively insensitive to path length and antenna heights [3]. Data from non-urban areas indicate that σ_L, the standard deviation of L in decibels, is approximated by

$$\sigma_L = \begin{cases} 6 + 0.55\sqrt{\dfrac{\Delta h}{\lambda}} - 0.004\,\dfrac{\Delta h}{\lambda}, & \dfrac{\Delta h}{\lambda} < 4700 \\ 24.9, & \dfrac{\Delta h}{\lambda} \geqslant 4700 \end{cases} \tag{1.2.2}$$

where Δh is the *interdecile range* of terrain elevations, which is the difference between the 90th and the 10th percentile terrain elevations.

For mobile communicators, the received signal power exhibits additional variation over short periods of time. This variation, which is called fading, is caused by time-varying multipath effects.

Because of the computational complexity of propagation models, an approximate equation for the median value of the received signal power is desirable. If R_s in (1.2.1) is interpreted as the median received signal power and the range of values of D_T is appropriately restricted, it is often possible to approximate the transmission losses by

$$L = L_{TR} D_T^2 \cong \frac{D_T^n}{K_T} \tag{1.2.3}$$

where n and K_T are independent of D_T, but are functions of other parameters such as λ and the antenna heights above the earth. Equations (1.2.1) and (1.2.3) yield

$$R_s = \frac{K_T P_T G_{TR} G_{RT} \lambda^2}{(4\pi)^2 D_T^n} \tag{1.2.4}$$

To use this equation, we need specific values for K_T and n. These values can be determined from experimental data or data calculated from a propagation model. Equation (1.2.4) indicates that n satisfies

$$n = -\frac{\partial \log R_s}{\partial \log D_T} \tag{1.2.5}$$

Thus, a consistent way to estimate n is to calculate this partial derivative from a graph of R_s in decibels *versus* D_T in decibels. Because this partial derivative usually tends to increase with D_T, the range must be subdivided into small enough pieces that each piece can be associated with a nearly constant value of n. Once n is determined for a range of D_T, the corresponding value of K_T can be calculated by fitting (1.2.4) to a single point of the R_s *versus* D_T curve within that range.

At optical and millimeter-wave frequencies, the severe atmospheric attenuation can be more accurately modeled by replacing (1.2.4) with

$$R_s = \frac{K_T P_T G_{TR} G_{RT} \lambda^2}{(4\pi)^2 D_T^n} \exp(-\alpha D_T) \tag{1.2.6}$$

where $\alpha \geqslant 0$. The parameters K_T, n, and α are probably best estimated by regression analysis.

The interference power that enters a receiver may be reduced by two major factors. First, there is a *polarization loss* due to the fact that the interference may not have the appropriate polarization. This relative polarization loss may be described by a coefficient p, which has the range $0 \leqslant p \leqslant 1$. A second power reduction may be caused by the receiver *bandpass filtering.* The effect of this filtering usually can be described by a function $g(B_R, B_I)$, where B_R is the bandwidth of the effective receiver bandpass filter, and B_I is the bandwidth of the interference signal. If the entire interference spectrum is included in the receiver passband,

$$g(B_R, B_I) = 1 \tag{1.2.7}$$

If the spectrum includes the entire receiver passband,

$$g(B_R, B_I) = \frac{B_R}{B_I} \tag{1.2.8}$$

The median value of the net interference power affecting the receiver can often be approximated by

$$R_i = \frac{K_I P_I G_{IR} G_{RI} \lambda^2 p g(B_R, B_I)}{(4\pi)^2 D_I^m} \exp(-\beta D_I) \tag{1.2.9}$$

where D_I is the distance between the interferer and the receiver, P_I is the average transmitted power, G_{IR} is the gain of the interferer's antenna in the direction of the receiver, and G_{RI} is the gain of the receiver antenna in the direction of the interferer. The parameters K_I, m, and β are defined analogously to K_T, n, and α, respectively. However, the former parameters may have different values from the latter ones because, for example, the communicators are on the earth while the interferer is airborne. If the average wavelengths of the interference and the desired signal are approximately equal, then an approximate median value of the interference-to-signal ratio is determined from (1.2.4) and (1.2.9) to be

$$\frac{R_i}{R_s} = \frac{P_I K_I G_{IR} G_{RI} p g(B_R, B_I)}{P_T K_T G_{TR} G_{RT}} \left(\frac{D_T^n}{D_I^m}\right) \exp(\alpha D_T - \beta D_I) \tag{1.2.10}$$

For specific locations of the communicators and the interferer, the interference-to-signal ratios may differ from the median value of (1.2.10) because of the particular terrain profiles.

Large propagation losses sometimes provide a natural countermeasure to interference. For example, if $n = m$, $\alpha = \beta$, and $D_T < D_I$, then (1.2.10) and calculus indicate that the interference-to-signal ratio decreases as n and α increase.

If R_i and R_s are measured in decibels relative to some level and are modeled as independent random variables over a geographical region, then the variance of the interference-to-signal ratio is equal to the sum of the variances of R_i and R_s. The variance of R_i or R_s due to propagation effects is given by (1.2.2), but there may be other sources of variance in a practical communication model

[4, 5]. For example, it may be a practical necessity to model an antenna radiation pattern as a random variable rather than as a deterministic function because of the complexity or uncertainty of the pattern.

According to information theory, the most destructive type of additive noise in a communication channel is *white Gaussian noise.* Thus, it is often desirable for a jammer to produce a facsimile of bandlimited white Gaussian noise, although it may be difficult to synthesize a waveform with the large voltage swings of this type of noise. To the extent that interference can be modeled as white Gaussian noise, well-known theoretical formulas can often be used to determine its impact on communications.

Although the effects of non-Gaussian interference are usually studied through computer simulations, an approximate mathematical analysis is sometimes feasible [6-8]. Although such analyses are necessarily limited in scope, they provide valuable insight into the characteristics of degraded communications and the countermeasures that can be used against the interference.

1.3 ANALOG COMMUNICATIONS

A general received waveform, which includes *amplitude-modulation* (AM), *phase-modulation* (PM), and *frequency-modulation* (FM) waveforms as special cases, is

$$s(t) = A(t)\cos[\omega_1 t + \phi_1(t)] \tag{1.3.1}$$

where $A(t)$ is the amplitude modulation, ω_1 is the carrier frequency, and $\phi_1(t)$ is the angle modulation. Consider an analog receiver with an initial bandpass filter that passes $s(t)$ with negligible distortion. A general form for the interference waveform at the receiver input is

$$i(t) = B(t)\cos[\omega_2 t + \phi_2(t)] \tag{1.3.2}$$

It is assumed that $i(t)$ passes the receiver bandpass filter with negligible distortion. Alternatively, $i(t)$ may be viewed as the description of the interference at the filter output.

Throughout the subsequent analysis, we neglect the effect of thermal noise for simplicity. Consequently, the signal at the output of the bandpass filter is $X(t) = s(t) + i(t)$. Using a trigonometric identity to expand the cosine term of (1.3.2), we obtain

$$\begin{aligned} X(t) = &\left\{A(t) + B(t)\cos[\omega_3 t + \phi_3(t)]\right\} \cos[\omega_1 t + \phi_1(t)] \\ &- B(t)\sin[\omega_3 t + \phi_3(t)]\sin[\omega_1 t + \phi_1(t)] \end{aligned} \tag{1.3.3}$$

where $\omega_3 = \omega_2 - \omega_1$ and $\phi_3 = \phi_2 - \phi_1$. Further trigonometry yields

$$X(t) = R(t)\cos[\omega_1 t + \phi_1(t) + \theta(t)] \tag{1.3.4}$$

where

$$R(t) = \left\{ A^2(t) + B^2(t) + 2A(t)B(t)\cos[\omega_3 t + \phi_3(t)] \right\}^{1/2} \tag{1.3.5}$$

and

$$\theta(t) = \tan^{-1} \left\{ \frac{B(t)\sin[\omega_3 t + \phi_3(t)]}{A(t) + B(t)\cos[\omega_3 t + \phi_3(t)]} \right\} \tag{1.3.6}$$

1.3.1 AM Systems

When an AM signal is transmitted, $\phi_1(t) = 0$, $\phi_3(t) = \phi_2(t)$, and the message is carried by $A(t)$. Consider a noncoherent system in which the receiver demodulates by means of an envelope detector. An ideal envelope detector produces an output proportional to the instantaneous amplitude, $R(t)$, if it is slowly varying relative to $\omega_1 t$. Assuming this ideal operation and $A(t) > 0$, the envelope detector output is proportional to

$$y(t) = A(t) \left\{ 1 + 2\frac{B(t)}{A(t)} \cos[\omega_3 t + \phi_2(t)] + \frac{B^2(t)}{A^2(t)} \right\}^{1/2} \tag{1.3.7}$$

We expand the square root as a Taylor series in the parameter $B(t)/A(t)$ about the origin. Only the first three terms of the expansion are retained. This truncation gives a reasonably small error if $A(t) \geqslant 2|B(t)|$ for most times of interest. The use of a trigonometric identity in the expansion yields

$$y(t) \approx A(t) + B(t)\cos[\omega_3 t + \phi_2(t)] + B(t) \left\{ \frac{B(t)}{4A(t)} - \frac{B(t)}{4A(t)} \cos[2\omega_3 t + 2\phi_2(t)] \right\} \tag{1.3.8}$$

If we set $\phi_2(t)$ equal to a constant, the effectiveness of the interference is not greatly impaired unless $\omega_3 \approx 0$ and $\phi_2 \approx \pi/2$. Furthermore, the spectral bandwidth of the interference is decreased, so that the receiver bandpass filter can block less power before it reaches the envelope detector. Thus, an AM signal is usually a suitable choice of waveform for jamming AM communications. If both $\phi_2(t)$ and $B(t)$ are constants, the energy in the interference terms of (1.3.8) can be significantly reduced by a blocking capacitor when $\omega_3 \approx 0$. Consequently, a *tone* (unmodulated carrier) is not a satisfactory jamming signal.

Ideal coherent demodulation of the intended signal is accomplished when $X(t)$ is multiplied by $2\cos(\omega_1 t)$ and the double-frequency terms are removed by a filter. From (1.3.3) with $\phi_1(t) = 0$, we find that the output is

$$y(t) = A(t) + B(t)\cos[\omega_3 t + \phi_2(t)] \tag{1.3.9}$$

This expression can be compared with (1.3.8) to see the effects of the highly nonlinear operation of envelope detection. The main problem with coherent demodulation is the need for a locally generated phase-coherent reference. The carrier synchronization system of the receiver is subject to degradation due to jamming. If the carrier component is transmitted, it is susceptible to detection by an opponent, who may be able to produce more effective jamming once the carrier frequency has been determined.

1.3.2 PM Systems

When a PM signal is transmitted, we have $A(t) = A$, a positive constant. The message is carried by $\phi_1(t)$. The output of an ideal PM discriminator with $X(t)$ as an input is proportional to $\phi_1(t) + \theta(t)$ if the instantaneous amplitude, $R(t)$, is slowly varying as a function of time, which may require that the discriminator be preceded by a bandpass limiter. From (1.3.4) and (1.3.6), we find that the ideal discriminator output is proportional to

$$y(t) = \phi_1(t) + \tan^{-1}\left\{\frac{B(t)\sin[\omega_3 t + \phi_3(t)]}{A + B(t)\cos[\omega_3 t + \phi_3(t)]}\right\} \tag{1.3.10}$$

We expand the arctangent as a Taylor series in the parameter $B(t)/A$ about the origin and retain the first three terms. This truncation gives a reasonably small error if $A \geqslant 2|B(t)|$ for most times of interest. Simple trigonometry yields

$$y(t) \approx \phi_1(t) + \frac{B(t)}{A}\sin[\omega_3 t + \phi_3(t)] - \frac{1}{2}\left[\frac{B(t)}{A}\right]^2 \sin[2\omega_3 t + 2\phi_3(t)] \tag{1.3.11}$$

If ω_3 exceeds the message bandwidth, the post-detection filter removes most of the power in the last two terms. Thus, although the bandwidth of the initial band-pass filter may be greater than in the AM case, it is important for a jammer to accurately estimate the receiver center frequency. Assuming that ω_3 does not exceed the message bandwidth, there appears to be little loss in jamming effectiveness if $\phi_2(t)$ is a constant, because the interference terms in (1.3.11) are still phase modulated by $\phi_1(t)$ and $\phi_3 = \phi_2 - \phi_1$. We conclude that a PM jamming waveform offers no particular advantage in the jamming of a PM communication system.

If we set both $B(t)$ and $\phi_2(t)$ equal to constants, the jamming effectiveness does not appear to be seriously impaired. Thus, a tone is often a satisfactory jamming signal against a PM communication system.

1.3.3 FM Systems

An FM transmission may be described by (1.3.1) with $A(t) = A$, a positive constant. The message is carried by the derivative of the phase function, which is denoted by $\phi_1'(t)$. Under the same assumptions made in the PM case, the output of an ideal FM discriminator is proportional to $\phi_1'(t) + \theta'(t)$, where the second term is the derivative of (1.3.6). A straightforward calculation yields

$$y(t) = \phi_1'(t) + \frac{[\omega_3 + \phi_3'(t)]\left\{AB(t)\cos[\omega_3 t + \phi_3(t)] + B^2(t)\right\}}{A^2 + B^2(t) + 2AB(t)\cos[\omega_3 t + \phi_3(t)]} \tag{1.3.12}$$

As usual, we approximate the second term in this expression by the first three terms of a Taylor series, assuming that $A \geqslant 2|B(t)|$ for most times of interest. The result is

$$y(t) \approx \phi_1'(t) + [\omega_3 + \phi_3'(t)]\,\frac{B(t)}{A}\left\{\cos[\omega_3 t + \phi_3(t)] - \frac{B(t)}{A}\cos[2\omega_3 t + 2\phi_3(t)]\right\} \tag{1.3.13}$$

By reasoning similar to that used in the PM case, we draw analogous conclusions. An FM jamming waveform offers no compelling advantage in the jamming of an FM communication system; a tone is often satisfactory.

Equation (1.3.13) indicates that the interference at the receiver output increases with the frequency offset ω_3. Comparing (1.3.11) and (1.3.13) reveals that interference is more damaging to FM systems than to PM systems when the frequency offset is large but less than the bandwidth of the post-detection filter. However, if we use a *de-emphasis filter* with a transfer function that decreases as ω^{-1} for large frequencies, the effect of the interference does not increase with frequency offset. To compensate for the distortion of the message due to the presence of the de-emphasis filter, the message should be modified by a *pre-emphasis filter* before transmission, as shown in Figure 1.2. The pre-emphasis filter should have a transfer function equal to the reciprocal of the transfer function of the de-emphasis filter so that the demodulated message is unchanged [9, 10].

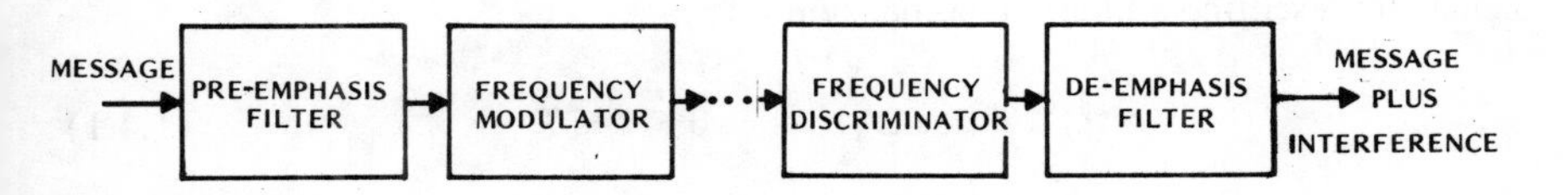

Figure 1.2 FM system that resists interference.

Careful design is required when this pair of filters is used. The pre-emphasis filter amplifies high-frequency spectral components of the message, thereby increasing the bandwidth required for transmission.

It has been shown in this section that jamming effectiveness against analog communications is not dependent upon a similarity between the intended signal and the jamming waveform. An AM jamming waveform is usually effective against AM, PM, and FM systems. A tone is often a satisfactory choice for jamming PM and FM systems.

As the jamming power is increased, angle-modulation systems and noncoherent AM systems are susceptible to sudden disruptions due to well-known threshold effects [9, 10]. When the jamming has the appropriate form and its power is so great that the receiver responds to the jamming rather than to the intended signal, the receiver is said to have been *captured.* Generally, complete disruption of angle-modulation systems requires less power than complete disruption of AM systems.

Analog communication systems are declining in importance relative to digital communication systems for military applications. The two most important reasons are probably the proliferation of the digital computer and the increased security provided by cryptographic digital communications.

1.4 DIGITAL COMMUNICATIONS

To analyze the performance of various digital communication systems, it is usually necessary to model the interference as either noise interference or tone interference. *Noise interference* is interference that approximates stationary Gaussian noise, whereas *tone interference* approximates a tone (unmodulated carrier).

1.4.1 Noncoherent FSK Systems

A noncoherent demodulator for a binary *frequency-shift-keying* (FSK) system is illustrated in Figure 1.3. At the demodulator input, the two possible intended signals representing a binary channel symbol are

$$s_1(t) = A \cos \omega_1 t \,, \qquad 0 \leqslant t \leqslant T_s \tag{1.4.1}$$

$$s_2(t) = A \cos \omega_2 t \,, \qquad 0 \leqslant t \leqslant T_s \tag{1.4.2}$$

where A is a constant amplitude, T_s is the symbol duration, and ω_1 and ω_2 are the center frequencies of bandpass filter 1 and bandpass filter 2, respectively. Each of these spectrally disjoint filters has a passband that accommodates one of the possible intended signals with negligible distortion, while completely rejecting the other. The filters have the same spectral shape and bandwidth, but differ in center frequency. The channel symbol may be an information bit or a binary code symbol.

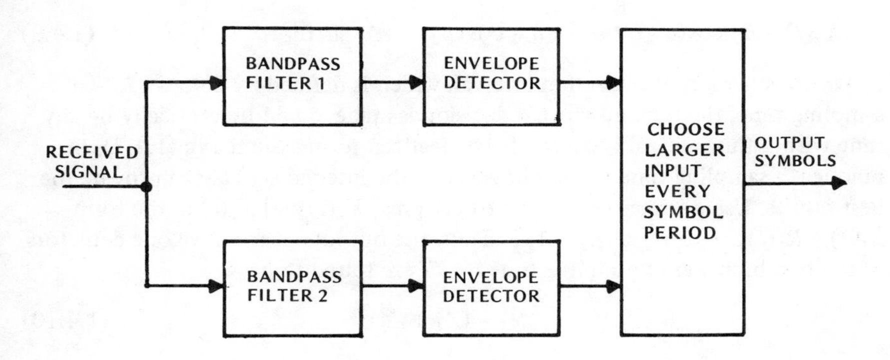

Figure 1.3 Noncoherent frequency-shift-keying demodulator.

Let N_1 represent the power of wide-sense-stationary Gaussian noise at the output of bandpass filter 1, and let N_2 be defined similarly. Some of the noise power is due to noise interference. The remainder is due to thermal and background noise, which is assumed to have the same power, denoted by N_t, at each filter output. Because the noise interference is statistically independent of the thermal and background noise,

$$N_1 = N_t + N_{i1} \tag{1.4.3}$$

$$N_2 = N_t + N_{i2} \tag{1.4.4}$$

where N_{i1} and N_{i2} are interference powers.

At the output of bandpass filter i, the stationary Gaussian noise has the narrowband representation (Appendix A):

$$n_j(t) = n_{cj}(t)\cos\omega_j t - n_{sj}(t)\sin\omega_j t\,, \qquad j = 1, 2 \tag{1.4.5}$$

where ω_j is the center frequency and $n_{cj}(t)$ and $n_{sj}(t)$ are Gaussian processes with noise powers equal to N_j. We assume that interference signals of the form

$$i_j(t) = B_j(t)\cos[\omega_j t + \phi_j(t)]\,, \qquad j = 1, 2 \tag{1.4.6}$$

emerge from the bandpass filters. Suppose that a binary symbol represented by

$$s_1(t) = A\cos\omega_1 t \tag{1.4.7}$$

is received and passes through bandpass filter 1 with negligible distortion. The total signals at the outputs of the two bandpass filters are

$$X_1(t) = A_1 \cos \omega_1 t + B_1 \cos(\omega_1 t + \phi_1) + n_{c1} \cos \omega_1 t - n_{s1} \sin \omega_1 t \qquad (1.4.8)$$

$$X_2(t) = B_2 \cos(\omega_2 t + \phi_2) + n_{c2} \cos \omega_2 t - n_{c2} \sin \omega_2 t \qquad (1.4.9)$$

We consider a typical symbol interval, which is defined by $0 \leqslant t \leqslant T_s$. The sampling time, the time at which a decision is made, could theoretically be any time within this interval because of the idealized form assumed in (1.4.7). In practice, a sampling time at the midpoint of the interval is likely to provide the best results. Using trigonometry, we can express $X_1(t)$ and $X_2(t)$ in the form $X_j(t) = R_j(t)\cos(\omega_j t + \psi_j), j = 1, 2$. Thus, the outputs of the envelope detectors of the two branches at sampling time $t = T_1$ are found to be

$$R_1 = (Z_1^2 + Z_2^2)^{1/2} \qquad (1.4.10)$$

$$R_2 = (Z_3^2 + Z_4^2)^{1/2} \qquad (1.4.11)$$

where the following definitions are made for notational convenience:

$$Z_1 = A + B_1(T_1)\cos[\phi_1(T_1)] + n_{c1}(T_1) \qquad (1.4.12)$$

$$Z_2 = B_1(T_1)\sin[\phi_1(T_1)] + n_{s1}(T_1) \qquad (1.4.13)$$

$$Z_3 = B_2(T_1)\cos[\phi_2(T_1)] + n_{c2}(T_1) \qquad (1.4.14)$$

$$Z_4 = B_2(T_1)\sin[\phi_2(T_1)] + n_{s2}(T_1) \qquad (1.4.15)$$

Since the $n_i(t)$ are assumed to be zero-mean processes, all the noise variables in (1.4.12) to (1.4.15) are zero mean. Denoting the expected value of Z_j by M_j for $j = 1, 2, 3, 4$, we have

$$M_1 = A + B_1(T_1)\cos[\phi_1(T_1)] \qquad (1.4.16)$$

$$M_2 = B_1(T_1)\sin[\phi_1(T_1)] \qquad (1.4.17)$$

$$M_3 = B_2(T_1)\cos[\phi_2(T_1)] \qquad (1.4.18)$$

$$M_4 = B_2(T_1)\sin[\phi_2(T_1)] \qquad (1.4.19)$$

Assuming that $B_1(T_1)$ and $\phi_1(T_1)$ are given, the joint probability density function of Z_1 and Z_2 is

$$g_1(z_1, z_2) = \frac{1}{2\pi N_1} \exp\left[- \frac{(z_1 - M_1)^2 + (z_2 - M_2)^2}{2N_1} \right] \qquad (1.4.20)$$

which follows from the statistical independence of $n_{c1}(T_1)$ and $n_{s1}(T_1)$ (Appendix A). Let R_1 and Θ be implicitly defined by $Z_1 = R_1 \cos\Theta$ and $Z_2 = R_1 \sin\Theta$. It follows that the joint density of R_1 and Θ is

$$g_2(r_1, \theta) = \frac{r_1}{2\pi N_1} \exp\left(-\frac{r_1^2 - 2r_1M_1\cos\theta - 2r_1M_2\sin\theta + M_1^2 + M_2^2}{2N_1}\right)$$

$$r_1 \geqslant 0, \ |\theta| \leqslant \pi \tag{1.4.21}$$

The density of the envelope R_1 is obtained by integration over θ. Because the modified Bessel function of the first kind and zero order satisfies

$$I_0(x) = \frac{1}{2\pi}\int_0^{2\pi} \exp[x\cos(u+v)]\, du \tag{1.4.22}$$

regardless of the value of v, the integral of (1.4.21) over θ can be reduced, after suitable trigonometric manipulation, to the density

$$f_1(r_1) = \frac{r_1}{N_1}\exp\left(-\frac{D_1^2 + r_1^2}{2N_1}\right) I_0\left(\frac{D_1 r_1}{N_1}\right), \quad r_1 \geqslant 0 \tag{1.4.23}$$

where

$$D_1^2 = M_1^2 + M_2^2 = A^2 + B_1^2(T_1) + 2AB_1(T_1)\cos[\phi_1(T_1)] \tag{1.4.24}$$

Similarly, the output at $t = T_1$ of the envelope detector in the other branch of the FSK demodulator has the density

$$f_2(r_2) = \frac{r_2}{N_2}\exp\left(-\frac{B_2^2 + r_2^2}{2N_2}\right) I_0\left(\frac{B_2 r_2}{N_2}\right), \quad r_2 \geqslant 0 \tag{1.4.25}$$

Because $s_1(t)$ has been transmitted, an error occurs in a demodulated symbol if $R_2 > R_1$. A straightforward derivation proves that the cross-covariance of $X_1(t)$ and $X_2(t)$ is zero because the bandpass filters are disjoint. Because $X_1(T_1)$ and $X_2(T_1)$ are Gaussian random variables, they are statistically independent. Hence, the envelopes R_1 and R_2 are statistically independent and the probability of an error is

$$P(E/1) = \int_0^{\infty} f_1(r_1)\left[\int_{r_1}^{\infty} f_2(r_2)\, dr_2\right] dr_1 \tag{1.4.26}$$

Substituting (1.4.23) and (1.4.25) into (1.4.26), we obtain

$$P(E/1) = \int_0^\infty q\left(\frac{D_1}{\sqrt{N_1}}, x\right) Q\left(\frac{B_2}{\sqrt{N_2}}, \frac{\sqrt{N_1}\; x}{\sqrt{N_2}}\right) dx \tag{1.4.27}$$

where

$$q(\alpha, x) = x \exp\left(-\frac{x^2 + \alpha^2}{2}\right) I_0(\alpha x) \tag{1.4.28}$$

and the *Q-function* is defined by

$$Q(\alpha, \beta) = \int_\beta^\infty q(\alpha, x)\, dx \tag{1.4.29}$$

The integral in (1.4.27) can be evaluated by using the identity [11]

$$\int_0^\infty q(a, x) Q(b, rx)\, dx = Q(v_2, v_1) - \frac{r^2}{1 + r^2} \exp\left[-\frac{a^2 r^2 + b^2}{2(1 + r^2)}\right] I_0\left(\frac{abr}{1 + r^2}\right) \tag{1.4.30}$$

where

$$v_1 = \frac{ar}{\sqrt{1 + r^2}}, \qquad v_2 = \frac{b}{\sqrt{1 + r^2}} \tag{1.4.31}$$

Carrying out the algebra, we obtain

$$P(E/1) = Q\left(\frac{B_2}{\sqrt{N_1 + N_2}}, \frac{D_1}{\sqrt{N_1 + N_2}}\right) - \frac{N_1}{N_1 + N_2} \times \exp\left[-\frac{B_2^2 + D_1^2}{2(N_1 + N_2)}\right] I_0\left(\frac{B_2 D_1}{N_1 + N_2}\right) \tag{1.4.32}$$

This expression gives $P(E/1)$ for fixed values of $B_1(T_1)$, $B_2(T_1)$, and $\phi_1(T_1)$. From symbol interval to symbol interval, these parameters generally vary in value. If these parameters are modeled as random variables, an aggregate $P(E/1)$ can be calculated by integrating the product of (1.4.32) and the joint density of the parameters. To obtain reasonably simple results, we assume that narrowband angle-modulated interference is present so that for $j = 1, 2$, $B_j(t) = B_j(T_1) = B_j$, a constant. If $\phi_1(t)$ is nonsynchronous with the carrier frequency of $s_1(t)$, it is logical to model $\phi_1(T_1)$ as uniformly distributed from 0 to 2π radians. Thus, the aggregate probability of error, given that $s_1(t)$ was transmitted, is

$$\bar{P}(E/1) = \frac{1}{2\pi} \int_0^{2\pi} P(E/1)\, d\phi_1 \tag{1.4.33}$$

If the interference is a tone (continuous-wave interference), then $\phi_1(t) = \Delta_1 t + \psi_1$, where Δ_1 is the frequency offset of the tone and ψ_1 is a constant.

When a symbol represented by

$$s_2(t) = A \cos \omega_2 t \tag{1.4.34}$$

is received, the error probabilities can be determined by an analogous procedure. Defining

$$D_2^2 = A^2 + B_2^2(T_1) + 2AB_2(T_1) \cos[\phi_2(T_1)] \tag{1.4.35}$$

we obtain

$$P(E/2) = Q\left(\frac{B_1}{\sqrt{N_1 + N_2}}, \frac{D_2}{\sqrt{N_1 + N_2}}\right) - \frac{N_2}{N_1 + N_2} \times \exp\left[-\frac{B_1^2 + D_2^2}{2(N_1 + N_2)}\right] I_0\left(\frac{B_1 D_2}{N_1 + N_2}\right) \tag{1.4.36}$$

If for $j = 1, 2$, $B_j(t) = B_j(T_1) = B_j$, a constant, and $\phi_2(T_1)$ is uniformly distributed, then

$$\bar{P}(E/2) = \frac{1}{2\pi} \int_0^{2\pi} P(E/2)\, d\phi_2 \tag{1.4.37}$$

If the transmission of $s_1(t)$ or $s_2(t)$ is equally likely, the probability of a symbol error is

$$P_e = \frac{1}{2} \bar{P}(E/1) + \frac{1}{2} \bar{P}(E/2) \tag{1.4.38}$$

Substitution of the previous equations into (1.4.38) yields

$$P_e = \frac{1}{4\pi}\int_0^{2\pi} dx \left\{ Q\left[\frac{B_2}{\sqrt{N_1+N_2}}, \frac{D_1(x)}{\sqrt{N_1+N_2}}\right] + Q\left[\frac{B_1}{\sqrt{N_1+N_2}}, \frac{D_2(x)}{\sqrt{N_1+N_2}}\right] - \frac{N_1}{N_1+N_2} \times \exp\left[-\frac{B_2^2+D_1^2(x)}{2(N_1+N_2)}\right] I_0\left[\frac{B_2 D_1(x)}{N_1+N_2}\right] - \frac{N_2}{N_1+N_2} \times \exp\left[-\frac{B_1^2+D_2^2(x)}{2(N_1+N_2)}\right] I_0\left[\frac{B_1 D_2(x)}{N_1+N_2}\right]\right\} \tag{1.4.39}$$

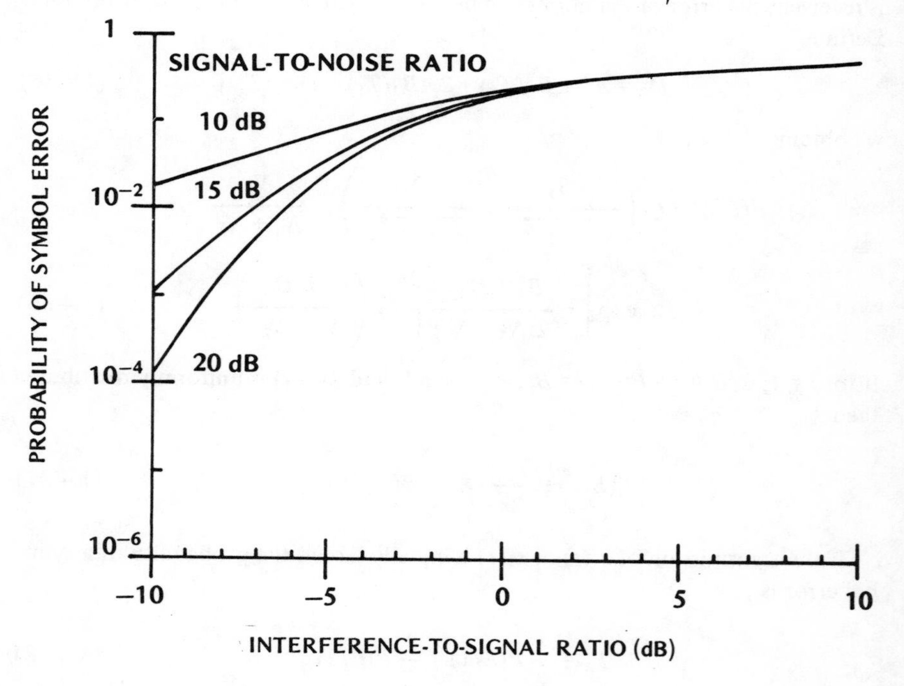

Figure 1.4 Symbol error probability for noise interference.

where

$$D_j^2(x) = A^2 + B_j^2 + 2AB_j \cos x\,, \qquad j = 1, 2 \tag{1.4.40}$$

If only noise interference is present, (1.4.39) greatly simplifies. We set $B_1 = B_2 = 0$ and $A = \sqrt{2R_s}$, where R_s is the average power of the intended signal. Using $I_0(0) = 1$ and

$$Q(0,\beta) = \exp\left(-\frac{\beta^2}{2}\right) \tag{1.4.41}$$

and substituting (1.4.3) and (1.4.4), we obtain the symbol error probability in the presence of noise interference:

$$P_s = \frac{1}{2}\exp\left(-\frac{R_s}{2N_t + N_{i1} + N_{i2}}\right) \tag{1.4.42}$$

Because P_s is a function of the sum $N_{i1} + N_{i2}$, the distribution of the total interference power between the two different filter passbands or channels is irrelevant. Figure 1.4 plots P_s *versus* the interference-to-signal ratio, $(N_{i1} + N_{i2})/R_s$, for various fixed values of the signal-to-noise ratio, R_s/N_t.

If only narrowband angle-modulated or tone interference is present [7], then $N_{i1} = N_{i2} = 0$ and $N_1 = N_2 = N_t$. It is convenient to refer to the passband of each bandpass filter as a channel. Let S_2 denote the symbol error probability given that an equal amount of interference power, R_i, enters both channels. Setting $B_1 = B_2 = \sqrt{2R_i}$ and $A = \sqrt{2R_s}$ in (1.4.39) and (1.4.40), we obtain the symbol error probability in the presence of tone interference:

$$S_2 = \frac{1}{2\pi}\int_0^{2\pi} dx \left\{ Q\left[\sqrt{\frac{R_i}{N_t}}, \frac{D(x)}{\sqrt{2N_t}}\right] - \frac{1}{2}\exp\left[-\frac{2R_i + D^2(x)}{4N_t}\right] I_0\left[\frac{\sqrt{2R_i}\,D(x)}{2N_t}\right]\right\} \tag{1.4.43}$$

where

$$D^2(x) = 2R_s + 2R_i + 4\sqrt{R_s R_i}\,\cos x \tag{1.4.44}$$

Figure 1.5 depicts S_2 *versus* the interference-to-signal ratio per channel, R_i/R_s, for various fixed values of R_s/N_t.

The symbol error probability given that tone interference enters a single channel is denoted by S_1 and is determined by setting $B_1 = \sqrt{2R_i}$, $B_2 = 0$, $A = \sqrt{2R_s}$, and $N_1 = N_2 = N_t$ in (1.4.39) and (1.4.40). Equations (1.4.22) and

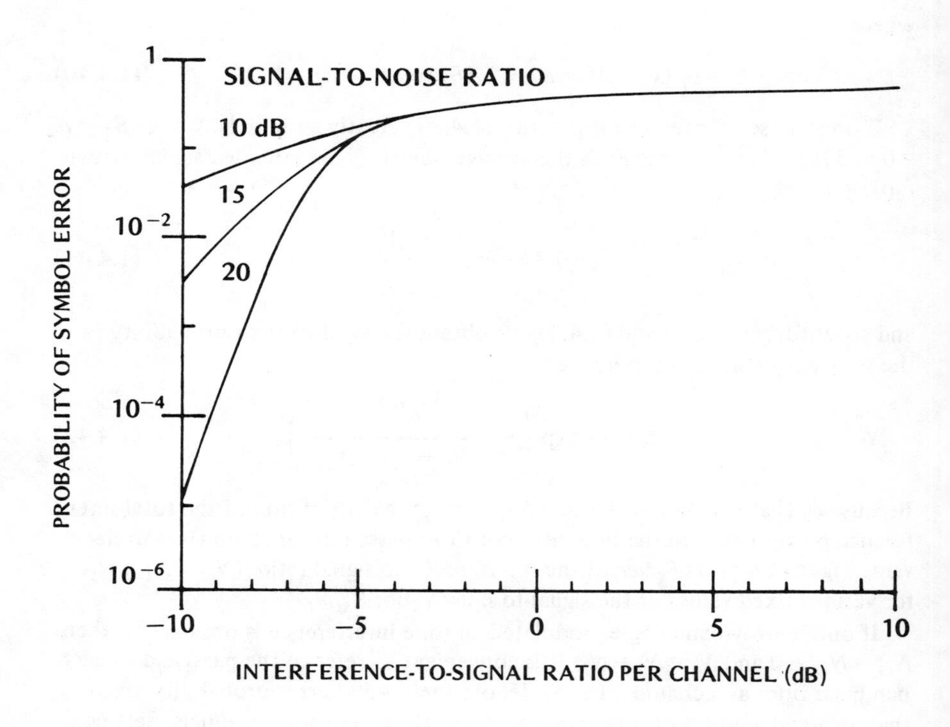

Figure 1.5 Symbol error probability for tone interference in both channels.

(1.4.41) help to simplify the result, which is

$$S_1 = \frac{1}{2} Q\left(\sqrt{\frac{R_i}{N_t}}, \sqrt{\frac{R_s}{N_t}}\right) \tag{1.4.45}$$

Figure 1.6 depicts S_1 *versus* the interference-to-signal ratio, R_i/R_s.

Figures 1.4 to 1.6 indicate that noise interference is more damaging than tone interference unless the symbol error probability is on the order of 10^{-1} or more. Since most practical communication systems do not operate satisfactorily when the symbol error probability exceeds 10^{-1}, it is usually not necessary for a jammer to drive the symbol error probability beyond this point. Thus, noise jamming is usually preferable to the jammer who does not know the jamming-to-signal ratio at the receiver. Figures 1.7 and 1.8 compare the effects of noise and tone interference when a single channel or both channels receive interference, respectively.

A comparison of Figures 1.5 and 1.6 indicates that for tone jamming of fixed total power, the jamming of both channels is usually preferable to the jamming of a single channel. Only when the symbol error probability exceeds 10^{-1} is the

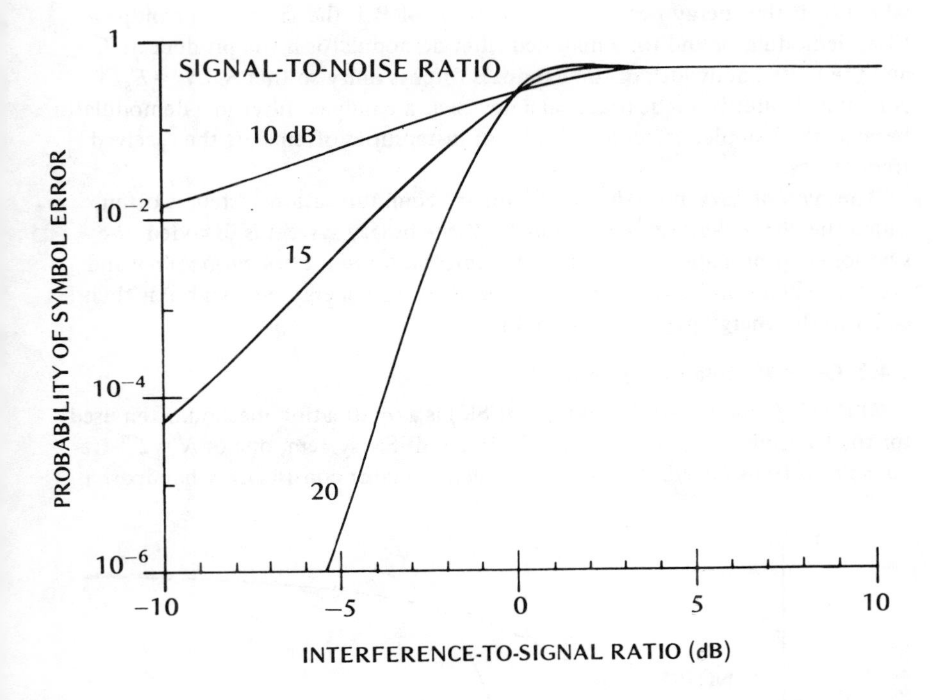

Figure 1.6 Symbol error probability for tone interference in a single channel.

jamming of a single channel more effective. Figure 1.9 compares the effects of tone interference in a single channel and in both channels as a function of the total interference-to-signal ratio.

An ideal noncoherent FSK demodulator has *matched filters* in place of the bandpass filters in Figure 1.3. The matched filters are matched to the signal described by (1.4.1) and (1.4.2). For orthogonal signaling, the two transmission frequencies satisfy $f_d = (\omega_1 - \omega_2)/2\pi = k/2T_s$, where k is a nonzero integer, or $|f_d| \gg 1/T_s$. Suppose that noise interference at each matched-filter input can be modeled as bandlimited white Gaussian noise over the filter passband. Let $N_0/2$ denote the two-sided power spectral density of the thermal and background noise. Let $N_{01}/2$ and $N_{02}/2$ denote the two-sided power spectral densities of the noise interference at matched filters 1 and 2, respectively. A derivation following along classical lines [10], but preserving the distinction between N_{01} and N_{02}, yields the symbol error probability

$$P_s = \frac{1}{2} \exp\left(- \frac{E_s}{2N_0 + N_{01} + N_{02}} \right) \tag{1.4.46}$$

where E_s is the energy per symbol. The value of P_s is the same for a bandpass-filter demodulator and for a matched-filter demodulator if the product of T_s and the noise bandwidth of the bandpass filter is unity so that $R_s/N_t = E_s/N_0$. A matched filter is not as practical a device as a bandpass filter in a demodulator because the Doppler effect and oscillator instabilities often shift the received frequencies.

The *symbol error probability* of a binary communication system is often called the *channel-bit error probability.* If the binary system is uncoded, the symbol error probability is equal to the *information-bit error probability* and is often called simply the *bit error probability.* The energy per symbol is then equal to the energy per information bit.

1.4.2 Other Modulation Systems

Multiple frequency-shift keying (MFSK) is a modulation method often used for the transmission of m-bit symbols. In an MFSK system, one of $N = 2^m$ frequencies is transmitted. A noncoherent demodulator consists of N bandpass

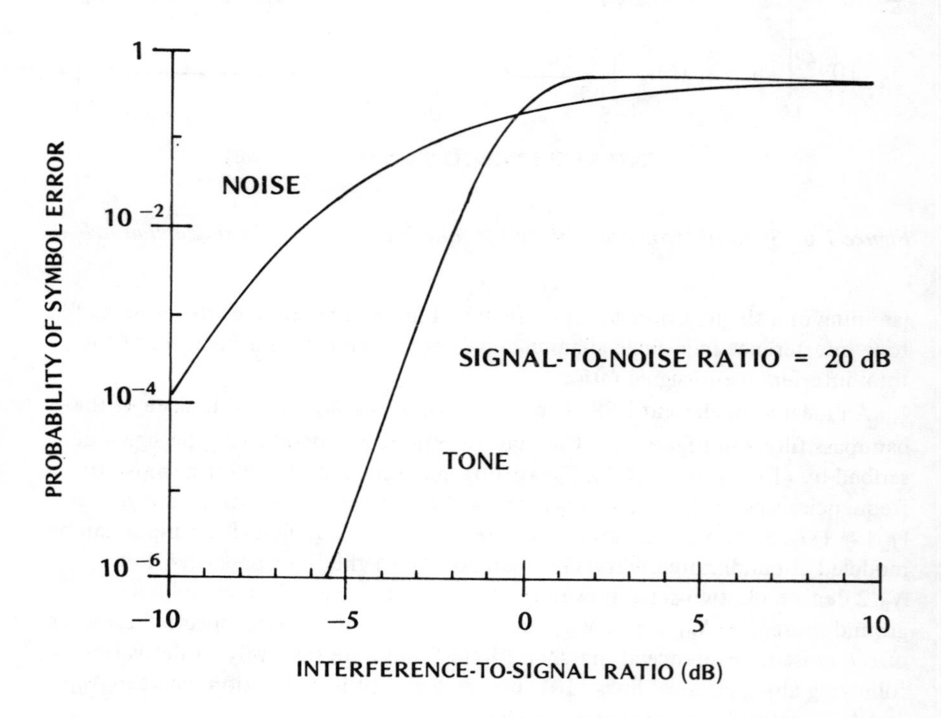

Figure 1.7 Comparison of error probabilities for noise and tone interference when single channel receives interference.

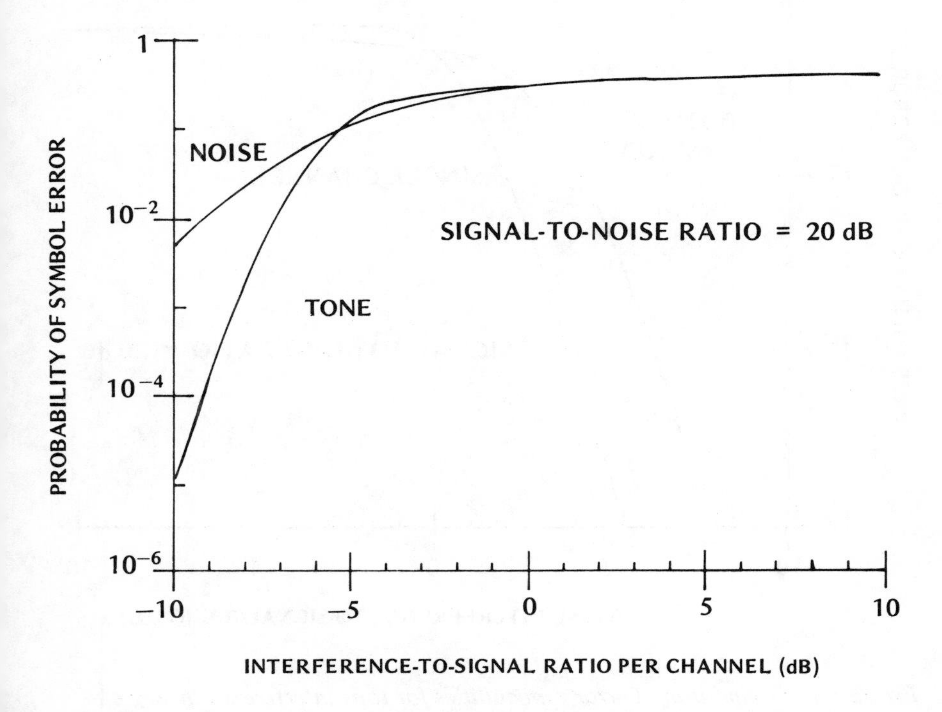

Figure 1.8 Comparison of error probabilities for noise and tone interference when both channels receive interference.

filters and envelope detectors in parallel. The demodulator decides in favor of a particular symbol if the corresponding envelope detector output in the largest. If only white Gaussian noise is present, a simple expression for the symbol error probability can be derived. A symbol error is made if the correct envelope detector output does not exceed the $N - 1$ other envelope detector outputs. Reasoning similar to that leading to (1.4.26) gives the generalization

$$P_s = 1 - \int_0^\infty f_1(r_1) \left[\int_0^{r_1} f_2(r_2)\, dr_2 \right]^{N-1} dr_1 \qquad (1.4.47)$$

where $f_1(r_1)$ and $f_2(r_2)$ are determined by setting $B_1 = B_2 = 0$ and $N_1 = N_2 = N_t$ in (1.4.23) to (1.4.25):

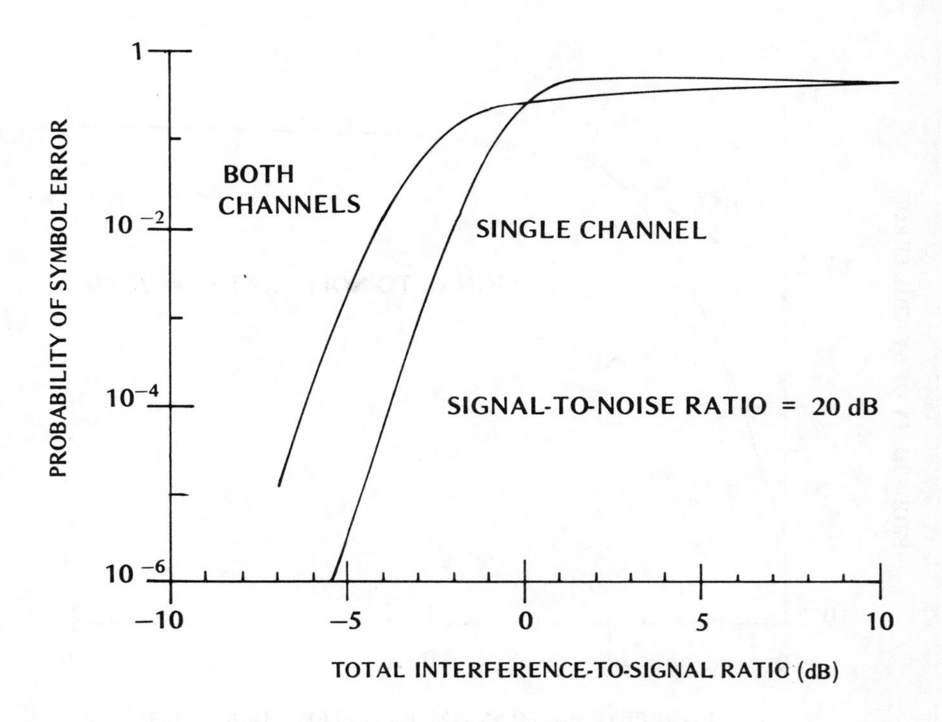

Figure 1.9 Comparison of error probabilities for tone interference in single channel and both channels.

$$f_1(r_1) = \frac{r_1}{N_t} \exp\left(-\frac{A^2 + r_1^2}{2N_t}\right) I_0\left(\frac{Ar_1}{N_t}\right), \quad r_1 \geqslant 0 \qquad (1.4.48)$$

$$f_2(r_2) = \frac{r_2}{N_t} \exp\left(-\frac{r_2^2}{2N_t}\right), \quad r_2 \geqslant 0 \qquad (1.4.49)$$

Evaluating the inner integral of (1.4.47) and using the binomial expansion, we obtain

$$P_s = 1 - \int_0^\infty f_1(r_1)\left[\sum_{i=0}^{N-1} (-1)^i \binom{N-1}{i} \exp\left(-\frac{ir_1^2}{2N_t}\right)\right] dr_1 \qquad (1.4.50)$$

Changing variables and using $Q(\alpha,0) = 1$ and $A = \sqrt{2R_s}$, we obtain

$$P_s = \sum_{i=1}^{N-1} \frac{(-1)^{1+i}}{1+i} \binom{N-1}{i} \exp\left[-\frac{iR_s}{(1+i)N_t}\right] \tag{1.4.51}$$

Because each channel symbol represents m bits, the bandwidth of each of the 2^m bandpass filters can be reduced by a factor of m relative to a binary FSK system. Therefore, N_t is reduced by a factor of m and R_s/N_t is increased by a factor of m. The total bandwidth required is $2^m/m$ times that required by a binary FSK system.

The symbol error probability for communications in white Gaussian noise can often be described at least approximately over a large range of values of P_s by an equation of the form

$$P_s = b \exp\left(-\frac{\xi E_s}{N_0}\right) \tag{1.4.52}$$

or the form

$$P_s = b \operatorname{erfc}\left(\sqrt{\frac{\xi E_s}{N_0}}\right) \tag{1.4.53}$$

where $N_0/2$ is the two-sided noise power spectral density, b and ξ are parameters, and the *complementary error function* is defined by

$$\operatorname{erfc}(x) = \frac{2}{\sqrt{\pi}} \int_x^\infty \exp(-y^2)\,dy \tag{1.4.54}$$

When intersymbol interference is disregarded, (1.4.52) holds for *differential phase-shift keying* (DPSK) if $b = 1/2$ and $\xi = 1$, and for noncoherent FSK with binary orthogonal signaling if $b = 1/2$ and $\xi = 1/2$. When intersymbol interference is disregarded and $b = 1/2$ and $\xi = 1$, (1.4.53) holds for *coherent phase-shift keying* (PSK) and for the binary output symbols of a *coherent quadriphase-shift keying* (QPSK) system.

The symbol error probability for *coherent binary FSK* is given by (1.4.53) with $b = 1/2$ and [1]

$$\xi = \frac{1}{2}(1 - \operatorname{sinc} 2f_d T_s) \tag{1.4.55}$$

where $f_d = (\omega_1 - \omega_2)/2\pi$ is the difference between the two possible frequencies and

$$\text{sinc}\, x = \frac{\sin(\pi x)}{\pi x} \tag{1.4.56}$$

For the usual orthogonal signaling, $2f_d T_s = k$ for some nonzero integer k, and hence $\xi = 1/2$. However, the minimum value of P_s results if

$$f_d T_s \cong 0.715 \tag{1.4.57}$$

which implies that $\xi \cong 0.61$. The value of E_s/N_0 necessary for a specified P_s is approximately 0.85 dB less than for orthogonal FSK.

The symbol error probability is not the ultimate performance measure of digital communication systems. For data communications, the *message error probability* is usually a more significant figure of merit. For voice communications, the message *intelligibility* is the performance measure of greatest interest to the users. Intelligibility is usually defined as the probability that a listener will correctly identify a word or phoneme drawn from a specific list. The exact method of measuring intelligibility is somewhat controversial. Whatever the method, the measured intelligibility is a sensitive function of the experience and ability of the listener, and thus is not easily reproducible. For both voice and data communications, the word error probability or the information-bit error rate is usually by far the most convenient performance measure to use.

1.5 DIGITAL FREQUENCY MODULATION

Digital modulations that produce compact spectra are desirable to reduce intersymbol interference in bandlimited channels and to limit crosstalk among communicators using carrier frequencies that are close to each other. The spectral sidelobes of FSK or other digital pulses can be reduced by filtering, or by generating approximately Gaussian or raised cosine pulses instead of rectangular pulses. However, nonlinearities in the final transmitter power amplifier and the propagation channel may considerably regenerate the spectral sidelobes so that the net benefit from the filtering or pulse shaping is modest. It is often impossible to implement a filter with the appropriate bandwidth and center frequency for spectral shaping of the signal after it emerges from the final power amplifier.

If an approximately constant-envelope signal is generated and filtering is not used, the spectral effect of the amplifier and channel nonlinearities is usually negligible. A class of constant-envelope signals with steep spectral roll-offs use *digital frequency modulation* (digital FM) and have continuous phase functions. Thus, the modulation is sometimes called *continuous phase modulation.* The general form of a digital FM signal is

$$s(t) = A \cos[2\pi f_0 t + \phi(t, \mathbf{a})] \tag{1.5.1}$$

where A is the amplitude, f_0 is the carrier or center frequency, and $\phi(t,\boldsymbol{\alpha})$ is the phase function, which carries the message. The phase function usually has the form

$$\phi(t,\boldsymbol{\alpha}) = 2\pi h \int_0^t \sum_{i=0}^{n} \alpha_i g(x - iT_s)\,dx + \phi_0 \tag{1.5.2}$$

where h is a constant called the *deviation ratio* or *modulation index*, $g(t)$ represents a pulse, T_s is the symbol duration, ϕ_0 is the phase at time $t = 0$, and the vector $\boldsymbol{\alpha}$ is a sequence of M-ary channel symbols. Each symbol α_i, $i = 0, 1, 2, \ldots, n$, takes one of M values; if M is even,

$$\alpha_i = \pm 1, \pm 3, \ldots, \pm (M-1)\,, \quad i = 0, 1, 2, \ldots, n-1 \tag{1.5.3}$$

To calculate power spectra, the phase function is usually represented in the idealized form

$$\phi(t,\boldsymbol{\alpha}) = 2\pi h \int_{-\infty}^t \sum_{i=-\infty}^{\infty} \alpha_i g(x - iT_s)\,dx \tag{1.5.4}$$

and it is assumed that the channel symbols are uncorrelated. Equations (1.5.2) and (1.5.4) indicate that the phase in any specified symbol interval depends upon the previous symbols. It is usually assumed that each channel symbol takes one of its M possible values with probability $1/M$. This assumption and the assumption that the channel symbols are uncorrelated are idealizations that are not always valid in practice.

The frequency pulse $g(t)$ is assumed to vanish outside an interval; that is,

$$g(t) = 0\,, \quad t < 0\,, \quad t > LT_s \tag{1.5.5}$$

where L is a positive integer and may be infinite. The presence of h as a multiplicative factor in the phase function makes it convenient to normalize $g(t)$ by assuming that

$$\int_0^{LT_s} g(x)\,dx = 1/2 \tag{1.5.6}$$

If $L = 1$, the digital FM modulation is called a *full-response modulation;* if $L > 1$, it is called a *partial-response modulation.* The normalization condition for a full-response modulation implies that the phase change over a symbol interval is equal to $h\pi\alpha_i$.

Continuous-phase frequency-shift keying (CPFSK) is a subclass of digital FM for which the instantaneous frequency is constant over each symbol interval.

Because of the normalization,

$$g(t) = \begin{cases} \dfrac{1}{2LT_s}, & 0 \leqslant t \leqslant LT_s \\ 0, & \text{otherwise} \end{cases} \tag{1.5.7}$$

A binary full-response CPFSK signal shifts between two frequencies separated by $f_d = h/T_s$.

Minimum-shift keying (MSK) is defined as full-response CPFSK with $h = 1/2$. *Generalized MSK* is defined as digital FM with $h = 1/2$. The class of generalized MSK signals includes signals with much faster spectral roll-offs than ordinary MSK signals. An example of a binary generalized MSK signal is *sinusoidal frequency-shift keying* (SFSK), for which the frequency pulse is

$$g(t) = \begin{cases} \dfrac{1}{2T_s}\left(1 - \cos\dfrac{2\pi t}{T_s}\right), & 0 \leqslant t \leqslant T_s \\ 0, & \text{otherwise} \end{cases} \tag{1.5.8}$$

To compare modulations that have different values of M, it is useful to define the energy per data bit

$$E_b = \frac{E_s}{\log_2 M} \tag{1.5.9}$$

and the duration of a data bit

$$T_b = \frac{T_s}{\log_2 M} \tag{1.5.10}$$

The data bit may be an information bit or a binary code symbol.

A measure of the spectral compactness of signals is provided by the *fractional out-of-band power* [12, 13], defined as

$$G(f) = 1 - \frac{\int_{-f}^{f} S(f')\,df'}{\int_{-\infty}^{\infty} S(f')\,df'}, \qquad f \geqslant 0 \tag{1.5.11}$$

where f is the frequency variable and $S(f)$ is the two-sided power spectral density of the *complex envelope* of the signal (Appendix A), which is often called the *equivalent lowpass waveform.* Figure 1.10 shows $G(f)$ for QPSK, binary MSK, and SFSK. The plots depict $G(f)$ in decibels as a function of f in units of $1/T_b$, where $T_s = 2T_b$ for QPSK. The rapid decrease in $G(f)$ for SFSK when $fT_b > 2$ occurs because the frequency has a continuous derivative, whereas the

MSK frequency is discontinuous at symbol transitions. The fractional power within a transmission channel of one-sided bandwidth B is given by

$$K_0 = 1 - G(B/2) \tag{1.5.12}$$

Usually, K_0 must exceed at least 0.9 to prevent a significant performance degradation in communications over a bandlimited channel.

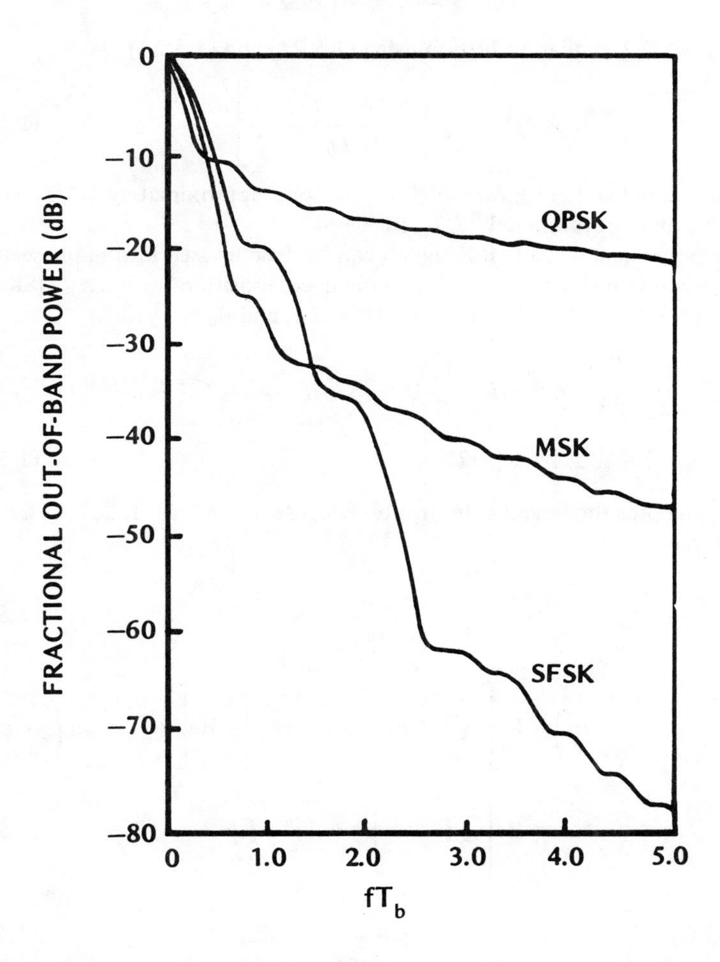

Figure 1.10 Fractional out-of-band power for equivalent lowpass waveforms.

The power spectral densities of QPSK and binary MSK signals are given by simple closed-form expressions if the data sequence is modeled as a random binary sequence. Expressing a QPSK signal in terms of its in-phase and quadrature components, and using (A.3.30) of Appendix A and (2.2.7) of Chapter 2, it is found that

$$S(f) = 2E_b \operatorname{sinc}^2(2T_b f) \tag{1.5.13}$$

For binary MSK, a similar derivation using (1.5.22) and (1.5.23) yields

$$S(f) = \frac{16E_b}{\pi^2}\left[\frac{\cos(2\pi T_b f)}{16T_b^2 f^2 - 1}\right]^2 \tag{1.5.14}$$

The transmission bandwidth for which $K_0 = 0.99$ is approximately $1.2/T_b$ for binary MSK, but approximately $8/T_b$ for QPSK.

Full-response generalized MSK signals can be decomposed into in-phase and quadrature components [14]. We derive the decomposition for binary MSK. Equations (1.5.2) and (1.5.7) with $h = 1/2, L = 1$, and $\phi_0 = 0$ yield

$$\phi(t,\boldsymbol{\alpha}) = \frac{\pi}{2}\sum_{i=0}^{2k-1} \alpha_i + \alpha_{2k}\frac{\pi}{2T_s}(t - 2kT_s), \quad 2kT_s \leqslant t \leqslant (2k+1)T_s\,,$$
$$k = 1, 2, \ldots, [n/2] \tag{1.5.15}$$

where $[x]$ denotes the largest integer in x. Because $\alpha_i = \pm 1, i = 1, 2, \ldots, 2k - 1$, it follows that

$$\sin\left(\frac{\pi}{2}\sum_{i=0}^{2k-1}\alpha_i\right) = 0 \tag{1.5.16}$$

$$\cos\left(\frac{\pi}{2}\sum_{i=0}^{2k-1}\alpha_i\right) = \operatorname{Re}\left[\prod_{i=0}^{2k-1}\exp\left(j\frac{\pi}{2}\alpha_i\right)\right] = \operatorname{Re}\left[\prod_{i=0}^{2k-1} j\sin\left(\frac{\pi}{2}\alpha_i\right)\right]$$
$$= \operatorname{Re}\left[\prod_{i=0}^{2k-1} j\alpha_i\right] = (-1)^k \prod_{i=0}^{2k-1}\alpha_i \tag{1.5.17}$$

$$\cos\left[\alpha_{2k}\frac{\pi}{2T_s}(t - 2kT_s)\right] = \cos\left[\frac{\pi}{2T_s}(t - 2kT_s)\right] \tag{1.5.18}$$

$$\sin\left[\alpha_{2k}\frac{\pi}{2T_s}(t - 2kT_s)\right] = \alpha_{2k}\sin\left[\frac{\pi}{2T_s}(t - 2kT_s)\right] \tag{1.5.19}$$

We define the symbols

$$d_{2k-1} = (-1)^k \prod_{i=0}^{2k-1} \alpha_i \tag{1.5.20}$$

$$d_{2k} = (-1)^k \prod_{i=0}^{2k} \alpha_i \tag{1.5.21}$$

These relations and a trigonometric expansion of (1.5.1) yield

$$s(t) = d_{2k-1} \cos\left[\frac{\pi}{2T_s}(t - 2kT_s)\right] \cos(2\pi f_0 t) - d_{2k} \sin\left[\frac{\pi}{2T_s}(t - 2kT_s)\right]$$

$$\times \sin(2\pi f_0 t), \qquad 2kT_s \leqslant t \leqslant (2k+1)T_s \tag{1.5.22}$$

Similarly, if $2k + 1 \leqslant n$, we can derive

$$s(t) = d_{2k+1} \cos\left[\frac{\pi}{2T_s}(t - (2k+2)T_s)\right] \cos(2\pi f_0 t) - d_{2k} \sin\left[\frac{\pi}{2T_s}(t - 2kT_s)\right]$$

$$\times \sin(2\pi f_0 t), \qquad (2k+1)T_s \leqslant t \leqslant (2k+2)T_s \tag{1.5.23}$$

Equations (1.5.22) and (1.5.23) explicitly indicate the modulations of the in-phase and quadrature components of the MSK signal. This form can be exploited in the implementation of the modulator and the demodulator. If the α_i are regarded as the true input symbols, then (1.5.20) and (1.5.21) indicate that the d_i are a form of differentially encoded symbols. Alternatively, the d_{2k} and the d_{2k+1}, $k = 0, 1, 2, \ldots$, can be independent symbol streams modulating the components of the MSK signal.

Since each MSK component uses antipodal signaling, ideal coherent demodulation of a binary MSK signal transmitted over a channel with white Gaussian noise yields the symbol error probability

$$P_s = \frac{1}{2}\,\text{erfc}\left(\sqrt{\frac{E_s}{N_0}}\right) \tag{1.5.24}$$

which is the same P_s provided by coherent PSK and QPSK systems. The in-phase and quadrature components of an MSK signal can be demodulated in parallel as in QPSK demodulation. Alternatively, a coherent MSK receiver can be designed to use serial symbol-by-symbol demodulation. There are other binary generalized MSK signals that provide nearly the same performance as ordinary binary MSK signals, but have steeper spectral roll-offs than SFSK [15].

There are full-response digital FM systems that provide both improved asymptotic spectral properties and a performance advantage relative to a binary MSK system [16]. Thus, M-ary CPFSK systems and certain other full-response systems can be designed to require a lower value of E_b/N_0 for the same P_s and K_0 as a binary MSK system, where E_b is given by (1.5.9). The performance advantage results partly from making each symbol decision based upon the processing of two or more symbols. The improved asymptotic spectral properties are achieved by using $M > 2$ and smoothing the phase at the symbol transitions.

Partial-response digital FM systems can improve the spectral properties relative to binary MSK even more than full-response systems without degrading the symbol error probability [17]. The improvement can be attributed to the memory in the modulation and demodulation processes. However, partial-response systems require complex implementations. Synchronization problems are substantial and grow as the system bandwidth decreases.

Noncoherent detection greatly simplifies the synchronization problems. With multisymbol noncoherent detection, full-response CPFSK systems can provide a better symbol error probability than coherent binary MSK systems. For r-symbol detection, where r is odd, the optimal receiver correlates the received waveform over all possible r-symbol patterns before making a decision about the middle symbol. However, when error-correcting codes are used with the CPFSK modulation, the improvement in the information-bit error rate appears to be less than the improvement when the coding is used with PSK or QPSK modulation [18]. Furthermore, the implementation complexity of multisymbol detection, even for three-symbol detection, is considerable.

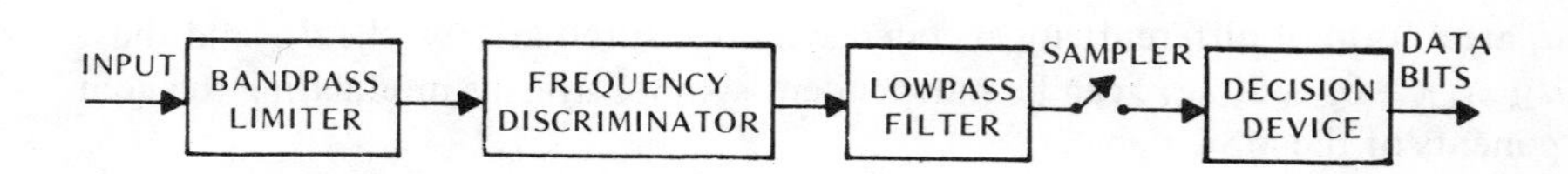

Figure 1.11 CPFSK demodulator.

Symbol-by-symbol noncoherent detection of binary full-response CPFSK signals can be inexpensively implemented by using a limiter and frequency discriminator, as illustrated in Figure 1.11. The theory of *limiter-discriminator* demodulation [19] provides complicated expressions for the symbol error probability in the presence of white Gaussian noise. However, the theoretical P_s can usually be approximated to within a few tenths of a decibel by the simple equation

$$P_s = \frac{1}{2} \operatorname{erfc}\left(\sqrt{\frac{\xi E_s}{N_0}}\right) \tag{1.5.25}$$

where $E_s = E_b$, $T_s = T_b$, and the parameter ξ depends upon h and the product of T_s and the noise bandwidth, B. Because the limiter-discriminator does not make optimal use of the signal phase, P_s exceeds the value that it has for ideal coherent FSK detection. Thus, (1.4.55) and $h = f_d T_s$ imply that ξ for the limiter-discriminator should satisfy

$$\xi \leqslant \frac{1}{2} (1 - \text{sinc } 2h) \tag{1.5.26}$$

If the bandpass limiter has a Gaussian filter, the lowpass filter is an integrator, and $BT_s = 1$, it is found that P_s is minimized when $h \cong 0.7$. The corresponding value of ξ, obtained by an approximate least-squares fit of (1.5.25) to the theoretical curve for $10^{-2} < P_s < 10^{-6}$, is approximately 0.56, which is consistent with (1.5.26). The least-squares fit for binary MSK requires that $\xi \approx 0.38$, which satisfies (1.5.26) because $h = 0.5$. To achieve $P_s = 10^{-4}$, CPFSK with $h = 0.7$ requires approximately 1.7 dB less in $E_s/N_0 = E_b/N_0$ than binary MSK. However, the 99 percent bandwidth of a CPFSK signal with $h = 0.7$ is approximately $1.8/T_s$, compared with $1.2/T_s$ for binary MSK.

Differential detection of binary full-response CPFSK signals provides a symbol error probability similar to that achieved with a limiter-discriminator [20]. If $h \geqslant 0.5$ and intersymbol interference is absent, a differential detector for binary CPFSK can be designed to provide [21]

$$P_s = \frac{1}{2} \exp\left(-\frac{R_s}{N_t}\right) \tag{1.5.27}$$

However, if the intersymbol interference is to cause a decibel or less increase in the required value of R_s/N_t to achieve $P_s \approx 10^{-4}$, then it is necessary that $BT_s > 2$.

1.6 PULSED JAMMING

In attempting to disrupt digital communications, it is often advantageous to concentrate the jamming energy in short pulses. *Pulsed jamming* can cause a substantial increase in the bit error rate relative to the rate caused by continuous jamming with the same average power. *Swept-frequency jamming,* which results from periodically sweeping the center frequency of a jamming signal over a frequency range, produces pulsed jamming in communication systems operating over part of the frequency range.

Let $N_0/2$ denote the two-sided power spectral density of the white Gaussian thermal and background noise. Suppose that the average transmitted jamming power is conserved as the pulse duration changes and that during a pulse the jamming can be modeled as an independent white Gaussian process. Let μ denote either the *pulse duty cycle*, which is the ratio of the pulse duration to the repetition period, or the probability of pulse occurrence if the pulses occur randomly. The jamming power spectral density during a pulse is $J_0/2\mu$, where $J_0/2$ is the two-sided power spectral density of continuous jamming with the same average power as the pulsed jamming. The power spectral density of the jamming plus noise during a jamming pulse is $J_0/2\mu + N_0/2$, whereas it is $N_0/2$ when no pulse occurs. Suppose that $P_s = g(E_s/N_0)$ is the symbol error probability of a communication system operating in white Gaussian thermal noise, where E_s is the symbol energy. If the jamming pulse duration exceeds the symbol duration, it is reasonable to approximate the symbol error probability by

$$P_s \cong \mu g\left(\frac{E_s}{N_0 + J_0/\mu}\right) + (1-\mu)g\left(\frac{E_s}{N_0}\right), \quad 0 \leqslant \mu \leqslant 1 \tag{1.6.1}$$

If $J_0 \gg N_0$, this equation reduces to

$$P_s \cong \mu g(\mu E_s/J_0) + (1-\mu)g(E_s/N_0), \quad 0 \leqslant \mu \leqslant 1 \tag{1.6.2}$$

To determine the optimal value of μ for a jammer, we differentiate (1.6.2) with respect to μ and equate the result to zero. If

$$g(E_s/N_0) = b \exp\left(-\frac{\xi E_s}{N_0}\right) \tag{1.6.3}$$

then we obtain

$$\exp\left(-\frac{\xi\mu E_s}{J_0}\right) - \exp\left(-\frac{\xi E_s}{N_0}\right) = \frac{\xi\mu E_s}{J_0}\exp\left(-\frac{\xi\mu E_s}{J_0}\right) \tag{1.6.4}$$

Because $J_0 \gg N_0$ and $\mu \leqslant 1$, the second term on the left-hand side is much smaller than the first term and can be ignored. It is then a simple matter to obtain

$$\mu_0 \approx \begin{cases} \dfrac{J_0}{\xi E_s}, & J_0 < \xi E_s \\ 1, & J_0 \geqslant \xi E_s \end{cases} \tag{1.6.5}$$

as the optimal value. Thus, a continuous jamming waveform is preferable to pulsed jamming if $J_0 \geqslant \xi E_b$.

By substituting $\mu = \mu_0$ and (1.6.5) into (1.6.2), we obtain an equation for the worst-case symbol error probability caused by optimal pulsed jamming when $J_0 \gg N_0$:

$$P_s \approx \begin{cases} \dfrac{b}{e\xi}\left(\dfrac{E_s}{J_0}\right)^{-1}, & J_0 < \xi E_s \\ b\exp\left(-\dfrac{\xi E_s}{J_0}\right), & J_0 \geqslant \xi E_s \end{cases} \tag{1.6.6}$$

In deriving this equation, it is assumed that the average jamming power is independent of the pulse duration. This assumption is valid only if the pulse duration exceeds some minimum value and if J_0/μ is less than some maximum value.

Equation (1.6.6) indicates that P_s varies inversely, rather than exponentially, with E_s/J_0 if this quantity is sufficiently large. For noncoherent binary FSK, the required value of E_s/J_0 for $P_s = 10^{-5}$ is raised by more than 30 dB when worst-case pulsed jamming is present instead of continuous jamming. Thus, some form of error correction is needed to mitigate the potential effect of pulsed jamming.

Similar results are obtained when

$$g(E_s/N_0) = b \operatorname{erfc}\left(\sqrt{\frac{\xi E_s}{N_0}}\right) \tag{1.6.7}$$

If we differentiate (1.6.2), equate the result to zero, and set $N_0 = 0$, then a numerical evaluation yields

$$\mu_0 \approx \begin{cases} \dfrac{0.7 J_0}{\xi E_s}, & J_0 < \dfrac{\xi E_s}{0.7} \\ 1, & J_0 \geqslant \dfrac{\xi E_s}{0.7} \end{cases} \tag{1.6.8}$$

as the value of μ corresponding to worst-case pulsed jamming.

1.6.1 Interleavers

An *interleaver* is a device that permutes the order of a sequence of symbols. A *deinterleaver* is the corresponding device that restores the original order of the sequence. An interleaver is used when it is desired to randomize the distribution of errors after reception.

The symbol errors induced by pulsed interference are clustered. If an error-correcting block code is used, but the pulse duration is comparable to the duration of a codeword, the error-correcting code may not be able to decrease the word error probability significantly. To reduce the clustering of errors, symbols from various codewords can be interleaved before transmission. After deinterleaving at the receiver, a burst of errors is spread over a number of different codewords, which facilitates the removal of the errors by the decoding.

A *block interleaver* performs identical permutations on successive blocks of symbols. Consider the $m \times n$ matrix representing the mn successive encoder output symbols that are stored in a buffer prior to transmission:

$$\begin{matrix} x_{11} & x_{12} & x_{13} & \cdots & x_{1n} \\ x_{21} & x_{22} & x_{23} & \cdots & x_{2n} \\ \cdot & \cdot & \cdot & & \cdot \\ \cdot & \cdot & \cdot & & \cdot \\ \cdot & \cdot & \cdot & & \cdot \\ x_{m1} & x_{m2} & x_{m3} & \cdots & x_{mn} \end{matrix}$$

Without interleaving, successive rows are transmitted. With interleaving, successive columns are transmitted so that the transmitted stream of symbols is

$$x_{11}x_{21}\ldots x_{m1}x_{12}x_{22}\ldots x_{m2}x_{13}x_{23}\ldots x_{mn}$$

For continuous interleaving, two buffers are needed. Symbols are written in one buffer while previous symbols are read out of the other buffer. In the deinterleaver, symbols are stored by column in a buffer while previous symbols are read out by rows from another buffer.

For a block code, the parameter n equals or exceeds the length of a codeword. When a burst of errors occurs in m or fewer consecutive symbols and there are no other errors in the stream, each codeword after deinterleaving has at most one error. Thus, a single-error-correcting code can eliminate all the errors. Similarly, a double-error-correcting code can be used to correct a single burst of errors spanning as many as $2m$ symbols. The parameter m should be chosen to accommodate any known repetition rates of periodic interference pulses. For convolutional codes, the randomization of errors is approximated only if n exceeds a constraint length.

A block interleaver is a type of *periodic interleaver,* which is one with a permutation function that repeats periodically. Another type of periodic interleaver is the *convolutional interleaver,* which has similar properties [22]. Periodic interleavers are potentially vulnerable to jamming. If the interleaving structure is known, a jammer can produce pulses spaced by small integer multiples of m symbols. If the pulses produce symbol errors, then multiple errors will lie within single codewords or constraint lengths after the deinterleaving, thereby nullifying the usefulness of the interleaving. To thwart this pulsed jamming threat, a different type of interleaver is needed.

A *pseudorandom interleaver* [22] permutes each block of symbols pseudorandomly. The desired permutation can be stored in a read-only memory (ROM) as a sequence of addresses. Each block of symbols to be interleaved can be written sequentially into a *random-access memory* (RAM). The symbols are interleaved by reading them out in the order dictated by the contents of the ROM.

If two RAMs are used, symbols can be written into one RAM while other symbols are being read from the other RAM.

Although this pseudorandom interleaver provides robustness against periodic pulsed jamming, it is vulnerable to a jammer with knowledge of the permutation. Thus, it is desirable to change the permutation for each block of symbols. One possibility is to store a large number of different permutations in the ROM, and then pseudorandomly select one of them for each block. The pseudorandom selection prevents a jammer from easily exploiting any knowledge of the set of permutations. Because a pseudorandom deinterleaver simply performs the inverse permutation, its implementation is similar to that of the interleaver.

1.7 ERROR-CORRECTING CODES

There are two methods for eliminating errors from a received digital message. One method, *forward error correction* (FEC), uses error-correcting codes that allow the automatic correction of at least some of the detected errors. The other method, *automatic repeat request* (ARQ), entails the detection of errors by the receiver and the subsequent sending of a request to the transmitter to repeat the message [23].

An ARQ system requires simpler decoding devices than an FEC system that uses the same code to correct errors that the ARQ system uses to detect errors. However, an ARQ system requires a two-way link, a retransmission protocol, and buffers to store messages while retransmissions are occurring. When strong interference is present, the receiver of an ARQ system may be forced to request many retransmissions, thereby significantly reducing the rate at which messages are correctly received relative to an FEC system. The feedback link for retransmission requests may be subjected to strong interference that causes unnecessary retransmissions or disrupts the feedback communications. Although a hybrid system combining ARQ and FEC may alleviate these problems, a pure FEC system appears to be usually preferable when strong interference or jamming is a threat.

In this section, some of the fundamental results of coding theory [22-25] are summarized and used to derive equations for the error probabilities produced by FEC systems. Coding makes possible some combination of an improved system performance, a decrease in transmitted power, a decrease in the antenna sizes, or an increase in the transmitted data rate.

1.7.1 Block Codes

An (n,k) *block code* is a set of *codewords*, each of which has n symbols and represents k information symbols. Each symbol is selected from an alphabet of q symbols. If $q = 2$, the two possible symbols are the digits 0 and 1. A block code is called a *linear block code* if its codewords form a k-dimensional subspace of the vector space of sequences with n symbols. Thus, the vector sum of two codewords of a linear block code is a codeword. If a binary block code is linear, the symbols of a codeword are modulo-two sums of information bits. The *code rate* is defined as the ratio k/n.

The *Hamming distance* between two sequences with an equal number of symbols is defined as the number of positions in which the symbol of one sequence differs from the corresponding symbol of the other sequence. The minimum Hamming distance between any two possible codewords is called the *minimum distance* of the code.

After the waveform representing a codeword is received and demodulated, the decoder uses the demodulator output to determine the information symbols corresponding to the codeword. If the demodulator produces a sequence of discrete symbols and the decoding is based on these symbols, the demodulator is said to make *hard decisions.* Conversely, if the demodulator produces analog or multilevel quantized samples of the waveform, the demodulator is said to make *soft decisions.* The advantage of soft decisions is that reliability or quality information is provided to the decoder, which can use this information to improve its performance.

When hard decisions are made, the demodulator output sequence is called the *received sequence* or the *received word.* A *complete decoder* assumes that the correct codeword is the one that is the smallest Hamming distance from the received sequence. Correct decoding is always possible if the number of incorrect symbols in the received sequence is bounded by

$$t_0 = [(d - 1)/2] \tag{1.7.1}$$

where d is the minimum distance between codewords and $[x]$ denotes the integer part of x. An *incomplete decoder* does not attempt to correct all received sequences.

A conceptual three-dimensional representation of the vector space of sequences of length n is depicted in Figure 1.12. Each codeword occupies the center of a *decoding sphere* with radius t in Hamming distance, where t is a positive integer such that $t \leqslant t_0$. A *bounded-distance decoder* corrects all received sequences with t or fewer symbol errors. If a received sequence lies within a decoding sphere, then the decoder output is the codeword at the center of the sphere. When more than t symbol errors occur, the received sequence may lie within a decoding sphere surrounding an incorrect codeword or it may lie in the interstices outside the decoding spheres. If the sequence lies within a decoding sphere, the decoder selects an incorrect codeword. Thus, the decoder produces an output word with an *undetected error.* If the sequence lies in the interstices, the decoder is unable to correct the errors but recognizes their existence. Thus, a *decoding failure* occurs and the decoder either reproduces the received sequence or erases it. Most practical decoders for block codes are bounded-distance decoders.

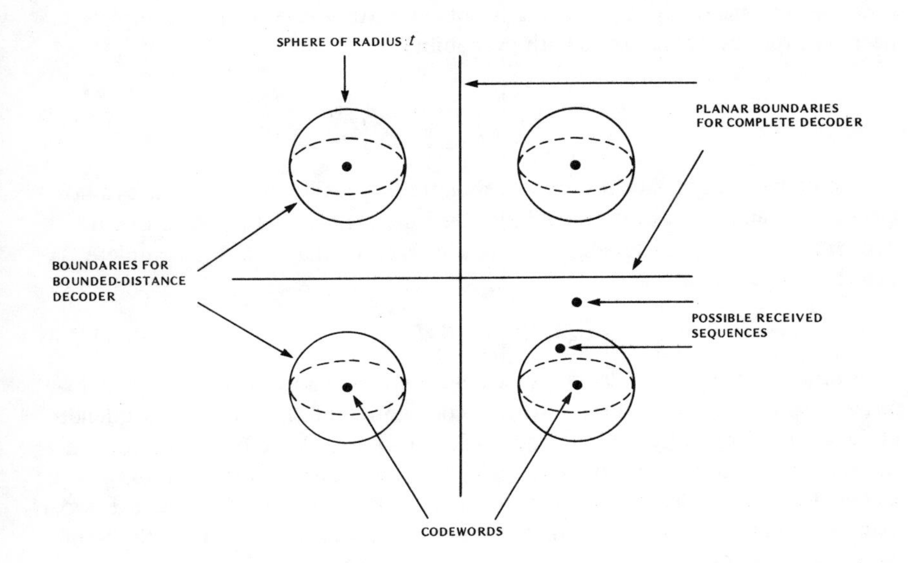

Figure 1.12 Conceptual representation of vector space of sequences of length n.

Let P_s denote the *channel-symbol error probability,* which is the probability of error in a demodulated code symbol and is assumed to be the same for all code symbols. It is assumed that the channel-symbol errors are statistically independent, which can usually be ensured, if necessary, by appropriate symbol interleaving. Let P_W denote the probability that one or more symbol errors occur in a decoded word, including the contributions due to both undetected errors and decoding failures. Since there are $\binom{n}{i}$ distinct ways in which i errors may occur among n symbols, a bounded-distance decoder produces word errors with probability

$$P_W = \sum_{i=t+1}^{n} \binom{n}{i} P_s^i (1 - P_s)^{n-i} \tag{1.7.2}$$

In many practical applications, $t = t_0$ and $d = 2t + 1$. However, if a block code is used for both error correction and error detection, a bounded-distance decoder is often designed with $t < t_0$. If a block code is used exclusively for error detection, then $t = 0$.

Conceptually, a complete decoder correctly decodes when the number of symbol errors exceeds t_0 if the received sequence lies within the planar boundaries associated with the correct codeword, as depicted in Figure 1.12. When a received sequence is equidistant from two or more codewords, a complete decoder selects one of them according to some arbitrary rule. Thus, a complete decoder produces word errors with probability

$$P_w \leqslant \sum_{i=t_0+1}^{n} \binom{n}{i} P_s^i (1 - P_s)^{n-i} \tag{1.7.3}$$

It is convenient to let the symbol t denote the maximum number of symbol errors that can *always* be corrected by a decoder. Thus, t is the radius of a decoding sphere if a bounded-distance decoder is used, and $t = t_0$ if a complete decoder is used. In all cases,

$$2t + 1 \leqslant d \tag{1.7.4}$$

A *perfect code* is a block code such that every n-symbol sequence is at a distance of at most t_0 from some n-symbol codeword, and the sets of all sequences at distance t_0 or less from each codeword are disjoint. Therefore, d is odd, and decoders can correct as many as $t_0 = (d - 1)/2$ errors, but not more than t_0 errors. A complete decoder for a perfect code is also a bounded-distance decoder. The only known perfect codes are the repetition codes of odd length, the Hamming codes, the Golay (23, 12) code, and the ternary Golay (11, 6) code.

Repetition codes have one information bit represented by n code symbols. For hard-decision decoding, n is odd, and the decoder decides the state of the information bit according to the states of the majority of the demodulated symbols. Thus, the decoder corrects all combinations of $t = (n - 1)/2$ or fewer errors, but no patterns of more errors. In this case, P_w is the probability of error in a decoded information bit, which is denoted by P_{ib}, and (1.7.2) becomes

$$P_{ib} = \sum_{i=(n+1)/2}^{n} \binom{n}{i} P_s^i (1 - P_s)^{n-i} \tag{1.7.5}$$

Although repetition codes are not efficient for white Gaussian noise channels, they can improve the system performance for fading channels if the number of repetitions is properly chosen.

A *Hamming* (n, k) *code* has $n = 2^{n-k} - 1$ and $d = 3$. Thus, a Hamming code is capable of correcting all single errors, and (1.7.2) holds with $t = 1$.

The *Golay* (23, 12) *code* has $d = 7$ and, thus, can correct three symbol errors. The *extended Golay* (24, 12) *code* is formed by adding an overall parity symbol to the perfect Golay (23, 12) code, thereby increasing the minimum distance to $d = 8$. As a result, some received sequences with four errors can be corrected by a complete decoder. The (24, 12) code is often preferable to the (23, 12) code because it has a code rate of exactly one-half, which simplifies the system timing.

Relatively simple encoding and hard-decision decoding techniques are known for binary codes belonging to the class of *Bose-Chaudhuri-Hocquenghem* (BCH) *codes* [22-25]. For any positive integers m and t such that $m \geqslant 3$ and $t < 2^{m-1}$, there exists a binary BCH code with $n = 2^m - 1$, $n - k \leqslant mt$, and $d \geqslant 2t + 1$. A BCH code is usually designed to allow the correction of t errors in a received sequence. However, because the minimum distance may exceed $2t + 1$, a BCH code sometimes allows the correction of more than t errors.

The most important nonbinary block codes belong to the class of *Reed-Solomon codes,* which provide the largest possible minimum distance of any linear code with specified values of n and k [22-24]. However, not all values of n and k have associated Reed-Solomon codes. The block length of a Reed-Solomon code is $n = q - 1$, where q is the number of symbols in the code alphabet. For convenience in implementation, q is usually chosen so that $q = 2^m$, where m is the number of bits per symbol. Thus, $n = 2^m - 1$ and the code provides correction of 2^m-ary symbols. The minimum distance is $d = n - k + 1$. Most Reed-Solomon decoders are bounded-distance decoders such that $d = 2t + 1$. Some incomplete decoders are designed to also correct some received sequences with $t + 1$ symbol errors.

The *information-bit error rate,* which is the expected fraction of decoded information bits that are erroneous, is a performance measure that is useful in comparing communication systems that use different codes. An exact equation relating the information-bit error rate to the channel-symbol error probability is a function of the code weight structure and is unknown for most codes. Thus, a simple approximation that does not depend upon the code structure is derived for block codes that use hard decisions. It is assumed that decoding failures cause the reproduction of the received sequence.

Let $P_{is}(j)$ denote the probability of an error in information symbol j at the decoder output. In general, it cannot be assumed that $P_{is}(j)$ is independent of j. The *average information-symbol error probability* is defined by

$$P_{is} = \frac{1}{k} \sum_{j=1}^{k} P_{is}(j) \tag{1.7.6}$$

The random variables $Z_j, j = 1, 2, \ldots, k$, are defined so that $Z_j = 1$ if information symbol j is in error and $Z_j = 0$ if it is correct. The expected number of information-symbol errors is

$$E[I] = E\left[\sum_{j=1}^{k} Z_j\right] = \sum_{j=1}^{k} E[Z_j] = \sum_{j=1}^{k} P_{is}(j) \tag{1.7.7}$$

where $E[x]$ denotes the expected value of x. The *information-symbol error rate* is defined as $E[I]/k$. Equations (1.7.6) and (1.7.7) imply that

$$P_{is} = \frac{E[I]}{k} \tag{1.7.8}$$

Thus, the average information-symbol error probability is equal to the information-symbol error rate.

Let $P_{ds}(j)$ denote the probability of an error in symbol j of the codeword chosen by the decoder or symbol j of the received sequence if a decoding failure occurs. The average decoded-symbol error probability is defined by

$$P_{ds} = \frac{1}{n} \sum_{j=1}^{n} P_{ds}(j) \tag{1.7.9}$$

A relation analogous to (1.7.8) can easily be established. A *systematic block code* is a code in which the information symbols appear unchanged in the codeword, which also has additional parity symbols. Thus, for systematic codes in which the first k values of j denote the information symbols, $P_{is}(j) = P_{ds}(j)$, where $j = 1, 2, \ldots, k$.

Let $P_{ds}(j|w)$ denote the probability of an error in symbol j of the codeword or sequence chosen by the decoder, given the event W that an incorrect codeword or sequence was chosen. The average decoded-symbol error probability given W is defined by

$$P_{dsw} = \frac{1}{n} \sum_{j=1}^{n} P_{ds}(j|w) \tag{1.7.10}$$

and is equal to the decoded-symbol error rate given W. Let $P_{is}(j|w)$ denote the probability of an error in information symbol j at the decoder output given W. The average information-symbol error probability given W is defined by

$$P_{isw} = \frac{1}{k} \sum_{j=1}^{k} P_{is}(j|w) \tag{1.7.11}$$

From the definition of a conditional probability and (1.7.6) and (1.7.11), it follows that

$$P_{is} = P_{isw} P_w \tag{1.7.12}$$

It is plausible that $P_{isw} \approx P_{dsw}$ for most useful systematic codes, and to a somewhat lesser degree for most useful nonsystematic codes. We make the weaker assumption that

$$P_{isw} \gtrsim P_{dsw} \tag{1.7.13}$$

Because a particular information symbol may be correct despite a word error, $P_{is}(j) \leqslant P_w$ and, hence, $P_{is} \leqslant P_w$. For perfect codes and other codes using a complete decoder, $P_{dsw} \geqslant d/n$, because an incorrectly chosen codeword is at least a distance d from the correct codeword. Thus, for a complete decoder, (1.7.12) and (1.7.13) imply that

$$\frac{d}{n} P_w \lesssim P_{is} \leqslant P_w \tag{1.7.14}$$

For a bounded-distance decoder, the possibility that a word error may be detected but remain uncorrected implies that $P_{dsw} \geqslant (t+1)/n$. Thus, for a bounded-distance decoder, (1.7.12) and (1.7.13) imply that

$$\frac{t+1}{n} P_w \lesssim P_{is} \leqslant P_w \tag{1.7.15}$$

Let $A(i)$ denote the event that i symbol errors occur in a received sequence of n symbols at the decoder input. Let $P_{is}(j\,|\,i)$ denote the probability of an error in decoded information symbol j given $A(i)$. The average information-symbol error probability given $A(i)$ is defined by

$$P_{isi} = \frac{1}{k} \sum_{j=1}^{k} P_{is}(j|i)\,, \quad i = 1, 2, \ldots, n \tag{1.7.16}$$

The theorem of total probability plus (1.7.6) and (1.7.16) yield

$$P_{is} = \sum_{i=t+1}^{n} P[A(i)]\, P_{isi} \tag{1.7.17}$$

where $P[A(i)]$ is the probability of $A(i)$. Let P_{dsi} denote the average decoded-symbol error probability given $A(i)$, which is defined analogously to P_{isi} in (1.7.16). It is plausible to assume that

$$P_{isi} \approx P_{dsi}\,, \quad i = t+1,\ t+2, \ldots, n \tag{1.7.18}$$

Consider a decoder with $d = 2t + 1$, where t is the maximum number of symbol errors that can always be corrected. Suppose that there are i symbol errors in a received sequence. For $d \leqslant i \leqslant n$, it is plausible to assume that the decoder, on the average, neither increases nor decreases the number of symbol errors but chooses a codeword with approximately i symbol errors so that $P_{dsi} \approx i/n$. For $t + 1 \leqslant i \leqslant d$, the decoder often selects a codeword that is a distance d from the correct codeword; thus, we assume that $P_{dsi} \approx d/n$. If the demodulated-symbol errors are independent,

$$P[A(i)] = \binom{n}{i} P_s^i (1 - P_s)^{n-i} \qquad (1.7.19)$$

Equations (1.7.17) to (1.7.19) and the preceding approximations lead to

$$P_{is} \approx \sum_{i=t+1}^{2t+1} \frac{2t+1}{n} \binom{n}{i} P_s^i (1 - P_s)^{n-i} + \sum_{i=2t+2}^{n} \frac{i}{n} \binom{n}{i} P_s^i (1 - P_s)^{n-i} \qquad (1.7.20)$$

This equation satisfies (1.7.14) and (1.7.15). The right-hand side probably gives an approximate upper bound on P_{is} for a complete decoder because some received sequences with more than $t = (d - 1)/2$ errors can be corrected and result in no decoded information-symbol errors. For binary codes, $P_{ib} = P_{is}$.

For repetition codes, $d = n$ and $t = (n - 1)/2$, so (1.7.20) reduces to (1.7.5), which is the exact expression in this case. A numerical comparison of the exact error rate with (1.7.20) for the Golay [23,12] code indicates that (1.7.20) produces an error of approximately seven percent or less [26].

In nonbinary communications, an information symbol represents m information bits. It is assumed that an incorrectly decoded information symbol is equally likely to be any of the remaining $2^m - 1$ symbols in the alphabet. Among those symbols, a given bit is incorrect in 2^{m-1} cases. Thus,

$$P_{ib} = \frac{2^{m-1}}{2^m - 1} P_{is} \qquad (1.7.21)$$

For Reed-Solomon codes, $n = 2^m - 1$ so (1.7.20) and (1.7.21) imply that

$$P_{ib} \approx \frac{n+1}{2n} \left[\sum_{i=t+1}^{2t+1} \frac{2t+1}{n} \binom{n}{i} P_s^i (1 - P_s)^{n-i} + \sum_{i=2t+2}^{n} \frac{i}{n} \binom{n}{i} P_s^i (1 - P_s)^{n-i} \right] \qquad (1.7.22)$$

Decoding failures occur often when bounded-distance decoders are used with Reed-Solomon codes. Because decoding failures are largely ignored in the heuristic derivation of (1.7.22), the right-hand side of this equation is an upper bound on P_{ib} for bounded-distance decoding at least when P_s is small enough that only the first term of the first summation is significant.

Let r denote the ratio of information bits to transmitted channel symbols. For binary codes, r is the code rate. For block codes with m information bits per symbol, $r = mk/n$. When coding is used but the information rate or the message duration is preserved, the duration of a channel symbol is changed relative to that of an information bit. Thus, the energy per received channel symbol is

$$E_s = rE_b = \frac{mk}{n} E_b \qquad (1.7.23)$$

where E_b is the energy per information bit. When $r < 1$, a code is potentially beneficial if its error-correcting capability is sufficient to overcome the degradation due to the reduction in the energy per received symbol. However, the communication channel must be capable of accommodating the increased bandwidth of the transmitted waveform. In the presence of white Gaussian noise, the channel-symbol error probability for a coherent binary PSK system is

$$P_s = \frac{1}{2}\,\text{erfc}\left(\sqrt{\frac{E_s}{N_0}}\right)$$

$$= \frac{1}{2}\,\text{erfc}\left(\sqrt{\frac{rE_b}{N_0}}\right) \tag{1.7.24}$$

When soft decisions are made, a number called the *metric* is associated with each possible codeword. The metric is a function of both the codeword and the demodulator output samples. A soft-decision decoder selects the codeword with the largest metric. The information bits are then recovered from this codeword.

Let $\mathbf{y}$ denote the n-dimensional vector of noisy output samples y_i, $i = 1, 2, \ldots, n$, produced by a demodulator that receives a sequence of n symbols. Let $\mathbf{x}_j$ denote the jth codeword vector with components x_{ji}, $i = 1, 2, \ldots, n$. Let $p(\mathbf{y}|\mathbf{x}_j)$ denote the *likelihood function,* which is the conditional probability density function of $\mathbf{y}$ given that $\mathbf{x}_j$ was transmitted. The *maximum-likelihood decoder* finds the value of j, $1 \leqslant j \leqslant 2^k$, for which the likelihood function is largest. If this value is j_0, the decoder decides that codeword j_0 was transmitted. Thus, if $p(\mathbf{y}|\mathbf{x}_j)$ is known, the metric is usually chosen to be a linear function of the logarithm of $p(\mathbf{y}|\mathbf{x}_j)$, which is called the *log-likelihood function.*

If the demodulator outputs are statistically independent and a single output corresponds to each code symbol, then the log-likelihood function can be expressed as

$$\ln p(\mathbf{y}|\mathbf{x}_j) = \sum_{i=1}^{n} \ln p(y_i\,|\,x_{ji}) \tag{1.7.25}$$

where $p(y_i\,|\,x_{ji})$ is the conditional probability density function of y_i given that code symbol x_{ji} was transmitted. For coherent binary PSK communications in white Gaussian noise with matched-filter demodulation [9, 10],

$$p(y_i|x_{ji}) = \frac{1}{\sqrt{\pi N_0}}\,\exp\left[-\,\frac{(y_i - x_{ji})^2}{N_0}\right] \tag{1.7.26}$$

and $x_{ji} = +\sqrt{E_s}$ or $-\sqrt{E_s}$. Because y_i^2 and x_{ji}^2 are independent of j, (1.7.25) and (1.7.26) indicate that a suitable metric for maximum-likelihood decoding is

$$m_0(j,n) = a \sum_{i=1}^{n} x_{ji} y_i + b \tag{1.7.27}$$

where a and b are constants chosen for computational convenience and do not affect the decoder performance.

The *Hamming weight* of a codeword is defined as the number of nonzero symbols in a codeword. For binary codes, the Hamming weight is the number of 1's in a codeword. The set of Hamming distances from a given codword to the other codewords is the same for all codewords of a linear block code [24]. Since the set of distances from the all-zero codeword is the same as the set of weights of the nonzero codewords, the set of weights is equivalent to the set of distances. The *weight distribution* of a code is a list of the number of codewords of each weight. Analytical expressions that can be used to compute the weight distribution are known for the Hamming and Reed-Solomon codes [22, 23]. The weight distributions of other codes can be determined by examining all valid codewords if the number of codewords is not too large for a computation. The weight distribution of the Golay codes is listed in Table 1-1.

Table 1-1

Weight Distribution of Golay Codes

Weight	Number of Codewords	
	(23, 12) Code	(24, 12) Code
0	1	1
7	253	0
8	506	759
11	1288	0
12	1288	2576
15	506	0
16	253	759
23	1	0
24	0	1

A fundamental property of a probability, called *countable subadditivity,* is that the probability of a finite or countable union of events $A_n, n = 1, 2, \ldots,$ satisfies

$$P\left(\bigcup_n A_n\right) \leqslant \sum_n P(A_n) \tag{1.7.28}$$

In communication theory, a bound obtained from this inequality is called a *union bound.*

Let $N(\delta)$ denote the number of codewords having weight δ. Let $Q(\delta)$ denote the probability that the metric for an incorrect codeword at distance δ from the correct codeword exceeds the metric for the correct codeword. It is assumed that the modulation is chosen so that $Q(\delta)$ is independent of the transmitted codeword. Thus, no generality is lost if it is assumed that the all-zero codeword is transmitted. The union bound and the relation between weights and distances imply that P_w for soft-decision decoding satisfies

$$P_w \leqslant \sum_{\delta=d}^{n} N(\delta)Q(\delta) \tag{1.7.29}$$

Metrics are defined so that $Q(\delta_1) \leqslant Q(\delta_2)$ if $\delta_1 \geqslant \delta_2$. Thus, in terms of the minimum distance, d, we have the much weaker bound

$$P_w \leqslant (N_c - 1)Q(d) \tag{1.7.30}$$

where N_c is the total number of codewords. This bound may be useful when the only known parameter of a code is the minimum distance between codewords.

For a repetition code of length $n, P_{ib} = Q(n)$, where n may be even or odd for soft-decision decoding. For other block codes, the union bound and the type of reasoning that led to (1.7.20) indicate that

$$P_{is} \lesssim \sum_{\delta=d}^{n} \frac{\delta}{n} N(\delta)Q(\delta) \tag{1.7.31}$$

Consider a binary code used with PSK modulation and transmitted over a channel that adds white Gaussian noise. For ideal coherent demodulation and an optimal metric, $Q(\delta)$ is easily derived from (1.7.26) and (1.7.27). Let $j = 1$ label the correct sequence and $j = 2$ label an incorrect one at distance δ. Then $Q(\delta)$ is the probability that $m_0(2,n) - m_0(1,n) > 0$. Because $m_0(2,n) - m_0(1,n)$ has a Gaussian distribution and $E_s = rE_b$, we find that

$$Q(\delta) = \frac{1}{2}\,\text{erfc}\left(\sqrt{\frac{\delta r E_b}{N_0}}\right) \tag{1.7.32}$$

Although it is theoretically advantageous, optimal soft-decision decoding cannot be efficiently implemented except for very short block codes, primarily because the number of codewords for which the metrics must be computed is prohibitively large. Among the suboptimal soft-decision decoding algorithms, the

Chase algorithm is one of the most useful [22]. The Chase algorithm generates a small set of candidate codewords that will almost always include the codeword with the largest metric. Various trial sequences are generated by first making hard decisions on each of the received symbols and then altering the least reliable symbols, which are determined from the soft-decision information. Hard-decision decoding of each trial sequence generates one of the candidate codewords. The decoder selects the candidate codeword with the largest metric.

For digital computations to be performed, soft-decision information must be quantized. Two levels of quantization correspond to hard decisions. More than two levels of quantization requires analog-to-digital conversion of the demodulator output samples. Because the optimal location of the levels is a function of the signal, thermal noise, and interference powers, automatic gain control is often necessary. For a white Gaussian noise channel, it is found that an eight-level quantization, which can be represented by three bits, and the appropriate uniform spacing between threshold levels cause no more than a few tenths of a decibel degradation in system performance relative to what could be achieved by processing unquantized analog voltages or using infinitely fine quantization.

The simplest practical soft-decision decoding entails the use of *erasures* to supplement hard-decision decoding [22]. A symbol is said to be erased when the demodulator instructs the decoder to ignore that symbol during the decoding. An erasure occurs when soft-decision information indicates that the symbol is unreliable or that interference is present. If a code has a minimum distance d and a received word contains ϵ erasures, then all codewords differ in at least $d - \epsilon$ of the unerased symbols. Hence, ν errors can be corrected if $2\nu + 1 \leqslant d - \epsilon$ and the maximum number of correctable errors is

$$t_m = \left[\frac{d - \epsilon - 1}{2} \right] \tag{1.7.33}$$

For coherent PSK modulation and a white Gaussian noise channel, decoding with optimal erasures provides only about a 0.2 dB performance improvement relative to hard-decision decoding. However, the use of erasures may be effective against sporadic or pulsed interference that is strong enough to trigger erasures when it occurs.

Calculations for specific communication systems and codes operating in white Gaussian noise have shown that optimal soft-decision decoding gives a superior performance relative to hard-decision decoding. Approximately 2 dB of additional signal power is required for a hard-decision receiver to produce the same information-bit error rate as the corresponding soft-decision receiver. However, soft-decision receivers are much more complex to implement and may be too slow for the processing of high information rates. For a given level of implementation complexity or expense, the length of a block code can usually be

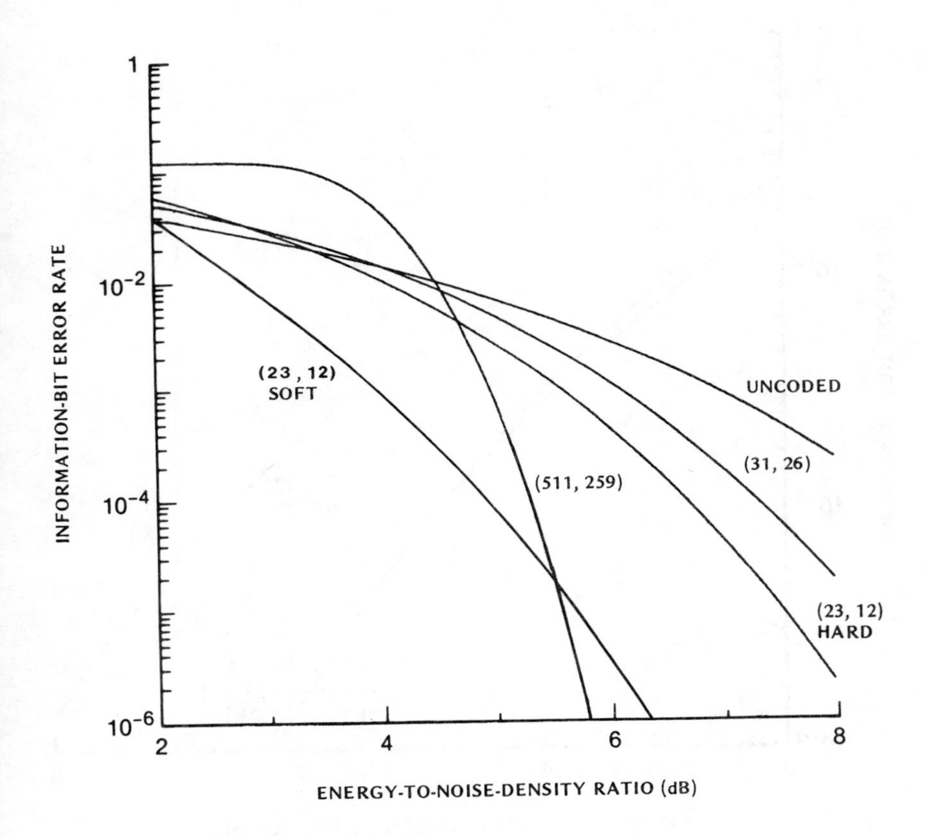

Figure 1.13 Information-bit error rate for binary block (n, k) codes and coherent PSK demodulation.

much longer when hard-decision decoding is used than when soft-decision decoding is used. Consequently, the hard-decision receiver may overcome the inherent advantage of the soft-decision receiver when the cost is fixed. In practice, soft-decision decoding other than erasures is seldom used with block codes of length greater than 50.

Figure 1.13 depicts P_{ib} *versus* E_b/N_0 for various binary block codes with $d = 2t + 1$, coherent PSK, and a white Gaussian noise environment. Equations (1.7.20) and (1.7.24) are used to compute the error rates of the BCH (511, 259)

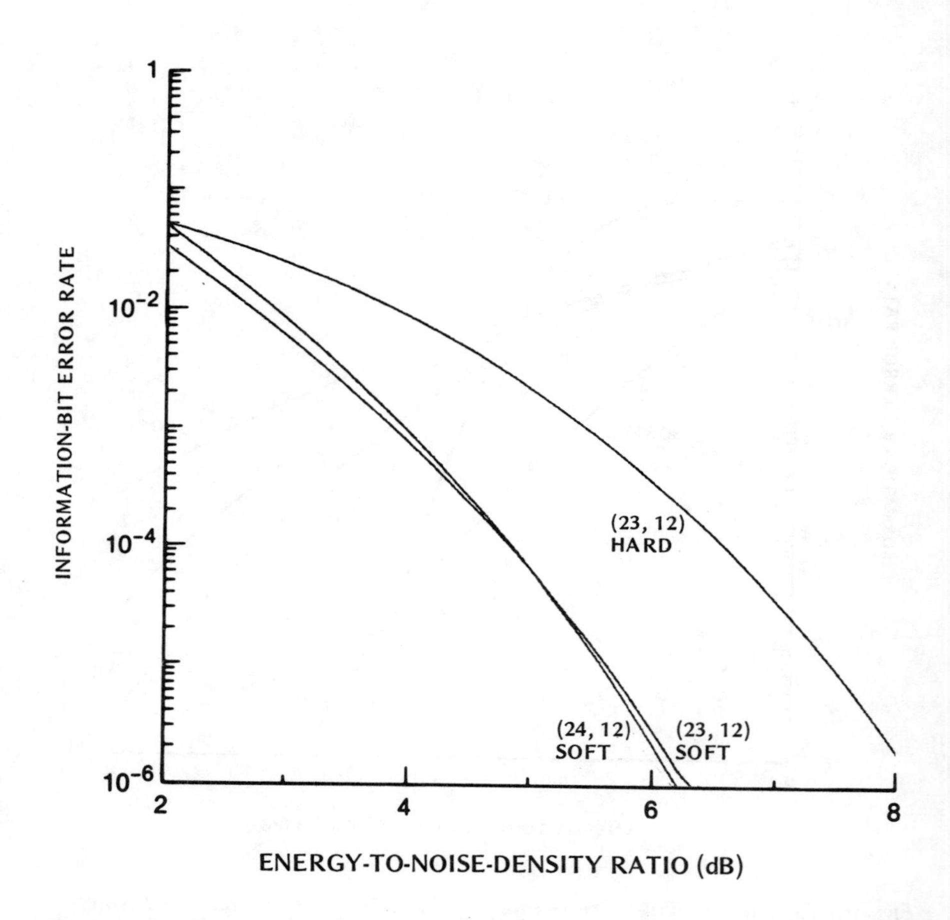

Figure 1.14 Information-bit error rate for Golay codes and coherent PSK demodulation.

code, the Hamming (31, 26) code, and the Golay (23, 12) code with hard decisions. Relations (1.7.31) and (1.7.32) and Table 1-1 are used to compute the performance of the Golay (23, 12) code with optimal soft decisions. Implementation losses from quantization and the possible use of the Chase algorithm are neglected. The curves illustrate the power of the soft-decision decoding. For the Golay (23, 12) code, soft-decision decoding requires approximately 2 dB less E_b/N_0 to achieve $P_{ib} = 10^{-5}$ than hard-decision decoding. The BCH (511, 259) code, which corrects $t = 30$ errors, requires approximately the same E_b/N_0 for

$P_{ib} = 10^{-5}$ as the soft-decision Golay (23, 12) code, which corrects only three errors. For $P_{ib} = 10^{-3}$, the soft decision Golay code provides a performance advantage of more than 0.8 dB relative to the BCH code. If $E_b/N_0 \leqslant 2$ dB, an uncoded system with coherent PSK provides a lower P_{ib} than a similar system that uses one of the block codes of Figure 1.13. A comparison of Golay codes is shown in Figure 1.14.

Figure 1.15 illustrates the performance of Reed-Solomon codes in white Gaussian noise when MFSK, noncoherent demodulation, and hard decisions are used. The error rates are computed from (1.7.22), (1.7.23), and (1.4.51) and the

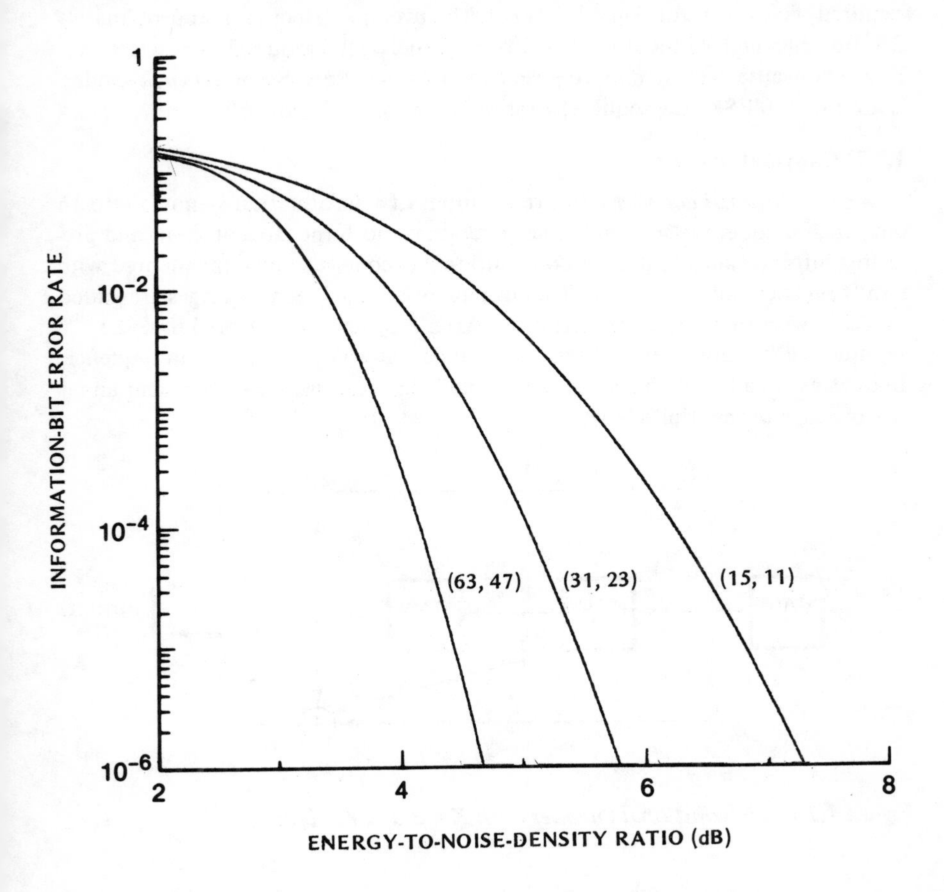

Figure 1.15 Information-bit error rate for Reed-Solomon (n, k) codes and noncoherent MFSK demodulation.

relations $d = n - k + 1 = 2t + 1$ and $N = 2^m = n + 1$. For the chosen values of n, the best performance at $P_{ib} = 10^{-5}$ is obtained if $r \cong 3/4$. The (31, 23) code provides a nearly 8 dB advantage over uncoded binary FSK at $P_{ib} = 10^{-5}$. Further gains result from increasing n and hence the implementation complexity. Reed-Solomon decoders can accommodate high data rates efficiently because operations are performed at the symbol rate rather than the higher information-bit rate. A major disadvantage of Reed-Solomon codes with MFSK is the bandwidth requirement. Let B denote the bandwidth required for an uncoded binary PSK signal. If the same data rate is accommodated by using uncoded binary FSK, the required bandwidth for demodulation with envelope detectors is approximately $2B$. For uncoded MFSK using $N = 2^m$ frequencies, the required bandwidth is $2^m B/m$ because each symbol represents m bits. If a Reed-Solomon (n, k) code is used with MFSK, the required bandwidth becomes $2^m nB/mk$.

1.7.2 Convolutional Codes

A *convolutional encoder* converts an input of k information symbols into an output of n code symbols that are a function of both the current input and preceding information symbols. A convolutional encoder can be implemented with a shift register and linear logic. The outputs of selected register stages are added modulo-two to form the code symbols. After k symbols are shifted into the register and k symbols are shifted out, n code symbols are read out in sequence. In contrast to a block codeword, a convolutional codeword can represent an entire message of indefinite length.

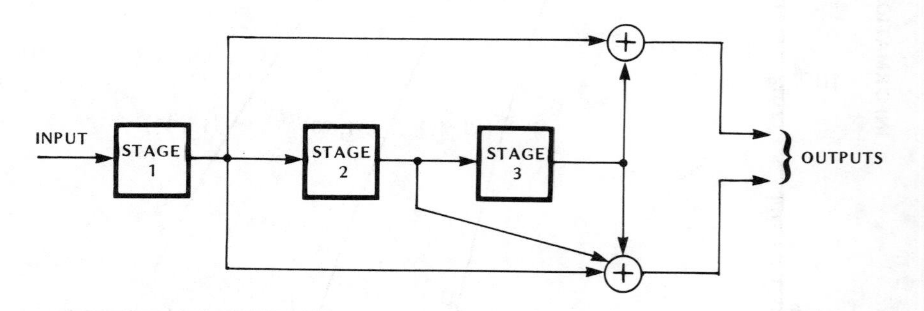

Figure 1.16 Convolutional encoder with $K = 3$ and $r = 1/2$.

A simple example of an encoder with $k = 1$ and $n = 2$ is shown in Figure 1.16. The shift register consists of three bistable devices (flip-flops). Information bits enter the shift register in response to clock pulses. After each clock pulse, the most recent information bit becomes the content and output of the first stage, the contents of the first two stages are shifted to the right, and the content of the third stage is shifted out of the register. The outputs of the modulo-two adders (Exclusive-OR gates) provide two code bits.

Several definitions of the *constraint length* of a convolutional code are in use. Here we define the constraint length, K, as the maximum number of shift-register stages that can influence an output symbol. Thus, $K = 3$ for the encoder of Figure 1.16.

If there are m bits per input symbol, then $b = mk$ bits enter the register with each iteration. Because b bits are dumped from the register as the new bits enter it, only the contents of $K - b$ stages affect the subsequent outputs. Therefore, the contents of these $K - b$ stages define the *state* of the encoder. The encoder of Figure 1.16 has $b = m = k = 1$ and the first $K - 1 = 2$ stages define its state.

The encoder generally starts in the zero state. After the message sequence has been encoded, $K - b$ *zeros* must be inserted into the encoder to terminate the codeword. These $K - b$ bits cause a decrease in the code rate and a loss in system performance. However, if the number of bits in the message is much greater than $K - b$, this effect is negligible and the code rate is $r \cong k/n$.

A *trellis diagram* depicts the structure of a convolutional code. A trellis diagram corresponding to the encoder of Figure 1.16 is shown in Figure 1.17. Each of the nodes in the column of a trellis diagram represents the state of the encoder

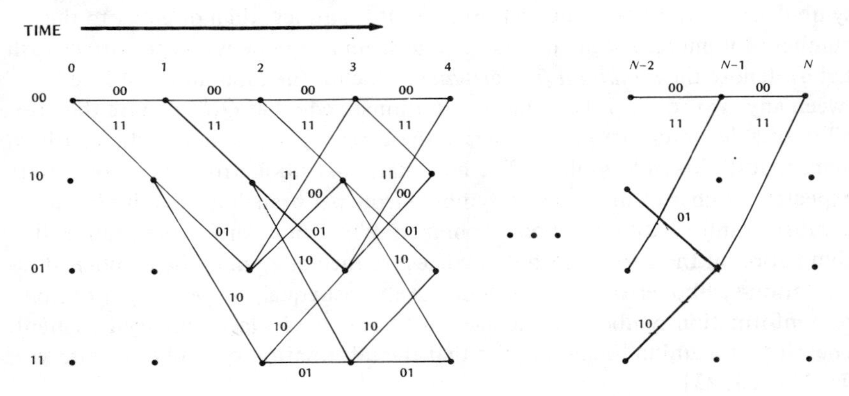

Figure 1.17 Trellis diagram for Fig. 1.16.

following a clock pulse. The first bit of a state represents the content of stage 1, while the second bit represents the content of stage 2. Branches connecting nodes represent possible changes of state. Each branch is labeled with the output bits or symbols produced following a clock pulse and the formation of a new encoder state. In the example of Figure 1.17, the first bit of a branch label refers to the upper output of the encoder. The upper branch leaving a node corresponds to a 0 input bit, while the lower branch corresponds to a 1.

If the encoder begins in the zero state, not all of the other states can be reached until the initial contents have been shifted out. The trellis diagram then becomes identical from column to column until the final $K - b$ input bits force the encoder back to the zero state. Every path from left to right through the trellis represents a possible codeword.

A maximum-likelihood decoder selects the codeword with the largest metric. Each branch of the trellis is associated with a branch metric, and the metric of a codeword is defined as the sum of the branch metrics for the path associated with the codeword. The *Viterbi decoder* implements maximum-likelihood decoding efficiently by sequentially eliminating many of the possible paths. At any node, only the path reaching that node with the largest partial metric is retained, for any other path into that node must ultimately produce a smaller total metric regardless of its subsequent trajectory.

To bound the information-symbol error rate for the Viterbi decoder, we assume that the all-zero codeword is transmitted. This assumption usually entails no loss of generality [24]. Let $a(\delta, i)$ denote the number of paths diverging at a node from the correct path, each having Hamming weight δ and i information-symbol errors over the unmerged segment; it is also less than or equal to the number of unmerged segments that merge again at a node with the correct path. Let d_f denote the *minimum free distance,* which is the minimum distance between any two codewords in the convolutional code. Let $Q(\delta)$ denote the probability of a decoding error in comparing the correct path segment with a path segment that differs in δ symbols. The information-symbol error rate is equal to the expected number of information-symbol errors per branch divided by k, the number of information symbols per branch. When an incorrect path causes the elimination of the correct path or a correct branch at a node, the number of information-symbol errors that are added is at most equal to the number of nonzero information symbols on the incorrect path over its last unmerged segment. Therefore, the union bound implies that the information-symbol error rate satisfies [22, 23, 25]

$$P_{is} \leqslant \frac{1}{k} \sum_{i=1}^{\infty} \sum_{\delta=d_f}^{\infty} i\, a(\delta, i) Q(\delta) \tag{1.7.34}$$

This bound is known to be very tight in most applications of interest. For binary codes, P_{is} is equal to the information-bit error rate, P_{ib}, which is equal to the average probability of an information-bit error.

Because the decoding complexity grows exponentially with constraint length, Viterbi decoders are limited to use with convolutional codes of short constraint lengths. *Sequential decoding* of convolutional codes is a suboptimal method that does not invariably provide maximum-likelihood decisions. However, because its implementation complexity is only weakly dependent upon the constraint length, very low error probabilities can be attained by using long constraint lengths [22, 23, 25]. The number of computations needed to decode a frame of data is fixed when Viterbi decoding is used, but is a random variable when sequential decoding is used. When strong interference is present, the excessive computational demands and consequent memory overflows of sequential decoding may result in a higher P_{ib} than for Viterbi decoding. Thus, Viterbi decoding appears to be preferable for most communication systems that are to resist strong interference and is assumed in the subsequent analysis of the performance of convolutional codes.

Among the convolutional codes of a given code rate and constraint length, one with favorable distance properties can sometimes be determined by a complete computer search. After elimination of the *catastrophic codes,* for which a finite number of demodulated symbol errors can cause an infinite number of decoded information-bit errors, the codes with the largest value of d_f are selected. If more than one code is selected, then the value of the *total information weight,*

$$A(\delta) = \sum_{i=1}^{\infty} i\,a(\delta, i), \qquad \delta \geqslant d_f \tag{1.7.35}$$

is examined for each of the selected codes. All codes that do not have the minimum value of $A(d_f)$ are eliminated. If more than one code remains, codes are eliminated on the basis of the minimal values of $A(d_f + 1), A(d_f + 2), \ldots,$ until one code remains. For binary codes of rates 1/2, 1/3, and 1/4 and constraint lengths up to 14, the codes with these favorable distance properties have been determined [27]. For these codes, Tables 1-2, 1-3, and 1-4 list the corresponding values of d_f and $A(d_f + i)$, $i = 0, 1, \ldots, 7$. In terms of $A(d_f + i)$, (1.7.34) becomes

$$P_{is} \leqslant \frac{1}{k} \sum_{i=0}^{\infty} A(d_f + i)\,Q(d_f + i) \tag{1.7.36}$$

If a binary code is used and the demodulator makes hard decisions an exact expression for $Q(\delta)$ can be determined. When a correct path segment is compared with an incorrect one, correct decoding results if the number of symbol errors in

Table 1-2
Parameter Values for Convolutional Codes with Rate = 1/2
and Favorable Distance Properties

K	d_f	$A(d_f)$	$A(d_f+1)$	$A(d_f+2)$	$A(d_f+3)$	$A(d_f+4)$	$A(d_f+5)$	$A(d_f+6)$	$A(d_f+7)$
3	5	1	4	12	32	80	192	448	1024
4	6	2	7	18	49	130	333	836	2069
5	7	4	12	20	72	225	500	1324	3680
6	8	2	36	32	62	332	701	2342	5503
7	10	36	0	211	0	1404	0	11,633	0
8	10	2	22	60	148	340	1008	2642	6748
9	12	33	0	281	0	2179	0	15,035	0
10	12	14	26	74	257	496	1378	4122	10,832
11	14	94	0	463	0	3783	0	26,711	0
12	15	76	180	374	1142	2783	6836	18,709	49,242
13	16	152	0	971	0	6933	0	45,436	0
14	16	22	99	218	608	1724	4404	11,108	30,438

Table 1-3
Parameter Values for Convolutional Codes with Rate = 1/3
and Favorable Distance Properties

K	df	$A(df)$	$A(df+1)$	$A(df+2)$	$A(df+3)$	$A(df+4)$	$A(df+5)$	$A(df+6)$	$A(df+7)$
3	8	3	0	15	0	58	0	201	0
4	10	6	0	6	0	58	0	118	0
5	12	12	0	12	0	56	0	320	0
6	13	1	8	26	20	19	62	86	204
7	15	11	16	19	28	55	96	169	338
8	16	1	0	24	0	113	0	287	0
9	18	11	0	32	0	195	0	564	0
10	20	29	0	91	0	246	0	954	0
11	22	53	0	92	0	347	0	1104	0
12	24	80	0	58	0	607	0	1563	0
13	24	27	0	74	0	228	0	794	0
14	26	41	0	165	0	319	0	1156	0

Table 1-4
Parameter Values for Convolutional Codes with Rate = 1/4
and Favorable Distance Properties

K	d_f	$A(d_f)$	$A(d_f+1)$	$A(d_f+2)$	$A(d_f+3)$	$A(d_f+4)$	$A(d_f+5)$	$A(d_f+6)$	$A(d_f+7)$
3	10	2	1	4	9	8	25	32	52
4	13	4	2	0	10	3	16	34	18
5	16	8	0	7	0	17	0	60	0
6	18	6	0	17	0	24	0	60	0
7	20	31	0	0	0	94	0	0	0
8	22	2	10	10	8	10	11	54	64
9	24	4	0	22	0	38	0	103	0
10	27	10	12	18	44	31	72	120	108
11	29	13	24	18	22	35	34	56	108
12	32	49	0	40	0	82	0	267	0
13	33	19	16	15	46	29	48	124	140
14	36	74	0	80	0	177	0	493	0

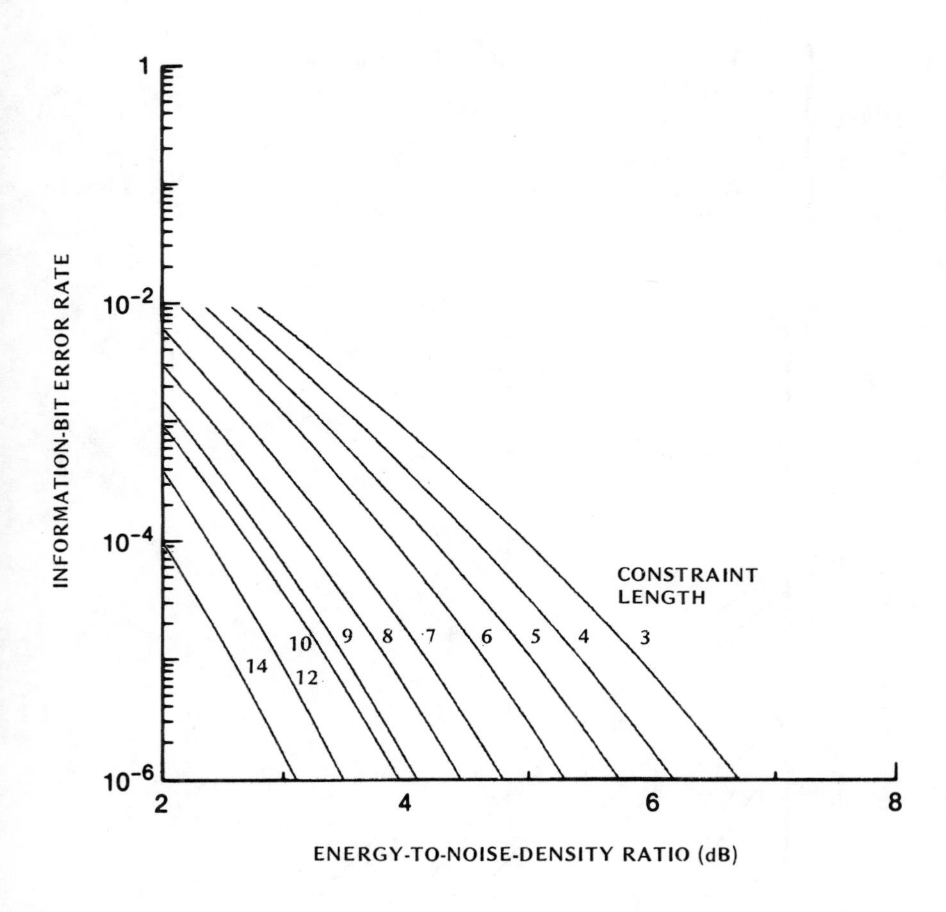

Figure 1.18 Information-bit error rate for rate-1/2 convolutional code and coherent PSK demodulation.

the demodulator output is less than half the number of symbols in which the two segments differ. If the number of symbol errors is exactly half the number of differing symbols, then either of the two segments is chosen with equal probability. Assuming the independence of symbol errors, it follows that

$$Q(\delta) = \begin{cases} \displaystyle\sum_{i=(\delta+1)/2}^{\delta} \binom{\delta}{i} P_s^i (1 - P_s)^{\delta-i}, & \delta \text{ is odd} \\ \displaystyle\sum_{i=\delta/2+1}^{\delta} \binom{\delta}{i} P_s^i (1 - P_s)^{\delta-i} + \frac{1}{2} \binom{\delta}{\delta/2} [P_s(1 - P_s)]^{\delta/2}, & \delta \text{ is even} \end{cases} \tag{1.7.37}$$

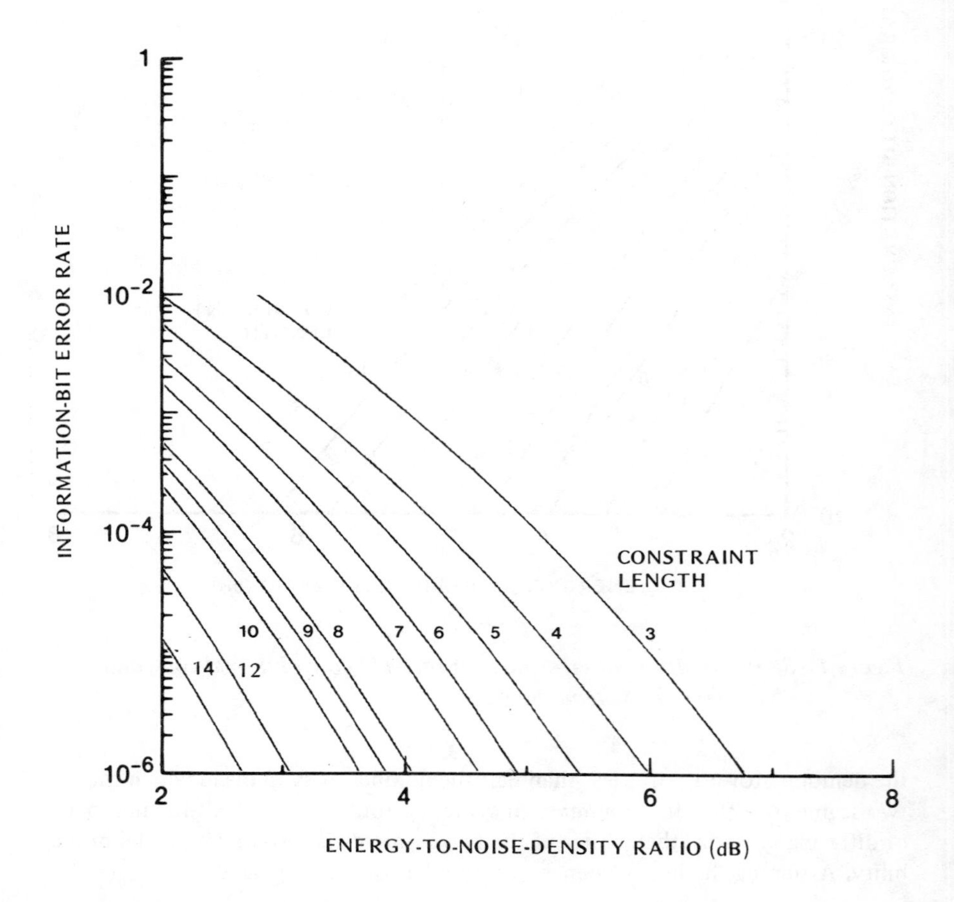

Figure 1.19 Information-bit error rate for rate-1/3 convolutional code and coherent PSK demodulation.

Soft-decision decoding typically provides a 2 dB power savings at $P_{ib} = 10^{-5}$ compared to hard-decision decoding for communications in white Gaussian noise. Because there is a minor decrease in implementation complexity and the loss due to three-bit quantization usually is 0.2 to 0.3 dB, soft-decision decoding is highly preferable for the Gaussian channel. Approximate upper bounds on P_{ib} for rate-1/2, rate-1/3, and rate-1/4 convolutional codes with coherent PSK, soft-decision decoding, and infinitely fine quantization are depicted in Figures 1.18 to 1.20.

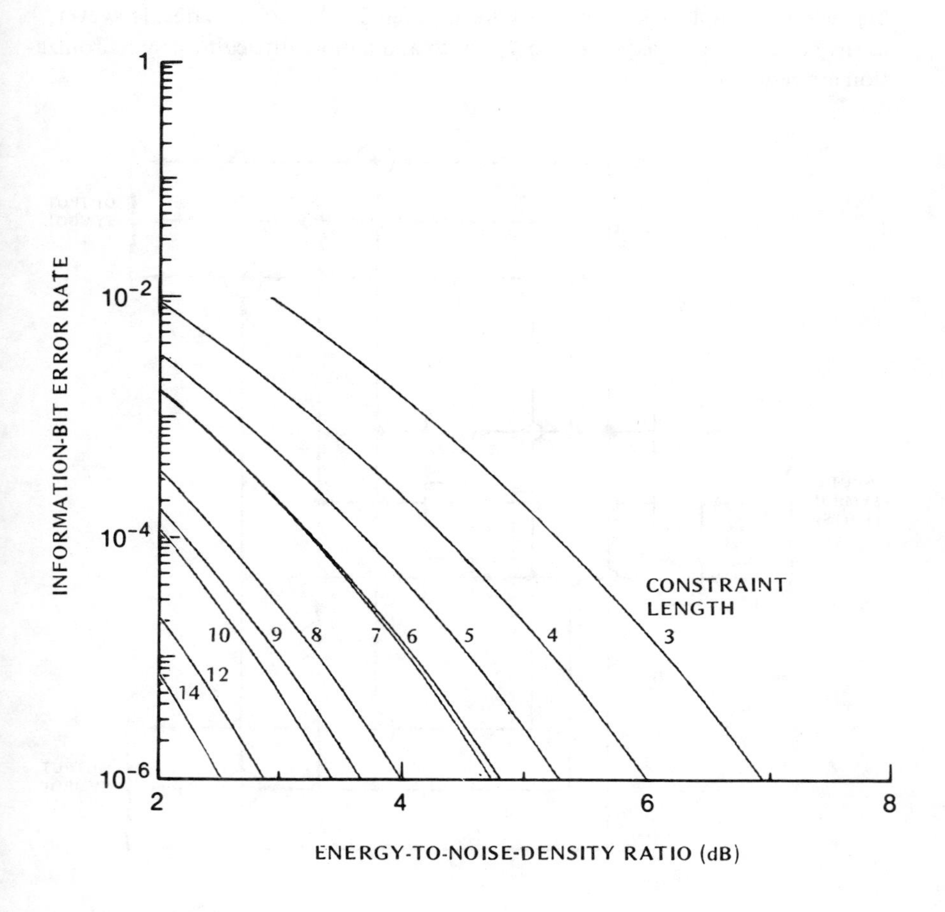

Figure 1.20 Information-bit error rate for rate-1/4 convolutional code and coherent PSK demodulation.

The curves are computed by using (1.7.32) and Tables 1-2 to 1-4 in (1.7.36) and truncating the series after eight terms. This truncation gives reasonably accurate results for $P_{ib} \leqslant 10^{-3}$. However, the truncation does not necessarily provide a close approximation to the upper bound on P_{ib} when $P_{ib} > 10^{-3}$ and the bound itself becomes looser as P_{ib} increases. The figures indicate that the code performance improves with increases in the constraint length. For a fixed value of K and $K \geqslant 4$, the rate-1/3 and rate-1/4 codes both exhibit a better performance than the rate-1/2 code. For a fixed value of K, the three codes require comparable decoder complexities because there are then 2^{K-1} encoder states. However, as the code rate decreases, more bandwidth and a more difficult bit synchronization are required.

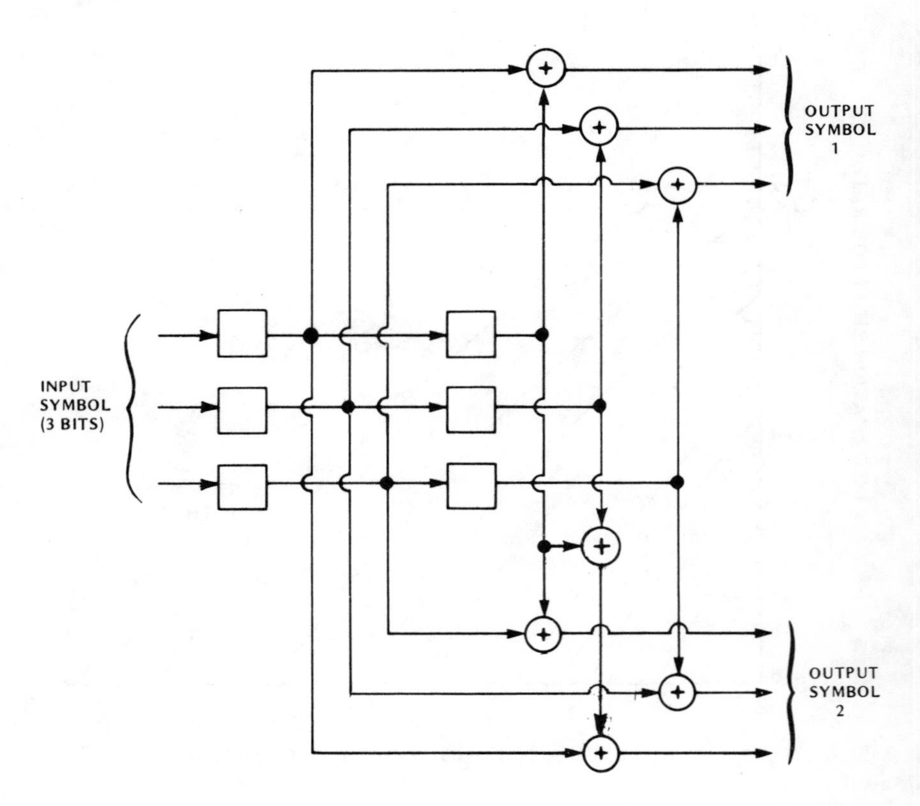

Figure 1.21 Rate-1/2, dual-3 convolutional encoder.

The codes with favorable distance properties are not necessarily the codes that require the smallest value of E_b/N_0 for a specified P_{ib}. For $K \leqslant 7$, other codes that require less energy to attain certain values of P_{ib} have been found [28]. However, the codes of Tables 1-2 and 1-3 require no more than an additional 0.1 dB in E_b/N_0, and the codes of Table 1-4 require no more than an additional 0.4 dB.

In principle, $A(\delta)$ can be determined from the *augmented generating function*, $T(D,I)$, which depends upon the structure of the convolutional code. An expansion of the augmented generating function has the form [22, 23, 25]

$$T(D,I) = \sum_{i=1}^{\infty} \sum_{\delta=d_f}^{\infty} a(\delta,i) D^{\delta} I^{i} \tag{1.7.38}$$

The derivative at $I = 1$ is

$$\left.\frac{\partial T(D,I)}{\partial I}\right|_{I=1} = \sum_{i=1}^{\infty} \sum_{\delta=d_f}^{\infty} i a(\delta,i) D^{\delta} = \sum_{i=0}^{\infty} A(d_f + i) D^{d_f+i} \tag{1.7.39}$$

Thus, the bound on P_{is}, given by inequality (1.7.36), is determined by substituting $Q(\delta)$ in place of D^{δ} in this expression and multiplying the result by $1/k$.

In many applications, it is possible to establish an inequality of the form

$$Q(\delta) \leqslant \alpha Z^{\delta} \tag{1.7.40}$$

where α and Z are independent of δ. It follows from (1.7.40), (1.7.39), and (1.7.36) that

$$P_{is} \leqslant \frac{\alpha}{k} \left.\frac{\partial T(D,I)}{\partial I}\right|_{I=1,\, D=Z} \tag{1.7.41}$$

For a white Gaussian noise channel, binary coding, and coherent PSK demodulation, $Q(\delta)$ is given by (1.7.32). Using the definition of erfc (y) given by (1.4.54) and changing variables, we obtain

$$\begin{aligned}\operatorname{erfc}(\sqrt{\nu+\beta}) &= \frac{2}{\sqrt{\pi}} \int_0^{\infty} \exp[-(y+\sqrt{\nu+\beta})^2]\,dy \\ &= \frac{2}{\sqrt{\pi}} \exp(-\beta) \int_0^{\infty} \exp[-(y+\sqrt{\nu})^2 + 2y(\sqrt{\nu}-\sqrt{\nu+\beta})]\,dy\end{aligned} \tag{1.7.42}$$

If $\nu \geqslant 0$ and $\beta \geqslant 0$, then $\sqrt{\nu} - \sqrt{\nu+\beta} \leqslant 0$ and

$$\operatorname{erfc}(\sqrt{\nu+\beta}) \leqslant \frac{2}{\sqrt{\pi}} \exp(-\beta) \int_0^{\infty} \exp[-(y+\sqrt{\nu})^2]\,dy \tag{1.7.43}$$

Thus, we obtain

$$\text{erfc}(\sqrt{\nu+\beta}) \leqslant \exp(-\beta)\text{erfc}(\sqrt{\nu}), \quad \nu \geqslant 0, \ \beta \geqslant 0 \tag{1.7.44}$$

This inequality and (1.7.32) give

$$Q(\delta) \leqslant \frac{1}{2} \text{ erfc}\left(\sqrt{\frac{d_f r E_b}{N_0}}\right) \exp[-(\delta - d_f) r E_b/N_0] \tag{1.7.45}$$

Thus, (1 7.40) holds with

$$\alpha = \frac{1}{2} \text{ erfc}\left(\sqrt{\frac{d_f r E_b}{N_0}}\right) \exp(d_f r E_b/N_0) \tag{1.7.46}$$

$$Z = \exp(-rE_b/N_0) \tag{1.7.47}$$

Closed-form expressions for $T(D,I)$ are known for the orthogonal and the dual-m convolutional codes. The encoder for an *orthogonal convolutional code* of constraint length K generates one of 2^K orthogonal binary sequences of length $n = 2^K$ for each input bit. Each nonzero sequence has weight $n/2$. For a binary transmission, the rate of the code is 2^{-K}. The augmented generating function can be shown to be [25]

$$T(D,I) = \frac{ID^{Kn/2}(1 - D^{n/2})}{1 - D^{n/2}\{1 + I[1 - D^{(n/2)(K-1)}]\}} \tag{1.7.48}$$

In the expansion of $T(D, I)$, the coefficient $a(\delta, i)$ represents the number of diverging paths that differ in δ binary code symbols and i information bits from the correct path.

By using a nonbinary modulation, the performance in white Gaussian noise is improved. If an orthogonal binary sequence is transmitted as a single code symbol, then the code rate is 1. In this case; let $a(\delta, i)$ represent the number of diverging paths that differ in δ nonbinary code symbols and i information bits from the correct path. Since each nonzero code symbol consists of bits with weight $n/2$, the augmented generating function is obtained by substituting D in place of $D^{n/2}$ in (1.7.48). The result is

$$T(D,I) = \frac{ID^K(1 - D)}{1 - D[1 + I(1 - D^{K-1})]} \tag{1.7.49}$$

Differentiation of this equation yields

$$\left.\frac{\partial T(D,I)}{\partial I}\right|_{I=1} = \frac{D^K(1 - D)^2}{(1 - 2D + D^K)^2} \tag{1.7.50}$$

Using $k = 1$ and $P_{is} = P_{ib}$ in (1.7.41), we obtain

$$P_{ib} \leqslant \alpha \frac{Z^K(1 - Z)^2}{(1 - 2Z + Z^K)^2} \tag{1.7.51}$$

A *dual-m code* is a nonbinary convolutional code. The encoder shifts one m-bit information symbol at a time into a shift register of $2m$ binary stages (two symbol stages). For each information symbol, a rate $1/n$ encoder generates n code symbols. Each code symbol is one of 2^m possible symbols and may be transmitted as one of 2^m modulated signals. A rate-1/2, dual-3 encoder is shown in Figure 1.21 [22].

Among the dual-m codes, the codes with the best distance properties can be shown to have [29]

$$T(D,I) = \frac{(2^m - 1)D^{2n}I}{1 - I[nD^{n-1} + (2^m - 1 - n)D^n]} \tag{1.7.52}$$

which implies that

$$\left.\frac{\partial T(D,I)}{\partial I}\right|_{I=1} = \frac{(2^m - 1)D^{2n}}{[1 - nD^{n-1} - (2^m - 1 - n)D^n]^2} \tag{1.7.53}$$

Relations (1.7.53), (1.7.41), and (1.7.21) with $k = 1$ give

$$P_{ib} \leqslant \frac{\alpha 2^{m-1} Z^{2n}}{[1 - nZ^{n-1} - (2^m - 1 - n)Z^n]^2} \tag{1.7.54}$$

To simplify the evaluation of (1.7.51) and (1.7.54) for the white Gaussian noise channel and noncoherent MFSK, we first derive the Chernoff bound.

1.7.3 Chernoff Bound

The *moment generating function* of the random variable X is defined as

$$M(s) = E[e^{sX}] = \int_{-\infty}^{\infty} \exp(sx)\, dF(x) \tag{1.7.55}$$

for all real s for which the integral is finite, where $F(x)$ is the distribution function of X. Let $P[\ ]$ denote the probability of the event in the brackets. For all nonnegative s,

$$P[X \geqslant 0] = \int_0^{\infty} dF(x) \leqslant \int_0^{\infty} \exp(sx)\, dF(x) \tag{1.7.56}$$

Comparing (1.7.55) and (1.7.56), we conclude that

$$P[X \geqslant 0] \leqslant M(s), \qquad 0 \leqslant s < s_1 \tag{1.7.57}$$

where s_1 is the upper limit of the interval in which $M(s)$ is defined. To make this bound as tight as possible, we choose the value of s that minimizes $M(s)$. Therefore,

$$P[X \geqslant 0] = \min_{0 \leqslant s < s_1} M(s) \tag{1.7.58}$$

The right-hand side of this inequality is called the *Chernoff bound.* It is potentially useful if it can be more easily evaluated than $P[X \geqslant 0]$. From (1.7.58) and (1.7.55), we obtain

$$P[X \geqslant b] = \min_{0 \leqslant s < s_1} M(s)\exp(-sb) \tag{1.7.59}$$

Because the moment generating function is finite in some neighborhood of $s = 0$, it can be shown that [30] we may differentiate under the integral sign in (1.7.55) to obtain the derivative of $M(s)$. The result is

$$M'(s) = \int_{-\infty}^{\infty} x \exp(sx)\, dF(x) \tag{1.7.60}$$

It follows that $M'(0) = E[X]$. Differentiating (1.7.60) gives the second derivative

$$M''(s) = \int_{-\infty}^{\infty} x^2 \exp(sx)\, dF(x) \tag{1.7.61}$$

This equation shows that $M''(s) \geqslant 0$, which implies that $M(s)$ is convex in its interval of definition. Suppose that

$$E(X) < 0, \qquad P(X > 0) > 0 \tag{1.7.62}$$

Then $M'(0) < 0$ by the first assumption, and $M(s) \to \infty$ as $s \to \infty$ by the second. Because $M(s)$ is convex, it has its minimum value at some positive $s = s_0$. We conclude that (1.7.62) is sufficient to ensure that the Chernoff bound is less than unity and $s_0 > 0$.

The Chernoff bound can be tightened [31] if X has a density function, $f(x)$, that satisfies

$$f(-x) \geqslant f(x), \qquad x \geqslant 0 \tag{1.7.63}$$

For s in A, where A is the open interval over which $M(s)$ is finite, (1.7.63) implies that

$$\begin{aligned} M(s) &= \int_{-\infty}^{\infty} \exp(sx) f(x)\,dx \\ &= \int_{0}^{\infty} \exp(sx) f(x)\,dx + \int_{-\infty}^{0} \exp(sx) f(x)\,dx \\ &\geqslant \int_{0}^{\infty} [\exp(sx) + \exp(-sx)]\, f(x)\,dx \\ &= \int_{0}^{\infty} 2\cosh(sx) f(x)\,dx \\ &\geqslant 2\int_{0}^{\infty} f(x)\,dx \\ &= 2P[X \geqslant 0] \end{aligned} \tag{1.7.64}$$

Thus, we obtain

$$P[X \geqslant 0] \leqslant \frac{1}{2} \min_{s \in A} M(s) \tag{1.7.65}$$

In this version of the Chernoff bound, the minimum value s_0 is not required to be nonnegative. However, if (1.7.62) holds, then the bound is less than 1/2, $s_0 > 0$, and

$$P[X \geqslant 0] \leqslant \frac{1}{2} \min_{0<s<s_1} M(s) \tag{1.7.66}$$

In soft-decision decoding, the sequence with the largest associated métric is converted into the decoded output. For block codes, the sequence is a codeword; for convolutional codes, the sequence is a path segment. Let $m_0(j,L)$ denote the value of the metric associated with sequence j of length L. Consider additive metrics having the form

$$m_0(j,L) = \sum_{i=1}^{L} m_1(j,i) \tag{1.7.67}$$

where $m_1(j,i)$ is the *symbol metric* determined from code symbol i. Let $j = 1$ label the correct sequence and $j = 2$ label an incorrect one at distance δ. By suitably relabeling the δ symbol metrics that may differ for the two sequences, we obtain

$$Q(\delta) \leqslant P[m_0(2,L) \geqslant m_0(1,L)]$$
$$= P\left[\sum_{i=1}^{\delta} [m_1(2,i) - m_1(1,i)] \geqslant 0\right] \tag{1.7.68}$$

where the inequality results because $m_0(2,L) = m_0(1,L)$ does not necessarily cause an error if it occurs. In all practical cases, (1.7.62) is satisfied for the random variable $X = m_0(2,L) - m_0(1,L)$. Therefore, the Chernoff bound implies that

$$Q(\delta) \leqslant \alpha \min_{0<s<s_1} E\left[\exp\left\{ s \sum_{i=1}^{\delta} [m_1(2,i) - m_1(1,i)] \right\}\right] \tag{1.7.69}$$

where s_1 is the upper limit of the interval over which the expected value is defined. If (1.7.63) is satisfied, then $\alpha = 1/2$; otherwise, $\alpha = 1$.

The assumption that the $m_1(2,i) - m_1(1,i)$, $i = 1, 2, \ldots, \delta$, are independent, identically distributed random variables and the definition

$$Z = \min_{0<s<s_1} E\left[\exp\left\{ s[m_1(2,i) - m_1(1,i)] \right\}\right] \tag{1.7.70}$$

yield

$$Q(\delta) \leqslant \alpha Z^{\delta} \tag{1.7.71}$$

which is the same as (1.7.40). As δ increases, the central-limit theorem implies that the distribution of $X = m_0(2,L) - m_0(1,L)$ approximates the Gaussian distribution. Thus, for large enough δ, (1.7.63) is satisfied when $E[X] < 0$, and we can set $\alpha = 1/2$ in (1.7.71). For small δ, (1.7.63) may be difficult to establish mathematically, but is often intuitively clear. In many applications of the Chernoff bound, whether $\alpha = 1/2$ or $\alpha = 1$ is insignificant and not worth pursuing if it is not obvious.

As a simple example of the application of (1.7.70), consider binary hard-decision decoding, which can be regarded as a special case of soft-decision decoding. From (1.7.27), it follows that a suitable symbol metric for maximum-likelihood decoding and coherent binary PSK is

$$m_1(j,i) = x_{ji} y_i \tag{1.7.72}$$

where $x_{ji} = +1$ if symbol i in candidate codeword j is a 1, $x_{ji} = -1$ if it is a 0, and y_i is the ith output sample of the demodulator. When hard decisions are made by the demodulator, the processing is equivalent to assigning the symbol metric

$$m_1(j,i) = x_{ji} g(y_i) \tag{1.7.73}$$

where $g(y_i) = +1$ if $y_i > 0$ and $g(y_i) = -1$ if $y_i < 0$. Therefore, $m_1(2,i) - m_1(1,i)$ is equal to -2 with probability $(1 - P_s)$ and is equal to $+2$ with probability P_s. It follows from (1.7.70) that

$$\begin{aligned} Z &= \min_{0<s} \left[(1 - P_s)\, e^{-2s} + P_s e^{2s}\right] \\ &= 2\left[P_s(1 - P_s)\right]^{1/2} \end{aligned} \tag{1.7.74}$$

for binary hard-decision decoding. Substituting (1.7.74) into (1.7.71), we obtain a simple upper bound on $Q(\delta)$, which is exactly given by (1.7.37).

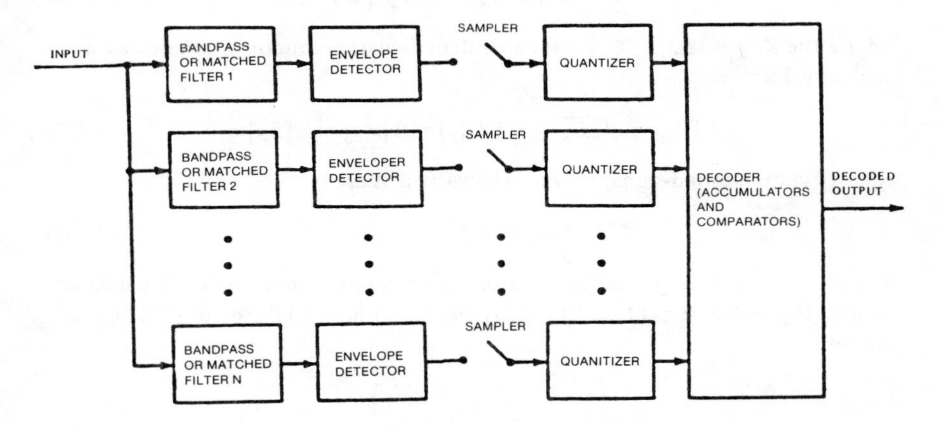

Figure 1.22 MFSK receiver with soft-decision decoding.

1.7.4 Noncoherent MFSK

A noncoherent MFSK demodulator usually consists of an array of bandpass filters, each followed by an envelope detector. For soft decisions, the detector outputs are sampled, quantized, and then used to compute the metrics, as indicated in Figure 1.22. We assume that the symbol metric is

$$m_1(j,i) = R_{ji}^2 \tag{1.7.75}$$

where R_{ji} is the sample value of the envelope-detector output that is associated with symbol i of candidate sequence j. In a white Gaussian noise environment, if the passbands of the two bandpass filters are disjoint, then the corresponding envelope detector outputs are statistically independent of each other (Section 1.4). Therefore, (1.7.70) gives

$$Z = \min_{0<s<s_1} E[\exp(sR_2^2)]\, E[\exp(-sR_1^2)] \tag{1.7.76}$$

where R_1 is the output of an envelope detector when a signal is present at the input, whereas R_2 is the output when a signal is absent. For a Gaussian random variable X with mean m and variance σ^2, a direct calculation yields

$$E[\exp(sX^2)] = \frac{1}{\sqrt{1-2s\sigma^2}} \exp\left(\frac{sm^2}{1-2s\sigma^2}\right), \quad s < \frac{1}{2\sigma^2} \tag{1.7.77}$$

From the analysis of Section 1.4, we have

$$R_1^2 = Z_1^2 + Z_2^2\,, \quad R_2^2 = Z_3^2 + Z_4^2 \tag{1.7.78}$$

where the $Z_j, j = 1, 2, 3, 4$, are independent Gaussian random variables with expected values

$$E[Z_1] = \sqrt{2R_s}\,, \quad E[Z_2] = E[Z_3] = E[Z_4] = 0 \tag{1.7.79}$$

and R_s is the desired-signal power. The variances are

$$\mathrm{VAR}(Z_j) = N_t, \quad j = 1, 2, 3, 4 \tag{1.7.80}$$

where N_t is the noise power at the input of each envelope detector. We assume that $R_s/N_t > 0$ so that (1.7.62) is satisfied. Combining (1.7.76) to (1.7.80), we obtain

$$Z = \frac{\exp\left[-\left(\frac{\lambda}{1+\lambda}\right)\left(\frac{R_s}{N_t}\right)\right]}{1-\lambda^2} \tag{1.7.81}$$

where

$$\lambda = -\left(\frac{1}{2} + \frac{R_s}{4N_t}\right) + \left[\left(\frac{1}{2} + \frac{R_s}{4N_t}\right)^2 + \frac{R_s}{2N_t}\right]^{1/2} \tag{1.7.82}$$

If the product of the symbol duration and the bandwidth of the bandpass filter is unity, then

$$\frac{R_s}{N_t} = \frac{rE_b}{N_0} \tag{1.7.83}$$

where r is the number of information bits per transmitted symbol, E_b is the received energy per information bit, and N_0 is the noise power spectral density.

Using (1.7.81) to (1.7.83), (1.7.54), and (1.7.51) and assuming that $\alpha = 1/2$, which is very tedious to verify, we obtain upper bounds on the information-bit error rate for orthogonal and dual-m convolutional codes with MFSK demodulation. Figure 1.23 illustrates the performance bounds of various orthogonal and rate-1/2 ($n = 2$) dual-m codes for the white Gaussian noise channel. More precise calculations indicate that the use of the Chernoff bound increases the upper bounds on P_{ib} by roughly 0.3 to 0.5 dB when $P_{ib} \leqslant 10^{-3}$.

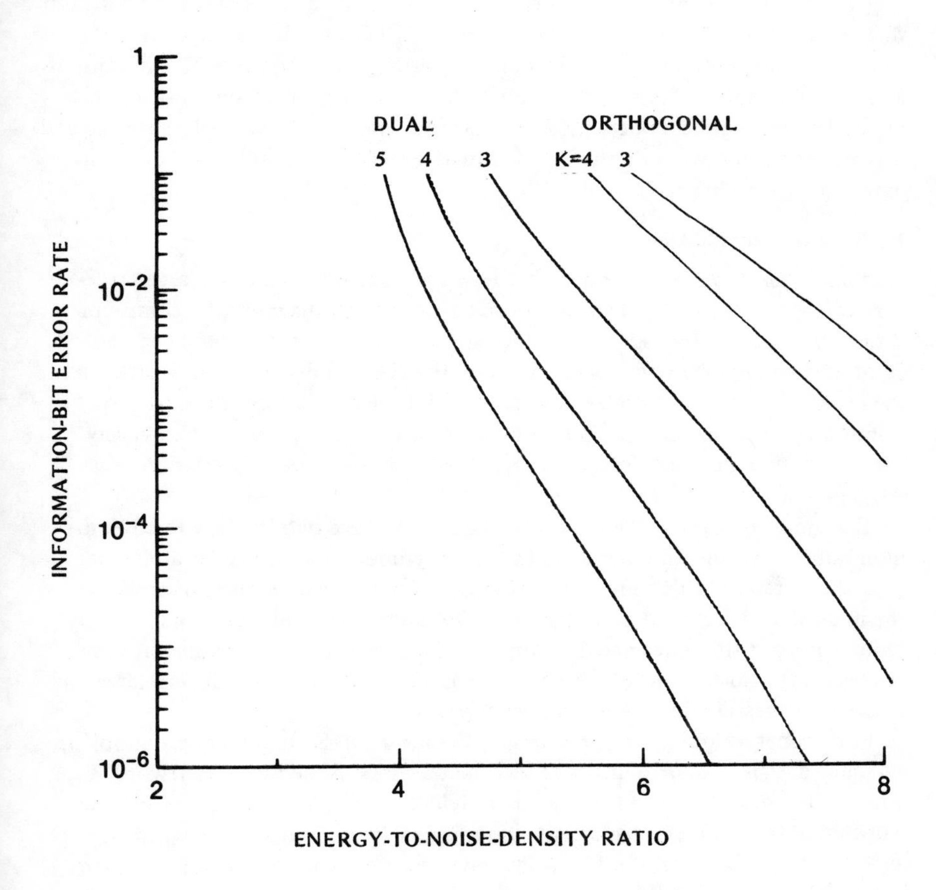

Figure 1.23 Information-bit error rate for rate-1/2 dual-m and orthogonal convolutional codes and noncoherent MFSK demodulation.

The bandwidth required to support a dual-m code and MFSK is approximately $2^m nB/m$, where $2B$ is the bandwidth required for uncoded binary FSK communications. Orthogonal convolutional codes require a bandwidth equal to $2^K B$. Thus, the rate-1/2 dual-4 code and the orthogonal code with $K = 3$ both require slightly less bandwidth than the Reed-Solomon (31, 23) code. However, even after allowing for the Chernoff-bound error, Figures 1.23 and 1.15 indicate that the dual-4 code requires a larger value of E_b/N_0 to achieve a specified P_{ib} than the Reed-Solomon code. Although the orthogonal code with $K = 3$ requires a much larger value of E_b/N_0 than the dual-4 code, it is much simpler to implement. The orthogonal code with $K = 3$ has four states and two paths entering and leaving each state, whereas the rate-1/2 dual-4 code has 16 states and 16 paths entering and leaving each state.

1.7.5 Concatenated Codes

A *concatenated code* uses multiple levels of coding to achieve a large error-correcting capability. Figure 1.24 is a functional block diagram of a communication system with two successive stages of coding. The inner interleaver and deinterleaver are often necessary to ensure the random distribution of errors at the input of the inner decoder. The outer deinterleaver, which requires a corresponding outer interleaver, is usually necessary to redistribute errors in the output of the inner decoder so that symbol errors are randomly distributed at the input of the outer decoder.

The most commonly used concatenated codes have two levels, a Reed-Solomon outer code and an inner code that uses symbols from a smaller alphabet than the outer-code alphabet. The outer decoder processes a burst of errors at the inner-decoder output as a single or a few outer-code symbol errors, thereby reducing P_{ib} at the outer-decoder output. The outer deinterleaver operates on outer-code symbols following a serial-to-parallel conversion and decorrelates errors in successive Reed-Solomon code symbols.

For a binary block inner code and m bits per Reed-Solomon code symbol, an efficient design sets the number of information bits in the block codeword, k_1, equal to an integer times m, and symbol deinterleaving prevents an inner-codeword error from affecting adjacent symbols in a Reed-Solomon codword. If $k_1 = m$, the probability of a Reed-Solomon code-symbol error, P_{s1}, is equal to the word error probability produced by the inner decoder. For hard demodulator decisions, (1.7.2) and (1.7.3) imply that

$$P_{s1} \leq \sum_{i=t_1+1}^{n_1} \binom{n_1}{i} P_s^i (1 - P_s)^{n-i} \tag{1.7.84}$$

where P_s is the channel-symbol error probability, equality holds for bounded-distance decoding, $t_1 = [(d_1 - 1)/2]$ for complete decoding, and t_1, n_1, and d_1 refer to the inner code. For soft demodulator decisions, (1.7.29) implies that

$$P_{s1} \leqslant \sum_{\delta=d_1}^{n_1} N(\delta)Q(\delta) \tag{1.7.85}$$

Because hard symbol decisions are presented to the outer decoder, an approximate upper bound on P_{ib} is determined from (1.7.85), (1.7.84), and (1.7.22) with P_{s1} substituted in place of P_s. In (1.7.22), the parameters t, n, and d refer to the outer code. The ratio of information bits to transmitted code symbols is

$$r = r_1 r_2 \tag{1.7.86}$$

where r_2 is the ratio of information bits to outer-code symbols and r_1 is the ratio of outer-code symbols to inner-code symbols. For coherent PSK demodulation, P_s is given by (1.7.24), and $Q(\delta)$ is given by (1.7.32).

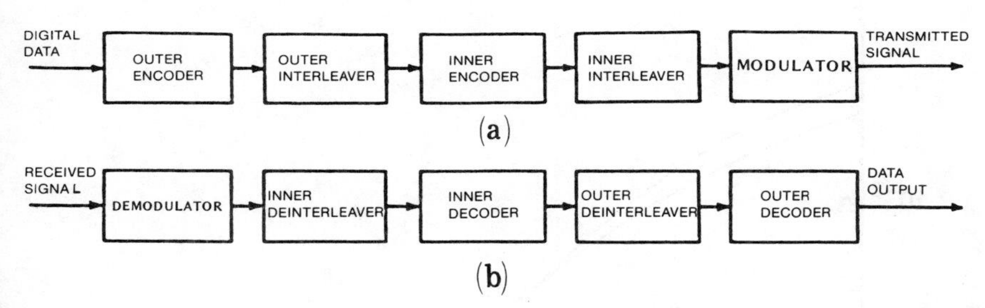

Figure 1.24 Concatenated coding in (a) transmitter and (b) receiver.

The combination of a high-rate Reed-Solomon outer code, a short binary block code with $d \leqslant 4$, and soft demodulator decisions is useful when a very high data rate must be accommodated [22]. Figure 1.25 depicts the approximate upper bound on P_{ib} for a Reed-Solomon (n, k) outer code, a Hamming (7,4) inner code, coherent PSK, and soft demodulator decisions in the presence of white Gaussian noise.

At the output of a convolutional inner decoder using the Viterbi algorithm, the bit errors occur over spans with an average length that depends upon the signal-to-noise ratio. Consequently, the associated outer deinterleaver should be designed to ensure that two input multiple-bit symbols at a distance less than the usual largest error span result in output Reed-Solomon symbols that do not belong to the same Reed-Solomon codeword. If we assume that the bit error rate is approximately one-half in spans with errors, then the bit error rate at the inner-decoder output is approximately one-half times P_{s1}, the error probability for a Reed-Solomon symbol. Thus, for convolutional inner codes,

$$P_{s1} \cong \frac{2}{k_1} \sum_{i=0}^{\infty} A(d_f + i)Q(d_f + i) \tag{1.7.87}$$

where all parameters and functions on the right-hand side are determined by the convolutional code. Assuming that the deinterleaving ensures independent symbol errors at the Reed-Solomon decoder input, an upper bound on P_{ib} is determined from (1.7.86), (1.7.87), and (1.7.22) with P_{s1} substituted in place of P_s. For coherent PSK demodulation with soft decisions, $Q(\delta)$ is given by (1.7.32); if hard decisions are made, (1.7.37) applies. Figure 1.26 depicts examples of the performance in white Gaussian noise for coherent PSK, soft demodulator decisions, an inner binary convolutional code with $K = 7$ and rate = 1/2, and various

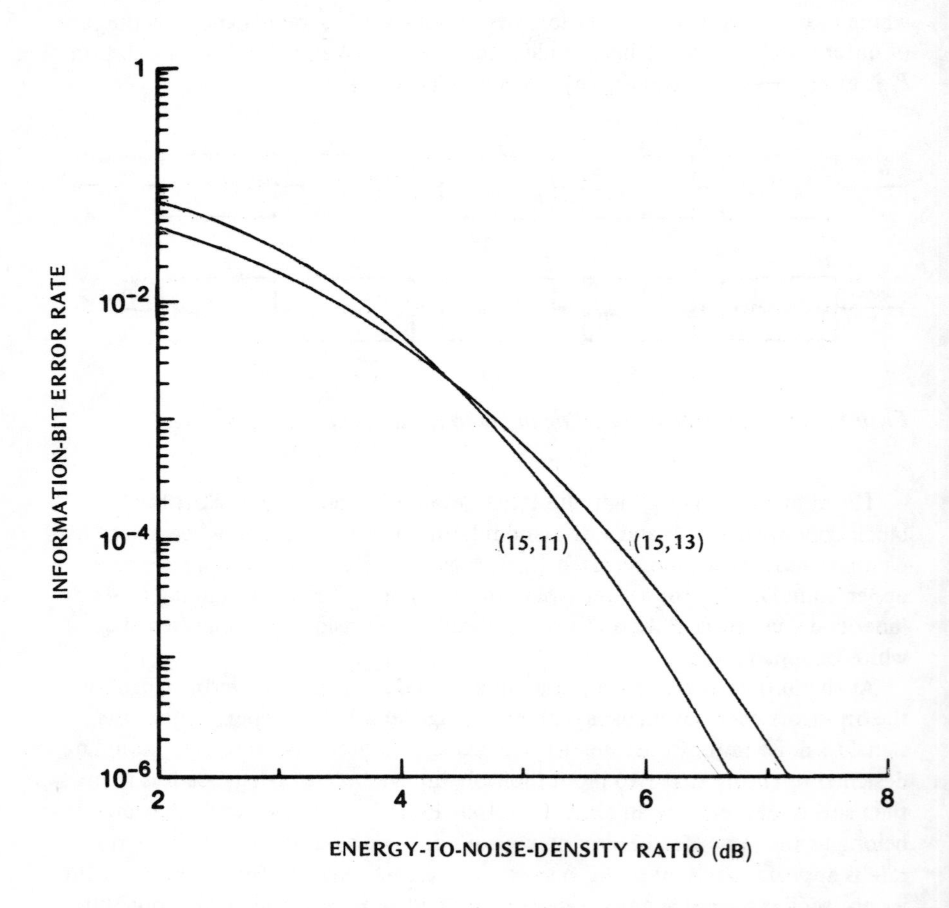

Figure 1.25 Information-bit error rate for concatenated codes with outer Reed-Solomon (n, k) code, inner Hamming (7,4) code, and coherent PSK demodulation.

Reed-Solomon outer codes. The bandwidth required by a concatenated code with a binary inner code is B/r, where B is the uncoded PSK bandwidth. Thus, the codes of Figure 1.26 require less bandwidth than the rate-1/3 convolutional codes of Figure 1.19.

1.7.6 Protection Against Pulsed Jamming

Metrics that are effective against white Gaussian noise are not necessarily effective against pulsed jamming that is optimized by an opponent. We examine [32]

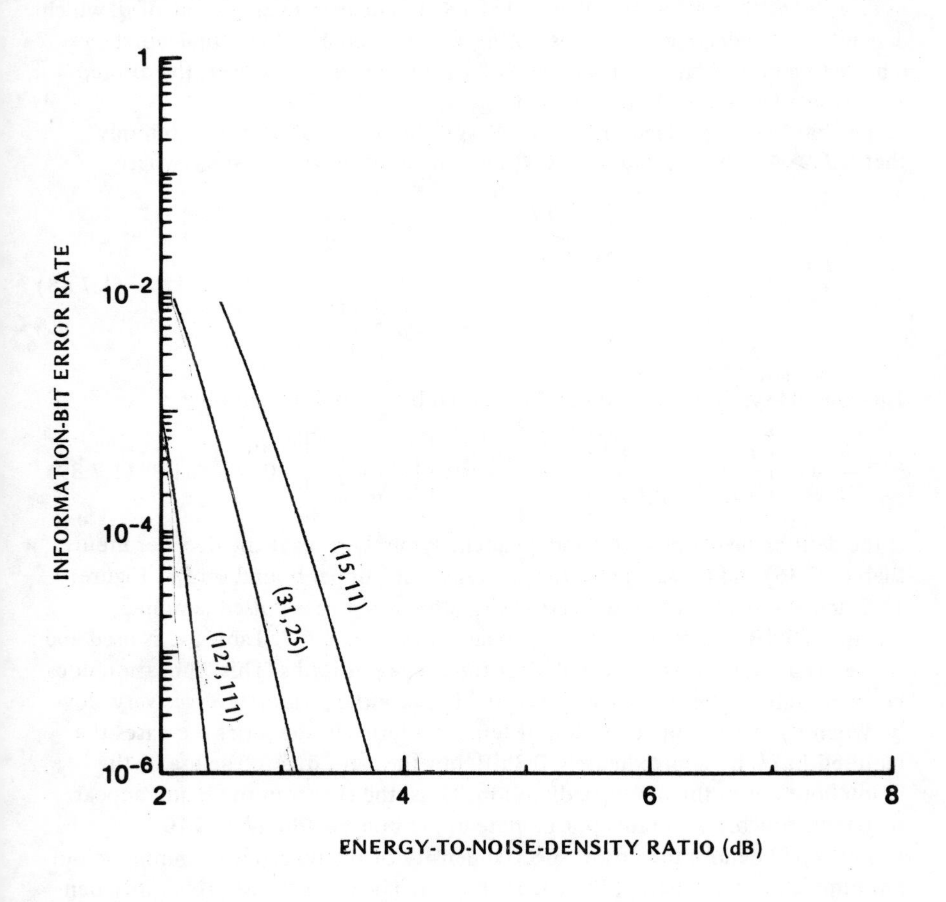

Figure 1.26 Information-bit error rate for concatenated codes with outer Reed-Solomon (n, k) code, inner convolutional (K = 7, rate = 1/2) code, and coherent PSK demodulation.

the performance of five different metrics against pulsed jamming when binary coherent phase-shift keying (PSK), interleaving, a convolutional code, and Viterbi decoding are used. According to the model of Section 1.6, the jamming signal is a Gaussian process when a pulse occurs, and the average jamming power is independent of the pulse duration. The power spectral density of the jamming plus noise is $J_0/2\mu + N_0/2$ during a jamming pulse and is $N_0/2$ when no pulse occurs, where μ denotes the probability that a pulse occurs. An upper bound on P_{ib} for worst-case pulsed jamming is obtained by maximizing the right-hand side of (1.7.36) with respect to μ, where $0 \leqslant \mu \leqslant 1$. The maximizing value of μ, which depends on the decoder metric, is not necessarily equal to the actual worst-case μ because a bound rather than an equality is maximized. However, the discrepancy is small when the bound is tight.

For hard-decision decoding, (1.7.24) and the results of Section 1.6 imply that if $J_0 \gg N_0$, then P_s and, hence, P_{ib} are maximized when $\mu \approx \mu_0$, where

$$\mu_0 = \begin{cases} \dfrac{0.7 J_0}{r E_b}, & J_0 < \dfrac{r E_b}{0.7} \\[2ex] 1, & J_0 \geqslant \dfrac{r E_b}{0.7} \end{cases} \qquad (1.7.88)$$

The channel-symbol error probability caused by the pulsed jamming is

$$P_s = \frac{\mu}{2} \operatorname{erfc}\left[\left(\frac{r E_b}{N_0 + J_0/\mu}\right)^{1/2}\right] + \frac{1-\mu}{2} \operatorname{erfc}\left[\left(\frac{r E_b}{N_0}\right)^{1/2}\right], \quad 0 \leqslant \mu \leqslant 1 \qquad (1.7.89)$$

If the deinterleaving produces independent symbol errors at the decoder input, then (1.7.36) and (1.7.37) give the corresponding upper bound on P_{ib}. Figure 1.27 depicts this upper bound *versus* E_b/J_0 for worst-case pulsed jamming, $E_b/N_0 = 20$ dB, and binary convolutional codes with $K = 7$. Table 1-2 is used and the series in (1.7.36) is truncated after the first eight terms. This truncation does not yield reliable results when $P_{ib} > 10^{-3}$ because the series converges very slowly. When $P_{ib} = 10^{-3}$, the inclusion of ten more terms in the series increases the required E_b/J_0 by approximately 0.3 dB, but the error due to the use of the union bound is in the opposite direction. Thus, the curves in the figure appear to give reasonably accurate approximate upper bounds when $P_{ib} \leqslant 10^{-3}$.

Let $N_{0i}/2$ denote the power spectral density of the noise plus jamming in output sample y_i of a coherent PSK demodulator. The conditional probability density of y_i given that code bit x_{ji} of sequence j was transmitted as

$$p(y_i \mid x_{ji}) = \frac{1}{\sqrt{\pi N_{0i}}} \exp\left[-\frac{(y_i - x_{ji})^2}{N_{0i}}\right], \quad i = 1, 2, \ldots, L \qquad (1.7.90)$$

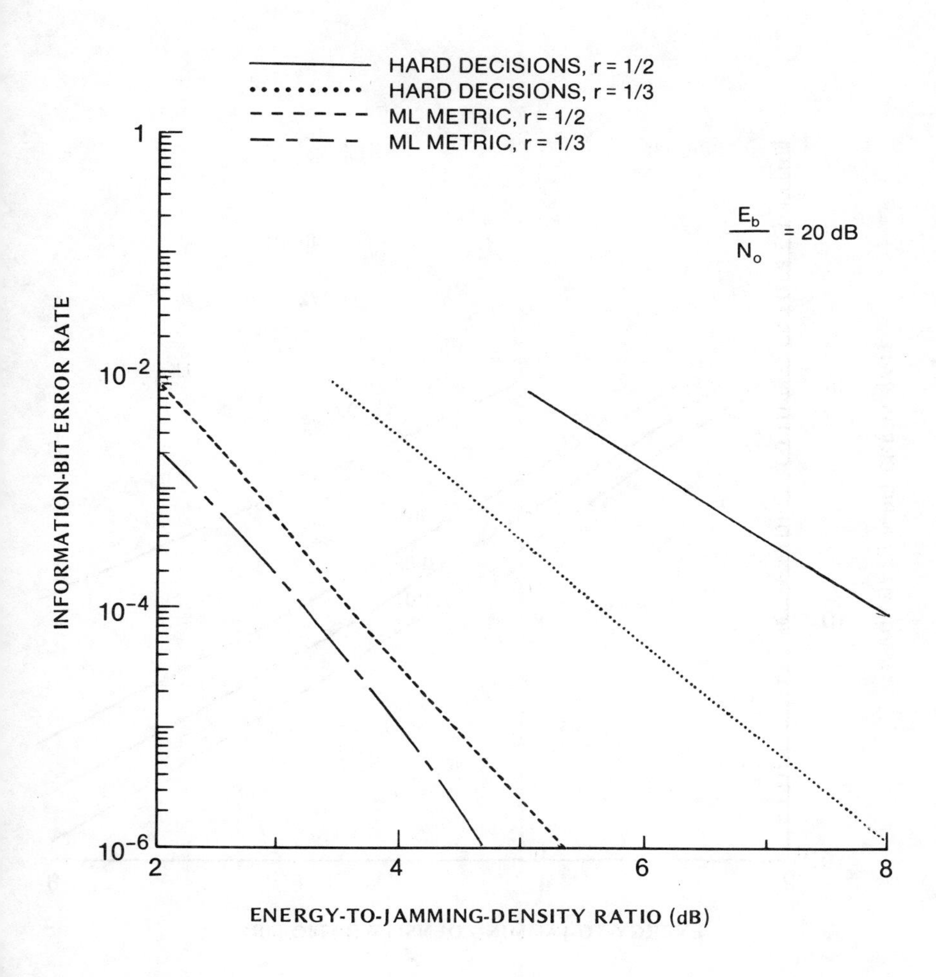

Figure 1.27 Worst-case performance against pulsed jamming for convolutional (K = 7) codes with hard decisions and with maximum-likelihood metric.

where $x_{ji} = +\sqrt{E_s}$ or $-\sqrt{E_s}$ and L is the sequence length. From the log-likelihood function and the statistical independence of the samples, it follows that when the values of $N_{01}, N_{02}, \ldots, N_{0L}$ are known, the optimal metric for maximum-likelihood decoding is

$$m_0(j,L) = \sum_{i=1}^{L} \frac{x_{ji} y_i}{N_{0i}} \tag{1.7.91}$$

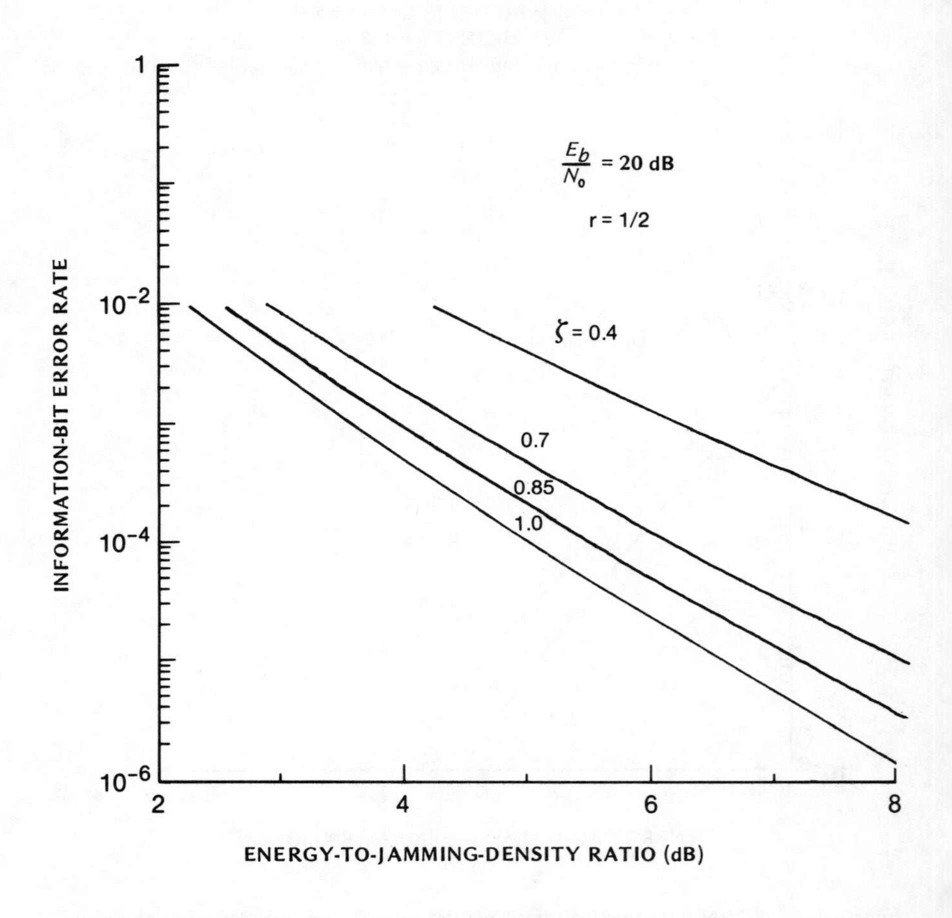

Figure 1.28 Performance against pulsed jamming for convolutional (K = 7, r = 1/2) code with white-noise metric.

This metric weights each y_i according to the level of the noise plus jamming. Because each y_i is assumed to be a Gaussian random variable, so is $m_0(j, L)$.

Let $j = 1$ label the correct sequence and $j = 2$ label an incorrect one at distance δ. We assume that there is no quantization of the sample values or that the quantization is infinitely fine. Therefore, the probability that $m_0(2, L) = m_0(1, L)$ is zero, and the probability of an error in comparing a correct sequence with an incorrect one that differs in δ symbols, $Q(\delta)$, is equal to the probability that $m_0(2,L) - m_0(1,L) > 0$. Let $Q(\delta \mid \nu)$ denote the conditional error probability given that a jamming pulse occurs during ν out of δ symbols and does not occur during δ-ν symbols. Because of the interleaving the probability that a symbol is jammed is statistically independent of the rest of the sequence and equals μ. Thus, (1.7.36) yields

$$P_{ib} \leqslant \sum_{i=0}^{\infty} A(d_f + i) \sum_{\nu=0}^{d_f+i} \binom{d_f + i}{\nu} \mu^{\nu}(1 - \mu)^{d_f+i-\nu} Q(d_f + i \mid \nu) \quad (1.7.92)$$

Because $m_0(2,L) - m_0(1,L)$ is a Gaussian random variable, $Q(\delta \mid \nu)$ is determined from the conditional means and variances of $m_0(2,L) - m_0(1,L)$. A straightforward calculation using (1.7.90) and (1.7.91) and the statistical independence of the samples yields

$$Q(\delta \mid \nu) = \frac{1}{2} \operatorname{erfc} \left\{ \sqrt{\frac{E_s}{N_0}} \left[\delta - \nu \left(1 + \frac{\mu N_0}{J_0} \right)^{-1} \right]^{1/2} \right\} \quad (1.7.93)$$

The upper bound on P_{ib} *versus* E_b/J_0 for worst-case pulsed jamming, $K = 7$, $E_b/N_0 = 20$ dB, and soft-decision decoding with the maximum-likelihood metric is shown in Figure 1.27. It is found that worst-case pulsed jamming causes very little degradation relative to continuous jamming. When $r = 1/2$, the *maximum likelihood metric* provides a performance that is approximately 4.5 dB superior at $P_{ib} = 10^{-4}$ to that provided by a hard-decision decoder; when $r = 1/3$, the advantage is approximately 2.5 dB. However, the physical realization of the maximum-likelihood metric presents severe difficulties.

Implementation of the maximum-likelihood metric entails knowledge of not only the jamming state, but also the jamming density level. Estimates of the N_{0i} might be based upon power measurements in adjacent frequency bands, but these measurements are reliable only if the jamming spectral density is fairly uniform over the desired-signal band and the adjacent bands. Any measurement of the power within the desired-signal band is contaminated by the presence of the desired signal. The average power of the desired signal is usually unknown since it fluctuates with time as the propagation conditions vary. Thus, a practical soft-decision decoder using an approximation to the maximum-likelihood metric can be expected to suffer a degradation relative to the performance predicted in Figure 1.27.

We analyze the performance of several metrics that are easier to implement than the maximum-likelihood metric for pulsed jamming. One candidate is the maximum-likelihood metric for white Gaussian noise. This *white-noise metric* is

$$m_0(j,L) = \sum_{i=1}^{L} x_{ji} y_i \quad (1.7.94)$$

Combining (1.7.94) and (1.7.90), we obtain

$$Q(\delta \mid \nu) = \frac{1}{2} \operatorname{erfc}\left\{\sqrt{\frac{E_s}{N_0}}\ \delta \left(\delta + \nu\ \frac{J_0}{\mu N_0}\right)^{-1/2}\right\} \tag{1.7.95}$$

The upper bound on P_{ib} is determined by (1.7.95) and (1.7.92). Figure 1.28 illustrates this upper bound *versus* E_b/J_0 for $K = 7$, $r = 1/2$, $E_b/N_0 = 20$ dB, and several values of $\zeta = \mu/\mu_0$. The figure demonstrates the vulnerability of soft-decision decoding with the white-noise metric to short high-power pulses if the jamming power is conserved. The high values of P_{ib} for $\zeta < 1$ are due to the domination of the metric by a few degraded symbol metrics.

Consider an *automatic-gain-control* (AGC) *device* that measures the average power within the frequency band of the desired signal during each channel-symbol interval. The AGC device then weights the demodulator outputs in proportion to the inverse of the measured power to form the *AGC metric.* The average power during channel symbol i is $N_{0i}B + E_s/T_s$, where B is the noise bandwidth of the receiver's initial bandpass filter and T_s is the channel-symbol duration. If the power measurement is perfect and $BT_s \cong 1$, then the AGC metric is

$$m_0(j, L) = \sum_{i=0}^{L} \frac{y_i x_{ji}}{N_{0i} + E_s} \tag{1.7.96}$$

which is a Gaussian random variable. From (1.7.96) and (1.7.90), we obtain

$$Q(\delta \mid \nu) = \frac{1}{2} \operatorname{erfc}\left\{\sqrt{\frac{E_s}{N_0}}\ \frac{\delta (N_0 + E_s + J_0/\mu) - \nu J_0/\mu}{[\delta (N_0 + E_s + J_0/\mu)^2 - \nu (N_0 + J_0/\mu - E_s^2/N_0) J_0/\mu]^{1/2}}\right\} \tag{1.7.97}$$

Figure 1.29 illustrates the upper bound on P_{ib} *versus* E_b/J_0 for worst-case pulsed jamming, soft-decision decoding with the AGC metric, $K = 7$, and $E_b/N_0 = 20$ dB. The figure indicates that the potential performance of the AGC metric is only slightly inferior to that of the maximum-likelihood metric. When the jamming is absent, the AGC, maximum-likelihood, and white-noise metrics all provide the same performance.

The power measurement can be computed from an energy measurement made by a *radiometer.* Using the results of radiometer theory (Section 4.2), we find that an ideal radiometer provides an unbiased estimate of the energy received during a symbol interval. A full analysis of the effects of radiometer inaccuracy on P_{ib} is complicated and is not attempted here. A simple measure of the radiometer performance as an estimator of the energy received during symbol i is given by the ratio of the standard deviation of the radiometer output to the expected value of the output. If $BT_s = 1$, (4.2.37) and (4.2.38) yield the ratio

$$S_i = \frac{\left(\dfrac{2E_s}{N_{0i}} + 1\right)^{1/2}}{\dfrac{E_s}{N_{0i}} + 1} \tag{1.7.98}$$

This relation indicates that the energy measurements and, hence, the power measurements will entail large inaccuracies if $E_s/N_{0i} \leqslant 1$ during jamming pulses. Thus, the potential performance of the AGC metric is expected to be significantly degraded in practice. The degradation can be reduced if it is known that the jamming pulses extend over many channel symbols so that the energy can be measured over the corresponding interval.

Suppose that a coherent PSK demodulator erases its output and, hence, a received symbol whenever a jamming pulse occurs. The presence of the jamming pulse might be detected by examining a sequence of the demodulator outputs and determining which ones have inordinately large magnitudes compared to the others. Alternatively, the demodulator might decide that a jamming pulse has occurred if an output has a magnitude that exceeds a known upper bound for the desired signal. Consider an ideal demodulator that unerringly detects the jamming pulses and erases the corresponding received symbols. Following the deinterleaving of the demodulated symbols the decoder processes symbols that have a probability of being erased equal to μ. The unerased symbols are decoded by using the white-noise metric. The erasing of ν symbols causes two sequences that differ in δ symbols to be compared on the basis of δ-ν symbols where $0 \leqslant \nu \leqslant \delta$. Therefore, we find that

$$Q(\delta \mid \nu) = \frac{1}{2} \operatorname{erfc}\left[\sqrt{\frac{E_s}{N_0}} (\delta - \nu)^{1/2}\right] \tag{1.7.99}$$

The performance bound determined by (1.7.99) and (1.7.92) is illustrated in Figure 1.30 for $K = 7, r = 1/2, E_b/N_0 = 20$ dB, and several values of $\zeta = \mu/\mu_0$. In this example, erasures are not helpful in reducing the required E_b/J_0 for $P_{ib} = 10^{-5}$ if $\zeta > 0.85$, but are increasingly useful if $\zeta < 0.85$, which corresponds to $J_0/\mu > 1.68\, E_s$. Thus, an ideal demodulator activates erasures only when $\zeta \leqslant V$, where $V \approx 0.85$. If $\zeta = 0.85$, the required E_b/J_0 at $P_{ib} = 10^{-5}$ for erasure decoding is more than 2 dB less than for worst-case hard-decision decoding, which occurs when $\zeta \cong 1$. However, a practical demodulator will sometimes erroneously make erasures or fail to erase. Thus, the difference in worst-case performance between erasure decoding and hard-decision decoding may be much less in practice.

The error rates of the hard-decision, maximum-likelihood, white-noise, AGC, and erasure metrics against pulsed jamming have been analyzed and compared. The analysis has demonstrated the lack of robustness of soft-decision decoding with the usual white-noise metric when pulsed jamming is present. The maximum-likelihood metric is effective but difficult to accurately implement. The

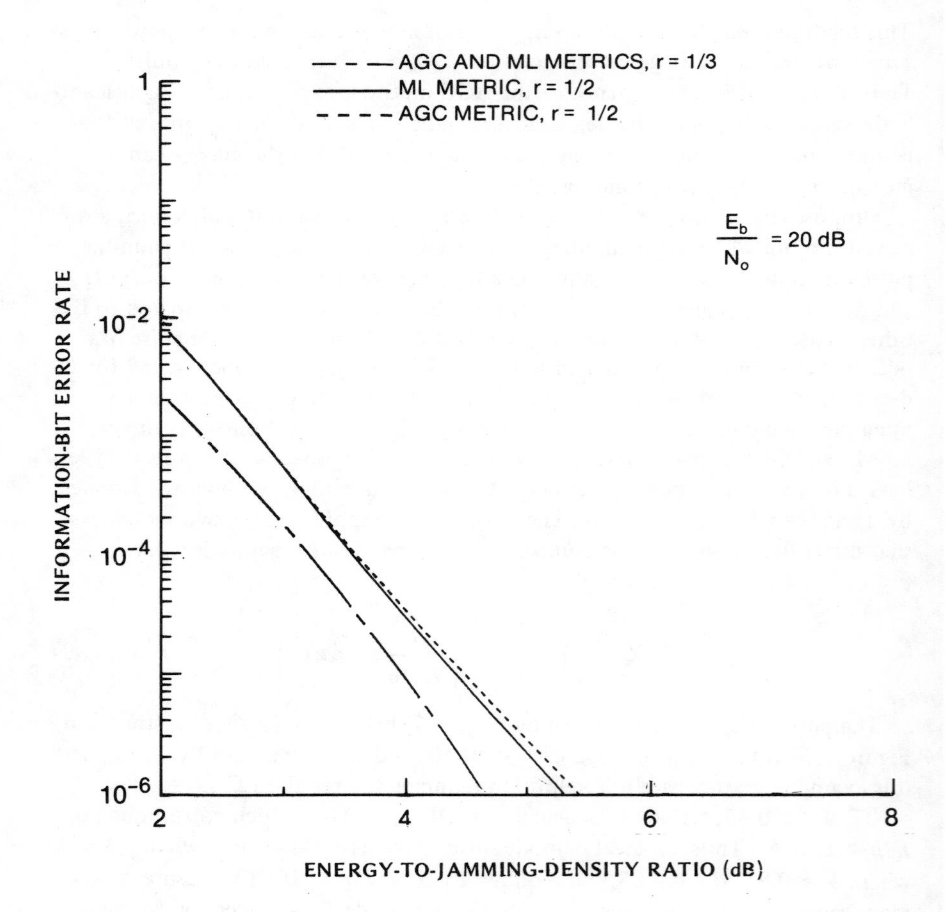

Figure 1.29 Worst-case performance against pulsed jamming for convolutional (K = 7) codes with AGC and maximum-likelihood metrics.

more practical AGC and erasure metrics theoretically offer better performance than hard-decision decoding. However, the performance advantage realizable in practice may sometimes not be sufficient to warrant the implementation complexity of soft-decision decoding.

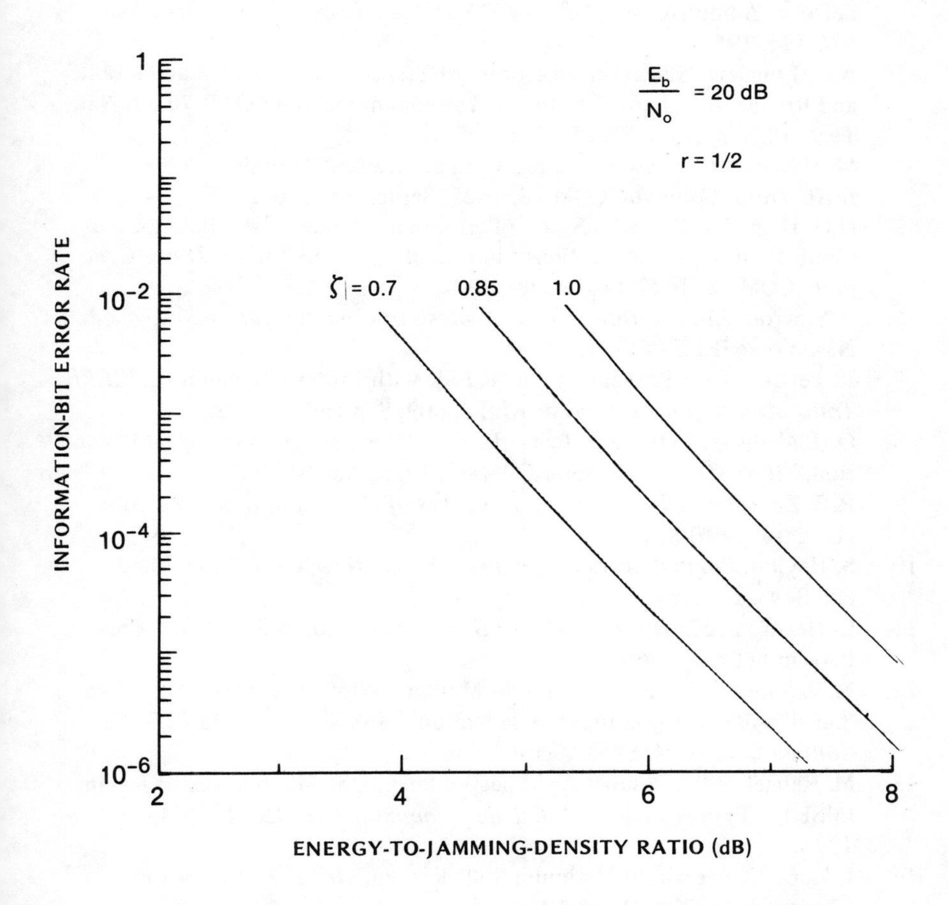

Figure 1.30 Performance against pulsed jamming for convolutional (K = 7, r = 1/2) code with erasures.

REFERENCES

1. R. Gagliardi, *Introduction to Communication Engineering.* New York: John Wiley and Sons, 1978.
2. A.G. Longley and P.L. Rice, "Prediction of Tropospheric Radio Transmission Loss over Irregular Terrain, A Computer Method," Environmental Sciences Administration ERL 79-ITS67, Nat. Tech. Inform. Serv., AD-676 874, 1968.
3. A.G. Longley, "Location Variability of Transmission Loss – Land Mobile and Broadcast Systems," Office of Telecommunications OTR 76-87, Nat. Tech. Inform. Serv., PB-254 472, 1976.
4. M. Marcus, "Analysis of Tactical Communications Jamming Problems," *IEEE Trans. Commun.* COM-28, 1625, September 1980.
5. G.H. Hagn, "VHF Radio System Performance Model or Predicting Communications Operational Ranges in Irregular Terrain," *IEEE Trans. Commun.* COM-28, 1637, September 1980.
6. P. Stavroulakis, ed., *Interference Analysis of Communication Systems.* New York: IEEE Press, 1980.
7. R. Pettit, "Error Probability for NCFSK with Linear FM Jamming," *IEEE Trans. Aerosp. Electron. Syst.* AES-3, 609, September 1972.
8. O.H. Shabsigh, "On the Effects of CW Interference on MSK Signal Reception," *IEEE Trans. Commun.* COM-30, 1925, August 1982.
9. R.E. Ziemer and W.H. Tranter, *Principles of Communications.* Boston: Houghton Mifflin, 1976.
10. S. Haykin, *Communication Systems,* 2nd ed., New York: John Wiley and Sons, 1983.
11. C. Helstrom, *Statistical Theory of Signal Detection,* 2nd ed., New York: Pergamon Press, 1968.
12. M.K. Simon, "A Generalization of Minimum-Shift-Keying (MSK) – Type Signaling Based Upon Input Data Symbol Pulse Shaping," *IEEE Trans. Commun.* COM-24, 845, August 1976.
13. M. Rabzel and S. Pasupathy, "Spectral Shaping in Minimum Shift Keying (MSK) – Type Signals," *IEEE Trans. Commun.* COM-26, 189, January 1978.
14. I. Korn, "Generalized Minimum Shift Keying," *IEEE Trans. Inform. Theory* IT-26, 234, March 1980.
15. F. Amoroso, "Experimental Results on Constant Envelope Signaling with Reduced Spectral Sidelobes," *IEEE Trans. Commun.* COM-31, 157, January 1983.
16. T. Aulin and C.E.W. Sundberg, "Continuous Phase Modulation – Part I: Full Response Signaling," *IEEE Trans. Commun.* COM-29, 196, March 1981.

17. T. Aulin, N. Rydbeck, and C.E.W. Sundberg, "Continuous Phase Modulation – Part II: Partial Response Signaling," *IEEE Trans. Commun.* COM-29, 210, March 1981.
18. R.F. Pawula and R. Golden, "Simulations of Convolutional Coding/Viterbi Decoding with Noncoherent CPFSK," *IEEE Trans. Commun.* COM-29, 1522, October 1981.
19. R.F. Pawula, "On the Theory of Error Rates for Narrow-Band Digital FM," *IEEE Trans. Commun.* COM-29, 1634, November 1981.
20. K.J.P. Fonseka and N. Ekanayake, "Differential Detection of Narrow-Band Binary FM," *IEEE Trans. Commun.* COM-33, 725, July 1985.
21. N. Ekanayake, "On Differential Detection of Binary FM," *IEEE Trans. Commun.* COM-32, 469, April 1984.
22. G.C. Clark and J.B. Cain, *Error-Correction Coding for Digital Communications.* New York: Plenum Press, 1981.
23. S. Lin and D.J. Costello, *Error Control Coding: Fundamentals and Applications.* Englewood Cliffs, NJ: Prentice-Hall, 1983.
24. R.E. Blahut, *Theory and Practice of Error Control Codes.* Reading, MA: Addison-Wesley, 1983.
25. A.J. Viterbi and J.K. Omura, *Principles of Digital Communication and Coding.* New York: McGraw-Hill, 1979.
26. D.J. Torrieri, "The Information-Bit Error Rate for Block Codes," *IEEE Trans. Commun.* COM-32, 474, April 1984.
27. J. Conan, "The Weight Spectra of Some Short Low-Rate Convolutional Codes," *IEEE Trans. Commun.* COM-32, 1050, September 1984.
28. P.J. Lee, "New Short Length, Rate 1/N Convolutional Codes Which Minimize the Required SNR for Given Desired Bit Error Rates," *IEEE Trans. Commun.* COM-33, 171, February 1985.
29. J.P. Odenwalder, "Dual-k Convolutional Codes for Noncoherently Demodulated Channels," International Telemetering Conference, 165, 1976.
30. P. Billingsley, *Probability and Measure.* New York: John Wiley and Sons, 1979.
31. I.M. Jacobs, "Probability-of-Error Bounds for Binary Transmission on Slowly Fading Rician Channel," *IEEE Trans. Inform. Theory* IT-12, 431, October 1966.
32. D.J. Torrieri, "The Performance of Five Different Metrics Against Pulsed Jamming," *IEEE Trans. Commun.* COM-34, 1986.

Chapter 2
Direct-Sequence Spread-Spectrum Systems

2.1 FUNDAMENTAL CONCEPTS

A *spread-spectrum system* is a system that produces a signal with a bandwidth much wider than the message bandwidth. Because a spread-spectrum system distributes the transmitted energy over a wide bandwidth, the signal-to-noise ratio at the receiver input is low. Nevertheless, the receiver is capable of operating successfully because the transmitted signal has distinct characteristics relative to the noise.

The generic forms of the transmitter and the receiver in a spread-spectrum system are shown in Figure 2.1. The *spreading waveform* is controlled by a *pseudonoise sequence* or *pseudonoise code,* which is a binary sequence that is apparently random but can be reproduced deterministically by intended users. The removal of the spreading waveform from the received signal to restore the modulated message is called the *despreading.*

Spread-spectrum systems are useful for secure communications because they make it difficult to detect the transmitted waveform, extract the message, or jam the intended receiver. Pseudonoise sequences give spread-spectrum systems identification and selective-calling capabilities. The most practical spread-spectrum methods known are *direct-sequence modulation, frequency hopping,* and hybrids of these two methods.

A *direct-sequence system,* which is sometimes called a *pseudonoise system,* spreads the transmitted spectrum by using the baseband pulses representing the pseudonoise sequence, which is produced by a pseudonoise code generator. Figure 2.2 shows the generic form of the transmitter of a direct-sequence system A pulse or bit of the pseudonoise sequence is called a *chip.*

Message privacy is provided by a direct-sequence system if a transmitted message cannot be recovered without knowledge of the pseudonoise sequence. Most analog message modulations are incompatible with message privacy. Amplitude modulation can usually be recovered by an envelope detector. Frequency modulation can usually be recovered by a squaring device and a frequency discriminator. Because of the lack of message privacy and synchronization problems, analog message modulations are rarely used in direct-sequence systems.

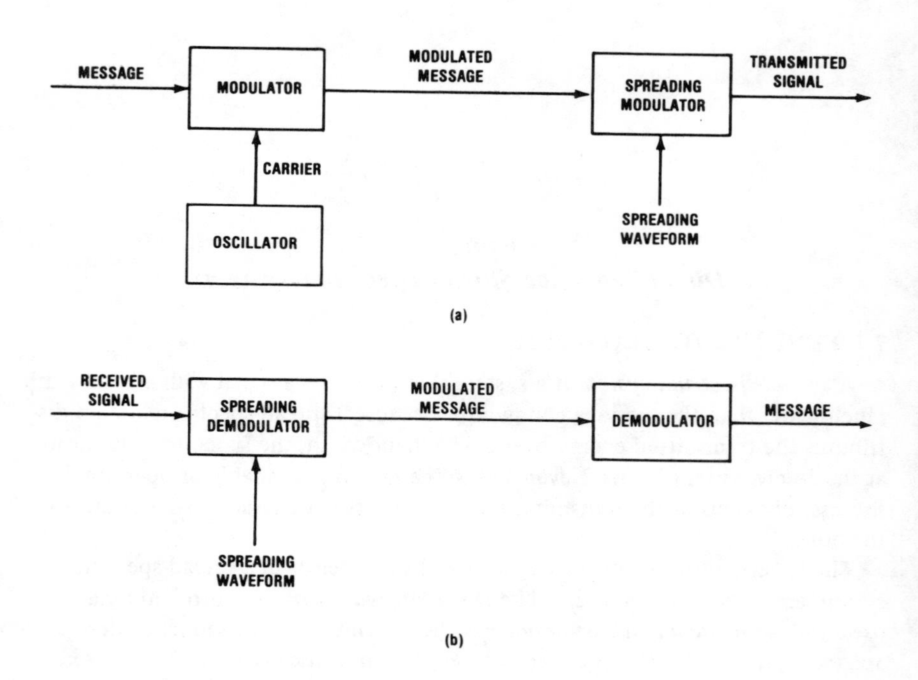

Figure 2.1 Generic spread-spectrum system: (a) transmitter and (b) receiver.

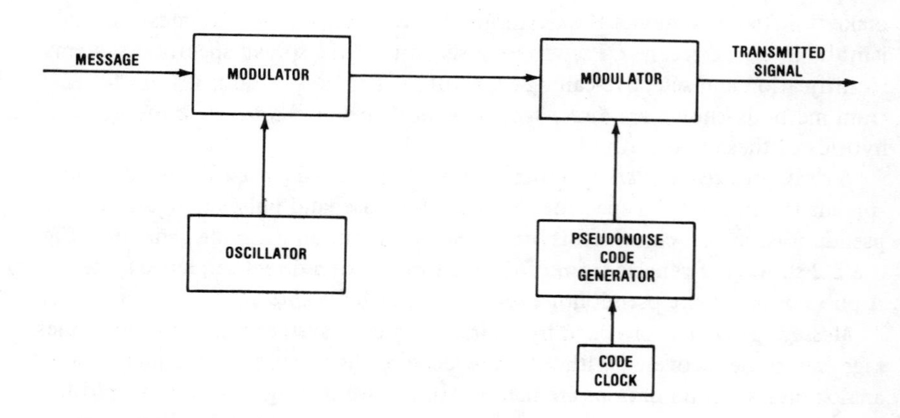

Figure 2.2 Generic form of direct-sequence system transmitter.

If the message is in digital form, but the data symbols are asynchronous with the code clock, then the symbol transitions do not coincide with the chip transitions, and message separation is possible. Thus, message privacy requires synchronization of the data symbol transitions with the code clock. This synchronization may be accomplished by either feeding the code clock back to the data source or providing for symbol storage. When the data and the code are synchronized at the transmitter, code synchronization in the receiver automatically gives symbol synchronization.

Figure 2.3 is a functional block diagram of a direct-sequence system with binary phase modulation. This system, which provides message privacy, is the most widely used direct-sequence implementation in practice. Synchronized data symbols, which may be information bits or binary code symbols, are modulo-two added to chips before the phase modulation. A coherent phase-shift keying (PSK) demodulator may be used in the receiver. Alternatively, if the uncertainty in the carrier frequency is sufficiently small, a differential phase-shift keying demodulator may be used.

The received spread-spectrum signal can be represented by

$$s(t) = Am(t)p(t)\cos(\omega_0 t + \theta) \tag{2.1.1}$$

where A is the amplitude, $m(t)$ is the data modulation, $p(t)$ is the spreading waveform, ω_0 is the carrier frequency, and θ is the phase angle at $t = 0$. The data modulation is a sequence of nonoverlapping rectangular pulses, each of which has an amplitude equal to +1 or −1. Each pulse of $m(t)$ represents a data symbol and has a duration of T_s. Each pulse of $p(t)$ represents a chip, is usually rectangular, and has a duration of T_c that is shorter than T_s. Because the transitions of the data symbol and the chips coincide on both sides of a symbol, the ratio of T_s to T_c is an integer. If W is the bandwidth of $s(t)$ and B is the bandwidth of $m(t)$ $\cos \omega_0 t$, the spreading due to $p(t)$ gives $W \gg B$. Usually, $p(t) = \pm 1$.

Assuming that code synchronization has been established, the received signal passes through the wideband filter and is multiplied by a local replica of $p(t)$. If $p^2(t) \cong 1$, this multiplication yields the *despread signal*

$$s_1(t) \cong Am(t) \cos(\omega_0 t + \theta) \tag{2.1.2}$$

at the input of the demodulator. Because $s_1(t)$ has the form of a PSK signal, the corresponding demodulation extracts $m(t)$.

The receiver reduces interference as qualitatively illustrated in Figure 2.4; quantitative results are given subsequently. Figure 2.4(a) shows the relative spectra of the desired signal and interference at the output of the wideband filter. Multiplication by the spreading waveform produces the spectra of Figure 2.4(b) at the demodulator input. The signal bandwidth is reduced to B, while the

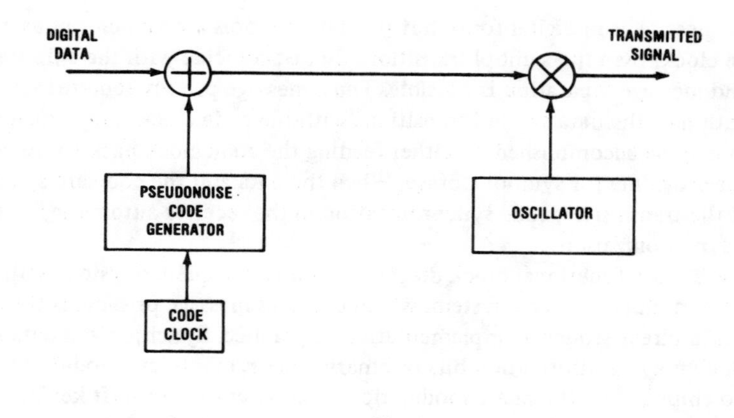

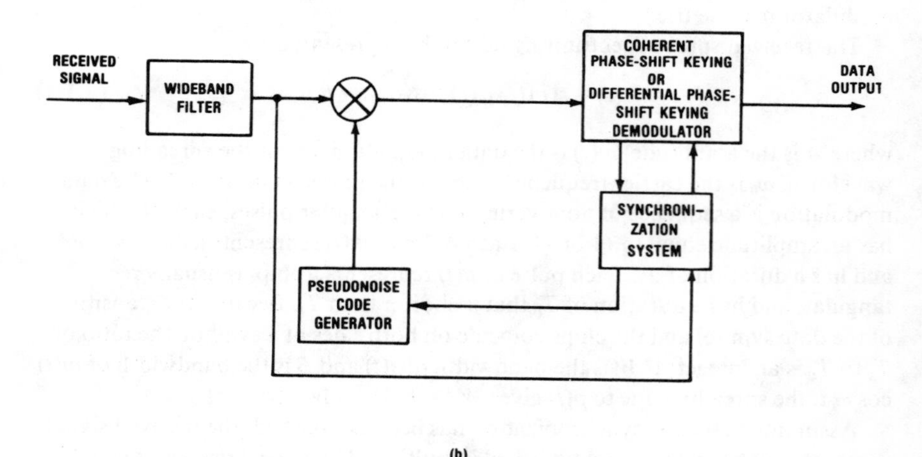

Figure 2.3 Direct-sequence system with binary phase modulation: (a) transmitter and (b) receiver. The data symbols may be information bits or binary code symbols.

interference energy is spread over a bandwidth exceeding W. The filtering action of the demodulator removes most of the interference spectrum that does not overlap the signal spectrum. Thus, most of the original interference energy is eliminated and does not affect the receiver performance. The *processing gain*, defined by

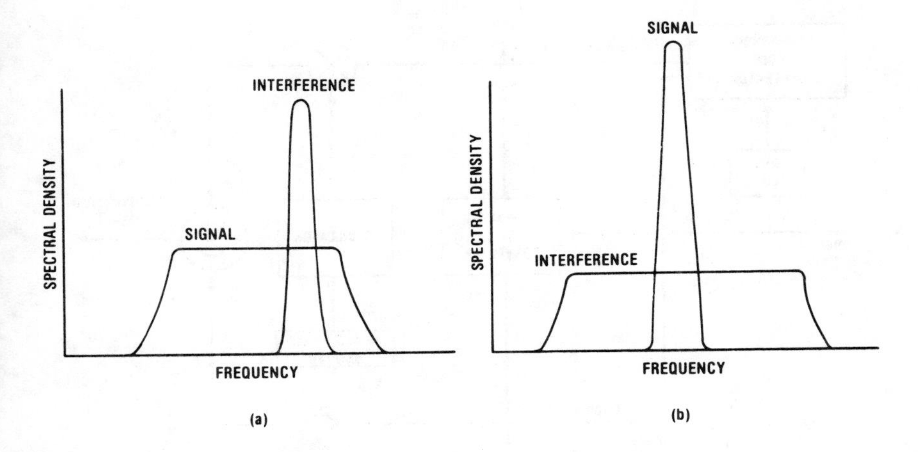

Figure 2.4 Spectra of desired signal and interference: (a) wideband filter output and (b) demodulator input.

$$G = \frac{T_s}{T_c} = \frac{W}{B} \tag{2.1.3}$$

is a measure of the interference rejection capability. It is equal to the number of chips in a symbol interval.

Two other direct-sequence systems with potential message privacy are diagrammed in Figures 2.5 and 2.6. For simplicity, the figures omit depictions of the synchronization systems.

Figure 2.5 shows a *quadriphase direct-sequence system,* which uses quadriphase-shift keying (QPSK). Two pseudonoise sequences, which may be derived from a single generator, are used with two quadrature carriers. Each member of each successive pair of data symbols is combined with one of the sequences and one of the quadrature carriers. Thus, the received signal can be represented by

$$s(t) = Am_1(t)p_1(t)\cos(\omega_0 t + \theta) + Am_2(t)p_2(t)\sin(\omega_0 t + \theta) \tag{2.1.4}$$

where $m_1(t)$ and $m_2(t)$ are the two data signals derived from the data symbols and $p_1(t)$ and $p_2(t)$ are the spreading waveforms. In each branch of the receiver, one of the sequences is removed, followed by coherent PSK demodulation. The outputs of the two branches are alternately sampled to reconstruct the data stream.

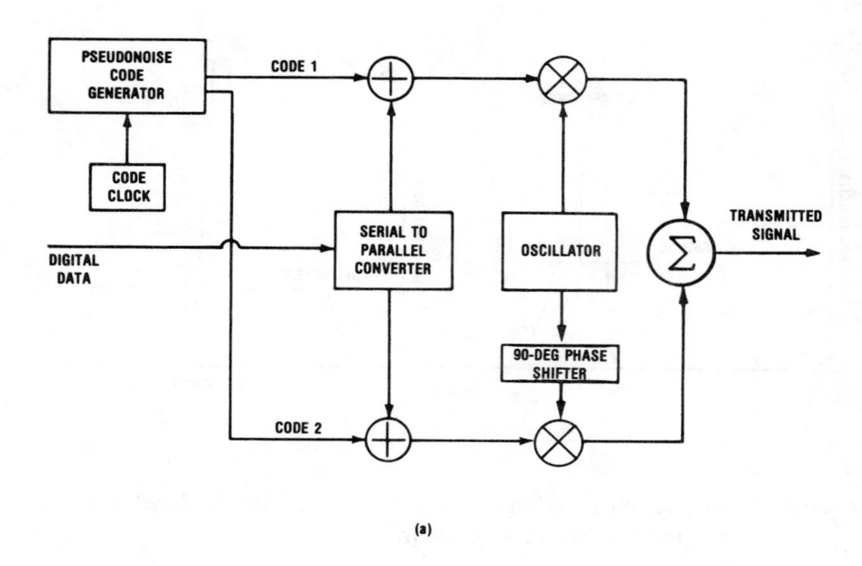

(a)

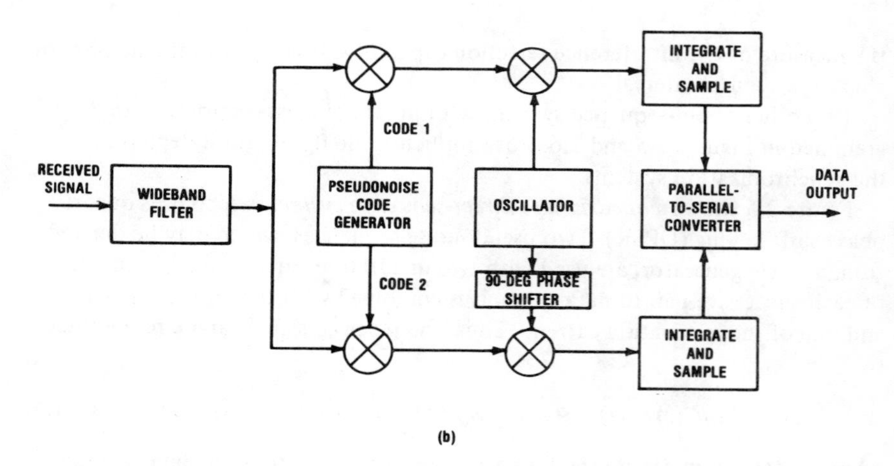

(b)

Figure 2.5 Quadriphase direct-sequence system: (a) transmitter and (b) receiver.

A direct-sequence system with offset QPSK or minimum-shift keying (MSK) is similar. A general *quaternary direct-sequence signal* can be represented by

$$s(t) = Am_1(t)p_1(t)\cos(\omega_0 t + \theta) + Am_2(t - t_0)p_2(t - t_0)\sin(\omega_0 t + \theta) \tag{2.1.5}$$

where t_0 is the relative delay between the in-phase and quadrature components of the signal. For offset QPSK and MSK, $|t_0| = T_c/2$. For MSK, the pulses of $p_1(t)$ and $p_2(t)$ are sinusoidal rather than rectangular, as indicated by (1.5.22) and (1.5.23). The use of MSK limits the spectral sidelobes of the direct-sequence signal, which may interfere with other signals, and limits the received signal power that falls outside the passband of the initial wideband filter.

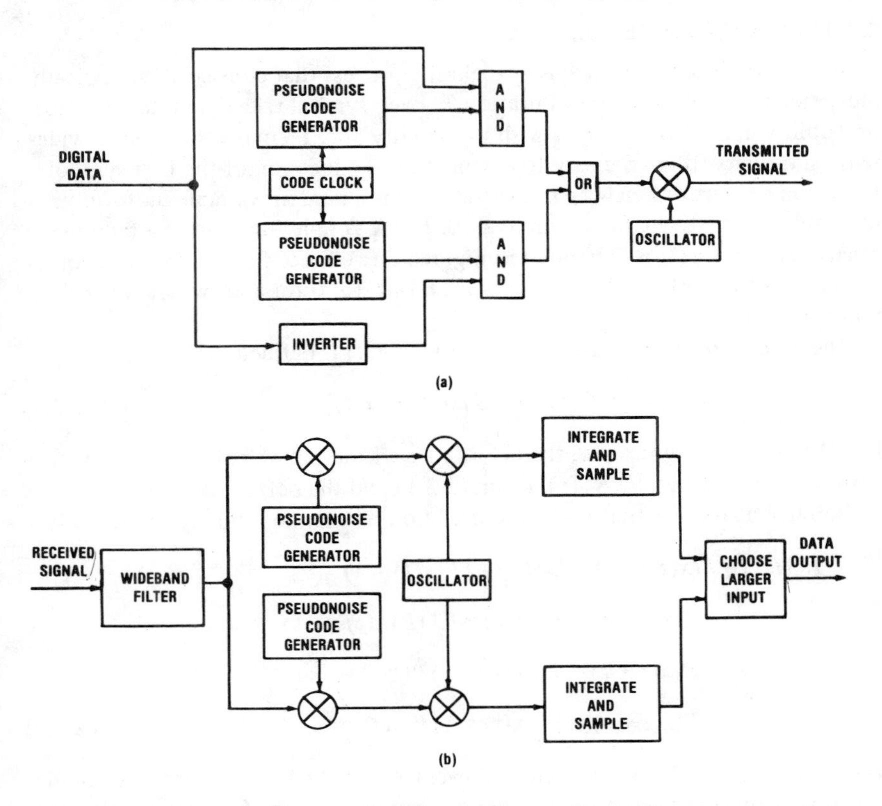

Figure 2.6 Direct-sequence system with binary code-shift keying: (a) transmitter and (b) receiver.

Figure 2.6 depicts a coherent direct-sequence system with binary *code-shift keying* (CSK). Depending upon the state of the binary data symbol, one or the other of two nearly orthogonal pseudonoise sequences is transmitted. In the absence of noise and interference, each sequence causes a significant output in only one of the two parallel branches in the receiver. Thus, the data is recovered by comparing the branch outputs. In a direct-sequence system with M-ary CSK, each group of n binary data symbols is encoded as one of $M = 2^n$ sequences chosen to have small cross correlations. Because they use orthogonal rather than antipodal signaling, binary CSK systems exhibit a relatively poor symbol error probability in the presence of white Gaussian noise. The M-ary systems provide improved error probabilities, but require complex implementations.

2.2 PSEUDONOISE SEQUENCES

A *random binary sequence* is a stochastic process that consists of statistically independent symbols, each of duration T. Each symbol takes the value +1 with probability 1/2 or the value −1 with probability 1/2. To make the process wide-sense stationary, it is necessary to assume that the location of the first symbol transition or start of a new symbol after $t = 0$ is a random variable uniformly distributed over the half-open interval $(0, T]$. A sample function of a random binary sequence $x(t)$ is illustrated in Figure 2.7(a). Since the amplitude is equally likely to equal +1 or −1 at any instant, $E[x(t)] = 0$ for all t, where $E[\ ]$ denotes the expected value.

The *autocorrelation* of a stochastic process $x(t)$ is defined as

$$R_x(t,\tau) = E[x(t)x(t+\tau)] \tag{2.2.1}$$

If $x(t)$ is a stationary process, then $R_x(t,\tau)$ is a function of τ alone, and the autocorrelation is denoted by $R_x(\tau)$. From (2.2.1) and the definition of a conditional probability, it follows that the autocorrelation of a random binary sequence is

$$\begin{aligned} R_x(t,\tau) = {} & P(x(t+\tau) = 1 \mid x(t) = 1)P(x(t) = 1) \\ & + P(x(t+\tau) = -1 \mid x(t) = -1)P(x(t) = -1) \\ & - P(x(t+\tau) = 1 \mid x(t) = -1)P(x(t) = -1) \\ & - P(x(t+\tau) = -1 \mid x(t) = 1)P(x(t) = 1) \end{aligned} \tag{2.2.2}$$

where $P(A)$ denotes the probability of event A and $P(A \mid B)$ denotes the conditional probability of event A given the occurrence of event B. From the theorem of total probability, it follows that

$$P(x(t+\tau) = i \mid x(t) = i) + P(x(t+\tau) = -i \mid x(t) = i) = 1\,, \qquad i = +1, -1 \tag{2.2.3}$$

Because the two amplitude values are equally likely,

$$P(x(t) = i) = \frac{1}{2}, \qquad i = +1, -1 \tag{2.2.4}$$

Because both of the following probabilities are equal to the probability that $x(t)$ and $x(t + \tau)$ differ,

$$P(x(t + \tau) = 1 \,|\, x(t) = -1) = P(x(t + \tau) = -1 \,|\, x(t) = 1) \tag{2.2.5}$$

Equations (2.2.2) to (2.2.5) yield

$$R_x(t,\tau) = 1 - 2P(x(t + \tau) = 1 \,|\, x(t) = -1) \tag{2.2.6}$$

If $|\tau| \geq T$, then $x(t)$ and $x(t + \tau)$ are independent random variables because t and $t + \tau$ are in different symbol intervals. Therefore, $P(x(t + \tau) = 1 \,|\, x(t) = -1) = P(x(t + \tau) = 1) = 1/2$ and (2.2.6) implies that $R_x(t,\tau) = 0$. If $|\tau| < T$, then $x(t)$ and $x(t + \tau)$ are independent only if a symbol transition occurs in the half-open interval $I_0 = (t, t + \tau]$. Consider any closed interval I_1 of length T that includes I_0. Exactly one transition occurs in I_1. Because of the uniform distribution of the first transition for $t > 0$, the probability that a transition in I_1 occurs in I_0 is $|\tau|/T$. If a transition occurs in I_0, then $x(t)$ and $x(t + \tau)$ are independent and differ with probability 1/2; otherwise, $x(t) = x(t + \tau)$. Consequently, $P(x(t + \tau) = 1 \,|\, x(t) = -1) = |\tau|/2T$ if $|\tau| < T$. Thus, (2.2.6) implies that the autocorrelation is

$$R_x(t,\tau) = R_x(\tau) = \Lambda\left(\frac{\tau}{T}\right) \tag{2.2.7}$$

where the triangular function is defined by

$$\Lambda(t) = \begin{cases} 1 - |t|, & |t| \leq 1 \\ 0, & |t| > 1 \end{cases} \tag{2.2.8}$$

Equation (2.2.7) and the fact that $E[x(t)] = 0$ indicate that the random binary sequence is wide-sense stationary. The autocorrelation of the random binary sequence is plotted in Figure 2.7(b).

A *pseudonoise* or *pseudorandom sequence* is a periodic binary sequence with an autocorrelation that resembles, over one period, the autocorrelation of a random binary sequence, which roughly resembles the autocorrelation of band-limited white noise. Although it is deterministic, a pseudonoise sequence has many characteristics, such as a nearly even balance of the digits 0 and 1, similar to those of a random binary sequence.

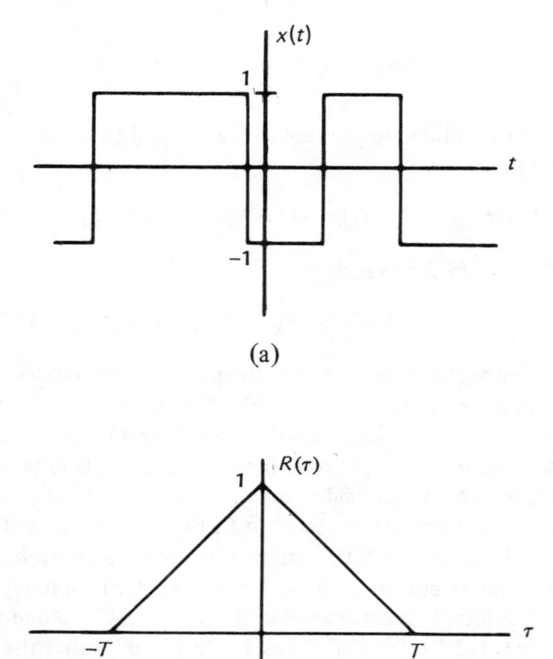

Figure 2.7 Random binary sequence: (a) sample function and (b) autocorrelation.

In general, pseudonoise sequences are generated by combining the outputs of *feedback shift registers.* A feedback shift register, which is diagrammed in Figure 2.8, consists of consecutive two-state memory or storage stages and feedback logic. Binary sequences are shifted through the shift register in response to clock pulses. The contents of the stages are logically combined to produce the input to the first stage. The initial contents of the stages and the feedback logic determine the successive contents of the stages. A feedback shift register and its output are called *linear* when the feedback logic consists entirely of modulo-two adders (Exclusive-OR gates).

An example of a linear feedback shift register with three stages is shown in Figure 2.9(a). The input to the first stage is the modulo-two sum of the contents of the second and third stages. The successive contents of the third stage provide the output sequence of the shift register. After each clock pulse, the contents of

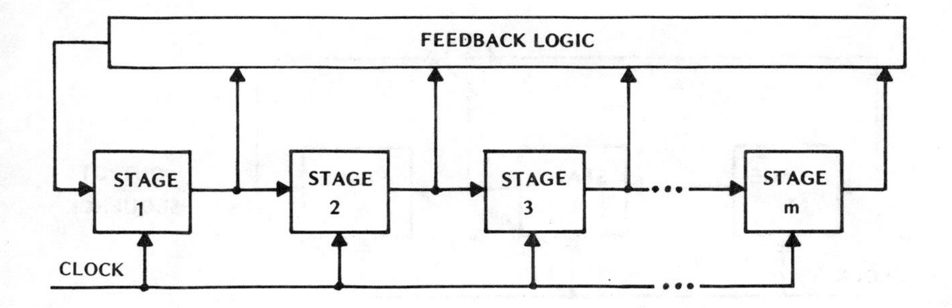

Figure 2.8 General feedback shift register with m stages.

the first two stages are shifted to the right, and the input to the first stage becomes its content. If the initial contents of the shift-register stages are 0 0 1, the subsequent contents after successive shifts are listed in Figure 2.9(b). After seven shifts, the shift register returns to its initial state. Thus, the periodic output sequence has a length of seven bits before repeating.

Let $s_j(i)$ denote the content of stage j after clock pulse i. The *state* of the shift register after clock pulse i is the vector

$$\mathbf{S}(i) = [s_1(i)\ s_2(i)\ \ldots\ s_m(i)], \quad i \geq 0 \tag{2.2.9}$$

The initial state is $\mathbf{S}(0)$. The *zero state* is the state for which all the contents are 0's. From the definition of a shift register, it follows that

$$s_j(i) = s_{j-1}(i-1), \quad i \geq 1, \quad 1 \leq j \leq m \tag{2.2.10}$$

where $s_0(i)$ denotes the input to stage 1 after clock pulse i. The output sequence is provided by the final stage. Thus, if a_i denotes the state of bit i of the output sequence, then $a_i = s_m(i)$. The state of a feedback shift register uniquely determines the subsequent sequence of states and the output sequence. Because the number of distinct states of an m-stage shift register is 2^m, the sequence of states and the output sequence must eventually become periodic with a period of at most 2^m.

If a linear feedback shift register reached the zero state at some time, it would always remain in the zero state and the output sequence would subsequently be all 0's. Since there are exactly $2^m - 1$ nonzero states, the period of a linear m-stage shift-register output sequence cannot exceed $2^m - 1$. A sequence of period $2^m - 1$ generated by a linear feedback shift register is called a *maximal-length* or *maximal sequence.* If a linear feedback shift register generates a maximal sequence, then all of its nonzero output sequences are maximal, regardless of the initial states.

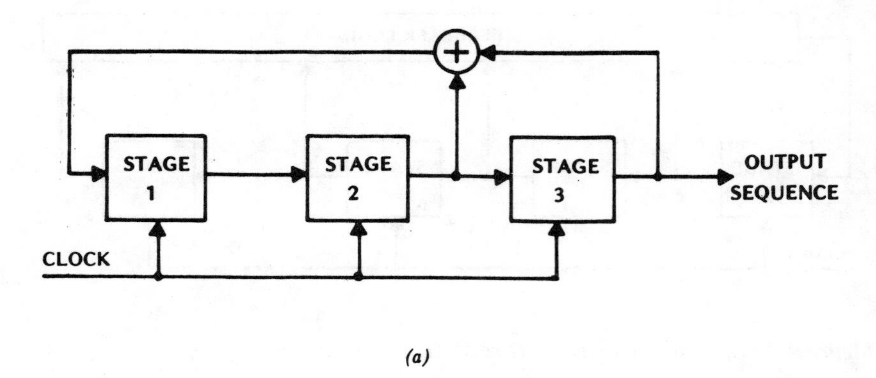

(a)

SHIFTS	CONTENTS			OUTPUT SEQUENCE
	STAGE 1	STAGE 2	STAGE 3	
INITIAL CONTENTS	0	0	1	1
1	1	0	0	0
2	0	1	0	0
3	1	0	1	1
4	1	1	0	0
5	1	1	1	1
6	0	1	1	1
7	0	0	1	1
8	1	0	0	0
9	0	1	0	0
10	1	0	1	1

(b)

Figure 2.9 (a) Three-stage linear feedback shift register and (b) contents and output after successive shifts.

Out of 2^m possible states, the content of the last stage, which is the same as the output bit, is a 0 in 2^{m-1} states. Thus, among the nonzero states, the output bit is a 0 in $2^{m-1} - 1$ states. We conclude that a maximal sequence contains exactly $2^{m-1} - 1$ *zeros* and 2^{m-1} *ones* per period.

2.2.1 Binary Arithmetic

Let the symbols 0 and 1 denote the elements of a set A. *Modulo-two addition* and *modulo-two multiplication* are binary operations defined by

$$\begin{gathered} 0+0=0\,,\quad 0+1=1\,,\quad 1+0=1\,,\quad 1+1=0 \\ 0\cdot 0=0\,,\quad 0\cdot 1=0\,,\quad 1\cdot 0=0\,,\quad 1\cdot 1=1 \end{gathered} \tag{2.2.11}$$

From these equations, it is easy to verify the following properties. The set A is closed under both modulo-two addition and modulo-two multiplication. Both operations are *associative* and *commutative.* Because -1 is defined as that element which when added to 1 yields 0, we have $-1 = 1$ and subtraction is the same as addition. Because $1 + 0 = 1$ and $0 + 0 = 0$, the additive identity element is 0. From $1 \cdot 1 = 1$ and $1 \cdot 0 = 0$ it follows that the multiplicative identity is 1 and the multiplicative inverse of 1 is $1^{-1} = 1$. By substituting all possible combinations, we can verify the *distributive laws:*

$$a(b+c) = ab + ac\,, \quad (b+c)a = ba + ca \tag{2.2.12}$$

where a, b, and c can each equal 0 or 1. The set of symbols 0 and 1, together with modulo-two addition and modulo-two multiplication, is called a *binary field* or *Galois field* of two elements and is denoted by $GF(2)$.

By definition of linear feedback shift register, the input to stage 1 satisfies

$$s_0(i) = \sum_{k=1}^{m} c_k s_k(i)\,, \qquad i \geqslant 0 \tag{2.2.13}$$

where the operations are modulo-two and the feedback coefficient c_k equals either 0 or 1, depending upon whether the output of stage k feeds a modulo-two adder. An m-stage shift register is defined to have $c_m = 1$; otherwise, the final stage would not contribute to the generation of the output sequence, but would only provide a one-shift delay. For example, Figure 2.9 gives $c_1 = 0$, $c_2 = c_3 = 1$, and $s_0(i) = s_2(i) + s_3(i)$. A general representation of a linear feedback shift register is shown in Figure 2.10(a). If $c_k = 1$, the corresponding switch is closed; if $c_k = 0$, it is open.

Equations (2.2.10) and (2.2.13) imply that the output bit $a_i = s_m(i)$ satisfies the linear recurrence relation

$$a_i = \sum_{k=1}^{m} c_k a_{i-k}, \qquad i \geqslant m \tag{2.2.14}$$

For $0 \leqslant i \leqslant m-1$, the output bits are determined solely by the initial state:

$$a_i = s_{m-i}(0), \qquad 0 \leqslant i \leqslant m-1 \tag{2.2.15}$$

Figure 2.10(a) is not necessarily the best way to implement the feedback shift register that produces the sequence satisfying Equation (2.2.14). Figure 2.10(b) illustrates an implementation that allows higher-speed operation. From this diagram, it follows that

$$s_j(i) = s_{j-1}(i-1) + c_{m-j+1}\, s_m(i-1), \quad i \geqslant 1, \quad 2 \leqslant j \leqslant m \tag{2.2.16}$$

$$s_1(i) = s_m(i-1), \qquad i \geqslant 1$$

These equations and the relations $c_m = 1$ and $a_i = s_m(i)$ imply (2.2.14). Thus, the two implementations can produce the same output sequence indefinitely if the first m output bits coincide. However, they require different initial states and have different sequences of states.

The sum of binary sequence $\{a_i\}$ and binary sequence $\{b_i\}$ is defined to be the binary sequence $\{a_i + b_i\}$, each bit of which is the modulo-two sum of the corresponding bits of $\{a_i\}$ and $\{b_i\}$. Thus, we can write

$$\{a_i\} + \{b_i\} = \{a_i + b_i\} \tag{2.2.17}$$

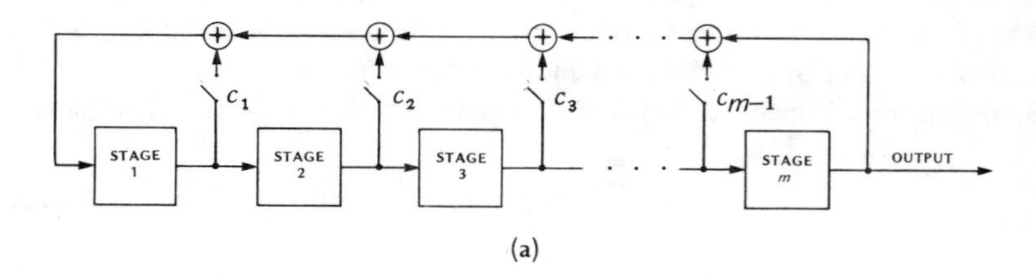

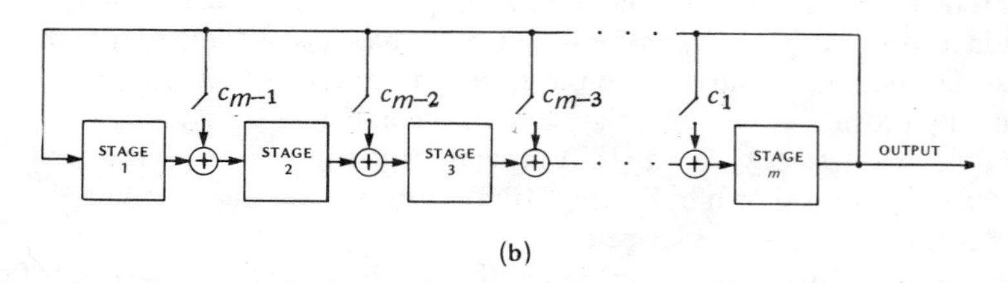

Figure 2.10 Linear feedback shift register: (a) standard representation and (b) high-speed form.

Suppose that $\{a_i\}$ and $\{b_i\}$ are generated by the same linear feedback shift register, but may differ because the initial states may be different. For the sequence $\{d_i\} = \{a_i + b_i\}$, (2.2.14) and the associative and distributive laws of binary fields imply that

$$d_i = \sum_{k=1}^{m} c_k a_{i-k} + \sum_{k=1}^{m} c_k b_{i-k}$$

$$= \sum_{k=1}^{m} (c_k a_{i-k} + c_k b_{i-k})$$

$$= \sum_{k=1}^{m} c_k (a_{i-k} + b_{i-k})$$

$$= \sum_{k=1}^{m} c_k d_{i-k} \tag{2.2.18}$$

This result indicates that the sequence $\{d_i\} = \{a_i + b_i\}$ can be generated by the same linear feedback logic as $\{a_i\}$ and $\{b_i\}$. Thus, if $\{a_i\}$ and $\{b_i\}$ are two output sequences of a linear feedback shift register, then $\{a_i + b_i\}$ is also. If $\{a_i\} = \{b_i\}$, then $\{a_i + b_i\}$ is the sequence of all 0's, which can be generated by any linear feedback shift register.

If $\{a_i\}$ is a maximal sequence and $j \neq 0$, modulo $2^m - 1$, then $\{a_i + a_{i+j}\} = \{a_i\} + \{a_{i+j}\}$ is not the sequence of all 0's. Because it is generated by the same shift register as $\{a_i\}$, $\{a_i + a_{i+j}\}$ must be a maximal sequence and, hence, some permutation of $\{a_i\}$. We conclude that the modulo-two sum of a maximal sequence and a cyclic shift of itself by j digits, where $j \neq 0$, modulo $2^m - 1$, produces another cyclic shift of the original sequence. This property is succinctly written as

$$\{a_i\} + \{a_{i+j}\} = \{a_{i+k}\}, \quad j \neq 0 \text{ (modulo } 2^m - 1) \tag{2.2.19}$$

In contrast, a nonmaximal linear sequence $\{a_i + a_{i+j}\}$ is not necessarily a permutation of $\{a_i\}$ and may not even have the same period.

2.2.2 Polynomials over the Binary Field

The use of polynomials allows one to concisely describe the dependence of the output sequence of a linear feedback shift register on its feedback coefficients and initial state. A *polynomial* over the binary field $GF(2)$ has the form

$$f(x) = f_0 + f_1 x + f_2 x^2 + \ldots + f_n x^n \tag{2.2.20}$$

where the coefficients $f_0, f_1, \ldots, f_n$ are elements of $GF(2)$ and the symbol x is an indeterminate introduced for convenience in calculations. The *degree* of a polynomial is the largest power of x with a nonzero coefficient. The *sum* of two polynomials, $f(x)$ and $g(x)$, over $GF(2)$ is another polynomial over $GF(2)$ defined by

$$f(x) + g(x) = \sum_{i=0}^{\max(n_1, n_2)} (f_i + g_i) x^i \qquad (2.2.21)$$

where the addition is modulo-two, n_1 is the degree of $f(x)$, n_2 is the degree of $g(x)$, and $\max(n_1, n_2)$ denotes the larger of n_1 and n_2. For example,

$$(1 + x^2 + x^3) + (1 + x^2 + x^4) = x^3 + x^4 \qquad (2.2.22)$$

The *product* of two polynomials over $GF(2)$ is another polynomial over $GF(2)$ defined by

$$f(x) g(x) = \sum_{i=0}^{n_1 + n_2} \left(\sum_{j=0}^{i} f_j g_{i-j} \right) x^i \qquad (2.2.23)$$

For example,

$$(1 + x^2 + x^3)(1 + x^2 + x^4) = 1 + x^3 + x^5 + x^6 + x^7 \qquad (2.2.24)$$

It is easily verified that associative, commutative, and distributive laws apply to polynomial addition and multiplication.

The *characteristic polynomial* associated with a linear feedback shift register of m stages is defined as

$$f(x) = 1 + \sum_{i=1}^{m} c_i x^i \qquad (2.2.25)$$

where $c_m = 1$ if we assume that all m stages contribute to the generation of the output sequence. The *generating function* associated with the output sequence is defined as

$$G(x) = \sum_{i=0}^{\infty} a_i x^i \qquad (2.2.26)$$

Substituting (2.2.14) into (2.2.26) and using the fact that modulo-two subtraction is equivalent to addition, we obtain

$$G(x) = \sum_{i=0}^{m-1} a_i x^i + \sum_{i=m}^{\infty} \sum_{k=1}^{m} c_k a_{i-k} x^i$$

$$= \sum_{i=0}^{m-1} a_i x^i + \sum_{k=1}^{m} c_k x^k \sum_{i=m}^{\infty} a_{i-k} x^{i-k}$$

$$= \sum_{i=0}^{m-1} a_i x^i + \sum_{k=1}^{m} c_k x^k [G(x) + \sum_{i=0}^{m-k-1} a_i x^i] \tag{2.2.27}$$

Rearranging this equation, using (2.2.25), and defining $c_0 = 1$ yield

$$G(x) f(x) = \sum_{i=0}^{m-1} a_i x^i + \sum_{k=1}^{m} c_k x^k \left(\sum_{i=0}^{m-k-1} a_i x^i \right)$$

$$= \sum_{k=0}^{m-1} c_k x^k \left(\sum_{i=0}^{m-k-1} a_i x^i \right) \tag{2.2.28}$$

Therefore,

$$G(x) = \frac{\sum_{k=0}^{m-1} c_k x^k \left(\sum_{i=0}^{m-k-1} a_i x^i \right)}{f(x)}, \quad c_0 = 1 \tag{2.2.29}$$

Because $a_i = s_{m-i}(0), i = 0, 1, \ldots, m - 1$, (2.2.29) and (2.2.15) explicitly indicate that the output sequence is completely determined by the initial state and the feedback coefficients $c_k, k = 1, 2, \ldots, m$. Because the numerator of (2.2.29) is of degree $m - 1$ or less and the denominator is of degree m, $G(x)$ cannot have a finite degree.

Equation (2.2.29) provides a convenient computational means of finding the output sequence of a linear feedback shift register. In the example of Figure 2.9, the feedback coefficients are $c_1 = 0, c_2 = 1$, and $c_3 = 1$, and the initial contents are $a_0 = 1, a_1 = 0$, and $a_2 = 0$. Therefore,

$$G(x) = \frac{1 + x^2}{1 + x^2 + x^3} \tag{2.2.30}$$

Performing the long division yields the output sequence listed in the figure. The long division follows the rules of binary arithmetic:

$$
\begin{array}{r l}
 & 1 + x^3 + x^5 + x^6 + x^7 + x^{10} + \ldots \\
\cline{2-2}
1 + x^2 + x^3 & \big) \; 1 + x^2 \\
 & \;\; 1 + x^2 + x^3 \\
\cline{2-2}
 & \;\; x^3 \\
 & \;\; x^3 + x^5 + x^6 \\
\cline{2-2}
 & \;\; x^5 + x^6 \\
 & \;\; x^5 \quad + x^7 + x^8 \\
\cline{2-2}
 & \;\; x^6 + x^7 + x^8 \\
 & \;\; x^6 \quad + x^8 + x^9 \\
\cline{2-2}
 & \;\; x^7 \quad + x^9 \\
 & \;\; x^7 \quad + x^9 + x^{10} \\
\cline{2-2}
 & \;\; x^{10}
\end{array}
$$

The numerator of (2.2.29) can be put into a different form:

$$
\begin{aligned}
\phi(x) &= \sum_{k=0}^{m-1} c_k x^k \left(\sum_{i=0}^{m-k-1} a_i x^i \right) = \sum_{k=0}^{m-1} \sum_{i=0}^{m-k-1} c_k a_i x^{k+i} \\
&= \sum_{k=0}^{m-1} \sum_{j=k}^{m-1} c_k a_{j-k} x^j = \sum_{j=0}^{m-1} \sum_{k=0}^{j} c_k a_{j-k} x^j \\
&= \sum_{i=0}^{m-1} x^i \left(\sum_{k=0}^{i} c_k a_{i-k} \right)
\end{aligned} \tag{2.2.31}
$$

Suppose that a polynomial $f(x)$ of degree $m \geqslant 1$ and a nonzero polynomial $g(x)$ of degree $m - 1$ or less are given. If $f_0 = 1$ and $f(x)$ is regarded as a characteristic polynomial, then the coefficients of $f(x)$ can be used to determine feedback coefficients c_k, $k = 1, 2, \ldots, m$. The coefficients of $g(x)$ can be compared to those of $\phi(x)$, as given by (2.2.31), to determine initial contents a_i, $i = 0, 1, \ldots, m - 1$. Thus, $f(x)$ and $g(x)$ uniquely determine an m-stage linear feedback shift register with an output sequence given by $G(x) = g(x)/f(x)$.

The polynomial $g(x)$ is said to *divide* the polynomial $f(x)$ if there is a polynomial $h(x)$ such that $f(x) = h(x)g(x)$. A polynomial $f(x)$ over $GF(2)$ of degree m is called *irreducible* if $f(x)$ is not divisible by any polynomial over $GF(2)$ of degree less than m but greater than zero. If $f(x)$ is irreducible over $GF(2)$, then $f(0) \neq 0$, for otherwise x would divide $f(x)$. If $f(x)$ has an even number of terms, then $f(1) = 0$ and the fundamental theorem of algebra implies that $x + 1$ divides $f(x)$. Therefore, an irreducible polynomial over $GF(2)$ must have an odd number of terms, but this condition may not be sufficient for irreducibility. For example, $1 + x + x^2$ is irreducible, but $1 + x + x^5 = (1 + x^2 + x^3)(1 + x + x^2)$ is not.

An irreducible polynomial over $GF(2)$ of degree m is called *primitive* if the smallest positive integer n for which $f(x)$ divides $1 + x^n$ is $n = 2^m - 1$. If $f(x)$ is the characteristic polynomial of (2.2.25), and $\{a_i\}$ is a nonzero sequence generated by the associated shift register, then $\{a_i\}$ is a maximal sequence if and only if $f(x)$ is primitive [1, 2]. For any positive integer $n > 1$, the *Euler function* $\psi(n)$ is the number of positive integers that are less than n and relatively prime to it. Thus, if n is a prime number, then $\psi(n) = n - 1$. For any positive integer m, the number of different primitive polynomials of degree m over $GF(2)$ is [1]

$$\lambda(m) = \frac{\psi(2^m - 1)}{m} \leqslant \frac{2^m - 2}{m} \tag{2.2.32}$$

The upper bound is achieved if $2^m - 1$ is a prime number. Because each primitive polynomial determines a distinct maximal sequence, the number of these sequences is large for moderate values of m. For example, if $m = 19$, then $\lambda(m) = 27{,}594$.

2.2.3 Autocorrelation of Spreading Waveform

The spreading waveform, which is controlled by a pseudonoise sequence, can be represented by

$$p(t) = \sum_{i=0}^{N-1} p_i\, \psi(t - iT_c)\,, \qquad 0 \leqslant t \leqslant NT_c \tag{2.2.33}$$

where N is the number of chips transmitted, $p_i = +1$ if the associated symbol a_i of the underlying pseudonoise sequence ia a 1, and $p_i = -1$ if a_i is a 0. The pulse $\psi(t)$, which is called the *chip waveform*, is usually confined to the interval $[0, T_c]$. We derive the autocorrelation of $p(t)$ assuming that we have a maximal sequence, a unit-amplitude chip waveform over $[0, T_c]$, and an ideal periodic spreading waveform of infinite extent.

The *autocorrelation* of a periodic function $x(t)$ with period T is defined as

$$R_x(\tau) = \frac{1}{T} \int_{-T/2}^{T/2} x(t)\, x(t + \tau)\, dt \tag{2.2.34}$$

where τ is the relative delay variable. It follows that $R_x(\tau)$ has period T. The autocorrelation of the spreading waveform, $R_p(\tau)$, varies linearly between the values that it assumes at the points $\tau = jT_c$, where j is an integer. Thus, only these points need to be considered. If the maximal sequence has length $K = 2^m - 1$, then $p(t)$ has period KT_c and (2.2.34) yields

$$R_p(jT_c) = \frac{1}{K} \sum_{i=1}^{K} p_i p_{i+j} = \frac{A_j - D_j}{K} \qquad (2.2.35)$$

where A_j denotes the number of terms with $p_i p_{i+j} = +1$ and D_j is the number of terms with $p_i p_{i+j} = -1$. The correspondence between p_i and a_i implies that A_j equals the number of 0's in a period of $\{a_i + a_{i+j}\}$. From (2.2.19), it follows that A_j equals the number of 0's in a maximal sequence if $j \neq 0$, modulo K. Thus, $A_j = (K - 1)/2$ and, similarly, $D_j = (K + 1)/2$ if $j \neq 0$, modulo K. We conclude that

$$R_p(jT_c) = -\frac{1}{K}, \qquad j \neq 0 \text{ (modulo } K) \qquad (2.2.36)$$

When $j = 0$, modulo K, $R_p(jT_c) = 1$. Therefore, over the interval

$$|\tau| \leqslant \frac{KT_c}{2}$$

the autocorrelation is

$$R_p(\tau) = \begin{cases} 1 - \left(\dfrac{K+1}{K} \dfrac{|\tau|}{T_c}\right), & |\tau| \leqslant T_c \\ -\dfrac{1}{K}, & |\tau| > T_c \end{cases} \qquad (2.2.37)$$

Because $R_p(\tau)$ has period KT_c, it is completely specified by this equation. The autocorrelation can be compactly expressed as

$$R_p(\tau) = -\frac{1}{K} + \frac{K+1}{K} \sum_{i=-\infty}^{\infty} \Lambda\left(\frac{\tau - iKT_c}{T_c}\right) \qquad (2.2.38)$$

It is plotted in Figure 2.11.

A straightforward calculation or the use of tables gives the Fourier transform of the triangular function:

$$F\left\{\Lambda\left(\frac{t}{T}\right)\right\} = \int_{-\infty}^{\infty} \Lambda\left(\frac{t}{T}\right) \exp(-j2\pi f t)\, dt$$

$$= T \operatorname{sinc}^2 fT \qquad (2.2.39)$$

where $j = \sqrt{-1}$ and $\operatorname{sinc} x = (\sin \pi x)/\pi x$. Because the summation in the right-hand side of (2.2.38) is a periodic function, it can be expressed as a complex exponential Fourier series. We take the Fourier transform of this series, express the Fourier coefficients as Fourier transforms, and use (2.2.39). The result is

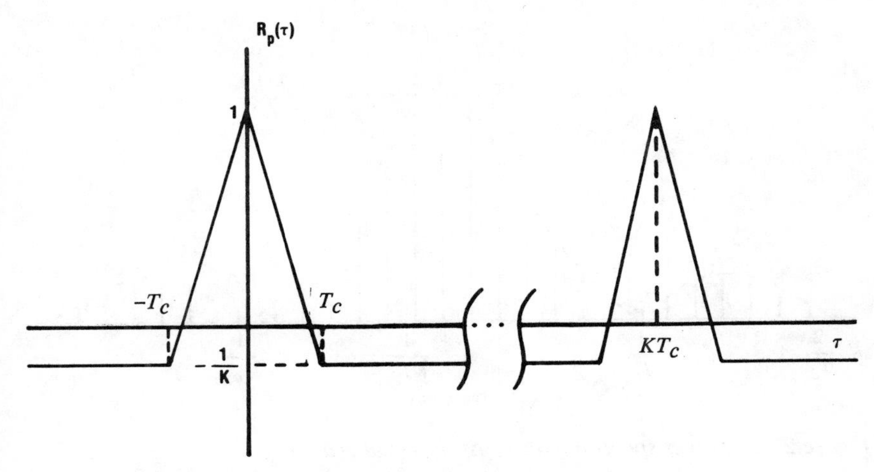

Figure 2.11 Autocorrelation of maximal sequence.

$$F\left\{\sum_{i=-\infty}^{\infty} \Lambda\left(\frac{t - iKT_c}{T_c}\right)\right\} = \frac{1}{K} \sum_{i=-\infty}^{\infty} \operatorname{sinc}^2\left(\frac{1}{K}\right)\delta\left(f - \frac{i}{KT_c}\right) \quad (2.2.40)$$

where $\delta(\ )$ is the Dirac delta function. We use the preceding results to determine $S_p(f)$, the power spectral density of $p(t)$, which is defined as the Fourier transform of $R_p(\tau)$. Taking the Fourier transform of (2.2.38), substituting (2.2.40), noting that the Fourier transform of a constant is a delta function, and rearranging the result, we obtain

$$S_p(f) = \frac{K+1}{K^2} \sum_{\substack{i=-\infty \\ i\neq 0}}^{\infty} \operatorname{sinc}^2\left(\frac{1}{K}\right)\delta\left(f - \frac{i}{KT_c}\right) + \frac{1}{K^2}\,\delta(f) \quad (2.2.41)$$

This function is plotted in Figure 2.12. It consists of delta functions separated by $1/KT_c$.

Some nonmaximal sequences have nearly even balances of 0's and 1's and have autocorrelations roughly similar to that of Figure 2.11. Thus, these sequences can be considered pseudonoise sequences. However, the autocorrelations of most nonmaximal sequences have minor peaks in addition to the major ones. Unless they are kept small, these minor peaks hinder code synchronization (Section 2.7).

Linear pseudonoise sequences are inherently susceptible to mathematical cryptanalysis (Chapter 6). Thus, if message security is desired when a linear sequence is used for spectrum spreading, then the data symbols must be enciphered before they are combined with the pseudonoise sequence.

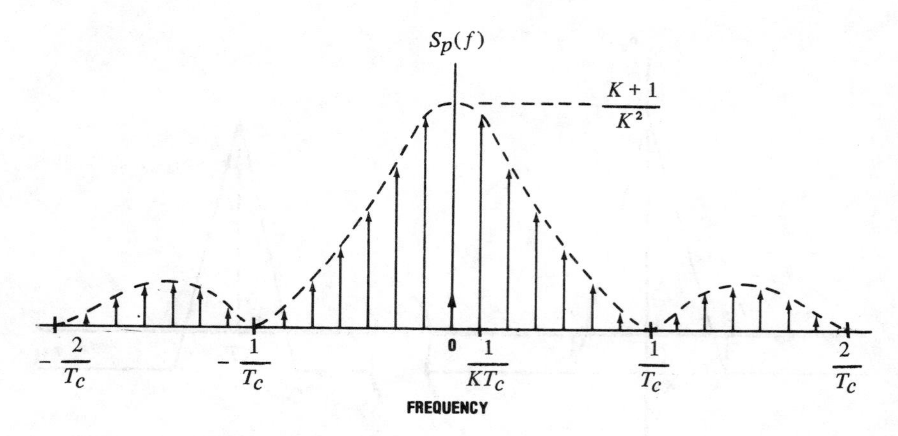

Figure 2.12 Power spectral density of maximal sequence.

A *long sequence* or *long code* is a sequence with a period that is much longer than a data-symbol duration. A *short sequence* or *short code* is a sequence with a period that is no more than a few times as large as a data-symbol duration. If short pseudonoise sequences are used, an opponent may be able to retransmit intercepted communications with enough power to capture the receiver during the interval between messages or during acquisition. Thus, long pseudonoise sequences are needed for secure communications. Even more is sometimes required. A sophisticated opponent might be capable of demodulating a direct-sequence signal, determining its pseudonoise sequence, and transmitting an appropriate sequence segment to capture the receiver. To counter this threat, the pseudonoise code generators in the transmitter and the receiver must be readily programmable and the sequences must be nonlinear (Chapter 6).

2.3 CONCEALMENT OF DIRECT-SEQUENCE WAVEFORMS

The presence of a signal is difficult to detect with a simple spectrum analyzer if the signal has a low power spectral density compared with that of the thermal and environmental noise and if the spectral density is a slowly varying function of frequency over a wide band. We derive an approximate necessary condition for a transmitted direct-sequence signal to have a slowly varying spectral density when a maximal sequence generates the spreading waveform. The results are similar when other pseudonoise sequences are used because they have similar autocorrelations.

Because the direct-sequence signal is a nonstationary process, we require a generalization of the autocorrelation function. Thus, the *average autocorrelation* of $x(t)$ is defined as

$$\bar{R}_x(\tau) = \lim_{T\to\infty} \frac{1}{2T} \int_{-T}^{T} R_x(t,\tau)\,dt \tag{2.3.1}$$

The limit exists and may be nonzero if $x(t)$ has finite power and infinite duration. If $x(t)$ is stationary, $\bar{R}_x(\tau) = R_x(\tau)$. The *average power spectral density*, denoted by $\bar{S}_x(f)$, is defined as the Fourier transform of the average autocorrelation.

The message, $m(t)$, is modeled as a stationary stochastic process with an autocorrelation $R_m(\tau)$. Suppose that θ in (2.1.1) is modeled as a random variable uniformly distributed over the interval $[0,2\pi]$ and statistically independent of $m(t)$. The autocorrelation of

$$m_1(t) = m(t)\cos(\omega_0 t + \theta) \tag{2.3.2}$$

is determined by applying (2.2.1) and using trigonometry. The result is independent of t, so we write

$$R_{m1}(\tau) = \frac{1}{2} R_m(\tau)\cos\omega_0\tau \tag{2.3.3}$$

Thus, the power spectral density of $m_1(t)$ is

$$S_{m1}(f) = \frac{1}{4}\,[S_m(f-f_0) + S_m(f+f_0)] \tag{2.3.4}$$

where $f_0 = \omega_0/2\pi$ and $S_m(f)$ is the power spectral density of $m(t)$.

By using (2.1.1) and (2.2.1) and noting that $p(t)$ is deterministic, the autocorrelation of $s(t)$ is determined to be

$$R_s(t,\tau) = A^2 p(t)\,p(t+\tau)R_{m1}(\tau) \tag{2.3.5}$$

which indicates that $s(t)$ is a nonstationary process. Because $p(t)$ is periodic, the definitions of (2.2.34) and (2.3.1) yield

$$\bar{R}_s(\tau) = A^2 R_p(\tau) R_{m1}(\tau) \tag{2.3.6}$$

Consequently, $\bar{S}_s(f)$, the average power spectral density of $s(t)$, is the convolution of $A^2 S_p(f)$ with $S_{m1}(f)$. Using (2.2.41) and (2.3.4), we obtain the average power spectrum of a transmitted direct-sequence signal when a maximal sequence is used:

$$\bar{S}_s(f) = \frac{A^2}{4K^2} S_m(f - f_0) + A^2 \frac{K+1}{4K^2} \sum_{\substack{i=-\infty \\ i \neq 0}}^{\infty} \text{sinc}^2 \left(\frac{i}{K}\right) S_m\left(f - f_0 - \frac{i}{KT_c}\right)$$

$$+ \frac{A^2}{4K^2} S_m(f + f_0) + A^2 \frac{K+1}{4K^2} \sum_{\substack{i=-\infty \\ i \neq 0}}^{\infty} \text{sinc}^2 \left(\frac{i}{K}\right) S_m\left(f + f_0 + \frac{i}{KT_c}\right) \tag{2.3.7}$$

If θ is modeled as a constant, rather than a random variable, (2.3.7) can be derived provided that

$$\lim_{T \to \infty} \frac{1}{2T} \int_{-T}^{T} p(t)p(t+\tau)\cos(2\omega_0 t + \omega_0 \tau + 2\theta)\, dt = 0 \tag{2.3.8}$$

This equation is satisfied if $\omega_0 \gg 1/T_c$ and in other cases.

To ensure that the spectrum is slowly varying, the spectral contributions of the terms in the summations of (2.3.7) must overlap. The center of the spectral contribution of a term is separated from the center of the spectral contribution of an adjacent term by $1/KT_c$. Thus, the bandwidth of $m_1(t)$ must satisfy $B \gg 2/KT_c$ if $\bar{S}_s(f)$ is to be slowly varying. Because $B \cong 2/T_s$, an approximate necessary condition for communication concealment from spectrum analysis is

$$KT_c \gg T_s \tag{2.3.9}$$

which states that the pseudonoise sequence period must equal or exceed a data-bit duration.

Since messages tend to be nearly random in character, it is plausible to model $m(t)$ as a random binary sequence with autocorrelation

$$R_m(\tau) = \Lambda \left(\frac{\tau}{T_s}\right) \tag{2.3.10}$$

Equation (2.2.39) gives the corresponding power spectral density:

$$S_m(f) = T_s \,\text{sinc}^2 f T_s \tag{2.3.11}$$

The fact that a signal is concealed does not mean that it cannot be detected. Suppose that $s(t)$ enters a wideband receiver and is squared. Because $m^2(t) = 1$, the output of the squaring device is

$$s^2(t) = A^2 p^2(t) \cos^2(\omega_0 t + \theta)$$
$$= \frac{A^2}{2} p^2(t)[1 + \cos(2\omega_0 t + 2\theta)] \qquad (2.3.12)$$

where $p^2(t) = 1$ when the chip waveform is rectangular. If $s^2(t)$ is applied to an integrator or a narrowband filter, the energy of the signal can often be detected, even if $\overline{S}_s(f)$ is far below the noise power spectral density. The double-frequency term of (2.3.12) can be applied to a separate device for estimation of the carrier frequency of the direct-sequence signal (Chapter 4). However, an interceptor cannot demodulate $s(t)$ without knowledge of $p(t)$.

The average power spectral density of a quadriphase direct-sequence signal has the same form as (2.3.7). The energy of a quaternary direct-sequence signal can be detected in the same manner as the energy of a binary signal. A device that raises the received signal of (2.1.5) to the fourth power produces a sinusoidal term with frequency $4\omega_0$. Thus, $s^4(t)$ can be used for the estimation of the carrier frequency of a quaternary direct-sequence signal. To extract $m_i(t)$, where $i = 1$ or 2, an interceptor must be able to produce $p_i(t)$.

2.4 ERROR PROBABILITIES FOR BINARY DIRECT-SEQUENCE SYSTEM

The essential elements of a binary direct-sequence receiver with coherent PSK demodulation are shown in Figure 2.13. After passage through the wideband filter, the total received signal is

$$r(t) = Am(t)p(t)\cos(\omega_0 t + \theta) + i(t) + n(t) \qquad (2.4.1)$$

where $i(t)$ denotes the interference and $n(t)$ denotes the zero-mean bandlimited white Gaussian noise. We assume that the spreading waveform consists of rectangular pulses and that perfect code, carrier phase, and symbol synchronization exist. Thus, the input to the demodulator is

$$r_1(t) = Am(t)\cos(\omega_0 t + \theta) + i(t)p(t) + n(t)p(t) \qquad (2.4.2)$$

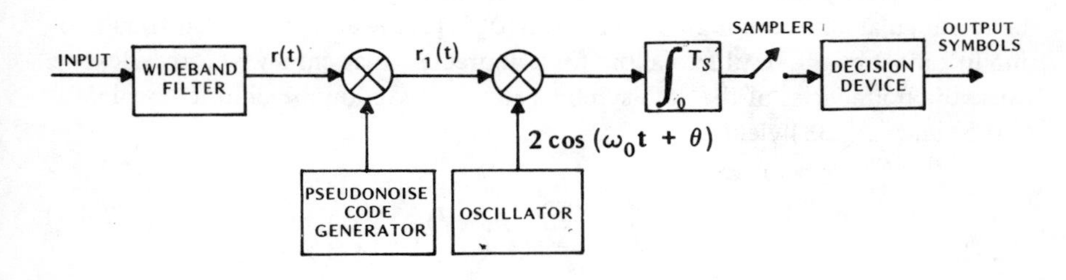

Figure 2.13 Basic elements of coherent receiver for direct-sequence system.

The factor $p(t)$ in this equation ensures that the interference energy is spread over a bandwidth at least equal to W. We consider a data-symbol interval that we take to be $[0, T_s]$ for convenience. The input sample applied to the decision device at the end of this interval is

$$L = \int_0^{T_s} 2r_1(t)\cos(\omega_0 t + \theta)\,dt \tag{2.4.3}$$

We assume that $f_0 = \omega_0/2\pi \gg W \geqslant 1/T_c \geqslant 1/T_s$. Therefore, (2.4.2) and (2.4.3) yield

$$L \cong A\int_0^{T_s} m(t)\,dt + L_1 + L_2$$

$$= \pm AT_s + L_1 + L_2 \tag{2.4.4}$$

where $m(t) = +1$ represents the symbol 1, $m(t) = -1$ represents the symbol 0, and

$$L_1 = \int_0^{T_s} 2i(t)p(t)\cos(\omega_0 t + \theta)\,dt \tag{2.4.5}$$

$$L_2 = \int_0^{T_s} 2n(t)p(t)\cos(\omega_0 t + \theta)\,dt \tag{2.4.6}$$

The decision device produces the symbol 1 if $L > 0$ and the symbol 0 if $L < 0$. An error occurs if $L < 0$ when $m(t) = +1$ or if $L > 0$ when $m(t) = -1$. The probability that $L = 0$ is zero.

Because $p(t)$ is deterministic and $n(t)$ is a Gaussian process, L_2 is a Gaussian random variable. If $i(t)$ is a Gaussian process, then L_1 is a Gaussian random variable; if $i(t)$ is not, L_1 may still approximate a Gaussian random variable under certain conditions.

For simplicity in the subsequent analysis, the chip waveform is assumed to be rectangular with a unit amplitude over $[0, T_c]$. However, it is not difficult to modify most of the results to allow for a nonrectangular chip waveform. Because the boundaries of the data symbols and the pseudonoise chips coincide, (2.4.5) and (2.2.33) yield

$$L_1 = \sum_{\nu=0}^{G-1} p_{\nu+l} J_\nu \tag{2.4.7}$$

where the processing gain G is equal to the number of chips in a symbol interval, p_l corresponds to the first chip in $[0, T_s]$, and

$$J_\nu = \int_{\nu T_c}^{(\nu+1)T_c} 2i(t)\cos(\omega_0 t + \theta)\,dt\,, \qquad \nu = 0, 1, \ldots, G-1 \tag{2.4.8}$$

We subsequently model $p(t)$ over the interval $[0, T_s]$ as a random binary sequence. This model is reasonable if

$$T_c \ll T_s \ll KT_c \tag{2.4.9}$$

which is satisfied in practical systems using long pseudonoise sequences.

To obtain a reasonably simple expression for the symbol error probability, further assumptions are necessary. Three models are presented. The stationary-process model is the most general and yields the simplest results, but it entails the strongest assumptions. The tone-interference and nearly exact models pertain to interference that can be modeled as a tone or set of tones.

2.4.1 Stationary-Process Model

If the J_ν are independent of each other, then L_1 is the sum of independent random variables. If the J_ν are also either uniformly bounded or identically distributed, then the central limit theorem [3] applies, and the distribution function of L_1 converges to a Gaussian distribution as G increases. If the J_ν are not independent, it is still reasonable to assume that L_1 has a roughly Gaussian distribution for large G. In the *stationary-process model,* the strong assumption is made that L_1 and, hence, L have approximately Gaussian distributions for large G.

Because they arise from different physical sources, it is assumed that $i(t)$, $n(t)$, $p(t)$, and θ are independent of each other. Because $p(t)$ is modeled as a random binary sequence, $E[L_1] = E[L_2] = 0$ and

$$E[L] = \pm AT_s \tag{2.4.10}$$

Because $E[n(t)] = 0$, the variance of L is equal to the sum of the variances of L_1 and L_2, Under the Gaussian assumption, a simple calculation gives the symbol error probability

$$P_s = \frac{1}{2}\,\mathrm{erfc}\left\{\left[\frac{E_s T_s}{\mathrm{VAR}(L_1) + \mathrm{VAR}(L_2)}\right]^{1/2}\right\} \tag{2.4.11}$$

where $E_s = A^2 T_s/2$ is the received energy per symbol, and the complementary error function is defined by

$$\mathrm{erfc}\, x = \frac{2}{\sqrt{\pi}} \int_x^\infty \exp(-y^2)\,dy \tag{2.4.12}$$

If we ignore the effect of the wideband filter and model $n(t)$ as ideal white noise, then a straightforward calculation gives

$$\mathrm{VAR}(L_2) = N_0 T_s \tag{2.4.13}$$

where $N_0/2$ is the two-sided noise-power spectral density. Thus, P_s can be calculated from (2.4.11) once the variance of L_1 has been determined.

We assume that $i(t)$ is a stationary stochastic process with autocorrelation $R_i(\tau)$. Because $p(t)$ is modeled as a random binary sequence, its autocorrelation is

$$R_p(\tau) = \Lambda\left(\frac{\tau}{T_c}\right) \tag{2.4.14}$$

Suppose that θ is an independent random variable uniformly distributed over $[0, 2\pi]$. Using (2.4.14) and (2.4.5) and the independence of $i(t)$ and $p(t)$, we can express the variance of L_1 as a double integral over autocorrelations. We obtain the approximation

$$\mathrm{VAR}(L_1) = 2\int_0^{T_s}\int_0^{T_s} R_i(t_1 - t_2)\Lambda\left(\frac{t_1 - t_2}{T_c}\right)\cos[\omega_0(t_1 - t_2)]\,dt_1 dt_2 \tag{2.4.15}$$

The use of (2.4.14) is an approximation because L_1 is defined over the fixed interval $[0, T_s]$ and the transitions of $p(t)$ occur at well-defined times, which is not true for a random binary sequence.

We change variables by using $\tau = t_1 - t_2$, $\sigma = t_1 + t_2$. The Jacobian of the transformation is 2. Evaluating one of the integrals leaves

$$\mathrm{VAR}(L_1) = 2\int_{-T_s}^{T_s}(T_s - |\tau|)R_i(\tau)\Lambda\left(\frac{\tau}{T_c}\right)\cos\omega_0\tau\,d\tau \tag{2.4.16}$$

If we assume that θ is a fixed constant instead of a random variable, we obtain an additional integral in the equation for $\mathrm{VAR}(L_1)$. However, since $\omega_0 \gg 1/T_s$, this integral is negligible.

Because $T_s \gg T_c$,

$$\mathrm{VAR}(L_1) \cong 2T_s\int_{-T_c}^{T_c} R_i(\tau)\Lambda\left(\frac{\tau}{T_c}\right)\cos\omega_0\tau\,d\tau \tag{2.4.17}$$

The limits in this integral can be extended to $\pm\infty$ because the integrand is truncated. Because $R_i(\tau)$ is an even function, the convolution theorem and (2.3.7) transform (2.4.17) into the alternative form

$$\mathrm{VAR}(L_1) \cong 2T_sT_c\int_{-\infty}^{\infty} S_i(f)\,\mathrm{sinc}^2[(f - f_0)T_c]\,df \tag{2.4.18}$$

where $S_i(f)$ is the power spectral density of the interference at the output of the wideband filter. If $S_i'(f)$ is the interference power spectral density at the input and $H(f)$ is the transfer function of the wideband filter, then

$$S_i(f) = S_i'(f)|H(f)|^2 \tag{2.4.19}$$

Because $f_0 \gg 1/T_c$, the integration over negative frequencies in (2.4.18) is usually negligible and

$$\mathrm{VAR}(L_1) \cong 2T_s T_c \int_0^{\infty} S_i(f)\,\mathrm{sinc}^2[(f - f_0)T_c]\,df \tag{2.4.20}$$

Several special cases are of particular interest. It is convenient to define the *interference factor, b*, implicitly by the equation

$$\mathrm{VAR}(L_1) = bR_i T_s T_c \tag{2.4.21}$$

where $R_i = R_i(0)$ is the average interference power. Because $\mathrm{sinc}\, x \leq 1$, (2.4.20) implies that $b \leq 1$.

Suppose that the interference has a flat spectrum over a band within the passband of the wideband filter so that

$$S_i(f) = \begin{cases} \dfrac{R_i}{2W_1}, & |f - f_1| \leq \dfrac{W_1}{2},\ |f + f_1| \leq \dfrac{W_1}{2} \\ 0, & \text{otherwise} \end{cases} \tag{2.4.22}$$

Equations (2.4.20) to (2.4.22) yield

$$b = \frac{1}{W_1} \int_{f_1 - W_1/2}^{f_1 + W_1/2} \mathrm{sinc}^2[(f_0 - f)T_c]\,df \tag{2.4.23}$$

Next, suppose that the interference has the form

$$i(t) = \sqrt{2R_i}\, q(t) \cos(\omega_1 t + \theta_1) \tag{2.4.24}$$

where $q(t)$ is a unit-power stationary process, carrier frequency $f_1 = \omega_1/2\pi$ is within the receiver passband, and θ_1 is an independent, uniformly distributed random variable. It follows that

$$R_i(\tau) = R_i R_q(\tau) \cos \omega_1 \tau \tag{2.4.25}$$

where $R_q(\tau)$ is the autocorrelation of $q(t)$. Combining (2.4.17), (2.4.21), and (2.4.25) and neglecting a double-frequency term that does not contribute significantly to the result, we obtain

$$b = \frac{1}{T_c} \int_{-T_c}^{T_c} R_q(\tau) \Lambda \left(\frac{\tau}{T_c}\right) \cos[(\omega_1 - \omega_0)\tau] d\tau \tag{2.4.26}$$

For tone (unmodulated-carrier) interference, $q(t) = 1$ so that $R_q(\tau) = 1$ and

$$b = \text{sinc}^2 [(f_1 - f_0) T_c] \tag{2.4.27}$$

For tone jamming at the carrier or center frequency, $f_1 = f_0$ so that $b = 1$.

Suppose that $q(t)$ in (2.4.24) is a pseudonoise sequence with a rectangular chip waveform, chip duration T_q, and length K_1. We neglect the effect of the wideband filter. If

$$T_q \ll T_s \ll K_1 T_q \tag{2.4.28}$$

and the cross correlation of $p(t)$ and $q(t)$ is small for all relative delays, it is reasonable to model $q(t)$ over the interval $[0, T_s]$ as an independent random binary sequence. The autocorrelation of $q(t)$ is

$$R_q(\tau) = \Lambda \left(\frac{\tau}{T_q}\right) \tag{2.4.29}$$

We define the parameters

$$T_0 = \min(T_c, T_q) \tag{2.4.30}$$

$$T_1 = \max(T_c, T_q) \tag{2.4.31}$$

Substituting (2.4.29) into (2.4.26) and using (2.2.8), the integral can be evaluated. For $T_0 \neq 0$, we obtain

$$b = \frac{2}{T_c} \left\{ \frac{\cos[2\pi(f_1 - f_0) T_0]}{(2\pi)^2 (f_1 - f_0)^2} \left(\frac{1}{T_1} - \frac{1}{T_0}\right) - \frac{2 \sin[2\pi(f_1 - f_0) T_0]}{(2\pi)^3 (f_1 - f_0)^3 T_1 T_0} + \frac{1}{(2\pi)^2 (f_1 - f_0)^2} \left(\frac{1}{T_1} + \frac{1}{T_0}\right) \right\}, \quad f_1 \neq f_0 \tag{2.4.32}$$

$$b = \frac{T_0}{T_c} \left(1 - \frac{T_0}{3T_1}\right), \quad f_1 = f_0 \tag{2.4.33}$$

If $T_c = T_q$, this equation becomes

$$b = \begin{cases} \dfrac{1 - \text{sinc}[2(f_1 - f_0) T_c]}{\pi^2 (f_1 - f_0)^2 T_c^2}, & f_1 \neq f_0, \ T_c = T_q \\[2ex] \dfrac{2}{3}, & f_1 = f_0, \ T_c = T_q \end{cases} \tag{2.4.34}$$

Equations (2.4.30) to (2.4.34) give b for interference due to a direct-sequence signal. Equation (2.4.26) indicates that b is greatest when $f_1 = f_0$.

If more than one statistically independent source of interference is present, the linearity of the demodulation allows calculation of the appropriate value of b. If R_i is interpreted as the total interference power,

$$b = \frac{1}{R_i} \sum_j b_j R_{ij} \tag{2.4.35}$$

where R_{ij} is the interference power due to source j, and b_j is the corresponding interference factor. If the $b_j \leqslant 1$, then $b \leqslant 1$.

If it is desired to account for the wideband filtering in the calculation of $\mathrm{VAR}(L_2)$, we may write

$$\mathrm{VAR}(L_2) = \beta N_0 T_s \tag{2.4.36}$$

where β is analogous to the interference factor. For an ideal rectangular bandpass filter of bandwidth W, application of the analogue of (2.4.23) yields

$$\beta = T_c \int_{f_0 - W/2}^{f_0 + W/2} \operatorname{sinc}^2 [(f - f_0) T_c] \, df \tag{2.4.37}$$

If $WT_c \geqslant 2$, then $0.89 \leqslant \beta \leqslant 1.0$.

Substituting (2.4.21) and (2.4.36) into (2.4.11), we obtain

$$P_s = \frac{1}{2} \operatorname{erfc} \left\{ \left[\frac{E_s}{\beta N_0 + R_i T_c b} \right]^{1/2} \right\} \tag{2.4.38}$$

Defining

$$N_{oe} = \beta N_0 + R_i T_c b = \beta N_0 + \left(\frac{R_i}{G} \right) T_s b \tag{2.4.39}$$

we can interpret $N_{0e}/2$ as the *equivalent two-sided power spectral density* of the interference plus noise. Equation (2.4.39) shows explicitly that increasing the processing gain is useful against interference for which R_i is fixed. *Increasing the processing gain by increasing W is not helpful against interference for which R_i increases proportionately with W, such as white noise.*

Figures 2.14 and 2.15 illustrate the dependence of b for direct-sequence interference on the *normalized frequency offset,*

$$\Delta = |(f_1 - f_0) T_c| \tag{2.4.40}$$

and the *chip rate ratio*

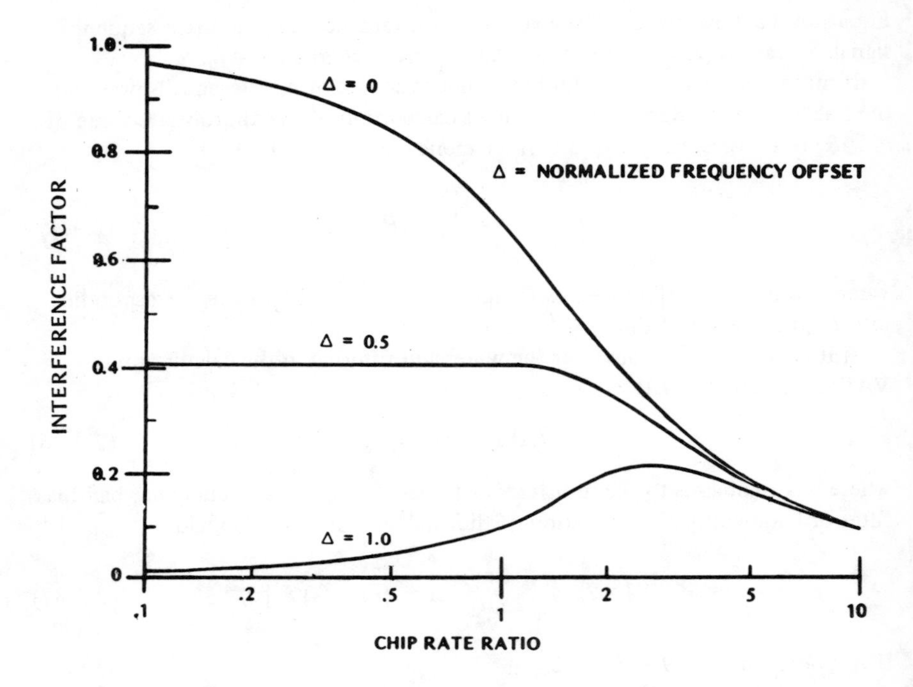

Figure 2.14 Interference factor for direct-sequence interference versus chip rate ratio.

$$\rho = \frac{1/T_q}{1/T_c} = \frac{T_c}{T_q} \tag{2.4.41}$$

Figure 2.15 also shows b for tone interference, as determined from (2.4.27).

A *multipath signal* can be modeled by (2.4.24) with

$$q(t) = m(t - T_m)p(t - T_m) \tag{2:4.42}$$

where T_m is the delay relative to the direct-path signal. Because the multipath signal is reflected before it reaches the receiver, it is reasonable to model θ_1 in (2.4.24) as a random variable uniformly distributed over $[0,2\pi]$. If $T_m > T_c$ and (2.4.28) is satisfied, then it is reasonable to model $q(t)$ as a random binary sequence that is independent of $p(t)$. Thus, Figures 2.14 and 2.15 can be used to determine the interference factor for multipath interference when $T_m > T_c$. In general, $\Delta \neq 0$ and $\rho \neq 1$ because of differences in the Doppler shifts of the direct-path and multipath signals. According to the stationary-process model, multipath power is reduced by a factor G/b before reaching the demodulator.

If $\Delta \cong 0$ and $\rho \cong 1$, then (2.4.34) indicates that $b \cong 2/3$.

If $|i(t)| \leqslant C$ over $[0, T_s]$, then (2.4.5) yields

$$|L_1| \leqslant 2T_sC \tag{2.4.43}$$

Therefore, the probability that $L_1 > 2T_sC$ is zero, whereas the Gaussian assumption assigns a nonzero probability to this event. It follows that the Gaussian assumption tends to cause an overestimate of P_s if E_s/R_iT_cb is sufficiently large.

2.4.2 Tone-Interference Model

Weaker approximations are sufficient to derive a reasonably simple expression for P_s when tone interference is present. Consider tone interference of the form

$$i(t) = \sqrt{2R_i}\, \cos(\omega_1 t + \theta_1) \tag{2.4.44}$$

where θ_1 is not initially assumed to be a random variable [4]. Substituting (2.4.44) into (2.4.8), evaluating the integral, assuming that $\omega_1 + \omega_0 \gg \omega_2 = \omega_1 - \omega_0$ so that terms involving $\omega_1 + \omega_0$ can be neglected, and using the identity $\sin A - \sin B = 2 \sin[(A - B)/2] \cos[(A + B)/2]$, we obtain

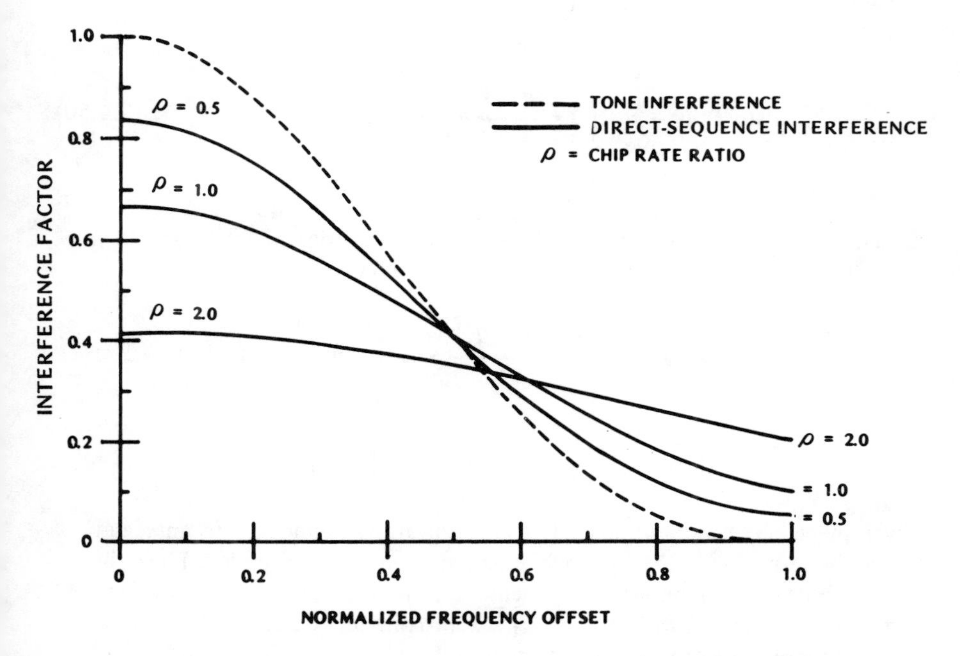

Figure 2.15 Interference factor versus normalized frequency offset.

$$J_\nu = T_c\sqrt{2R_i}\ \mathrm{sinc}(f_2T_c)\cos(\nu 2\pi f_2 T_c + \theta_2)\,, \qquad \nu = 0, 1, \ldots, G-1 \tag{2.4.45}$$

where $f_2 = \omega_2/2\pi$ and

$$\theta_2 = \theta_1 - \theta + \pi f_2 T_c \tag{2.4.46}$$

We model the p_ν in (2.4.7) as independent zero-mean random variables. Therefore,

$$E[p_{\nu+l}p_{j+l}] = 0\,, \qquad \nu \neq j \tag{2.4.47}$$

Consequently, (2.4.7) gives

$$\mathrm{VAR}(L_1) = \sum_{\nu=0}^{G-1} J_\nu^2 \tag{2.4.48}$$

Substituting (2.4.45) into (2.4.48) and expanding the squared cosine, we obtain

$$\mathrm{VAR}(L_1) = R_i T_c^2\ \mathrm{sinc}^2(f_2T_c)\left[G + \sum_{\nu=0}^{G-1} \cos(\nu 4\pi f_2 T_c + 2\theta_2)\right] \tag{2.4.49}$$

To evaluate the summation, we use the identity

$$\sum_{\nu=0}^{n-1} \cos(a + \nu b) = \cos\left(a + \frac{n-1}{2}\,b\right)\frac{\sin\left(\dfrac{nb}{2}\right)}{\sin\left(\dfrac{b}{2}\right)} \tag{2.4.50}$$

which is proved by using mathematical induction and trigonometric identities. Using (2.4.46), (2.4.21), and (2.1.3), we obtain the interference factor:

$$b(\phi) = \left[1 + \frac{\mathrm{sinc}(2f_2T_s)}{\mathrm{sinc}(2f_2T_c)}\cos\phi\right]\mathrm{sinc}^2(f_2T_c) \tag{2.4.51}$$

where

$$\phi = 2(\theta_1 - \theta) + 2\pi f_2 T_s \tag{2.4.52}$$

Given the value of ϕ, the J_ν in (2.4.45) are uniformly bounded constants, and hence the terms in (2.4.7) are independent and uniformly bounded. Thus, the central limit theorem implies that when G is large, it is reasonable to assume that the conditional distribution of L_1 is approximately Gaussian. The conditional symbol error probability is then given by

$$P(s|\phi) = \frac{1}{2} \text{ erfc} \left\{ \left[\frac{E_s}{\beta N_0 + R_i T_c b(\phi)} \right]^{1/2} \right\} \tag{2.4.53}$$

For the ensemble of interference waveforms with different values of θ_1, it is reasonable to model θ_1 as a random variable that is uniformly distributed over $[0,2\pi]$. If θ is either fixed or uniformly distributed, then the modulo-2π character of phase angles implies that ϕ is uniformly distributed over $[0, 2\pi]$. Thus, the symbol error probability, obtained by averaging $P(s|\phi)$ over ϕ, is

$$P_s = \frac{1}{4\pi} \int_0^{2\pi} \text{erfc} \left\{ \left[\frac{E_s}{\beta N_0 + R_i T_c b(\phi)} \right]^{1/2} \right\} d\phi \tag{2.4.54}$$

Because of the character of $\cos \phi$, this equation simplifies to

$$P_s = \frac{1}{2\pi} \int_0^{\pi} \text{erfc} \left\{ \left[\frac{E_s}{\beta N_0 + R_i T_c b(\phi)} \right]^{1/2} \right\} d\phi \tag{2.4.55}$$

When two or more interfering tones are present, the interference parameter is given by (2.4.35), where each of the b_j has the form of (2.4.51) with $\phi = \phi_j$. To determine P_s, a multidimensional integral over the ϕ_j must be evaluated.

We have derived two expressions for P_s when tone interference is present. Equations (2.4.55) and (2.4.51) give the expression derived under the weaker conditions of the tone-interference model. Equations (2.4.38) and (2.4.27) give an alternative expression derived from the stationary-process model, which we identify by an additional subscript:

$$P_{sa} = \frac{1}{2} \text{ erfc} \left\{ \left[\frac{E_s}{\beta N_0 + R_i T_c \, \text{sinc}^2(f_2 T_c)} \right]^{1/2} \right\} \tag{2.4.56}$$

To compare P_{sa} with P_s given by (2.4.55), we observe that the integrand of (2.4.55) is a convex function of $\cos \phi$ if $2E_s/3 > \beta N_0 + R_i T_c b(\phi)$ for all ϕ in $[0,\pi]$. If $g(\ )$ is a convex function over an interval containing the range of a random variable X, then *Jensen's inequality* states that [3]

$$g(E[X]) \leq E[g(X)] \tag{2.4.57}$$

provided that the indicated expected values exist. It follows from this inequality that $P_s \geq P_{sa}$ for sufficiently large values of E_s.

2.4.3 Nearly Exact Model

If $f_2 = 0$, a nearly exact equation for P_s can be derived without recourse to the central limit theorem. Substituting (2.4.44) with $\omega_1 = \omega_0$ into (2.4.5), assuming $\omega_0 T_s \gg 1$, and simplifying, we obtain the nearly exact result that

$$L_1 = \sqrt{2R_i} \cos \theta_2 \int_0^{T_s} p(t)\, dt \tag{2.4.58}$$

Let k_1 denote the number of chips in $[0, T_s]$ for which $p_\nu = +1$. The number of chips for which $p_\nu = -1$ is $G - k_1$. Using (2.4.58) and (2.2.33) and assuming that the chip waveform, $\psi(t)$, is rectangular with unit amplitude so that

$$\int_0^{T_c} \psi(t)\, dt = T_c \tag{2.4.59}$$

we obtain

$$L_1 = \sqrt{2R_i}\; T_c(2k_1 - G)\cos\theta_2 \tag{2.4.60}$$

We assume that the p_ν are independent and equally likely to be +1 or −1 and that it is equally likely that $m(t) = +1$ or -1. Given the value of θ_2, the conditional symbol error probability is

$$P(s|\theta_2) = \sum_{k_1=0}^{G} \binom{G}{k_1} \left(\frac{1}{2}\right)^G \left[\frac{1}{2} P(s|\theta_2, k_1, +1) + \frac{1}{2} P(s|\theta_2, k_1, -1)\right] \tag{2.4.61}$$

where $P(s|\theta_2, k_1, j)$ is the conditional symbol error probability given the values of θ_2 and k_1 and that $m(t) = j$ for $j = +1$ or -1. Under these conditions, L_1 is a constant and the conditional expected value of L is

$$E[L|\theta_2, k_1, j] = jAT_s + \sqrt{2R_i}\; T_c(2k_1 - G)\cos\theta_2, \qquad j = +1, -1 \tag{2.4.62}$$

The variance of L is equal to the variance of L_2, which is given by (2.4.36). Therefore, a straightforward calculation yields

$$P(s|\theta_2, k_1, j) = \frac{1}{2}\,\mathrm{erfc}\left[\frac{\sqrt{E_s T_s} + j\sqrt{R_i}\, T_c(2k_1 - G)\cos\theta_2}{\sqrt{\beta N_0 T_s}}\right], \qquad j = +1, -1 \tag{2.4.63}$$

Equations (2.4.63) and (2.4.61) give $P(s|\theta_2)$ and P_s is determined by integrating $P(s|\theta_2)$ over the distribution of θ_2, which is normally assumed to be uniform. These equations present computational difficulties for large values of G.

Figure 2.16 depicts the symbol error probability as a function of the signal-to-interference ratio, $R_s/R_i = E_s/T_s R_i$, for tone interference and $f_2 = 0$, $G = 50 = 17$ dB, and $E_s/\beta N_0 = 15$ dB. The three curves represent the results of the stationary-process model, the tone-interference model, and the nearly exact model. In this example, the tone-interference model is much more accurate than the stationary-process model over the displayed range of P_s. The difference between the results of the stationary-process and nearly exact models is approximately 1.9 dB at $P_s = 10^{-5}$.

In the preceding analysis, a long pseudonoise sequence satisfying (2.4.9) has been assumed. If sophisticated jamming is not a threat, then short pseudonoise sequences are attractive because some of them can provide a better performance than long sequences against unintentional interference. Tight bounds on the symbol error probability for short sequences in the presence of multitone interference can be calculated by first observing that given both the value of ϕ for each tone and the chip sequence during a symbol interval, the conditional distribution of L is Gaussian [5].

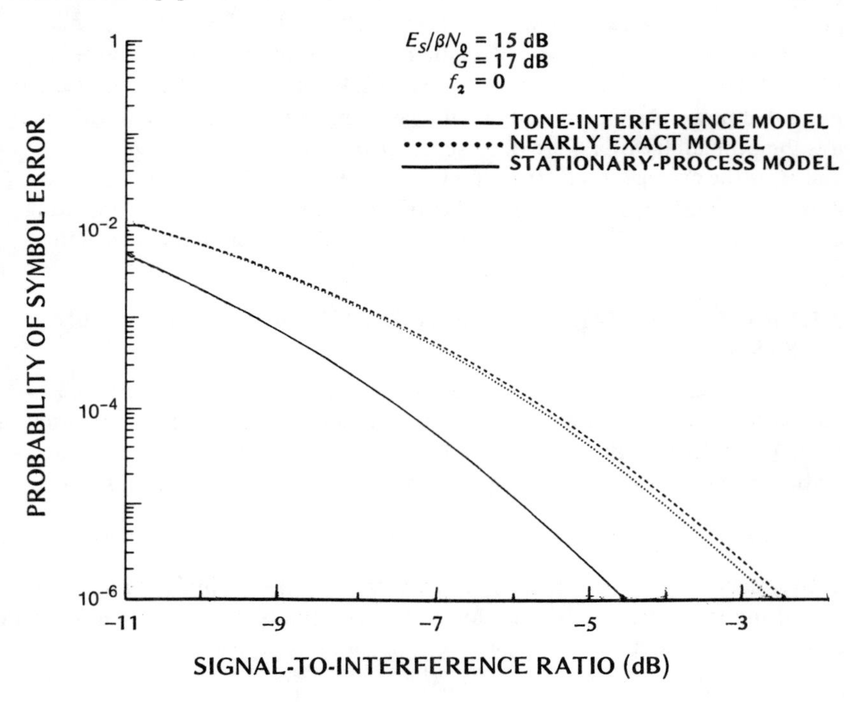

Figure 2.16 Symbol error probability of binary direct-sequence system in presence of tone interference at center frequency.

2.4.4 Error-Correcting Codes

When the total bandwidth and the information rate are fixed, the potential improvement due to error-correcting codes is partially counterbalanced by the decrease in the symbol error probability, which results from the increased rate of transmitted symbols. For binary encoding, the duration of a transmitted symbol is $T_s = rT_b$, where r is the code rate and T_b is the duration of an uncoded bit or an information bit. The energy per transmitted symbol is $E_s = rE_b$, where E_b is the energy per information bit. The processing gain of the encoded system is defined by $G = T_s/T_c = rG_u$, where G_u is the processing gain of the uncoded system. Once the channel-symbol error probability, P_s, has been determined, the information-bit error rate can be obtained from the equations of Section 1.7. For example, if the stationary-process model is applicable, then Figures 1.13 to 1.15,

1.18 to 1.20, 1.23, 1.25, and 1.26 illustrate the potential performance of a direct-sequence system if E_b/N_0 is replaced by E_b/N_{0e}, where N_{0e} is defined by (2.4.39).

When pulsed jamming is present, the interference factor depends upon the pulse duration. If the jamming is modeled as bandlimited white noise during a pulse, then the interference factor during a pulse is given by (2.4.23), where f_1 is the center frequency, and W_1 is the bandwidth of the jamming pulse if $W_1 \leqslant W$. The discussion of Section 1.7 and (2.4.39) imply that the equivalent two-sided power spectral density of the jamming during a pulse is $J_0/2\mu$, where μ is the probability of a pulse occurrence or the pulse duty cycle, $J_0 = R_i T_c b$, and R_i is the average interference power. Consequently, Figures 1.27 to 1.30 illustrate the information-bit error rate of a direct-sequence system operating against worst-case pulsed jamming if we set $J_0 = R_i T_c b$ and replace N_0 with βN_0.

2.5 ERROR PROBABILITIES FOR QUADRIPHASE DIRECT-SEQUENCE SYSTEMS

The preceding analysis of a binary direct-sequence system can easily be extended to the quadriphase direct-sequence system of Figure 2.5. Approximations based on the central limit theorem are again decisive in simplifying the mathematics. After passage through the wideband filter, the total received signal is

$$r(t) = Am_1(t)p_1(t)\cos(\omega_0 t + \theta) + Am_2(t)p_2(t)\sin(\omega_0 t + \theta) + i(t) + n(t) \quad (2.5.1)$$

Let T_{s1} denote the duration of the transmitted channel symbols of $m_1(t)$ and $m_2(t)$, and let T_s denote the duration of the data symbols before serial-to-parallel conversion. It is assumed that the synchronization is perfect and that

$$f_0 \gg W \geqslant \frac{1}{T_c} \geqslant \frac{1}{T_{s1}} \quad (2.5.2)$$

where $f_0 = \omega_0/2\pi$ is the carrier frequency, and T_c is the common chip duration of $p_1(t)$ and $p_2(t)$. Consequently, the sampled output of the upper integrator in Figure 2.5(b) at the end of a symbol interval is approximately

$$L = \pm AT_{s1} + L_1 + L_2 \quad (2.5.3)$$

where

$$L_1 = \int_0^{T_{s1}} 2i(t)p_1(t)\cos(\omega_0 t + \theta)\,dt \quad (2.5.4)$$

$$L_2 = \int_0^{T_{s1}} 2n(t)p_1(t)\cos(\omega_0 t + \theta)\,dt \quad (2.5.5)$$

Similarly, the output of the lower integrator at the end of a channel-symbol interval is approximately

$$Q = \pm AT_{s1} + Q_1 + Q_2 \tag{2.5.6}$$

where

$$Q_1 = \int_0^{T_{s1}} 2\,i(t)\,p_2(t)\sin(\omega_0 t + \theta)\,dt \tag{2.5.7}$$

$$Q_2 = \int_0^{T_{s1}} 2n(t)\,p_2(t)\sin(\omega_0 t + \theta)\,dt \tag{2.5.8}$$

In the subsequent analysis, unit-amplitude rectangular chip waveforms are assumed.

If $i(t)$ is a stationary stochastic process, the stationary-process model of Section 2.4 indicates that the probability of an error is the same for both inputs of the parallel-to-serial converter and hence for each symbol of the data output. Of the available desired-signal power, R_s, half is in each of the desired-signal components of (2.5.1); thus, $A = \sqrt{R_s}$. Because $T_{s1} = 2T_s$ and the average power of each component is half the total average power, the energy per symbol for each component is the same as the energy per symbol of the data sequence, E_s. According to (2.4.21) and (2.4.17) with T_{s1} in place of T_s, the quadriphase spreading does not usually cause a significant change in the interference factor. Thus, it is easy to verify that the probability of a symbol error in the data output is given by (2.4.38).

When the tone-interference model is used, the conditional symbol error probability given ϕ differs for each of the two symbol streams feeding into the parallel-to-serial converter because of the phase shifter in Figure 2.5(b). The average conditional symbol error probability for the data output is

$$P(s|\phi) = \frac{1}{4}\sum_{j=0}^{1} \operatorname{erfc}\left\{\left[\frac{E_s}{\beta N_0 + R_i T_c b_j(\phi)}\right]^{1/2}\right\} \tag{2.5.9}$$

where $b_0(\phi)$ and $b_1(\phi)$ are the interference factors for the two symbol streams. Because the channel-symbol duration is $T_{s1} = 2T_s$ and quadrature carriers are used, (2.4.51) and (2.4.52) imply that

$$b_j(\phi) = \operatorname{sinc}^2(f_2 T_c)\left[1 + \frac{\operatorname{sinc}(4f_2 T_s)}{\operatorname{sinc}(2f_2 T_c)}\cos(\phi + j\pi)\right], \quad j = 0, 1 \tag{2.5.10}$$

where $\phi = 2(\theta_1 - \theta) + 4\pi f_2 T_s$.

These equations indicate that $P(s|\phi)$ for a quadriphase direct-sequence system and the worst value of ϕ is usually lower than $P(s|\phi)$ for a binary direct-sequence system and the worst value of ϕ. After averaging $P(s|\phi)$ over ϕ, it is found that P_s for a quadriphase system in less than or equal to P_s for a binary system, but any difference is small when $f_2 > 1/T_s$. When $f_2 = 0$, the quadriphase and binary systems provide the same P_s.

In another version of the quadriphase direct-sequence system, the same data symbols are carried by the in-phase and quadrature components, which implies that $m_1(t) = m_2(t)$ in (2.5.1). Thus, although the spreading is done by QPSK, the data modulation is binary PSK. A receiver for this quadriphase system is shown in Figure 2.17. The synchronization system, which is assumed to operate perfectly in the subsequent analysis, is not shown for simplicity. Because of (2.5.2), the input to the decision device is approximately

$$L = \pm 2AT_s + L_1 + L_2 + Q_1 + Q_2 \tag{2.5.11}$$

where $T_s = T_{s1}$ and $A = \sqrt{R_s}$. If $p_1(t)$ and $p_2(t)$ can be approximated by independent random binary sequences, then L_1, L_2, Q_1, and Q_2 are uncorrelated with each other. Therefore, the variance of L is equal to the sum of the variances of L_1, L_2, Q_1, and Q_2. If $i(t)$ is modeled as a stationary stochastic process, an extension of the stationary-process model of Section 2.4 yields a symbol error probability given by (2.4.38), which also holds for a binary system.

If $i(t)$ is tone interference, an extension of the tone-interference model yields a $P(s|\phi)$ that is independent of ϕ. Therefore, $P_s = P(s|\phi)$ and

$$P_s = \frac{1}{2}\,\text{erfc}\left\{\left[\frac{E_s}{\beta N_0 + R_i T_c \,\text{sinc}^2(f_2 T_c)}\right]^{1/2}\right\} \tag{2.5.12}$$

This result is identical to that following from the stationary-process model. Jensen's inequality and the tone-interference model indicate that this quadriphase system provides a lower symbol error probability than a binary system when tone interference is present.

If $f_2 = 0$, the nearly exact model yields the conditional symbol error probability

$$P(s|\theta_2) = \sum_{k_1=0}^{G} \sum_{k_2=0}^{G} \binom{G}{k_1}\binom{G}{k_2}\left(\frac{1}{2}\right)^{2G}\left[\frac{1}{2}P(s|\theta_2,k_1,k_2,+1) + \frac{1}{2}P(s|\theta_2,k_1,k_2,-1)\right] \tag{2.5.13}$$

where k_1 and k_2 are the number of chips for which $p_1(t) = +1$ and $p_2(t) = +1$, respectively, and

$$P(s|\theta_2,k_1,k_2,j) = \frac{1}{2}\,\text{erfc}\left\{\frac{\sqrt{E_s T_s} + j\sqrt{R_i/2}\,T_c[(2k_1 - G)\cos\theta_2 - (2k_1 - G)\sin\theta_2]}{\sqrt{\beta N_0 T_s}}\right\}$$

$$j = +1, -1 \tag{2.5.14}$$

The derivation of these equations is analogous to that of (2.4.63) and (2.4.61). The symbol error probability P_s is determined by integrating $P(s|\theta_2)$ over the distribution of θ_2, which is normally assumed to be uniform.

Figure 2.18 depicts P_s as a function of R_s/R_i for tone interference and $f_2 = 0$, $G = 17$ dB, and $E_s/\beta N_0 = 15$ dB. The results for both the tone interference and nearly exact models indicate that the quadriphase direct-sequence system of Figure 2.17 provides a substantial advantage against tone interference relative to the binary direct-sequence system. Approximately 2.0 dB less in R_s/R_i is required for the quadriphase system to achieve $P_s = 10^{-5}$.

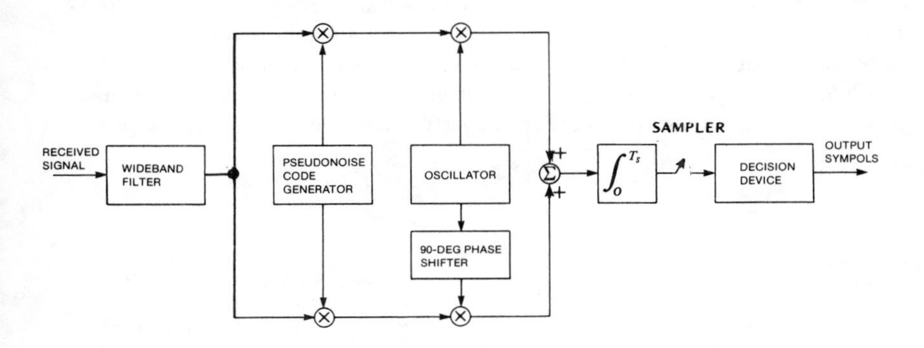

Figure 2.17 Receiver for quadriphase direct-sequence system with same message bits carried by both in-phase and quadrature components.

2.6 CODE-DIVISION MULTIPLE-ACCESS NETWORKS

A *multiple-access* communication system is one that is capable of coexisting with similar systems in a limited geographic area. A *code-division multiple-access* (CDMA) *system* is one that uses an identifying code to separate a desired signal at a receiver from other signals that simultaneously occupy the same general frequency band. A CDMA network consists of CDMA systems. In contrast with *frequency-division* or *time-division multiple-access networks,* a CDMA network accommodates an additional system without revising the signal formats, although some performance degradation results.

We consider CDMA networks of binary direct-sequence systems. At a particular receiver, the multiple-access interference signal from transmitter j has the form

$$i_j(t) = A_j m_j(t)\, p_j(t) \cos(\omega_0 t + \theta_j) \tag{2.6.1}$$

where $m_j(t)$ is the data modulation, $p_j(t)$ is the spreading waveform, and ω_0 is the same for all systems in a CDMA network. Equation (2.4.5) implies that the resulting interference term at the input of the decision device has the form

$$L_{1j} \cong A_j \cos(\theta - \theta_j) \int_0^{T_s} m_j(t)\, p_j(t)\, p(t)\, dt \tag{2.6.2}$$

where a double-frequency term is neglected and $p(t)$ is the spreading waveform of the desired signal. Suppose that the network is synchronized so that data-symbol transitions are aligned at each receiver. Then $m_j(t)$ is constant over $[0, T_s]$. If all the pseudonoise sequences have a common period equal to T_s, the common data-symbol duration, then L_{1j} is proportional to the cross-correlation of $p_j(t)$ and $p(t)$. If the pseudonoise sequences are also chosen to be orthogonal to each other, then $L_{1j} = 0$ for all j and the multiple-access interference is suppressed. However, network synchronization is seldom feasible because of changing path-length differences among the various communication links. Furthermore, short sequences are undesirable for secure communications.

When the symbol transitions of *asynchronous* multiple-access signals at a receiver are not simultaneous, the receiver performance depends upon the partial

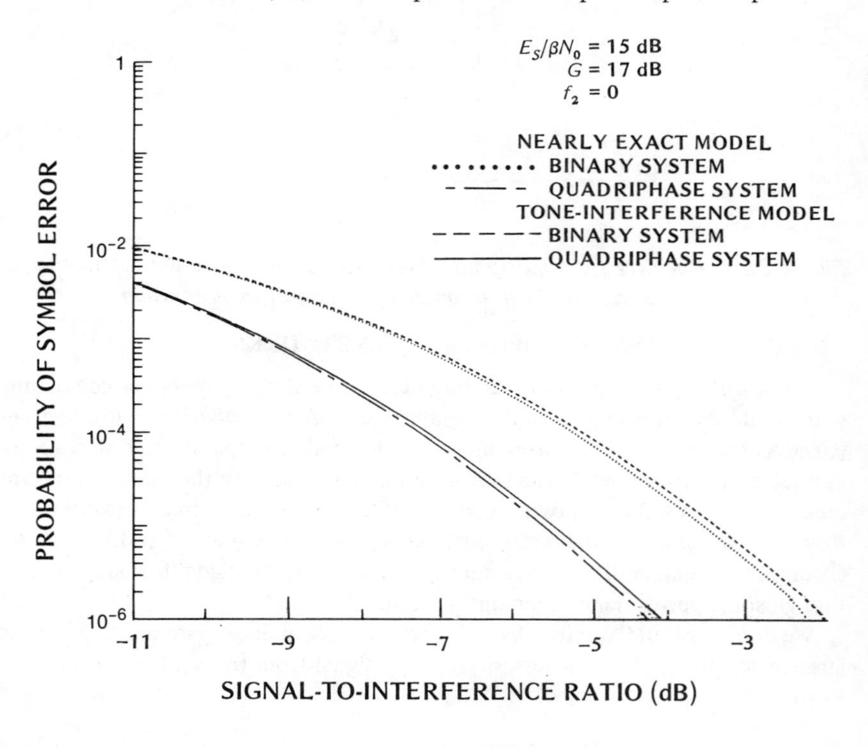

Figure 2.18 Symbol error probabilities of quadriphase direct-sequence system of Figure 2.17 and binary direct-sequence system. Tone interference is present at center frequency.

cross-correlations of the pseudonoise sequences. Small partial cross-correlations are desirable to limit the value of L_{1j}. There are large sets of nonmaximal sequences with small cross-correlations [6]. Carefully chosen subsets of these sets sometimes also exhibit small partial cross-correlations.

One such set consists of the *Gold sequences,* which may be generated by the modulo-two addition of certain maximal sequences. An example of a Gold code generator is shown in Figure 2.19. If each maximal code generator has m stages, a resulting Gold sequence has a period of $2^m - 1$. Different Gold sequences are obtained by selecting the initial contents of the maximal code generators. If n shifts of one of these generators is required to change its initial contents to the initial contents of the other generator, the two generators are said to have a relative displacement of n shifts. Since any relative displacement of the two maximal code generators from 0 to $2^m - 2$ shifts results in a different Gold sequence, $2^m - 1$ different nonmaximal Gold sequences can be produced by the system of Figure 2.19. Gold sequences identical to maximal sequences are produced by setting the contents of one of the maximal code generators to zero. Altogether, there are $2^m + 1$ different Gold sequences, each with a period of $2^m - 1$.

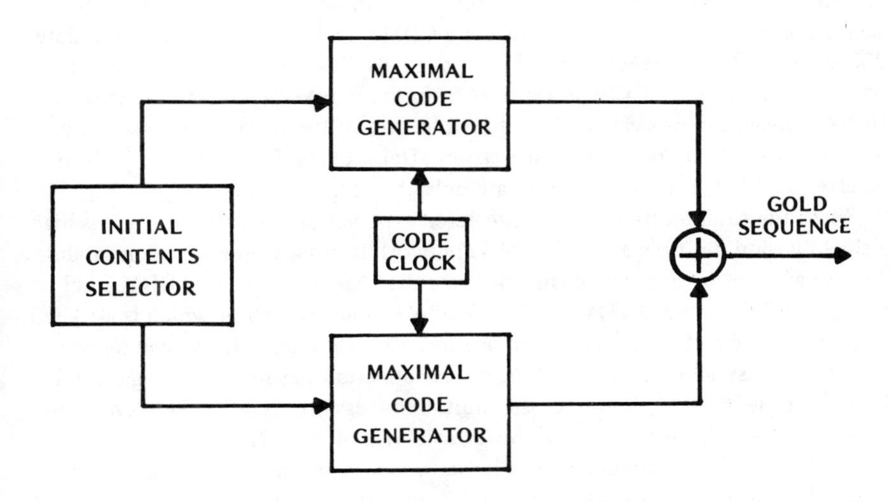

Figure 2.19 Gold code generator.

If all the pseudonoise sequences have a common period equal to a data-symbol duration, then the performance of CDMA systems can often be approximately evaluated or tightly bounded [7, 8]. By the proper selection of the sequences and their relative phases, one can obtain a performance better than that theoretically attainable with random sequences. However, if the sequences of a CDMA network have periods that exceed a data-symbol duration, then it is usually difficult to choose sets of them that provide a significant improvement over random sequences [9].

If all the pseudonoise sequences satisfy (2.4.28), then the symbol error probability in a receiver is approximately given by (2.4.38) and (2.4.35). Suppose that the Doppler shifts are negligible and that all the systems in a fixed network use the same carrier frequency and chip rate. Then (2.4.34) implies that $b_j = 2/3$ is the interference factor due to source j, and (2.4.35) gives $b = 2/3$. Thus, the stationary-process model yields

$$P_s = \frac{1}{2} \operatorname{erfc} \left\{ \left[\frac{E_s}{\beta N_0 + 2R_i T_c/3} \right]^{1/2} \right\} \tag{2.6.3}$$

where R_i is the total multiple-access interference power at the receiver. This equation indicates that P_s does not directly depend on the number of transmitting communicators in the network, but depends only on R_i.

A *near-far* problem occurs when interference arrives from a nearby network element with much more power than a desired signal that arrives from a more distant transmitter. This problem can result in a dramatic reduction in multiple-access capability. For example, consider a CDMA system that can accommodate 200 statistically independent equal-power signals. Suppose that the source of one of these signals, initially at range D from the receiver, moves to range $D/4$. If the propagation loss varies as the fourth power of the range, then this source introduces 24 dB more interference power after moving. Since 200 = 23 dB, the receiver can then at best accommodate only this single source.

When the total multiple-access interference power at network receivers is high, a time-division multiple-access (TDMA) network is usually able to accommodate a larger amount of this type of interference than can a comparable CDMA network. Consider an ideal TDMA network of N elements, each of which is assigned a separate time slot for transmission during each time frame. If we assume perfect network synchronization and negligible uncertainties in the propagation delays, the time division eliminates the multiple-access interference. However, the duration of a transmitted symbol must be decreased from T_s to T_s/N if the overall symbol rate is to remain equal to its value for a continuous transmission. If binary PSK modulation is used and interference from sources outside the network is absent, then the symbol error probability at a network element is

$$P_s = \frac{1}{2} \operatorname{erfc} \left[\left(\frac{R_{s1} T_s}{\beta N_0 N} \right)^{1/2} \right] \tag{2.6.4}$$

where R_{s1} is the received signal power from a transmitter during one of its time slots. This equation holds whether or not the network systems use spectral spreading. Increasing the signal power so that $R_{s1} = NR_s$ preserves the average received power at the level R_s that would exist without the time division. In this case, a comparison of (2.6.4) and (2.6.3) with $E_s = R_s T_s$ indicates that the TDMA network always performs better than a comparable CDMA network against multiple-access interference. However, if the transmitters are peak-power limited, the TDMA network performs better only if

$$N < \frac{R_{s1}}{R_s}\left(1 + \frac{2R_i T_c}{3\beta N_0}\right) \tag{2.6.5}$$

The main problem with time-division multiple access is the need for establishing accurate frame synchronization throughout the network.

Multiple-access interference can be alleviated without network coordination if an element uses a *time-hopping system,* which selects the time slots for transmission according to the state of a code generator. Figure 2.20 depicts a time-hopping direct-sequence system. The data symbols are temporarily stored for transmission at a high rate during the slot. After code synchronization has been established at the receiver, only signals corresponding to the desired portion of the frame pass through the initial switch. A time-hopping interference signal is blocked except for pseudorandom coincidences. The time hopping and the pseudonoise nature of the transmitted bursts are countermeasures to interception or jamming.

2.7 CODE ACQUISITION

A spread-spectrum receiver must generate a pseudonoise sequence that is synchronized with the received sequence; that is, the corresponding chips must precisely or nearly coincide. Any misalignment causes the signal amplitude at the demodulator output to fall in accordance with the autocorrelation or partial autocorrelation function. Range uncertainty, clock drifts, and the Doppler shift are the primary sources of synchronization errors.

Code synchronization consists of two operations, *acquisition* and *tracking.* Acquisition, also called *coarse synchronization,* is the operation by which the phase of the receiver-generated sequence is brought to within a fraction of a chip of the phase of the received sequence. After this condition is detected and verified, the tracking system is activated. Tracking, also called *fine synchronization,* is the operation by which synchronization errors are further reduced or at least maintained within certain bounds. The usual configuration of the acquisition and tracking systems in a direct-sequence receiver is shown in Figure 2.21.

When jamming is a threat, an acquisition system with the following features is desired.

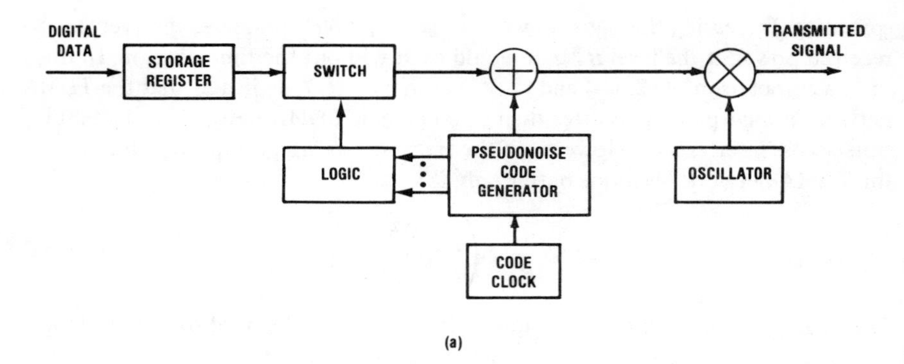

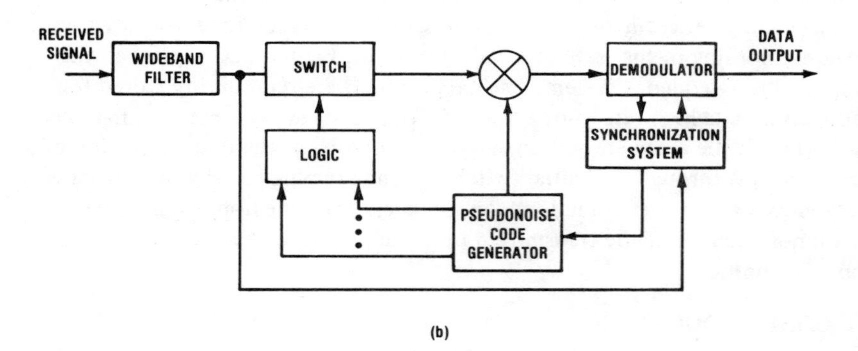

Figure 2.20 Time-hopping direct-sequence system: (a) transmitter and (b) receiver.

(a) Because successful jamming during acquisition completely disables a communication system, the acquisition system must have a strong capability to reject interference.

(b) The pseudonoise sequence used for acquisition must be programmable and sufficiently long for security.

(c) The acquisition should be rapid, so that a jammer must operate continuously to ensure jamming during acquisition: continuous operation reduces the amount of jamming power that can be produced.

Because of the first requirement, *matched-filter acquisition* and *serial-search acquisition* are the most effective techniques for most secure communications. Matched-filter acquisition provides potentially rapid acquisition when short sequences are used. Serial-search acquisition is more reliable and can easily accommodate long sequences.

In a benign environment, *sequential estimation* methods provide rapid acquisition [10]. Successive pseudonoise chips are demodulated and then loaded into the receiver's code generator to establish its initial state. However, because chip demodulation is required, the usual despreading mechanism cannot be used to suppress interference during acquisition.

2.7.1 Matched-Filter Acquisition

A filter is said to be *matched* to a signal $x(t)$ that is zero outside the interval $[0, T]$ if the impulse response of the filter is $h(t) = x(T - t)$. When $x(t)$ is the input to a filter matched to it, the output of the matched filter is

$$y(t) = \int_{-\infty}^{\infty} x(u)\,h(t - u)\,du = \int_{-\infty}^{\infty} x(u)\,x(u + T - t)\,du$$

$$= \int_{\max(t-T,0)}^{\min(t,T)} x(u)\,x(u + T - t)\,du \tag{2.7.1}$$

The autocorrelation of a deterministic signal with finite energy is defined by

$$R_x(\tau) = \int_{-\infty}^{\infty} x(u)\,x(u + \tau)\,du \tag{2.7.2}$$

Therefore, the response of a matched filter to the matched signal is

$$y(t) = R_x(T - t) = R_x(t - T) \tag{2.7.3}$$

The peak value of the response occurs at $t = T$ and is equal to $R_x(0)$, the signal energy.

Consider a filter matched to $p_1(t)\cos(\omega_0 t + \theta_1)$, where $p_1(t)$ is an integral number of periods of the spreading waveform $p(t)$ and is zero outside $[0,T]$. If the filter input is $p(t)\cos(\omega_0 t + \theta)$ over $[0,T_1]$, $T_1 > T$, and $\omega_0 T/2\pi \gg 1$, then the matched-filter output over $[T,T_1]$ is

$$y(t) = \int_{-\infty}^{\infty} p(u)\cos(\omega_0 u + \theta)\,p_1(u + T - t)\cos(\omega_0 u + \omega_0 T - \omega_0 t + \theta_1)\,du$$

$$\cong \frac{1}{2}\left[\int_{t-T}^{t} p(u)\,p(u + T - t)\,du\right]\cos(\omega_0 t - \omega_0 T + \theta - \theta_1)$$

$$= \frac{T}{2}\,R_p(T - t)\cos(\omega_0 t - \omega_0 T + \theta - \theta_1)$$

$$= \frac{T}{2}\,R_p(t - T)\cos(\omega_0 t + \theta_0)\,, \quad T \leqslant t \leqslant T_1 \tag{2.7.4}$$

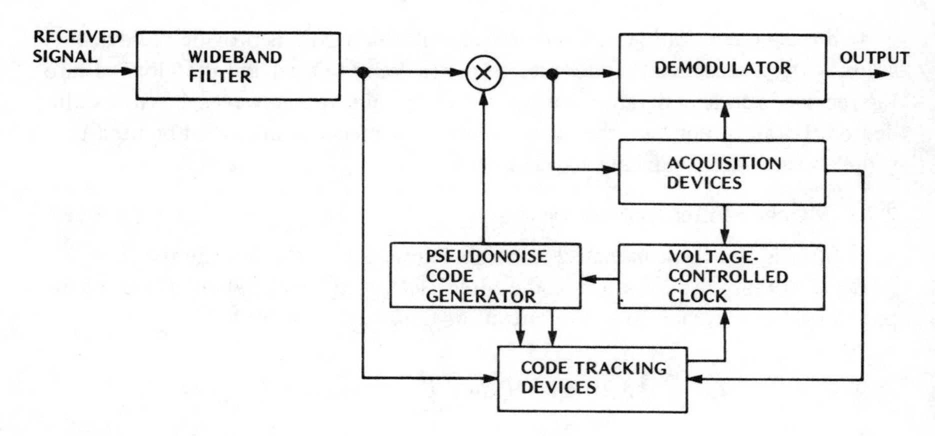

Figure 2.21 Configuration of code acquisition and tracking systems.

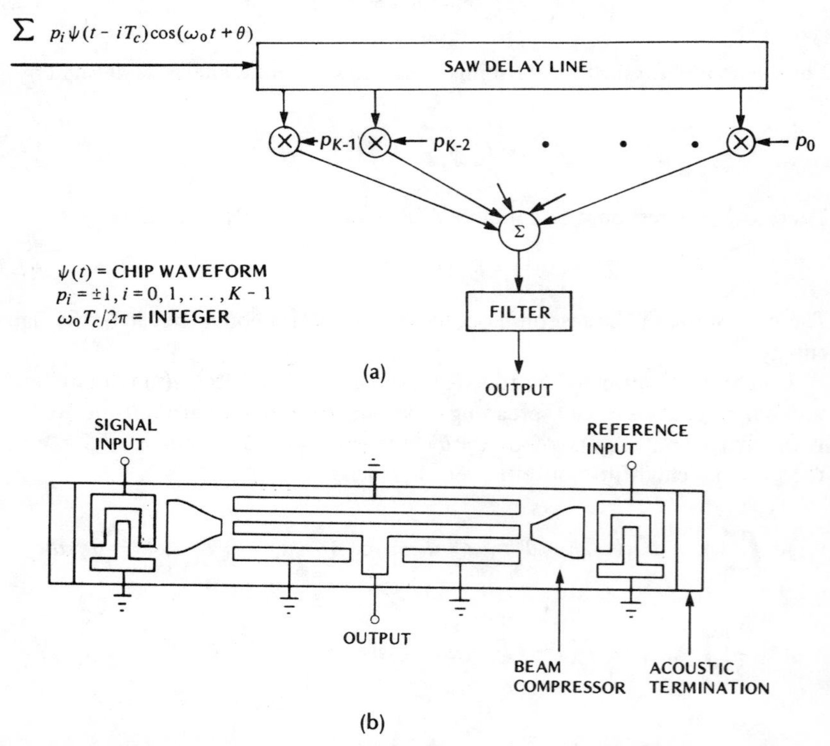

Figure 2.22 Matched filters: (a) SAW transversal filter and (b) SAW elastic convolver.

where $\theta_0 = \theta - \theta_1 - \omega_0 T$ and $R_p(\tau)$ is the periodic autocorrelation of the periodic spreading waveform of infinite extent (Section 2.2.3). Thus, the envelope of the matched-filter output is proportional to a delayed version of the autocorrelation of the spreading waveform. When the chip waveform is rectangular, this envelope is a series of pulses of the form of Figure 2.11.

The matched filter in an acquisition system is matched to the spreading waveform, which is transmitted without modulation during acquisition. The filter output or its envelope is applied to a threshold detector. The occurrence of the threshold crossing provides the receiver with the timing information used to complete the acquisition.

In many applications, the amplitude of the received signal is unknown and has a wide dynamic range. If the threshold level is set to accommodate a small amplitude, then a signal with a large amplitude may cause erroneous threshold crossings that coincide with the sidelobes of the autocorrelation. One solution to this problem is to use more than one threshold level. Then the crossing of the highest threshold during a period of the spreading waveform determines the acquisition.

A passive matched filter can be implemented as an analog or digital transversal filter that essentially stores a replica of the underlying pseudonoise sequence. The basic form of a surface-acoustic-wave (SAW) transversal filter is shown in Figure 2.22(a). The SAW delay line consists primarily of a piezoelectric substrate, which serves as the acoustic propagation medium, and interdigital transducers, which serve as the taps and the input transducer. If the SAW filter is matched to one period of the spreading waveform $p(t)\cos(\omega_0 t + \theta)$, where $p(t)$ is given by (2.2.33) with $N = K$, then the propagation delay between taps is T_c, and $\omega_0 T_c/2\pi$ is an integer. The filter following the summer is matched to $\psi(t)\cos(\omega_0 t + \theta)$. It is easily verified that the impulse response of the SAW filter is that of a filter matched to the spreading waveform over $0 \leqslant t \leqslant KT_c$.

An active, readily programmable matched filter can be implemented with a surface-acoustic-wave or an acoustooptic *convolver*, which produces the convolution of two input signals. When used as a direct-sequence matched filter, a convolver uses a receiver-generated, recirculating, time-reversed replica of more than one period of the spreading waveform as a reference waveform. The top view of a SAW elastic convolver is depicted in Figure 2.22(b). The received signal and the reference waveform are applied to interdigital transducers at opposite ends of the substrate. The two resulting acoustic waves travel in opposite directions. The beam compressors, which consist of thin metallic strips, focus the acoustic energy to increase the convolver's efficiency. When the acoustic waves overlap beneath the central electrode, a nonlinear piezoelectric effect causes a surface charge distribution that is spatially integrated by the electrode. The envelope of the filtered convolver output is proportional to a delayed version of $R_p(2t)$, where the factor of two results from the counterpropagation of the two acoustic waves. The acoustic terminations suppress reflected acoustic waves.

If the pseudonoise-sequence period is equal to the data-bit duration, a matched-filter output can be used for message demodulation, thereby eliminating the need for a separate code synchronization subsystem. However, the short sequence that must be used with a practical matched filter is susceptible to false correlation with interference and reproduction by a jammer. Thus, when jamming is a threat, matched-filter acquisition is used primarily for burst communications and for continuous communications when serial-search acquisition with a long sequence fails or takes too long. The short sequence used with the matched filter may be concealed by embedding it within the long sequence. The short sequence may be a subsequence of the long sequence that is presumed to be ahead of the received sequence and is stored in a programmable matched filter.

Figure 2.23 depicts the configuration of a matched filter for short-sequence acquisition and a serial-search system for long-sequence acquisition. The reception of the short sequence is detected when the matched-filter output or its envelope crosses a threshold. The threshold detector output starts a long-sequence generator at a predetermined initial state. The long sequence is used for verifying the acquisition by serial search and for despreading the received direct-sequence signal. A number of matched filters in parallel can be used to expedite the process [11].

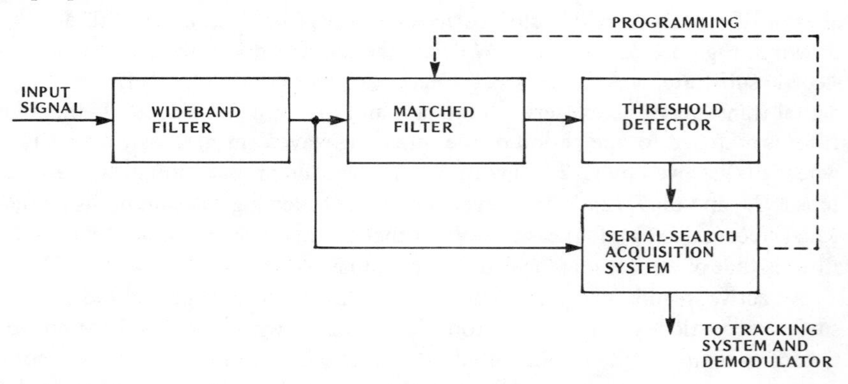

Figure 2.23 Configuration of matched-filter and serial-search systems.

2.7.2 Serial-Search Acquisition

Serial-search acquisition consists of a search, usually in discrete steps, of the possible time alignments of a receiver-generated sequence relative to a received pseudonoise sequence [12-15]. The time uncertainty region is usually quantized into a finite number of search positions or *cells.* The cells are serially tested until it is determined that a particular cell corresponds to the alignment of the two sequences to within a fraction of a chip. The *step size* or separation between cells is typically one-half of a chip.

A serial-search acquisition system is shown in Figure 2.24. The received direct-sequence signal is multiplied by the receiver-generated sequence. The energy of the product signal is measured and compared to a threshold. If the sequences are not aligned, the bandpass filter blocks most of the power in the product signal. If the threshold is not exceeded, the cell under test is rejected and phase of the local sequence is retarded or advanced, possibly by generating an extra clock pulse or by blocking one. A new cell is then tested. If the sequences are nearly aligned, the product signal is a tone or narrowband signal that passes through the filter with little loss of power. If the threshold is exceeded, the search is stopped, and the two sequences run in parallel at some fixed phase offset. Subsequent tests verify that the correct cell has been identified. If a cell fails the verification tests, the search is resumed. If a cell passes, the two sequences are assumed to be coarsely synchronized, and the tracking system is activated. The threshold detector output continues to be monitored so that any subsequent loss of synchronization can be detected and the serial search initiated.

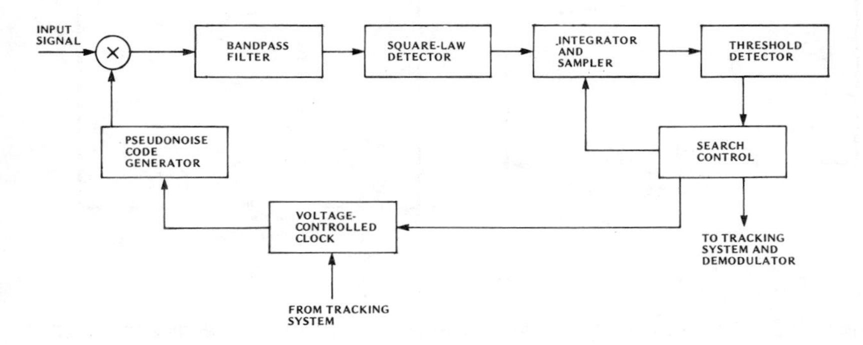

Figure 2.24 Serial-search acquisition system.

Diagrams of two representative serial-search strategies are shown in Figure 2.25. The correct cell is the one that results in the closest time alignment of the two sequences. Figure 2.25(a) depicts a *uniform search* over the uncertainty region. The *expanding-window search*, illustrated in Figure 2.25(b), is appropriate when *a priori* information makes part of the time uncertainty region more likely to contain the correct cell than the rest of the region. *A priori* information may be derived from a short preamble. If the sequences are synchronized with the time of day, then the receiver's estimate of the transmitter range combined with the time of day provide the *a priori* information.

Acquisition can be expedited by assigning several parallel acquisition systems to search different sections of the time uncertainty region. To reject noise and interference and to prevent false detections, the bandwidth of the bandpass filter of an acquisition system should be as small as possible. If Doppler shifts and oscillator drifts can cause a large uncertainty in the carrier frequency, it may be desirable to use an array of filters, detectors, and integrators in parallel so that the bandwidth of each filter can be kept small while the frequency uncertainty is accommodated by the array.

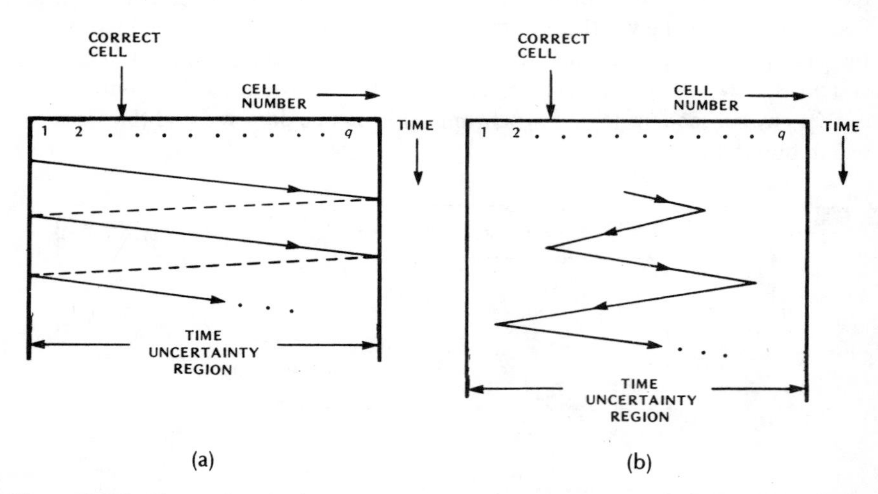

Figure 2.25 Trajectories of search positions: (a) uniform search and (b) expanding-window search.

Depending upon the step size, there may be several cells that potentially yield a correct detection by the acquisition system. However, if the sequences are not completely aligned, the detected energy is reduced from its potential peak value. This reduction degrades the performance somewhat. If the step size is one-half of a chip, then one of the cells corresponds to an alignment within one-fourth of a chip. On the average, the misalignment of this cell is one-eighth of a chip, which may cause a negligible degradation. As the step size is decreased, the potential detected energy during acquisition increases. However, the increase in the number of cells to be searched eventually causes an increase in the time it takes for acquisition.

The *dwell time* is the amount of time required for testing a cell and is approximately equal to the length of the integration interval. If a single test determines whether a cell is accepted as the correct one and the dwell time is fixed, then an acquisition system is called a *single-dwell-time system.* If verification testing occurs before acceptance, the system is called a *multiple-dwell-time system.* The dwell times may be either fixed or variable but bounded by some maximum value. When fixed dwell times are used, the dwell time for the initial test of a cell is usually much shorter than the dwell times for verification testing. The purpose of this approach is to expedite the acquisition by quickly eliminating the bulk of the incorrect cells. In any serial-search system, the dwell time allotted to a test is limited by the Doppler shift, which causes the received and receiver-generated chip rates to differ. As a result, an initial alignment of the two sequences may disappear by the end of the dwell time. Thus, the dwell time should be much smaller than the inverse of the maximum difference between the chip rates.

A flow graph for a *double-dwell time system* is shown in Figure 2.26. For the initial test of a cell, the dwell time is τ_{D1} seconds. If the threshold is not exceeded, the cell fails the test, and the next cell is tested. If it is exceeded, the cell passes the test, the search is stopped, and the system enters the *verification mode.* The same cell is tested again, but the dwell time is changed to τ_{D2} seconds, and the threshold may be changed. If this threshold is not exceeded, the system returns to the search mode. If it is exceeded, the code tracking is activated, and the system enters the *lock mode.* In the lock mode, the dwell time may be changed again. Tests continue to verify that code synchronization is maintained. Under certain conditions, the system decides that synchronization has been lost. *Reacquisition* begins in the search mode. The first cell to be tested is determined by the general search strategy and any *a priori* information.

Acquisition time, T_a, is the time it takes an acquisition system to locate the correct cell and initiate the code tracking system. To simplify the derivation of the probability density function of T_a, we assume that only one cell can give a correct detection. In the subsequent analysis, the search strategy is described by a single functional relation, as proposed by Meyr and Polzer [12].

If an incorrect cell is accepted, the receiver eventually recognizes the mistake and reinitiates the search at the next cell. The wasted time expended in code tracking is called the *penalty time,* T_p. Let L denote the number of times the correct cell is tested before it is accepted and acquisition terminates. We assume that the results of one test are independent of the results of another test. Therefore, the probability that $L = i$ is

$$p_L(i) = P_D(1 - P_D)^{i-1} \tag{2.7.5}$$

where P_D is the probability that the correct cell is detected when it is tested during a scan. Let C denote the number of the correct cell and $p_c(j)$ denote the

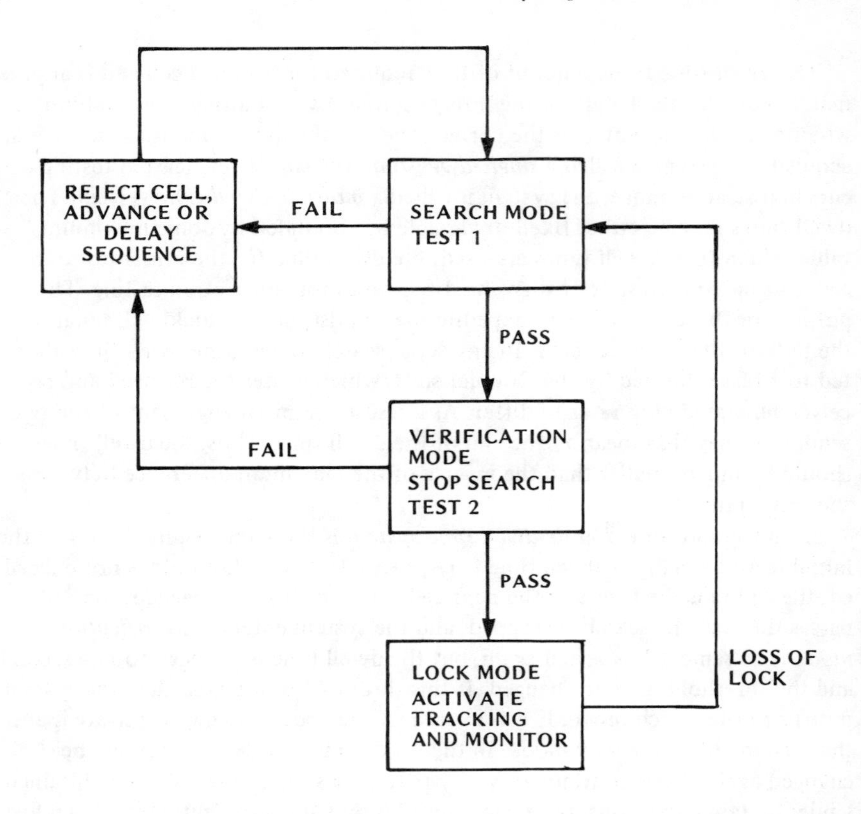

Figure 2.26 Flow graph for double-dwell-time acquisition system.

probability that $C = j$. The conditional density of T_a given that $L = i$ and $C = j$ is denoted by $f_a(t \mid i, j)$. The probability density function of T_a is

$$f_a(t) = \sum_{i=1}^{\infty} \sum_{j=1}^{q} p_L(i)\, p_c(j)\, f_a(t \mid i, j) \tag{2.7.6}$$

where q is the number of cells in the uncertainty region. Substituting (2.7.5) into (2.7.6), we obtain

$$f_a(t) = P_D \sum_{i=1}^{\infty} (1 - P_D)^{i-1} \sum_{j=1}^{q} p_c(j)\, f_a(t \mid i, j) \tag{2.7.7}$$

In the absence of *a priori* information, it is reasonable to set

$$p_c(j) = 1/q\,, \qquad 1 \leqslant j \leqslant q \tag{2.7.8}$$

Because an incorrect cell is always ultimately rejected, there are only three events that occur during the serial search. Either an incorrect cell is dismissed after T_{11} seconds, a correct cell is falsely dismissed after T_{12} seconds, or a correct cell is accepted after T_{22} seconds, where the first subscript is 1 if dismissal occurs and the second subscript is 1 if the cell is incorrect. An incorrect cell is temporarily accepted after T'_{21} seconds with false-alarm probability P_F. After penalty time T_p, the search resumes. An incorrect cell is dismissed without temporary acceptance after T'_{11} seconds with probability $1 - P_F$. Thus, either $T_{11} = T'_{21} + T_p$ or $T_{11} = T'_{11}$. Each of these decision times is a random variable that is identically distributed for each cell to which it applies, but T_{12}, T_{22}, T'_{11}, and T'_{21} may be constants in special cases.

Let $\nu(i,j)$ denote the total number of incorrect cells tested during the acquisition process when $C = j$ and $L = i$. The function $\nu(i,j)$ depends on the search strategy. For example, if the cells in Figure 2.25(a) are labeled consecutively from left to right, then

$$\nu(i,j) = (i-1)(q-1) + j - 1 \tag{2.7.9}$$

Given $L = i$, let $T_r(i)$ denote the total *rewinding time* it takes the system to go discontinuously from one part of the uncertainty region to another. In Figure 2.25(a), the rewinding is indicated by the broken lines. If the rewinding time associated with each broken line is T_r, then

$$T_r(i) = (i-1)\,T_r \tag{2.7.10}$$

If the uncertainty region covers an entire sequence period, then the cells at the two edges are actually adjacent and $T_r = 0$.

When acquisition has been completed, $\nu(i,j)$ incorrect cells have been dismissed, $i - 1$ correct cells have been dismissed, and one correct cell has been accepted. Therefore, it follows from (2.7.7) that the mean acquisition time is

$$\begin{aligned} E[T_a] &= \int_0^\infty t f_a(t)\,dt \\ &= P_D \sum_{i=1}^{\infty} (1 - P_D)^{i-1} \sum_{j=1}^{q} p_c(j) \\ &\quad \times \left\{ \nu(i,j)E[T_{11}] + (i-1)E[T_{12}] + E[T_{22}] + T_r(i) \right\} \end{aligned} \tag{2.7.11}$$

We assume that the time it takes to decide whether a cell is correct is statistically independent of the times for the other cells. Therefore,

$$E[T_a^2] = P_D \sum_{i=1}^{\infty} (1 - P_D)^{i-1} \sum_{j=1}^{q} p_c(j)$$
$$\times \Big\{\big\{\nu(i,j)E[T_{11}] + (i-1)E[T_{12}] + E[T_{22}] + T_r(i)\big\}^2$$
$$+ \nu(i,j)\,\mathrm{VAR}(T_{11}) + (i-1)\,\mathrm{VAR}(T_{12}) + \mathrm{VAR}(T_{22})\Big\} \qquad (2.7.12)$$

The variance of T_a is

$$\sigma_a^2 = E[T_a^2] - \{E[T_a]\}^2 \qquad (2.7.13)$$

Assuming that T'_{21} and T_p are statistically independent, we obtain

$$E[T_{11}] = (1 - P_F)E[T'_{11}] + P_F\{E[T'_{21}] + E[T_p]\} \qquad (2.7.14)$$

$$E[T_{11}^2] = (1 - P_F)[\mathrm{VAR}(T'_{11}) + \{E[T'_{11}]\}^2]$$
$$+ P_F[\mathrm{VAR}(T'_{21}) + \mathrm{VAR}(T_p) + \{E[T'_{21}] + E[T_p]\}^2] \qquad (2.7.15)$$

The variance of T_{11} is

$$\mathrm{VAR}(T_{11}) = E[T_{11}^2] - \{E[T_{11}]\}^2 \qquad (2.7.16)$$

Let $*$ denote the convolution operation and $[f(t)]^{*n}$ denote the n-fold convolution of the density $f(t)$ with itself. We define $[f(t)]^{*0} = 1$ and $[f(t)]^{*1} = f(t)$. Using this notation, we obtain

$$f_a(t \mid i,j) = [f_{11}(t)]^{*\nu(i,j)} * [f_{12}(t)]^{*(i-1)} * [f_{22}(t)] \qquad (2.7.17)$$

where $f_{11}(f)$, $f_{12}(t)$, and $f_{22}(t)$ are the densities associated with T_{11}, T_{12}, and T_{22}, respectively. If one of the decision times is a constant, then the associated density is a delta function. Because T'_{21} and T_p are independent, $f_{11}(t)$ can be decomposed into

$$f_{11}(t) = (1 - P_F)f'_{11}(t) + P_F[f'_{21}(t) * f_p(t)] \qquad (2.7.18)$$

where $f'_{11}(t)$, $f'_{21}(t)$, and $f_p(t)$ are the densities associated with T'_{11}, T'_{21}, and T_p, respectively. Equations (2.7.6), (2.7.17), and (2.7.18) give $f_a(t)$ in terms of various other densities.

Given that $L = i$ and $C = j$, the acquisition time is the sum of independent random variables. Thus, to simplify the computation of $f_a(t \mid i,j)$, it is reasonable to approximate this density by a Gaussian density with mean

$$\mu_{ij} = \nu(i,j)E[T_{11}] + (i-1)E[T_{12}] + E[T_{22}] + T_r(i) \qquad (2.7.19)$$

and variance

$$\sigma_{ij}^2 = \nu(i,j)\,\mathrm{VAR}(T_{11}) + (i-1)\,\mathrm{VAR}(T_{12}) + \mathrm{VAR}(T_{22}) \qquad (2.7.20)$$

To further simplify numerical computations, the Gaussian density may be truncated to $\pm 3\sigma_{ij}$ or some other interval. In most cases of interest, P_D is large and the infinite series in (2.7.7) converges rapidly. Therefore, the series can be accurately approximated by its first few terms.

In some applications, the serial-search acquisition must be completed within a specified period of duration T_{max}. If the acquisition time exceeds T_{max}, it is terminated and special measures are undertaken, for example, matched-filter acquisition of a short sequence. Thus, the probability that $T_a \leqslant T_{max}$ is an important performance measure. It can be calculated from $f_a(t)$. Alternatively, it can be bounded by using Chebyshev's inequality. Assuming that $T_{max} > E[T_a]$, we obtain

$$\begin{aligned} P(T_a \leqslant T_{max}) &= P(T_a - E[T_a] \leqslant T_{max} - E[T_a]) \\ &\geqslant P(|T_a - E[T_a]| \leqslant T_{max} - E[T_a]) \\ &\geqslant 1 - \frac{\sigma_a^2}{(T_{max} - E[T_a])^2} \end{aligned} \tag{2.7.21}$$

where $P(A)$ denotes the probability of the event A. This lower bound is only useful if $T_{max} - E[T_a] > \sigma_a$.

As an important application, we consider the uniform search of Figure 2.25(a) and a uniform *a priori* distribution for the location of the correct cell. We use (2.7.8) to (2.7.12) and the following identities:

$$\sum_{i=0}^{\infty} r^i = \frac{1}{1-r}, \quad \sum_{i=1}^{\infty} ir^i = \frac{r}{(1-r)^2}, \quad \sum_{i=1}^{\infty} i^2 r^i = \frac{r(1+r)}{(1-r)^3}$$

$$\sum_{i=1}^{n} i = \frac{n(n+1)}{2}, \quad \sum_{i=1}^{n} i^2 = \frac{n(n+1)(2n+1)}{6} \tag{2.7.22}$$

where $0 \leqslant |r| < 1$. Defining

$$\alpha = (q-1)E[T_{11}] + E[T_{12}] + T_r \tag{2.7.23}$$

we obtain

$$E[T_a] = (q-1)\left(\frac{2-P_D}{2P_D}\right) E[T_{11}] + \left(\frac{1-P_D}{P_D}\right)(E[T_{12}] + T_r) + E[T_{22}] \tag{2.7.24}$$

$$E[T_a^2] = (q-1)\left(\frac{2-P_D}{2P_D}\right)\text{VAR}(T_{11}) + \left(\frac{1-P_D}{P_D}\right)\text{VAR}(T_{12}) + \text{VAR}(T_{22})$$

$$+ \frac{(2q+1)(q+1)}{6}(E[T_{11}])^2 + (q+1)\left\{\frac{\alpha(1-P_D)}{P_D} + E[T_{22}] - E[T_{11}]\right\}$$

$$\times E[T_{11}] + \frac{\alpha^2(1-P_D)(2-P_D)}{P_D^2} + \frac{2\alpha(1-P_D)}{P_D}(E[T_{22}] - E[T_{11}])$$

$$+ (E[T_{22}] - E[T_{11}])^2 \tag{2.7.25}$$

Let τ_D denote the fixed dwell time of a single-dwell-time system. By definition, $T_{12} = T_{22} = T'_{11} = T'_{21} = \tau_D$ and $\text{VAR}(T_{12}) = \text{VAR}(T_{22}) = \text{VAR}(T'_{11}) = \text{VAR}(T'_{21}) = 0$. Equations (2.7.24) and (2.7.14) yield

$$E[T_a] = \frac{(q-1)(2-P_D)(\tau_D + P_F E[T_p]) + 2\tau_D + 2(1-P_D)T_r}{2P_D} \tag{2.7.26}$$

An equation for σ_a^2 follows from (2.7.25), (2.7.23), (2.7.26), and (2.7.13) to (2.7.16). For a quick estimate, we observe that as $q \to \infty$,

$$\sigma_a^2 \to q^2 \left(\frac{1}{P_D^2} - \frac{1}{P_D} + \frac{1}{12}\right)(\tau_D + P_F E[T_p])^2 \tag{2.7.27}$$

The serial-search system of Figure 2.24 uses a radiometer to test the cells. A performance analysis of the radiometer is given in Section 4.2. To apply this analysis, we assume that as a result of the autocorrelation properties of a pseudonoise sequence, negligible signal energy passes through the bandpass filter when an incorrect cell is tested. If data modulation is present, we assume that any symbol transition causes a negligible loss in the signal energy that passes through the bandpass filter when the correct cell is tested. If we assume that the synchronization is nearly perfect when the correct cell is tested, then (4.2.46) and (4.2.47) imply that

$$P_F = \frac{1}{2}\,\text{erfc}\left[\frac{V_T - N_0\tau_D W}{(2N_0^2\tau_D W)^{1/2}}\right], \qquad \tau_D W \gg 1 \tag{2.7.28}$$

$$P_D = \frac{1}{2}\,\text{erfc}\left[\frac{V_T - N_0\tau_D W - E_D}{(2N_0^2\tau_D W + 4N_0 E_D)^{1/2}}\right], \qquad \tau_D W \gg 1 \tag{2.7.29}$$

where W is the bandwidth of the bandpass filter, V_T is the threshold level, E_D is the signal energy emerging from the bandpass filter during the observation interval, and $N_0/2$ is the two-sided power spectral density of the noise. Given the values of E_D/N_0, $\tau_D W$, and $E[T_p]$, the value of V_T can be chosen to minimize $E[T_a]$, which is given by (2.7.26), (2.7.28), and (2.7.29).

The statistics of the penalty time depend upon the details of the operation of the acquisition system in the lock mode. Given that code synchronization is maintained, the time that elapses before the system incorrectly leaves the lock mode is called the *hold-in time.* It is desirable to have a large mean hold-in time and a small mean penalty time, but the realization of one of these goals tends to impede the realization of the other.

As an example, we assume that during the lock mode, the radiometer monitors the code synchronization and the dwell time is τ_L. A single missed detection, which occurs with probability $1 - P_{D1}$, causes the acquisition system to lose lock and initiate a search. Therefore, the mean hold-in time is

$$E[T_H] = \sum_{i=1}^{\infty} i\tau_L (1 - P_{D1}) P_{D1}^{i-1}$$

$$= \frac{\tau_L}{1 - P_{D1}} \qquad (2.7.30)$$

The penalty time expires unless false alarms, each of which occur with probability P_{F1}, continue to occur every τ_L seconds. Therefore, the mean penalty time is

$$E[T_p] = \sum_{i=1}^{\infty} i\tau_L (1 - P_{F1}) P_{F1}^{i-1}$$

$$= \frac{\tau_L}{1 - P_{F1}} \qquad (2.7.31)$$

The probabilities P_{D1} and P_{F1} are given by Equations (2.7.28) and (2.7.29) with τ_L in place of τ_D, $E_{D1} = E_D \tau_L / \tau_D$ in place of E_D, and V_{T1} in place of V_T. If V_{T1} is small, then P_{D1} and, hence, $E[T_H]$ are large. However, P_{F1} is also large, which causes $E[T_p]$ to be large. Thus, the selection of V_{T1} depends upon the importance of a large $E[T_H]$ relative to a small $E[T_p]$.

The use of a single dwell time to verify the lock condition makes the acquisition system vulnerable to deep fades and pulsed interference. Thus, a preferable strategy is for the lock mode to be maintained until a number of consecutive or cumulative misses occur during a series of tests. The performance of this type of lock detection can be analyzed by using Markov chains [13].

Suppose that a double-dwell-time system uses fixed dwell times in the search mode and performs a uniform search. We assume a uniform *a priori* distribution of the correct cell location. Let τ_{Di}, P_{Di}, and P_{Fi} denote the dwell time, detection probability, and false alarm probability for dwell i, where $i = 1$ or 2. From the flow graph of Figure 2.26, we derive

$$P_D = P_{D1}P_{D2} \tag{2.7.32}$$

$$E[T_{22}] = \tau_{D1} + \tau_{D2} \tag{2.7.33}$$

$$\begin{aligned} E[T_{12}] &= \tau_{D1}(1 - P_{D1}) + P_{D1}(\tau_{D1} + \tau_{D2}) \\ &= \tau_{D1} + P_{D1}\tau_{D2} \end{aligned} \tag{2.7.34}$$

$$\begin{aligned} E[T_{11}] &= (1 - P_{F1})\tau_{D1} + P_{F1}(1 - P_{F2})(\tau_{D1} + \tau_{D2}) + P_{F1}P_{F2} \\ &\quad \times (\tau_{D1} + \tau_{D2} + E[T_p]) \\ &= \tau_{D1} + P_{F1}(\tau_{D2} + P_{F2}E[T_p]) \end{aligned} \tag{2.7.35}$$

These equations and (2.7.24) yield

$$\begin{aligned} E[T_a] = (2P_{D1}P_{D2})^{-1}\Big\{&(q - 1)(2 - P_{D1}P_{D2})(\tau_{D1} + P_{F1}\tau_{D2} + P_{F1}P_{F2}E[T_p]) \\ &+ (1 - P_{D1}P_{D2})(2\tau_{D1} + 2P_{D1}\tau_{D2} + 2T_r)\Big\} + \tau_{D1} + \tau_{D2} \end{aligned} \tag{2.7.36}$$

An expression for σ_a^2 can be derived similarly. As $q \to \infty$, we find that

$$\sigma_a^2 \to q^2\left(\frac{1}{P_D^2} - \frac{1}{P_D} + \frac{1}{12}\right)(\tau_{D1} + P_{F1}\tau_{D2} + P_{F1}P_{F2}E[T_p])^2 \tag{2.7.37}$$

where P_D is given by (2.7.32).

Systems with variable dwell times and sequential detection potentially can provide a smaller mean acquisition time than those with fixed dwell times but require an increased implementation complexity [13].

2.8 CODE TRACKING

The *delay-locked loop* and the *tau-dither loop* are the predominant types of code-tracking loops used in direct-sequence systems. Either of these loops may be designed for coherent or noncoherent operation. The noncoherent versions of these loops are more common because the signal-to-noise ratio at the receiver input is usually too low for carrier synchronization prior to code synchronization and the subsequent despreading of the received signal.

Figure 2.27(a) is a block diagram of the main components of a delay-locked loop. The code generator produces three sequences, one of which is the reference sequence used for acquisition and demodulation. The other two sequences are advanced and delayed, respectively, by δT_c relative to the reference sequence. The product δT_c is usually equal to the acquisition step size, and thus usually $\delta = 1/2$, but other values are plausible [16]. The advanced and delayed sequences are multiplied by the received direct-sequence signal in separate branches. Each multiplier output is filtered and then applied to a square-law device or, alternatively, an envelope detector. The difference between the two branch outputs is filtered and then applied to a voltage-controlled clock. The filtered difference signal alters the clock frequency, which ideally causes the reference sequence to converge toward alignment with the received pseudonoise sequence.

The following approximate analysis indicates in more detail how the loop works in the absence of noise. If the received direct-sequence signal is given by (2.1.1), then the signal portion of the upper-branch multiplier output is

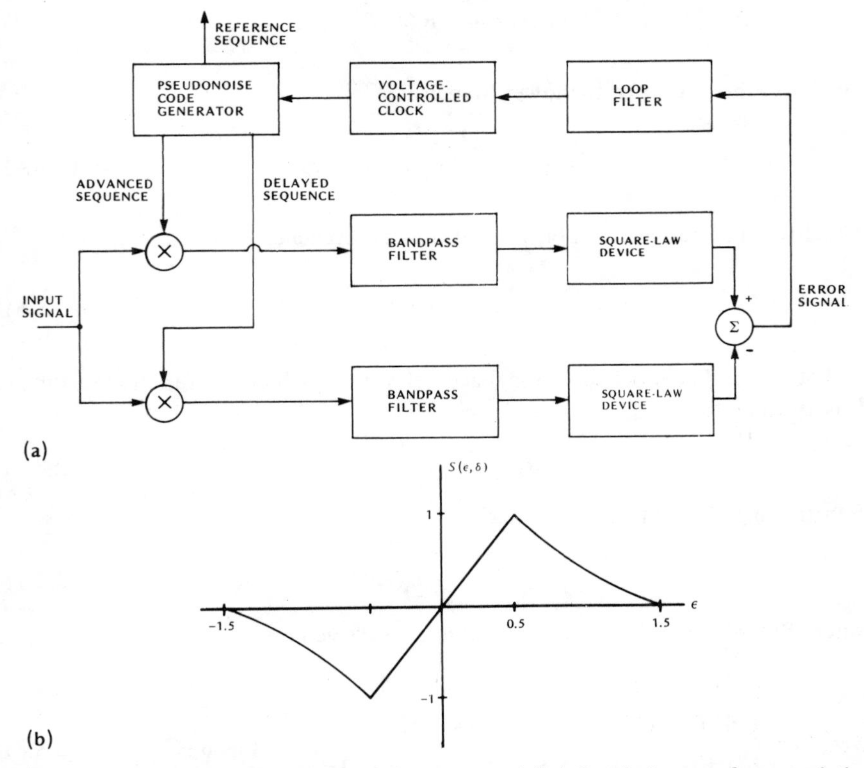

Figure 2.27 Delay-locked loop: (a) system and (b) discriminator characteristic for δ = 1/2.

$$s_{u1}(t) = Am(t)p(t)p(t + \delta T_c - \tau_e)\cos(\omega_0 t + \theta) \tag{2.8.1}$$

where τ_e is the loop's error in estimating the code timing and equals the delay of the reference sequence relative to the received sequence. Although τ_e is a function of time because of the loop dynamics, the time dependence is suppressed for notational convenience. It is assumed that the upper-branch bandpass filter distorts $m(t)$ negligibly, but approximately integrates the product $p(t)p(t + \delta T_c - \tau_e)$, which has a wider spectrum. Thus, the filter output is

$$s_{u2}(t) \cong Am(t)\hat{R}_p(\delta T_c - \tau_e)\cos(\omega_0 t + \theta) \tag{2.8.2}$$

where $\hat{R}_p(\tau)$ is an estimate of the full or partial autocorrelation function of the pseudonoise sequence. If the bandpass filter produced an exact integral over the interval $[t - T_s, t]$ and the pseudonoise sequence had a period equal to T_s, then $\hat{R}_p(\tau)$ would equal $R_p(\tau)$. Any double-frequency component produced by the square-law device is ultimately suppressed by the loop filter and thus is ignored. Since $m^2(t) = 1$, the upper-branch output is

$$s_{u3}(t) = \frac{A^2}{2} \hat{R}_p^2(\delta T_c - \tau_e) \tag{2.8.3}$$

Similarly, the output of the lower branch is

$$s_{l3}(t) = \frac{A^2}{2} \hat{R}_p^2(-\delta T_c - \tau_e) \tag{2.8.4}$$

The difference between the outputs of the two branches is the error signal

$$s_e(t) = \frac{A^2}{2} [\hat{R}_p^2(\delta T_c - \tau_e) - \hat{R}_p^2(-\delta T_c - \tau_e)] \tag{2.8.5}$$

Let $\epsilon = \tau_e / T_c$ denote the normalized delay error, which is a function of time. It is assumed that

$$\hat{R}_p(\tau) \approx \Lambda\left(\frac{\tau}{T_c}\right) \tag{2.8.6}$$

Substituting (2.8.6) into (2.8.5) yields

$$s_e(t) = \frac{A^2}{2} S(\epsilon, \delta) \tag{2.8.7}$$

where $S(\epsilon, \delta)$ is the *discriminator characteristic* given by

$$S(\epsilon,\delta) = \begin{cases} 4\epsilon(1-\delta), & 0 \leqslant \epsilon \leqslant \delta, \\ 4\delta(1-\epsilon), & \delta \leqslant \epsilon \leqslant 1-\delta, \\ 1+(\epsilon-\delta)(\epsilon-\delta-2), & 1-\delta \leqslant \epsilon \leqslant 1+\delta, \\ 0, & 1+\delta \leqslant \epsilon, \end{cases} \quad 0 \leqslant \delta \leqslant 1/2 \tag{2.8.8}$$

$$S(\epsilon,\delta)=\begin{cases}4\epsilon(1-\delta), & 0\leqslant\epsilon\leqslant 1-\delta,\\ 1+(\epsilon-\delta)(\epsilon-\delta+2), & 1-\delta\leqslant\epsilon\leqslant\delta,\\ 1+(\epsilon-\delta)(\epsilon-\delta-2), & \delta\leqslant\epsilon\leqslant 1+\delta,\\ 0, & 1+\delta\leqslant\epsilon,\end{cases}\qquad 1/2\leqslant\delta\leqslant 1 \tag{2.8.9}$$

$$S(-\epsilon,\delta)=-S(\epsilon,\delta) \tag{2.8.10}$$

The discriminator characteristic is plotted for $\delta = 1/2$ in Figure 2.27(b).

When $\epsilon(t) > 0$, the reference sequence is delayed relative to the received sequence. Because $S(\epsilon,\delta)$ and, hence, $s_e(t)$ are positive, the clock rate is increased and $\epsilon(t)$ decreases. Equations (2.8.7) to (2.8.9) indicate that $s_e(t) \to 0$ as $\epsilon(t) \to 0$. Similarly, when $\epsilon(t) < 0$ we find that $\epsilon(t)$ tends to increase and $s_e(t) \to 0$ as $\epsilon(t) \to 0$. Thus, the delay-locked loop tracks the received code timing once the acquisition system has finished the coarse alignment.

The main components of a noncoherent tau-dither loop are depicted in Figure 2.28(a). The dither generator controls a switch that alternately passes an advanced or delayed receiver-generated sequence. Let $d(t)$ denote the *dither signal,* which alternates between +1 and −1 as shown in Figure 2.28(b). In the absence of noise, the output of the switch can be represented by

$$s_1(t)=\left[\frac{1+d(t)}{2}\right]p(t+\delta T_c-\tau_e)+\left[\frac{1-d(t)}{2}\right]p(t-\delta T_c-\tau_e) \tag{2.8.11}$$

where the two factors within brackets are orthogonal functions of time and alternate between +1 and 0. Only one of the factors is nonzero at any instant. The received direct-sequence signal is multiplied by $s_1(t)$, filtered, and then applied to a square-law device. If the bandpass filter has the appropriate response, we find in a manner similar to the derivation of (2.8.3) that the device output is

$$s_2(t)=\frac{A^2}{2}\left[\frac{1+d(t)}{2}\right]\hat{R}_p^2(\delta T_c-\tau_e)+\frac{A^2}{2}\left[\frac{1-d(t)}{2}\right]\hat{R}_p^2(-\delta T_c-\tau_e) \tag{2.8.12}$$

Multiplication of $s_2(t)$ by $d(t)$ produces the loop filter input

$$s_3(t)=\frac{A^2}{2}\left[\frac{1+d(t)}{2}\right]\hat{R}_p^2(\delta T_c-\tau_e)-\frac{A^2}{2}\left[\frac{1-d(t)}{2}\right]\hat{R}_p^2(-\delta T_c-\tau_e) \tag{2.8.13}$$

which is a rectangular wave if the time variation of τ_e is ignored. The loop filter has a narrow bandwidth so that its output is approximately the direct-current component of $s_3(t)$, which is the average value of $s_3(t)$. Because τ_e is a slowly varying function of time, the filter output is found to be

$$s_4(t)\cong\frac{A^2}{4}\left[\hat{R}_p^2(\delta T_c-\tau_e)-\hat{R}_p^2(-\delta T_c-\tau_e)\right] \tag{2.8.14}$$

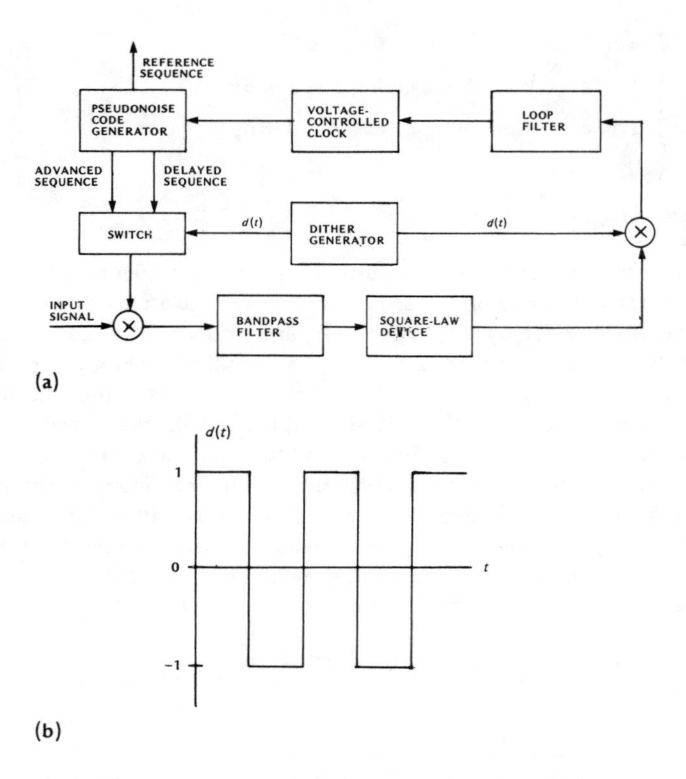

Figure 2.28 Tau-dither loop: (a) system and (b) dither signal.

If (2.8.6) is valid, then the clock input is

$$s_4(t) = \frac{A^2}{4} S(\epsilon, \delta) \tag{2.8.15}$$

where the discriminator characteristic is given by (2.8.8) to (2.8.10). Thus, the tau-dither loop can track the code timing in a manner similar to that of the delay-locked loop. The tau-dither loop requires less hardware than the delay-locked loop, but a detailed analysis indicates that the tau-dither loop provides less accurate code tracking [13].

2.9 BURST-COMMUNICATION SYSTEMS

Burst communications are short and infrequent communications, which are primarily used for interrogation or identification among network elements. Burst-communication systems can establish synchronization by including preamble symbols for this purpose in each transmitted burst [17]. However, if the bursts are short enough, it may be desirable to omit the preambles to reduce the susceptibility to interception.

A *direct-sequence burst-communication system* [18] that does not require preambles is depicted in Figure 2.29. Binary CSK is used as the data modulation. A programmable switching matrix connects each tap of a delay line with a pair of code generators or matched filters. In the transmitter; the delay line is excited by a pulse. As the pulse traverses the delay line, a corresponding pulse is applied to a different pair of pseudonoise code generators every symbol interval. Depending upon the state of the corresponding data symbol, each symbol selector, which may consist of logic circuitry similar to that shown in Figure 2.6(a), passes either the mark or space sequence to the phase modulator. Thus, M data symbols produce M contiguous pseudonoise sequences.

If tap i of the transmitter delay line is connected to code generators j, then tap $M - i + 1$ of the receiver delay line is connected to matched filters j. Thus, the received waveform does not produce significant matched-filter outputs until it completely fills the delay line. Shortly afterwards, either the mark or space matched filter produces a correlation peak. Each comparator output corresponds to its largest input. The correlation peaks occur simultaneously if the tap separations correspond to a symbol duration and if the matched filters cause identical propagation delays. If the mark and space sequences are nearly orthogonal to each other, one output of each pair of matched filters is negligible compared to the other output. Thus, the sum of the matched-filter outputs is a signal that has a peak amplitude equal to M times the peak amplitude of the output of a single matched filter. This sum has a peak power equal to M^2 times the peak power of a matched-filter output. If the noise outputs of the matched filters are nearly uncorrelated with each other, the noise power entering the threshold detector is $2M$ times the noise power of the output of a single matched filter. Thus, the signal-to-noise ratio at the threshold detector input is $M/2$ times the signal-to-noise ratio at the output of one of the matched filters. As a result, the detector output pulses provide the symbol timing with a precision that increases with M. For large M, the receiver performance approaches that of a synchronized CSK receiver despite the absence of *a priori* synchronization information. In a practical receiver, envelope detection is a necessary part of the matched-filter operation.

The programmable switching matrices permit the selection of $M!$ different permutations of the code pairs. Consequently, the burst-communication system allows selective addressing and resists deceptive jamming. When the value of M is sufficiently large and the reference level of the threshold detector is appropriately chosen, incorrect received sequences are rejected with high probability.

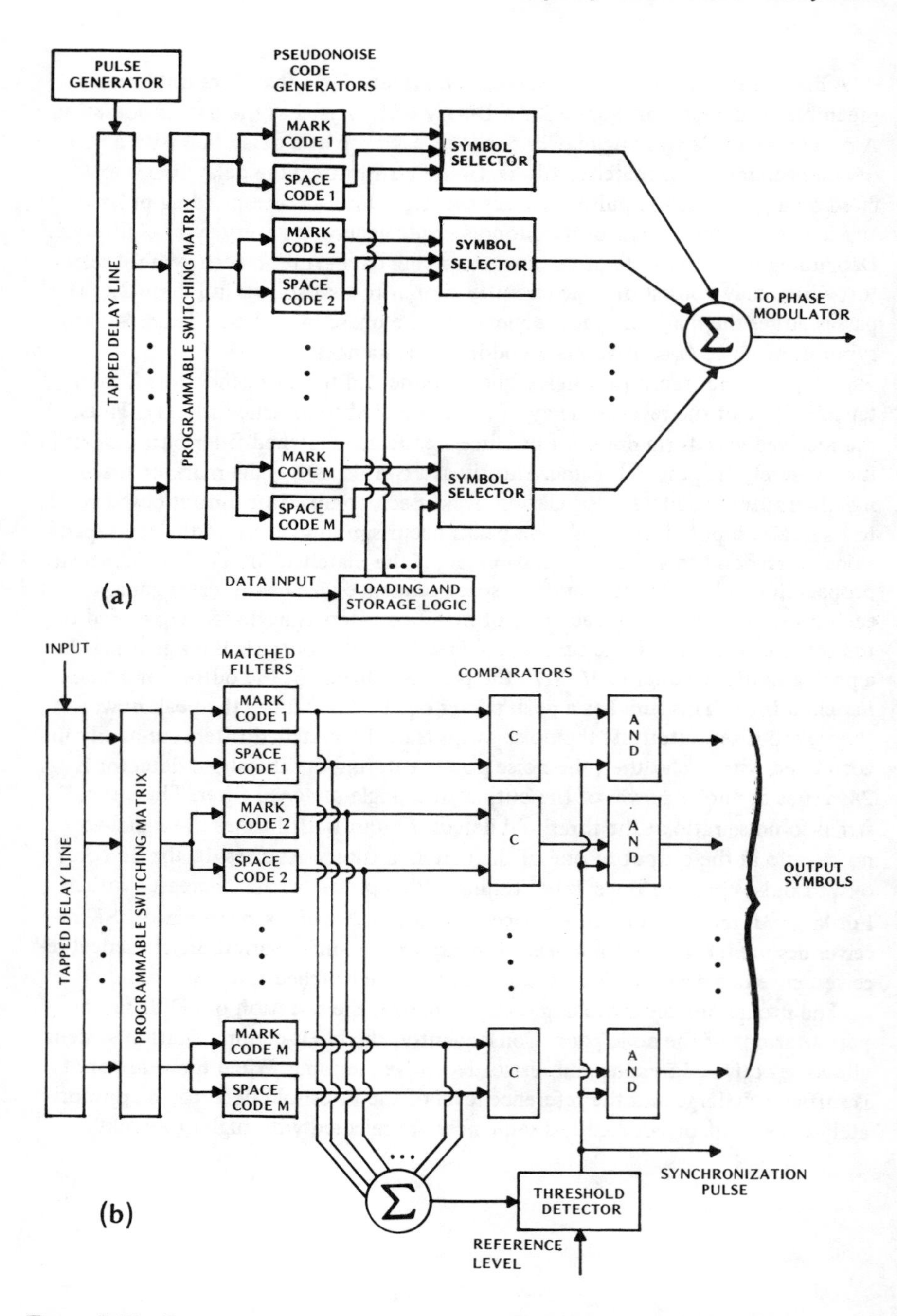

Figure 2.29 Burst-communication system: (a) transmitter and (b) receiver.

REFERENCES

1. S.W. Golomb, *Shift Register Sequences,* revised ed. Laguna Hills, California: Aegean Park Press, 1982.
2. H. Beker and F. Piper, *Cipher Systems, The Protection of Communications.* New York: John Wiley and Sons, 1982.
3. P. Billingsley, *Probability and Measure.* New York: John Wiley and Sons, 1979.
4. R.S. Lunayach, "Performance of a Direct Sequence Spread-Spectrum System with Long Period and Short Period Code Sequences," *IEEE Trans. Commun.* COM-31, 412, March 1983.
5. R.H. Dou and L.B. Milstein, "Error Probability Bounds and Approximations for DS Spread-Spectrum Communication Systems with Multiple Tone or Multiple-Access Interference," *IEEE Trans. Commun.* COM-32, 493, May 1984.
6. D.V. Sarwate and M.B. Pursley, "Crosscorrelation Properties of Pseudorandom and Related Sequences," *Proc. IEEE* 68, 593, May 1980.
7. M.B. Pursley, "Spread-Spectrum Multiple-Access Communications," in *Multi-User Communication Systems,* G. Longo, ed., New York: Springer-Verlag, 1981.
8. M.B. Pursley, D.V. Sarwate, and W.E. Stark, "Error Probability for Direct-Sequence Spread-Spectrum Multiple-Access Communications – Part I: Upper and Lower Bounds," *IEEE Trans. Commun.* COM-30, 975, Mary 1982.
9. D.V. Sarwate, M.B. Pursley, and T.U. Basar, "Partial Correlation Effects in Direct-Sequence Spread-Spectrum Multiple-Access Communication Systems," *IEEE Trans. Commun.* COM-32, 567, May 1984.
10. R.B. Ward and K.P. Yiu, "Acquisition of Pseudonoise Signals by Recursion-Aided Sequential Estimation," *IEEE Trans. Commun.* COM-25, 784, August 1977.
11. J.W. Mark and I.F. Blake, "Rapid Acquisition Techniques in CDMA Spread-Spectrum Systems," *IEE Proc.,* Part F, 223, April 1984.
12. H. Meyr and G. Polzer, "Performance Analysis for General PN-Spread-Spectrum Acquisition Techniques," *IEEE Trans. Commun.* COM-31, 1317, December 1983.
13. J.K. Holmes, *Coherent Spread Spectrum Systems.* New York: John Wiley and Sons, 1982.
14. A. Polydoros and C.L. Weber, "A Unified Approach to Serial Search Spread-Spectrum Code Acquisition – Part I: General Theory," *IEEE Trans. Commun.* COM-32, 542, May 1984.
15. W.R. Braun, "Performance Analysis for the Expanding Search PN Acquisition Algorithm," *IEEE Trans. Commun.* COM-30, 424, March 1982.

16. A. Polydoros and C.L. Weber, "Analysis and Optimization of Correlative Code-Tracking Loops in Spread-Spectrum Systems," *IEEE Trans. Commun.* COM-33, 30, January 1985.
17. M. Kowatsch, "Application of Surface-Acoustic-Wave Technology to Burst-Format Spread-Spectrum Communications," *IEE Proc.*, Part F, 734, December 1984.
18. M.G. Unkauf, "Surface Wave Devices in Spread Spectrum Systems," in *Surface Wave Filters,* H. Mathews, ed. New York: John Wiley and Sons, 1977.
19. R.C. Dixon, *Spread Spectrum Systems,* 2nd ed. New York: John Wiley and Sons, 1984.
20. C.R. Cahn *et al., Spread-Spectrum Communication,* NATO Advisory Group AGARD-LS-58, Nat. Tech. Inform. Serv. AD-766-914, 1973.

Chapter 3
Frequency Hopping

3.1 FUNDAMENTAL CONCEPTS

Frequency hopping is the periodic changing of the frequency or frequency set associated with a transmission. Successive frequency sets are determined by a pseudonoise sequence. If the data modulation is *multiple frequency-shift keying,* two or more frequencies are in the set that changes at each hop. For other data modulations, a single center or carrier frequency is changed at each hop.

A frequency-hopping signal may be regarded as a sequence of modulated pulses with pseudorandom carrier frequencies. Hopping occurs over a frequency band that includes a number of *frequency channels.* Each channel is defined as a spectral region with a center frequency that is one of the possible carrier frequencies and a bandwidth large enough to include most of the power in a pulse with the corresponding carrier frequency. The bandwidth of a frequency channel is often called the *instantaneous bandwidth.*

When only a single carrier frequency and frequency channel are used between hops, digital data modulation is called *single-channel modulation.* Figure 3.1 depicts the general form of a system with single-channel data modulation. Figure 3.2(a) illustrates the frequency changes with time for frequency hopping with this type of modulation. The time duration between hops is called the *hop duration* or the *hopping period* and is denoted by T_h. The *total hopping bandwidth* and the instantaneous bandwidth are denoted by W and B, respectively. The residual signal after the effects of the frequency hopping have been removed from the received signal is called the *dehopped signal.* If the frequency pattern produced by the receiver synthesizer in Figure 3.1(b) is synchronized with the frequency pattern of the received signal, then the mixer output is a dehopped signal at a fixed difference frequency. Before demodulation, the dehopped signal is applied to a bandpass filter that excludes double-frequency components and power that originated outside the appropriate frequency channel. However, as long as a signal occupies a particular channel, the noise and interference in that channel are translated in frequency so that they enter the demodulator.

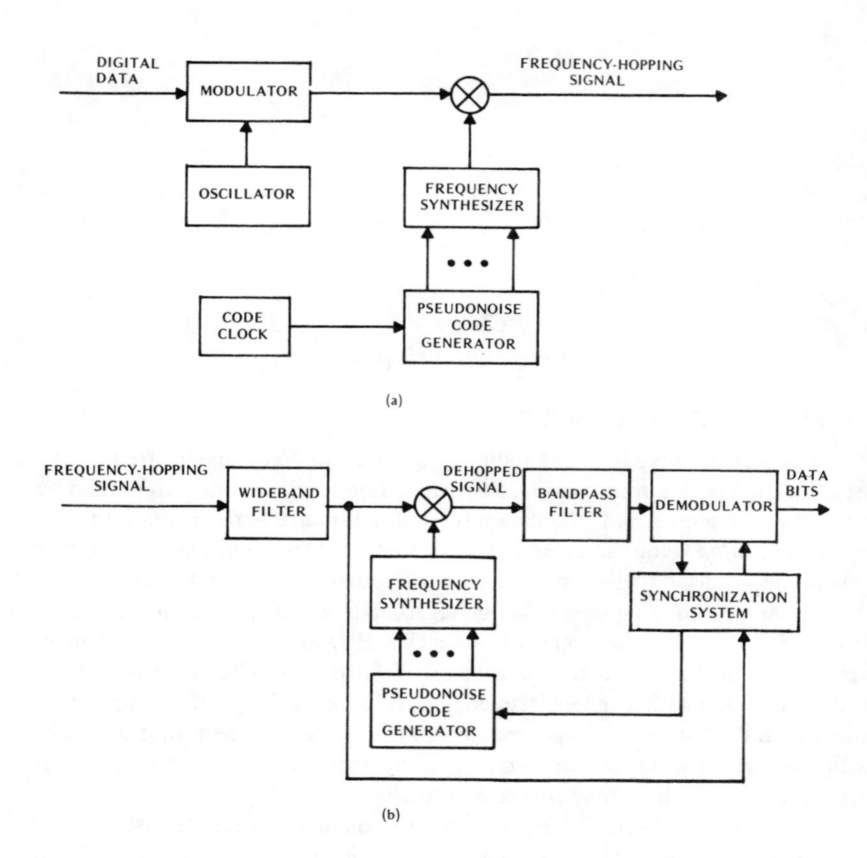

Figure 3.1 General form of frequency-hopping system with single-channel modulation: (a) transmitter and (b) receiver.

Frequency hopping may be classified as fast or slow. *Fast frequency hopping* occurs if there is a frequency hop for each transmitted symbol. Thus, for binary communications fast frequency hopping implies that the hopping rate equals or exceeds the information-bit rate. *Slow frequency hopping* occurs if two or more symbols are transmitted in the time interval between frequency hops.

When binary frequency-shift keying (FSK) is used, each symbol is transmitted as one of two frequencies, where one frequency represents the symbol 1 (mark), and the other, the symbol 0 (space). The pair of possible frequencies changes

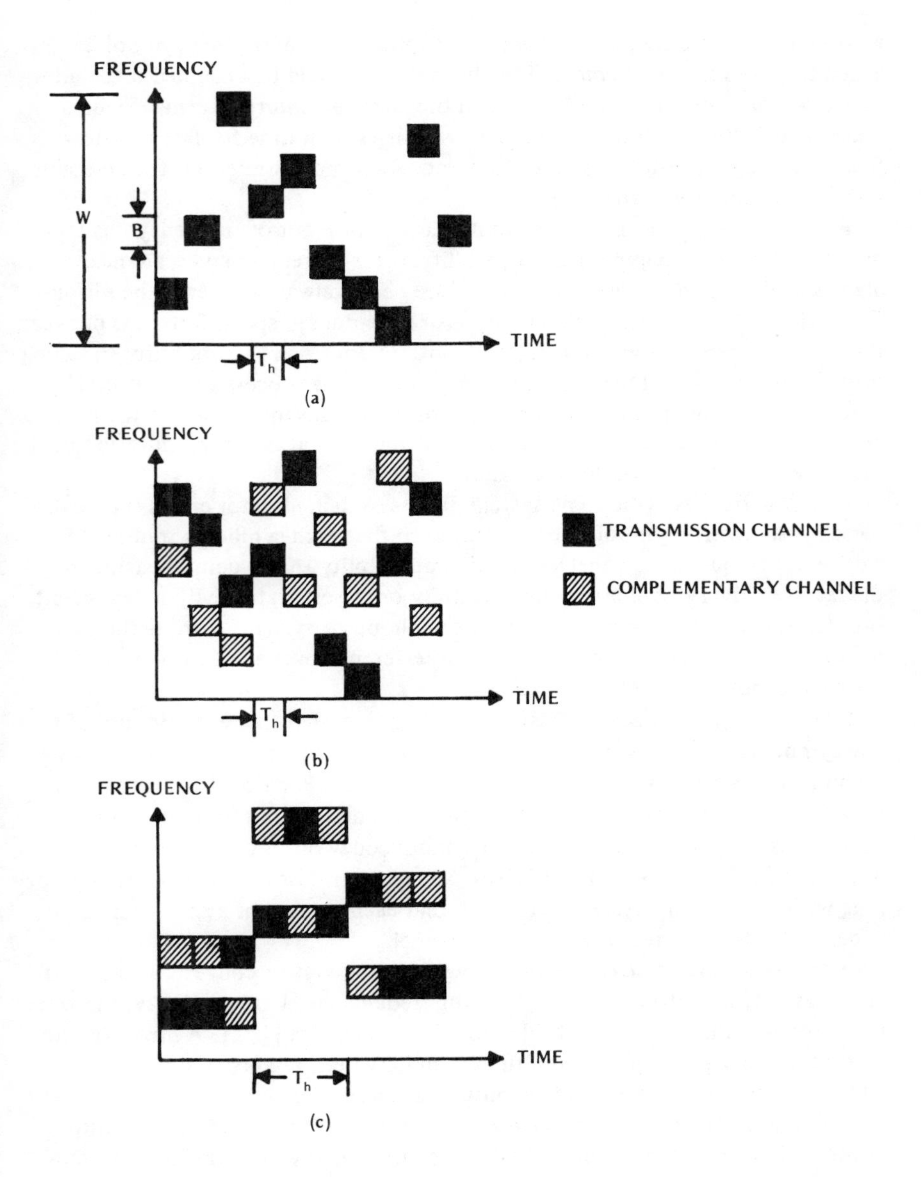

Figure 3.2 Frequency changes with time for (a) frequency hopping with single-channel data modulation, (b) fast hopping with FSK, and (c) slow hopping with FSK.

with each hop. The frequency channel occupied by a transmitted symbol is called the *transmission channel.* The channel that would be occupied if the alternative symbol were transmitted is called the *complementary channel.* Figures 3.2(b) and 3.2(c) illustrate the frequency changes with time for fast and slow frequency hopping with FSK data modulation and non-contiguous transmission and complementary channels.

Frequency hopping allows communicators to hop out of frequency channels with interference. To exploit this capability, error-correcting codes are nearly always used, and the frequency channels are nearly always chosen to be disjoint. The disjoint channels may be contiguous or have unused spectral regions between them. Some spectral regions with steady interference or a susceptibility to fading may be omitted from the set of frequency channels, a process called *spectral notching.* Although frequency hopping provides no advantage against white noise, it does provide a form of frequency diversity that limits the duration of the effects of frequency-selective fading.

In a slow frequency-hopping system, bursts of demodulator errors can be dispersed if successive symbols are interleaved so that each symbol is transmitted in a different frequency channel with a high probability. After deinterleaving in the receiver, the symbol errors are then fully or almost statistically independent. Interleaving in both fast and slow frequency-hopping systems permits the correction of bursts of errors due to pulsed interference over a large fraction of the total bandwidth.

Figure 3.2(a) can represent fast or slow hopping such that each symbol of a five-symbol codeword is transmitted in a different frequency channel. For fast hopping, one symbol is transmitted each T_h seconds. For slow hopping with interleaving, five symbols, each from a different codeword, are transmitted each T_h seconds. The transmission of a five-symbol codeword is completed after $5T_h$ seconds. Figures 3.2(b) and 3.2(c) can represent fast hopping and slow hopping with interleaving, respectively, such that each symbol of a three-symbol codeword is transmitted in a different channel.

A *frequency synthesizer* in a frequency-hopping system converts a stable reference frequency into the various hopping frequencies. A frequency synthesizer may use direct, indirect, or hybrid methods of synthesis [1, 2]. A *direct frequency synthesizer* uses frequency multipliers and dividers, mixers, bandpass filters, and electronic switches to produce output signals at the desired frequencies. A typical section of a direct synthesizer is illustrated in Figure 3.3(a). The output of the digital frequency generator, which is one of many possible tones, is multiplied by the switch output, which is one of several possible tones at frequencies $f_a, f_b, \ldots$. The multiplier (mixer) output is filtered to suppress undesired components and the filter output is ideally a single sine wave at the desired frequency. Both the switch and the digital frequency generator are controlled by bits produced by a pseudonoise code generator. The reference signal is usually the output of a frequency divider fed by a stable frequency source, such as an atomic or crystal oscillator. The signals applied to the switch are ultimately derived from the stable frequency source.

The *digital frequency generator* contains memory and a digital-to-analog converter. The sample values of a sine wave over one period are stored in the memory, read out at a rate equal to the reference frequency provided by the reference signal, and converted into an analog sine wave. A waveform with a frequency equal to n times the lowest generated frequency is produced when every nth

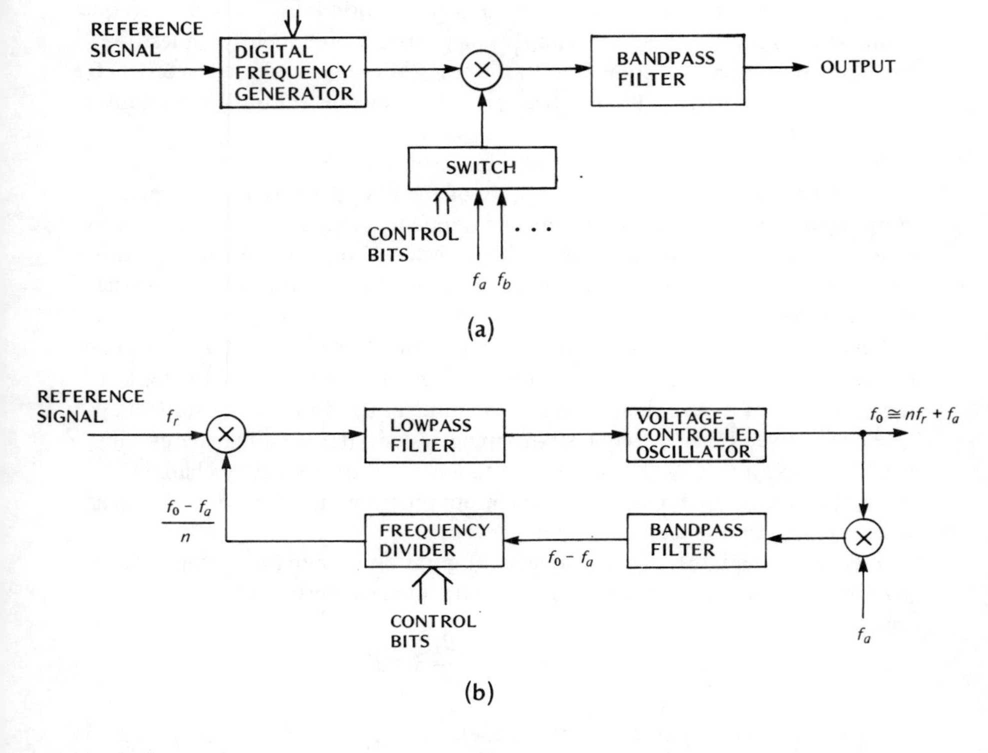

Figure 3.3 Sections of frequency synthesizers: (a) direct synthesizer and (b) indirect synthesizer.

sample value is read out at the constant rate. The digital frequency generator usually cannot be used alone as a frequency synthesizer because of practical limitations on the maximum frequency of its output signal.

An *indirect frequency synthesizer* uses a voltage-controlled oscillator and feedback. The principal components of a typical section of an indirect synthesizer, which is similar in operation to a phase-locked loop, are depicted in Figure 3.3(b). The feedback loop forces the frequency of the divider output, $(f_0 - f_a)/n$, to closely approximate the reference frequency, f_r. Consequently, the output of the voltage controlled oscillator is a sine wave with frequency $f_0 \cong nf_r + f_2$, where n is the divider factor. The control bits, which determine the value of n, are supplied by a pseudonoise code generator. Indirect synthesizers usually require less hardware than comparable direct ones, but require more time to switch from one frequency to another.

It is difficult to maintain phase coherence from hop to hop between frequency synthesizers in the transmitter and the receiver, primarily because of frequency-dependent multipath effects and Doppler shifts. Consequently, unless the hopping rate is very low compared to the transmitted symbol rate, practical frequency-hopping systems almost always require noncoherent or differentially coherent demodulators.

The effectiveness of frequency hopping against a sophisticated jamming threat depends upon the unpredictability of the hopping pattern. Hence, a *chirp signal* that continuously sweeps its frequency across a band cannot provide secure communications. It is relatively easy to intercept and analyze the chirp signal and then transmit chirp jamming to capture a receiver or degrade communications.

The benefits of frequency hopping are potentially neutralized by a *repeater jammer,* also known as a *follower jammer,* which is a device that intercepts a signal, processes it, and then transmits jamming at the same center frequency. To be effective against a frequency-hopping system, the jamming energy must reach the victim receiver before it hops to a new set of frequency channels. Thus, the greater the hopping rate, the more protected the frequency-hopping system is against a repeater jammer.

Figure 3.4 depicts the geometrical configuration of communicators and a jammer. For the repeater jamming to be effective, we must have

$$\frac{d_2 + d_3}{c} + T_{pr} \leqslant \frac{d_1}{c} + \eta T_d \tag{3.1.1}$$

where c is the velocity of an electromagnetic wave, T_{pr} is the processing time required by the repeater, η is a fraction, and T_d is the *dwell time,* which is the duration of a frequency-hopping pulse and is less than or equal to the hop duration. This equation states that the arrival-time delay of the jamming relative to the desired signal must not exceed a certain fraction of the dwell time if the jamming is to be effective. The value of η is determined by the jamming power at the receiver, the number of symbols per frequency-hopping pulse, and the details of the receiver design.

A rearrangement of (3.1.1) yields

$$d_2 + d_3 \leqslant (\eta T_d - T_{pr})\, c + d_1 \tag{3.1.2}$$

If the right-hand side of this inequality is regarded as a constant, then equating the two sides defines an ellipse with the transmitter and the receiver at the two foci. If the repeater jammer is located outside this ellipse, the jamming cannot be effective. Figure 3.4 shows a jammer located on the boundary of the ellipse.

In addition to the geometric restrictions, there are other problems entailed in using a repeater jammer. A large portion of the frequency-hopping band must usually be monitored. If many signals are simultaneously intercepted in different frequency channels, the repeater jammer must either divide its transmitted power among these signals or attempt to isolate a small number of signals to be jammed. In the presence of many communicators, the repeater jammer may be faced with a formidable signal-sorting problem.

Suppose that a jammer always distributes its power uniformly over a frequency-hopping band of contiguous frequency channels. To maintain a certain jamming power level in each frequency channel, the total transmitted jamming power

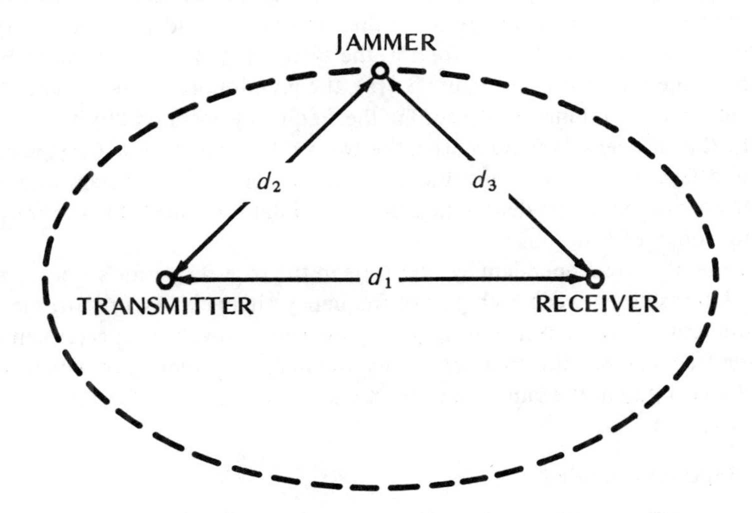

Figure 3.4 Geometrical configuration of communicators and jammer.

must increase linearly with the number of channels. For this reason, the number of available channels is often called the *processing gain* of a frequency-hopping system. However, this terminology is not recommended because it is misleading when the jamming power is not uniformly distributed or spectral notching is used.

In the next four sections, the performance of frequency-hopping systems with various types of modulation and coding are evaluated. Perfect code synchronization and the suppression of intermodulation products are always assumed. Hopping patterns are modeled as random rather than pseudorandom. It is found that the performance depends upon the signal-to-interference ratio and the signal-to-noise ratio. It is reasonable to assume that the average signal-to-interference ratio is nearly independent of the carrier frequency since both the signal and the interference use similar propagation channels. However, if W is large, the thermal and environmental noise do not necessarily vary with frequency in the same manner as the signal power. Thus, the average signal-to-noise ratio may have a significant frequency dependence unless the frequency-hopping system can be designed to make an appropriate compensation. However, to ensure the tractability of the mathematical analysis, it is always assumed that both power ratios are independent of the carrier frequency.

3.2 BINARY FSK

A frequency-hopping system with binary FSK and noncoherent demodulation may have the form of Figure 3.1 if the transmission and complementary channels are contiguous. If they are not, the system may have the form of Figure 3.5. In the transmitter of Figure 3.5(a), the pseudonoise sequence and the digital input are combined to determine the frequency generated by the synthesizer. In the receiver of Figure 3.5(b), the two synthesizers produce frequencies that are offset from the two possible received frequencies by constant intermediate frequencies. After the dehopping, the demodulation is the same as for ordinary noncoherent FSK.

The use of two independent synthesizers in the receiver permits a non-constant relationship between each pair of frequency channels. As a result, the measurement of the transmitted frequency does not provide a repeater jammer with the frequency of the complementary channel, the jamming of which is more damaging than the jamming of the transmission channel, as shown subsequently.

3.2.1 Repeater Jamming

Consider jamming by a repeater that responds rapidly enough to interfere with the reception of the same symbol that is intercepted. Suppose that the jamming power enters only the transmission channel. If the jamming waveform can be modeled as wide-sense-stationary Gaussian noise over the transmission

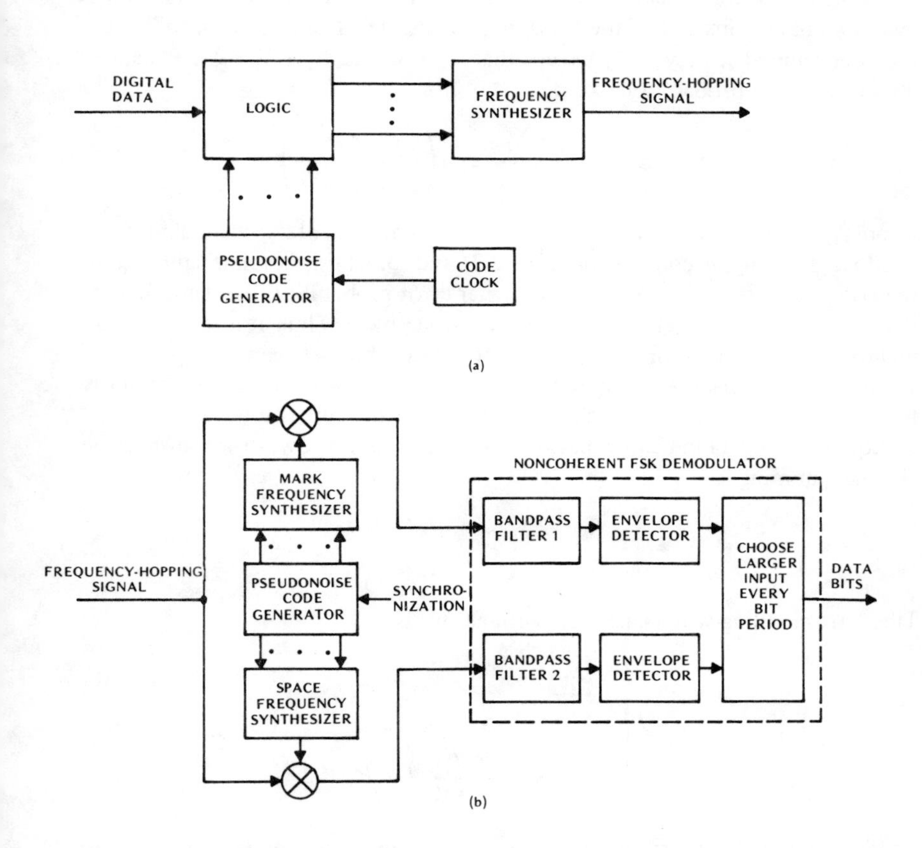

Figure 3.5 Frequency-hopping system with FSK: (a) transmitter and (b) receiver.

channel, then the symbol error probability is determined from (1.4.33), (1.4.32), and (1.4.24) with $B_1 = B_2 = 0$, $A = \sqrt{2R_s}$, $N_1 = N_t + N_i$, and $N_2 = N_t$. Using (1.4.41), we obtain the error probability

$$S_t = \frac{N_t}{2N_t + N_i} \exp\left(- \frac{R_s}{2N_t + N_i} \right) \tag{3.2.1}$$

where N_t is the average thermal-noise power common to both frequency channels, N_i is the average jamming power in the transmission channel after bandpass filtering, and R_s is the average power in the desired signal at the receiver.

Suppose that the repeater can concentrate the jamming power in the complementary channel instead of the transmission channel. The error probability is then determined as previously except that $N_1 = N_t$ and $N_2 = N_t + N_i$. Thus, we obtain the error probability

$$S_c = \frac{N_t + N_i}{2N_t + N_i} \exp\left(-\frac{R_s}{2N_t + N_i}\right) \tag{3.2.2}$$

where N_i is the average jamming power in the complementary channel after bandpass filtering. A comparison of (3.2.2) and (3.2.1) shows that jamming in the complementary channel causes a higher error probability than jamming in the transmission channel with the same average power. Thus, it is desirable to maintain a pseudorandom or nonconstant relationship between each pair of frequency channels. If the relationship remains unknown, a repeater jammer is forced to transmit the jamming in the transmission channel alone.

Equation (3.2.1) indicates that the optimal jamming power for transmission-channel jamming is

$$N_i = \begin{cases} R_s - 2N_t, & R_s > 2N_t \\ 0, & R_s \leqslant 2N_t \end{cases} \tag{3.2.3}$$

The corresponding worst-case error probability is

$$S_t = \begin{cases} \dfrac{N_t}{eR_s}, & R_s > 2N_t \\ \dfrac{1}{2} \exp\left(-\dfrac{R_s}{2N_t}\right), & R_s \leqslant 2N_t \end{cases} \tag{3.2.4}$$

If the jamming waveform that enters the transmission channel is modeled as a tone that is undistorted by the receiver bandpass filter, then the error probability is determined from (1.4.33), (1.4.32), and (1.4.24) with $B_1 = \sqrt{2R_i}$, $B_2 = 0, A = \sqrt{2R_s}$, and $N_1 = N_2 = N_t$. Using (1.4.41) and (1.4.22), we obtain

$$S_t = \frac{1}{2} \exp\left(-\frac{R_s + R_i}{2N_t}\right) I_0\left(\frac{\sqrt{R_s R_i}}{N_t}\right) \tag{3.2.5}$$

where R_i is the average jamming power in the tone. If the complementary channel is jammed by a tone, but the transmission channel is not, then a similar derivation gives

$$S_c = Q\left(\sqrt{\frac{R_i}{N_t}}, \sqrt{\frac{R_s}{N_t}}\right) - \frac{1}{2} \exp\left(-\frac{R_s + R_i}{2N_t}\right) I_0\left(\frac{\sqrt{R_s R_i}}{N_t}\right) \tag{3.2.6}$$

Use of the asymptotic expression for the Bessel function,

$$I_0(x) \cong \frac{\exp(x)}{\sqrt{2\pi x}}, \qquad x \gg 1 \tag{3.2.7}$$

in (3.2.5) yields

$$S_t \cong \left(\frac{N_t}{8\pi\sqrt{R_s R_i}} \right)^{1/2} \exp\left[- \frac{(\sqrt{R_s} - \sqrt{R_i})^2}{2N_t} \right], \qquad \sqrt{R_s R_i} \gg N_t \tag{3.2.8}$$

It follows that the optimal jamming power is $R_i \cong R_s$. The corresponding worst-case error probability is

$$S_t \cong \left(\frac{N_t}{8\pi R_s} \right)^{1/2}, \qquad R_i = R_s \gg N_t \tag{3.2.9}$$

As an example, Figure 3.6 displays S_t *versus* N_i/R_s or R_i/R_s, as determined from (3.2.1) and (3.2.5). The signal-to-noise ratio is R_s/N_t = 13 dB. The figure

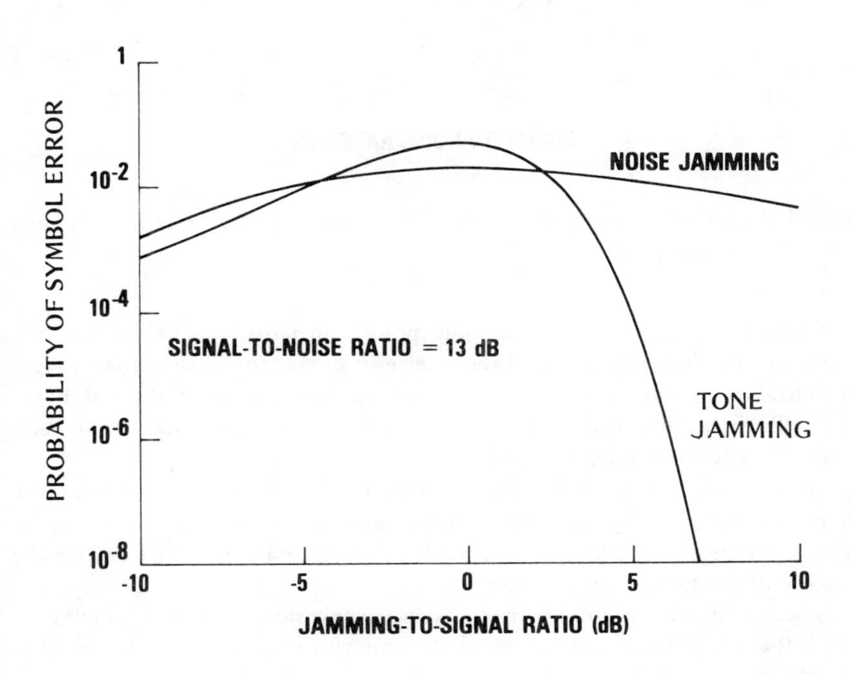

Figure 3.6 Symbol error probability for FSK modulation and repeater jamming.

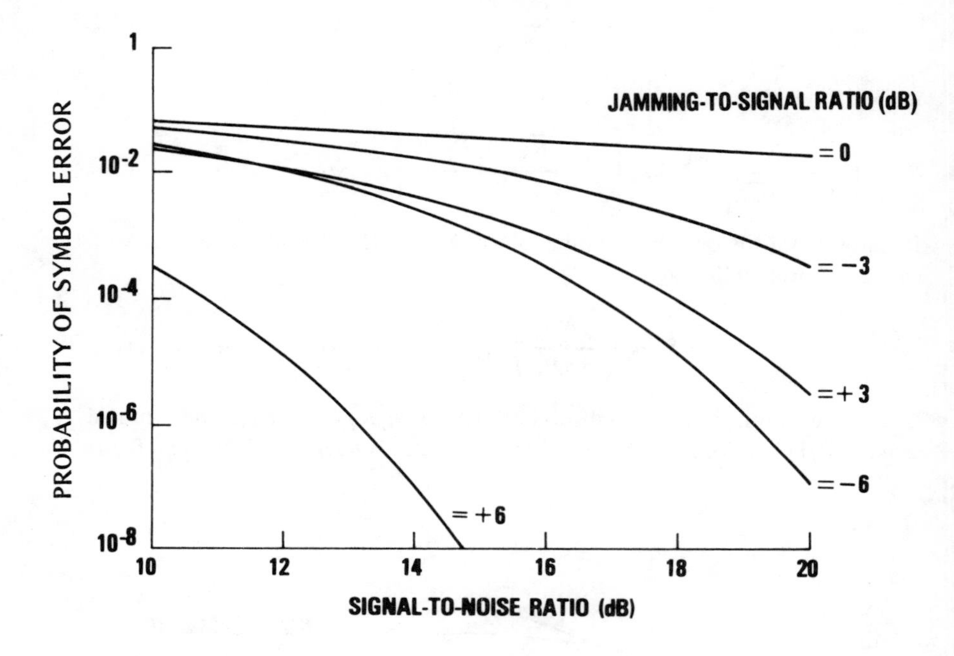

Figure 3.7 Symbol error probability for FSK modulation and tone jamming by repeater.

demonstrates that excessive tone jamming power can actually be helpful to the communicators. The reason is that the receiver responds to the strong jamming, which has the same form and enters the same channel as the desired signal. Figure 3.7 illustrates how S_t decreases with the signal-to-noise ratio for various fixed values of the tone jamming-to-signal ratio.

Equations (3.2.4) and (3.2.9) indicate that if R_s/N_t is large, then the channel-symbol error probability may be sufficiently small that an error-correcting code and interleaving can produce an acceptable information-bit error rate despite the presence of worst-case repeater jamming.

Repeater jamming is potentially damaging against slow frequency hopping even if the repeater cannot respond rapidly enough to interfere with the reception of the same symbol that is intercepted. Suppose that there are n symbols during the dwell time of a pulse. If the jamming energy reaches the receiver too late to affect the first m symbols and $n > m$, then the jamming energy may enter either the transmission or the complementary channels of each of the final $n - m$ symbols. Thus, the error probability for one of these symbols is given by (1.4.45) for tone jamming and (1.4.42) with $N_{i1} + N_{i2} = N_i$ for noise jamming.

3.2.2 Partial-Band Interference

As a signal hops among the M available frequency channels that remain after any spectral notching, the transmission channel, the complementary channel, or both channels may contain significant interference power. It is convenient to refer to the channels with interference as *jammed channels,* even though the interference may be unintentional. We assume that the interference-to-signal ratio is the same in all J jammed channels, where $J \leqslant M$. Suppose that the jammed channels are randomly distributed among the available frequency channels. Consider the pair of frequency channels associated with a transmitted symbol. We define

$$j_0 = \max(0, J - M + 2) \tag{3.2.10}$$

$$j_1 = \min(2, J) \tag{3.2.11}$$

If $j_0 \leqslant j \leqslant j_1$, there are $\binom{2}{j}$ ways to choose j jammed channels out of the pair. There are then $\binom{M-2}{J-j}$ ways to choose $J - j$ jammed channels out of the $M - 2$ channels that are not associated with the transmitted symbol. There are $\binom{M}{J}$ ways to choose J channels out of M total channels. Thus, the symbol error probability is

$$P_s = \sum_{j=j_0}^{j_1} \frac{\binom{2}{j}\binom{M-2}{J-j}}{\binom{M}{J}} S_j \tag{3.2.12}$$

where S_j is the conditional symbol error probability given that j channels associated with a symbol contain interference.

For the system of Figure 3.5, we assume that the transmission and complementary channels are not only distinct but also are both randomly chosen for each transmitted symbol. If the J distinct jammed channels are regarded as fixed, rather than randomly distributed, then a combinatorial argument and the easily proved identity that

$$\frac{\binom{J}{j}\binom{M-J}{2-j}}{\binom{M}{2}} = \frac{\binom{2}{j}\binom{M-2}{J-j}}{\binom{M}{J}} \tag{3.2.13}$$

lead to (3.2.12). Thus, this equation is valid for two different models of the frequency-hopping system and the jamming.

The S_j for tone interference are determined from (1.4.42) to (1.4.45), which give

$$S_0 = \frac{1}{2} \exp\left(-\frac{R_s}{2N_t}\right) \tag{3.2.14}$$

$$S_1 = \frac{1}{2} Q\left(\sqrt{\frac{R_i}{N_t}}, \sqrt{\frac{R_s}{N_t}}\right) \tag{3.2.15}$$

$$S_2 = \frac{1}{2\pi} \int_0^{2\pi} dx \left\{ Q\left[\sqrt{\frac{R_i}{N_t}}, \left(\frac{R_s + R_i + 2\sqrt{R_s R_i}\cos x}{N_t}\right)^{1/2}\right] - \frac{1}{2} \exp\left[-\frac{R_s + 2R_i + 2\sqrt{R_s R_i}\cos x}{2N_t}\right] \times I_0\left[\frac{\sqrt{R_i}\,(R_s + R_i + 2\sqrt{R_s R_i}\cos x)^{1/2}}{N_t}\right]\right\} \tag{3.2.16}$$

For noise interference, it follows from (1.4.42) that

$$S_j = \frac{1}{2} \exp\left(-\frac{R_s}{2N_t + jN_i}\right), \qquad j = 0, 1, 2 \tag{3.2.17}$$

When error-correcting codes are used, the word and average information-bit error probabilities can be determined from (3.2.12) and the equations of Section 1.7. If fast hopping or slow hopping with interleaving is used, then it is reasonable to assume that the channel-symbol errors are independent, which is necessary to apply the results of Section 1.7. For an (n, k) block code and hard-decision decoding, the word error probability satisfies

$$P_w \leqslant \sum_{i=t+1}^{n} \binom{n}{i} P_s^i (1 - P_s)^{n-i} \tag{3.2.18}$$

where P_s is given by (3.2.12) and equality holds if the code is perfect or if a bounded-distance decoding algorithm is used.

The disadvantage of randomly selecting pairs of channels is that the same channel may be used for more than one symbol of a codeword, thereby reducing the capability of a block decoder to correct errors caused by a few interfering signals. An alternative strategy is to randomly select the set of $2n$ distinct transmission and complementary channels associated with a codeword of n symbols, where $2n \leq M$. To determine the effectiveness of this strategy, let $A(j,m,q)$ denote the event that the transmission channels of j symbols, the complementary channels of m symbols, and both channels of q symbols are jammed. The probability of $A(j,m,q)$ is denoted by P_5. The probability of i symbol errors, given $A(j,m,q)$, is denoted by $P(i \mid j,m,q)$. From these definitions, it follows that

$$P_W \leq \sum_{i=t+1}^{n} \sum_{j} \sum_{m} \sum_{q} P(i \mid j,m,q) P_5 \tag{3.2.19}$$

The summation needs to be carried out only over those index values for which $P(i \mid j,m,q) P_5$ is nonzero. From the definitions of the probabilities, we obtain the following bounds for the index values:

$$\begin{aligned} &0 \leq q \leq m \leq n, \qquad 0 \leq q \leq j \leq n, \qquad m+j \leq J, \\ &m+j-q \leq n, \qquad n-j \leq M-J, \qquad J-j-m \leq M-2n \end{aligned} \tag{3.2.20}$$

Other inequalities required for consistency are implied by those above.

Probability P_5 can be evaluated by combinatorial analysis. Out of a total of M possible channels, n transmission channels and n complementary channels are associated with each codeword. We may regard the $2n$ different channels associated with a particular codeword as fixed and the interference as introduced into J randomly chosen channels. Alternatively, we may consider the J jammed channels as fixed and the transmission and complementary channels as randomly chosen for each codeword. In either case, we can derive the same formula for P_5. However, the former approach yields a simpler derivation.

There are $\binom{M}{J}$ ways to choose the J jammed channels out of the M total channels. The number of ways in which the event $A(j,m,q)$ can occur may be determined by specifying a four-step process. There are $\binom{n}{j}$ ways to choose the j jammed transmission channels of a codeword. Having chosen these channels, there are $\binom{j}{q}$ ways to choose those channels that have jammed complementary channels associated with them. There are $\binom{n-j}{m-q}$ ways to choose $m-q$ complementary channels that are jammed but are not associated with jammed transmission channels. The final step in the process is to select the $J-j-m$ jammed channels out of the $M-2n$ channels that are not associated with the codeword. This selection can be accomplished in $\binom{M-2n}{J-j-m}$ ways. Thus, the probability of event $A(j,m,q)$ is

$$P_5 = \frac{\binom{n}{j}\binom{j}{q}\binom{n-j}{m-q}\binom{M-2n}{J-j-m}}{\binom{M}{J}} \tag{3.2.21}$$

Let $B(\alpha,\beta,\gamma)$ denote the event that α errors occur in the $j - q$ symbols for which only the transmission channel is jammed, β errors occur in the $m - q$ symbols for which only the complementary channel is jammed, and γ errors occur in the q symbols for which both associated channels are jammed. The probability of $B(\alpha,\beta,\gamma)$ given $A(j,m,q)$ is denoted by $P(\alpha,\beta,\gamma \mid j,m,q)$. The probability of i symbol errors given the event $A(j,m,q) \cap B(\alpha,\beta,\gamma)$ is denoted by P_4. From these definitions, it follows that

$$P(i \mid j,m,q) = \sum_{\alpha} \sum_{\beta} \sum_{\gamma} P(\alpha,\beta,\gamma \mid j,m,q) P_4 \tag{3.2.22}$$

The summation needs to be carried out only over those index values for which $P(\alpha,\beta,\gamma \mid j,m,q)P_4$ is nonzero. From the definitions of the probabilities, we obtain the following bounds for the index values:

$$\begin{aligned} &0 \leqslant \alpha \leqslant j - q, \qquad 0 \leqslant \beta \leqslant m - q, \\ &0 \leqslant \gamma \leqslant q, \qquad \alpha + \beta + \gamma \leqslant i, \\ &n - (j - q) - (m - q) - q \geqslant i - (\alpha + \beta + \gamma) \end{aligned} \tag{3.2.23}$$

From its definition, P_4 is equal to the probability that there are $i - \alpha - \beta - \gamma$ errors among the $n - j - m + q$ symbols for which there is no interference in either associated channel. Since the channels associated with these symbols are all different, it is plausible to assume the independence of errors among these $n - j - m + q$ symbols. It then follows that

$$P_4 = \binom{n - j - m + q}{i - \alpha - \beta - \gamma} S_0^{i-\alpha-\beta-\gamma} \, (1 - S_0)^{n-j-m+q-i+\alpha+\beta+\gamma} \tag{3.2.24}$$

where S_0 is the probability of a bit error when neither associated channel is jammed and is given by (3.2.14). Similar assumptions of the independence of errors among various sets of symbols lead to

$$P(\alpha,\beta,\gamma \mid j,m,q) = P_1 P_2 P_3 \tag{3.2.25}$$

where

$$P_1 = \binom{j - q}{\alpha} S_t^{\alpha} (1 - S_t)^{j-q-\alpha} \tag{3.2.26}$$

$$P_2 = \binom{m - q}{\beta} S_c^{\beta} (1 - S_c)^{m-q-\beta} \tag{3.2.27}$$

$$P_3 = \binom{q}{\gamma} S_2^{\gamma}(1 - S_2)^{q-\gamma} \tag{3.2.28}$$

The conditional probabilities S_t, S_c, and S_2 are given by (3.2.1), (3.2.2), and (3.2.17) for noise interference, and by (3.2.5), (3.2.6), and (3.2.16) for tone interference.

Combining the above definitions, equations, and inequalities, we obtain the upper bound on the word error probability for either fast hopping or slow hopping with interleaving:

$$P_w \leqslant \sum_{i=t+1}^{n} \sum_{j=j_0}^{j_1} \sum_{m=m_0}^{m_1} \sum_{q=q_0}^{q_1} \sum_{\alpha=0}^{\alpha_1} \sum_{\beta=0}^{\beta_1} \sum_{\gamma=\gamma_0}^{\gamma_1} P_1P_2P_3P_4P_5 \tag{3.2.29}$$

where equality holds for bounded-distance decoding and

$$\begin{aligned}
j_0 &= \max(0, n + J - M), & q_1 &= \min(m, j),\\
j_1 &= \min(n, J), & \alpha_1 &= \min(i, j - q),\\
m_0 &= \max(0, 2n + J - M - j), & \beta_1 &= \min(i - \alpha, m - q),\\
m_1 &= \min(n, J - j), & \gamma_0 &= \max(0, i - \alpha - \beta - n + j + m - q),\\
q_0 &= \max(0, j + m - n), & \gamma_1 &= \min(i - \alpha - \beta, q)
\end{aligned} \tag{3.2.30}$$

The summation limits ensure that all the binomial coefficients, $\binom{a}{b}$, are well defined.

The symbol error probability can be determined by setting $n = 1$ and $t = 0$ in (3.2.29), which becomes an equality. Observing that

$$S_1 = \frac{1}{2}(S_t + S_c) \tag{3.2.31}$$

because it is equally likely that the transmission or the complementary channel is jammed, we again obtain (3.2.12).

If the message duration is preserved after encoding, the duration of a transmitted symbol is reduced relative to an uncoded bit and the channel bandwidths must be increased. Thus, if the total bandwidth is not changed, the number of available channels for frequency hopping, M, is reduced relative to the number of channels, M_u, that would be available in the absence of coding. The thermal noise power, N_t, is increased relative to the thermal noise power, N_{tu}, that would be present in the absence of coding. Specifically,

$$M = \left\lfloor \frac{M_u k}{n} \right\rfloor \tag{3.2.32}$$

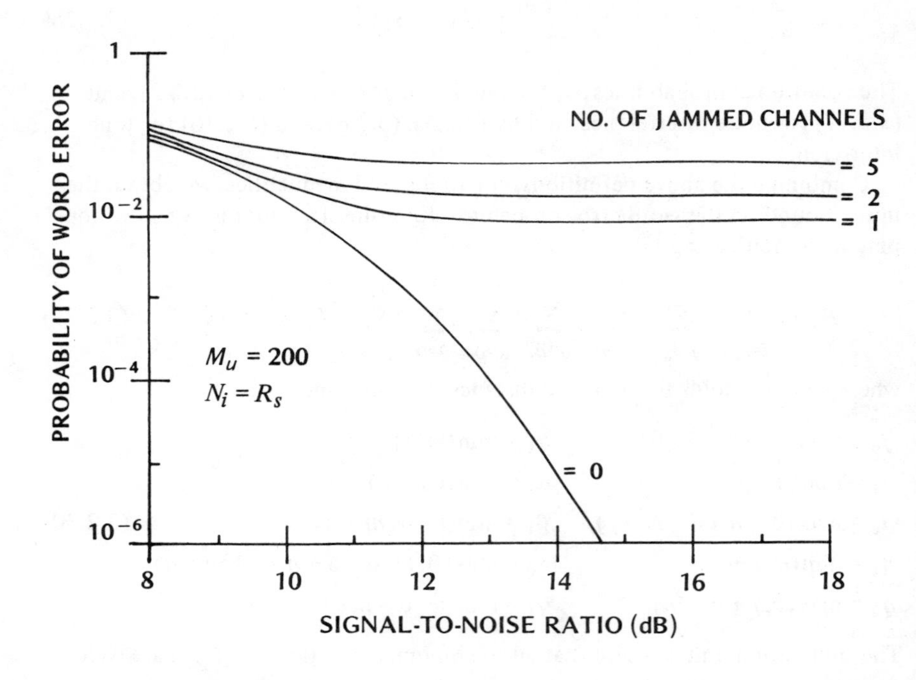

Figure 3.8 Word error probability for fast hopping or slow hopping with bit interleaving, four-bit uncoded word, randomly selected channel set, and FSK modulation.

$$N_t = \frac{N_{tu} n}{k} \tag{3.2.33}$$

where $[x]$ is the largest integer contained in x. The coding is effective when its error-correcting capability is sufficient to overcome the degradation implied by these equations.

As an example, we determine the word error probability for four-bit words, $M_u = 200$, and either fast hopping or slow hopping with interleaving when noise interference is present. The results are similar for tone interference. Each jammed channel is assumed to have interference power $N_i = R_s$ at the receiver. In this example, the upper bounds on P_w become equalities. Figures 3.8 to 3.10 depict P_w as a function of R_s/N_{tu}, the signal-to-noise ratio for uncoded communications. Figure 3.8 shows P_w for uncoded words, as calculated from (3.2.18) with $t = 0$. Figure 3.9, which is calculated from (3.2.29), shows the greatly improved performance when a Hamming (7,4) code is used and we still have $N_i = R_s$ in each jammed channel. Figure 3.10 depicts P_w calculated from (3.2.18). A comparison of Figures 3.9 and 3.10 illustrates the advantage of randomly selecting a set of $2n$ different channels for transmitting a codeword instead of randomly selecting each pair of channels. Although the implementation of randomly selected channel sets is more complicated, their advantage is usually significant when $J \leqslant t$ or when $t < J \leqslant 2t$ and n/M is sufficiently large.

The use of a repetition code provides a form of frequency diversity that is often effective against partial-band interference. To determine the probability of an information-bit error, we set $t = (n - 1)/2$ in (3.2.29), which is an equality in this case, and $k = 1$ in (3.2.32) and (3.2.33). As an example, suppose that $M_u = 1000$ and there are three interfering tones, each with power $R_i = R_s$ in a separate frequency channel. Figure 3.11 depicts the probability of an information-bit error *versus* the uncoded signal-to-noise ratio, R_s/N_{tu}, for $n = 1, 3, 5,$

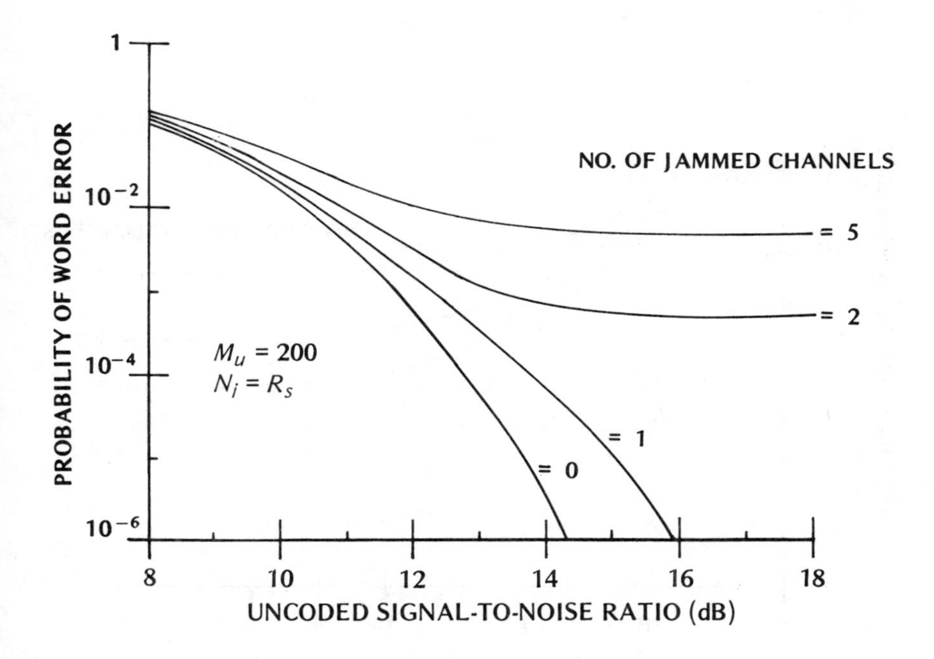

Figure 3.9 Word error probability for fast hopping or slow hopping with bit interleaving, (7,4) block code, randomly selected channel set, and FSK modulation.

and 7. For slow frequency hopping, the interleaving ensures that each code symbol is transmitted with a different carrier frequency. The figure indicates that increasing the amount of repetition is helpful only if the signal power is sufficiently large.

If the available jamming power is less than MR_s, then it is usually advantageous for a jammer to concentrate the jamming power in part of the total hopping band. It is intuitively reasonable that the highest information-bit error rate occurs when the jamming power is distributed so that the power in each jammed channel is approximately equal to the signal power, since increasing the jamming power beyond this level does not significantly increase the probability of a bit error when the communicators hop into a jammed channel. The effects of partial-band jamming, which are analogous to the effects of pulsed jamming (Section 1.6), are analyzed in a more general setting in the next section. Since a jammer can usually ensure that $J > 2t$, the use of randomly selected channel sets is not considered.

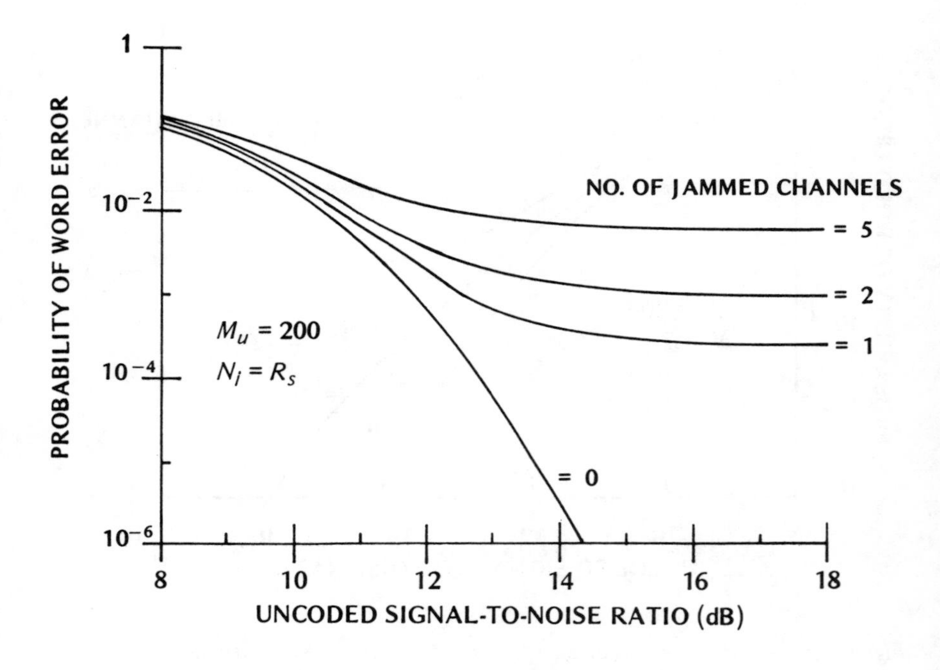

Figure 3.10 Word error probability for fast hopping or slow hopping with bit interleaving, (7,4) block code, randomly selected channel pairs, and FSK modulation.

3.3 MFSK AND PARTIAL-BAND JAMMING

Multiple frequency-shift keying (MFSK) entails choosing one of N frequencies as the carrier or center frequency for each transmitted symbol in a communication system. When frequency hopping is superimposed on MFSK, resulting in an FH/MFSK system, the set of N possible frequencies changes with each hop.

In this section, hard-decision decoding is considered [3]; soft-decision decoding is treated in Section 3.5. Soft-decision decoding is preferable in principle, but hard-decision decoding with its simpler implementation eliminates significant problems in the acquisition of reliable soft-decision information.

In a standard implementation of an FH/MFSK system, the N frequency channels of an MFSK set are contiguous, but have a randomly chosen position within the total hopping bandwidth. Figure 3.12 depicts the main elements of a noncoherent receiver with hard-decision decoding. After the dehopping, separate symbol decisions are made and the demodulated symbols are applied to a decoder, which produces the decoded message. The N frequencies are separated

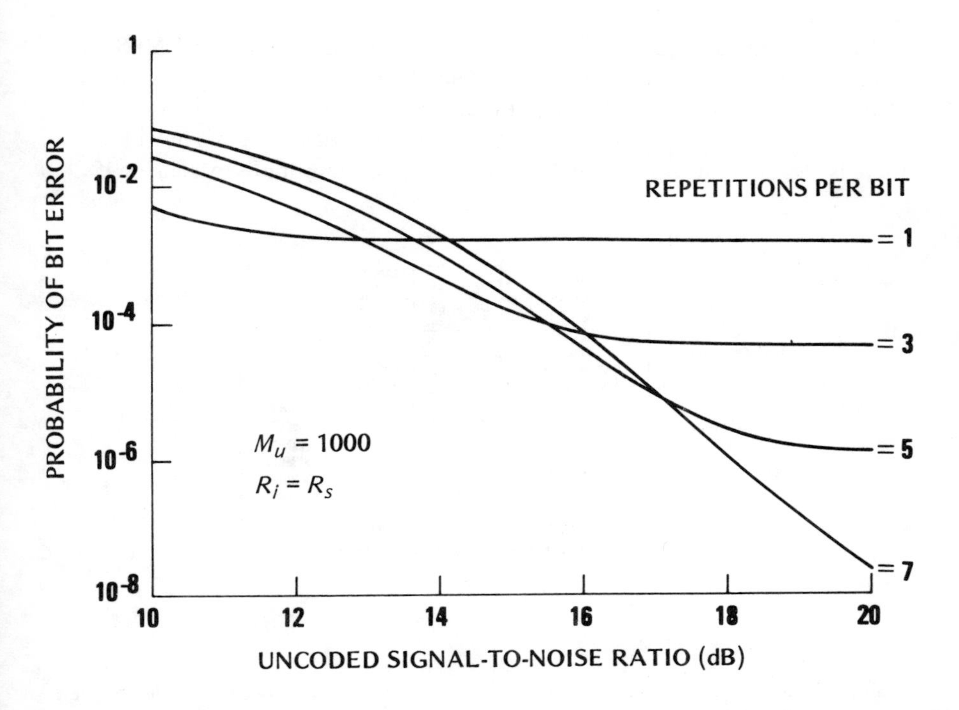

Figure 3.11 Bit error probability for repetition code, randomly selected channel set, FSK modulation, and three interfering tones.

enough so that the received signal produces negligible responses in the incorrect bandpass filters.

When the MFSK channels are contiguous, then whether noise or tone interference is generated, it is not advantageous to the jammer to transmit the jamming in a single continuous band since only one frequency channel of each MFSK set needs to be jammed to cause a symbol error. By randomly selecting the jammed channels, the number of MFSK sets that are affected by the jamming is increased and the practical difficulty of spectral notching is greatly increased.

The judicious spacing of jamming signals can exploit the uniform structure of contiguous MFSK frequency sets. If the sets are nonoverlapping, one jamming tone every MFSK set is more damaging than randomly placed tones. To preclude this type of sophisticated jamming, to facilitate the use of narrow spectral notches, and to increase the resistance of the receiver to repeater jamming, each frequency of the frequency sets can be randomly selected for each successive hop. Figure 3.13 depicts the main elements of the receiver appropriate for this strategy.

In the subsequent analysis of the error-rate performance of an FH/MFSK system, it is assumed that the errors in the demodulated symbols are independent. This assumption is reasonable for fast frequency hopping, in which there is a frequency hop for every MFSK symbol. For slow frequency hopping, in which there is less than one hop per MFSK symbol, the assumption is reasonable if

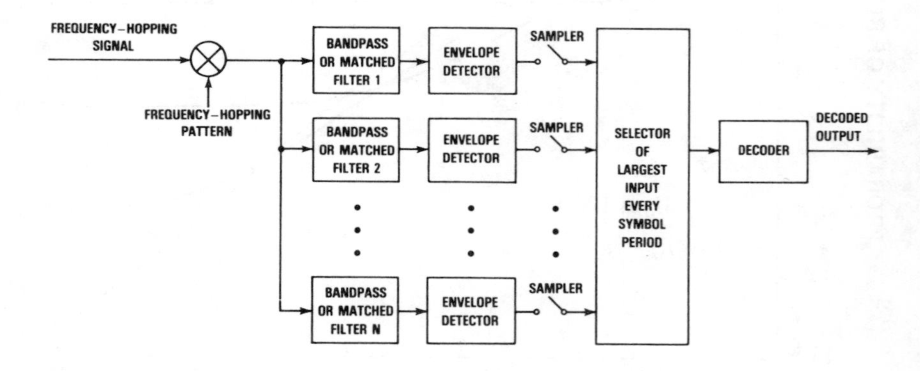

Figure 3.12 FH/MFSK receiver with hard-decision decoding and single frequency-hopping pattern.

symbols are interleaved over a sufficient number of hops. The deinterleaving in the receiver disperses bursts of errors, thereby facilitating the removal of errors by the decoder.

3.3.1 Block Codes

When a block code is used, each group of m information bits is encoded as a symbol, and then each group of k symbols is encoded as a codeword of n code symbols that are transmitted. Since a codeword represents mk information bits, there are 2^{mk} possible codewords. In an FH/MFSK system, one of $N = 2^m$ frequencies is selected as the carrier for each transmitted symbol.

Consider decoders of binary block codes such that $d = 2t + 1$, where d is the minimum distance between codewords and t is the number of symbol errors that can always be corrected (Section 1.7). When the channel-symbol errors are independent, the information-bit error rate for binary codes is often well-approximated by (1.7.20), which gives

$$P_{ib} \approx \sum_{i=t+1}^{2t+1} \frac{2t+1}{n} \binom{n}{i} P_s^i (1 - P_s)^{n-i} + \sum_{i=2t+2}^{n} \frac{i}{n} \binom{n}{i} P_s^i (1 - P_s)^{n-i} \quad (3.3.1)$$

For the nonbinary Reed-Solomon codes, the approximate expression is

$$P_{ib} \approx \frac{n+1}{2n} \left[\sum_{i=t+1}^{2t+1} \frac{2t+1}{n} \binom{n}{i} P_s^i (1 - P_s)^{n-i} + \sum_{i=2t+2}^{n} \frac{i}{n} \binom{n}{i} P_s^i (1 - P_s)^{n-i} \right] \quad (3.3.2)$$

where $n = 2^m - 1$ and $d = n - k + 1$.

Because an exact expression for P_s is complicated when partial-band jamming causes different power levels in the N receiver filters, we use a union bound. Referring to Figures 3.12 and 3.13, a code-symbol error occurs if the correct envelope sample does not exceed the other $N - 1$ envelope samples. Since the probability that a particular envelope sample exceeds the correct one is a function of only these two envelope samples, it is the same as P_{s2}, the channel-error probability of a frequency-hopping system with binary frequency-shift keying and the same energy per symbol. Thus, the union bound is

$$P_s \leqslant (N - 1) P_{s2}, \qquad N \geqslant 2 \quad (3.3.3)$$

If $N > 2$, the inequality is strict and becomes tighter as $(N - 1) P_{s2}$ decreases. If $N = 2$, we have $P_s = P_{s2}$. When (3.3.3) is substituted into (3.3.2), we obtain an approximate upper bound for P_{ib}.

Figures 1.7 and 1.8 indicate that noise jamming and tone jamming cause nearly the same P_s when the jamming-to-signal ratio per jammed channel is high. Thus, noise jamming is expected to be nearly as effective as tone jamming when optimal partial-band jamming exists. Because noise jamming is described by simpler equations, it is assumed in the subsequent examples. The noise jamming in each jammed frequency channel is modeled as a Gaussian process that is wide-sense stationary. We assume either that the N possible frequency channels are chosen randomly for each hop, which is true by definition for the system of Figure 3.13, or that the jammed channels are randomly chosen. For either assumption, (3.2.10) to (3.2.12) and (3.2.17) yield

$$P_{s2} = \frac{1}{2} \sum_{j=j_0}^{j_1} \frac{\binom{2}{j}\binom{M-2}{J-j}}{\binom{M}{J}} \exp\left(-\frac{R_s}{2N_t + jN_i}\right) \tag{3.3.4}$$

where

$$j_0 = \max(0, J+2-M), \quad j_1 = \min(2, J) \tag{3.3.5}$$

The parameters M, N_t, J, and N_i depend upon the coding. Let M_u denote the number of channels that are available for frequency hopping if an uncoded binary modulation is used. When MFSK and coding are used but the information rate (or the message duration) and the total hopping bandwidth are preserved, the number of available channels becomes approximately

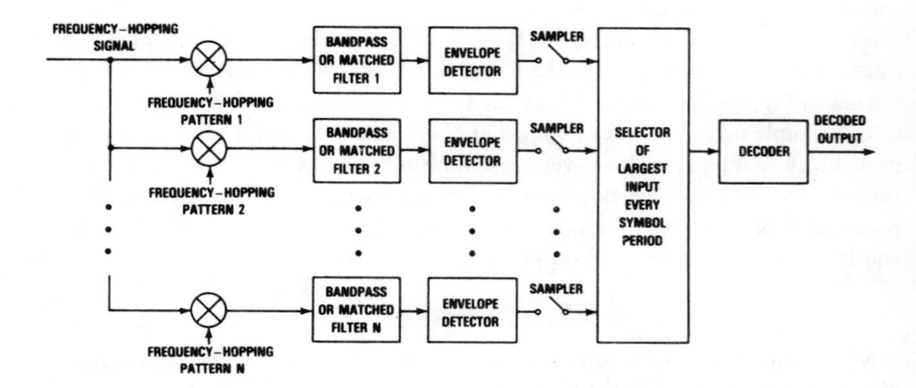

Figure 3.13 FH/MFSK receiver with hard-decision decoding and multiple frequency-hopping patterns.

$$M = [rM_u] \tag{3.3.6}$$

where r is the ratio of information bits to transmitted code symbols. For block codes, $r = mk/n$. The thermal noise power becomes

$$N_t = \frac{N_{tu}}{r} \tag{3.3.7}$$

where N_{tu} is the thermal noise power for uncoded binary modulation. The number of jammed channels is approximately

$$J = [\mu M] \tag{3.3.8}$$

where μ is the fraction of the total hopping band that contains jamming. If the jamming power is uniformly distributed and the channels are disjoint, then

$$N_i = \frac{N_{jt}}{J} \tag{3.3.9}$$

where N_{jt} denotes the total jamming power distributed in the total hopping band.

In the following examples, we assume that $M_u = 1000$. Figures 3.14 and 3.15 depict P_{ib} *versus* μ for uncoded binary communications and the Golay (23,12) code, respectively. The improvement due to the coding and the sharpness of the peaks in P_{ib} increase with increases in the signal-to-noise ratio, R_s/N_{tu}, and decrease with increases in the total jamming-to-signal ratio, N_{jt}/R_s. The optimal band occupancy for the jammer increases as N_{jt}/R_s increases. The potential effectiveness of partial-band jamming is evident in the difference between P_{ib} at $\mu = 1$ and the maximum value of P_{ib}. Figures 3.16 to 3.18 depict the maximum value of P_{ib} as a function of N_{jt}/R_s, assuming the optimal band occupancy for the jammer, $R_s/N_{tu} = 20$ dB, and various block codes with $d = 2t + 1$. Thus, these figures illustrate the worst-case performance of an FH/MFSK system operating against partial-band jamming. Instead of using N_{jt}/R_s, the abscissas can be expressed in terms of E_b/N_{0j}, where E_b is the energy per information bit and N_{0j} is the power spectral density that corresponds to uniform jamming over the total hopping band. A straightforward evaluation gives (in decibels)

$$E_b/N_{0j}\ (\text{dB}) = -N_{jt}/R_s\ (\text{dB}) + M_u\ (\text{dB}) + B_uT_{ib}\ (\text{dB}) \tag{3.3.10}$$

where B_u is the frequency-channel bandwidth for uncoded binary modulation and T_{ib} is the information-bit duration.

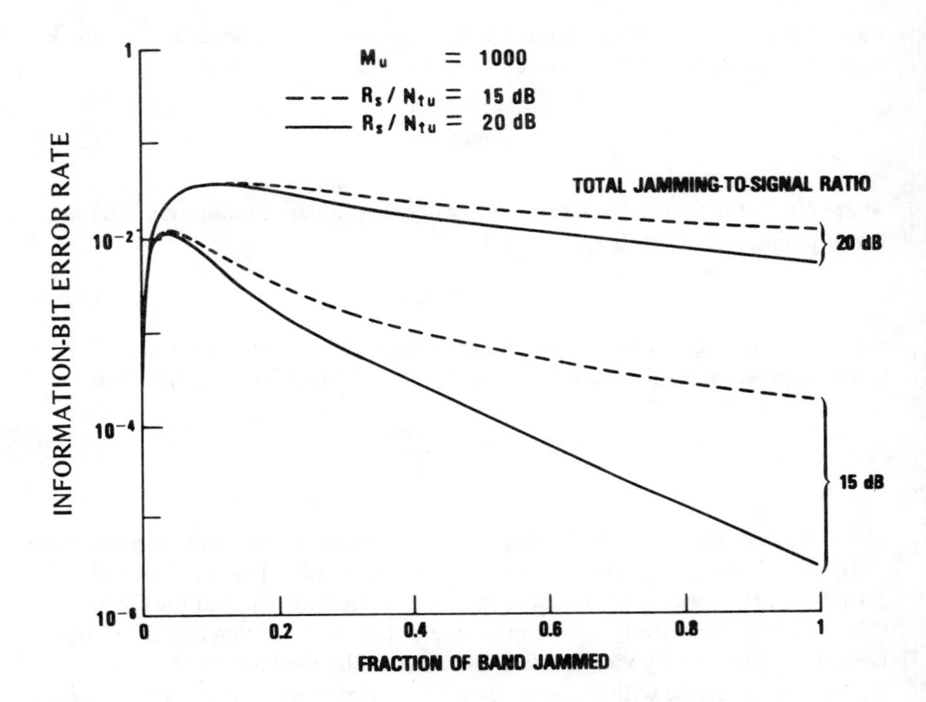

Figure 3.14 Information-bit error rate for no coding, binary hard decisions, and partial-band jamming.

The Reed-Solomon codes of Figure 3.18 are among those providing the best performance against worst-case partial-band jamming for given values of $N = n + 1$. The best codes tend to have rates between one-half and one-third. Since (3.3.3) is used, the curves of Figure 3.18 are actually approximate upper bounds for the worst-case performance. Assuming that the union-bound approximation causes a 1 dB error or less for $N \leqslant 16$ and $P_{ib} \cong 10^{-5}$, a comparison of Figures 3.16 and 3.18 shows that the Reed-Solomon codes do not offer a significant performance advantage over the Golay (23,12) code unless the required P_{ib} is 10^{-5} or less and $N \geqslant 32$, which requires a complex system implementation.

Communication systems are often required to provide an information-bit error rate below a specified value. Thus, a measure of performance for these systems is the signal-to-total-jamming ratio required to achieve the specified

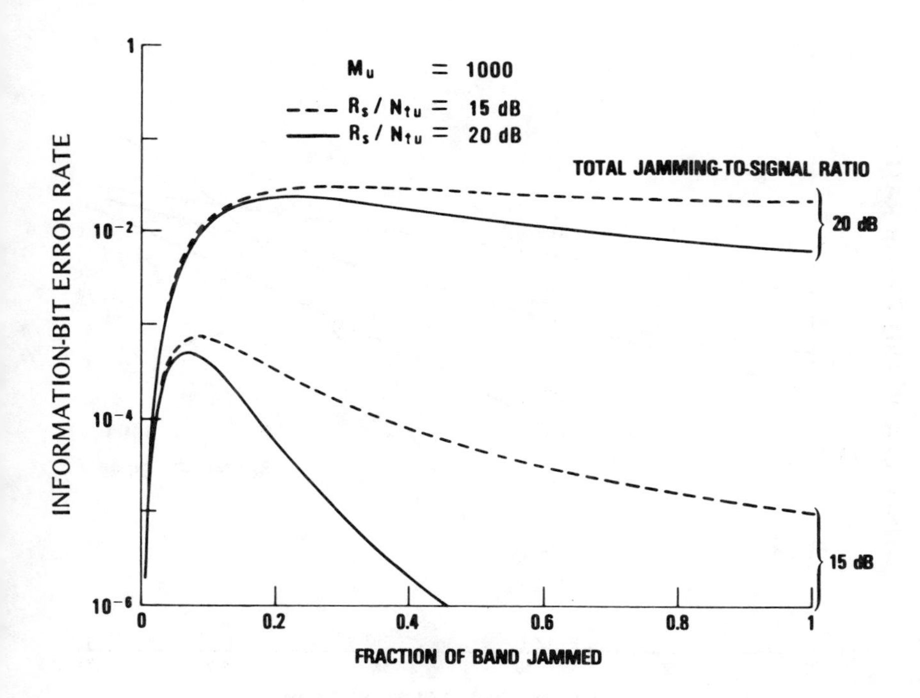

Figure 3.15 Information-bit error rate for Golay (23,12) code, hard-decision decoding, and partial-band jamming.

P_{ib}. For example, if an FH/MFSK system with repetition coding must provide $P_{ib} = 10^{-3}$, then Figure 3.17 indicates that $n = 7$ is the optimal choice for minimizing the required value of R_s/N_{jt} or, equivalently, maximizing the value of N_{jt}/R_s that can be tolerated.

3.3.2 Convolutional Codes

For binary convolutional codes of rate $r = k/n$ and Viterbi decoding, (1.7.36) gives

$$P_{ib} \leqslant \frac{1}{k} \sum_{i=0}^{\infty} A(d_f + i)\, Q(d_f + i) \tag{3.3.11}$$

where $Q(\delta)$ is the probability of error in comparing the correct path segment with a path segment that differs in δ symbols, $A(\delta)$ is the total information

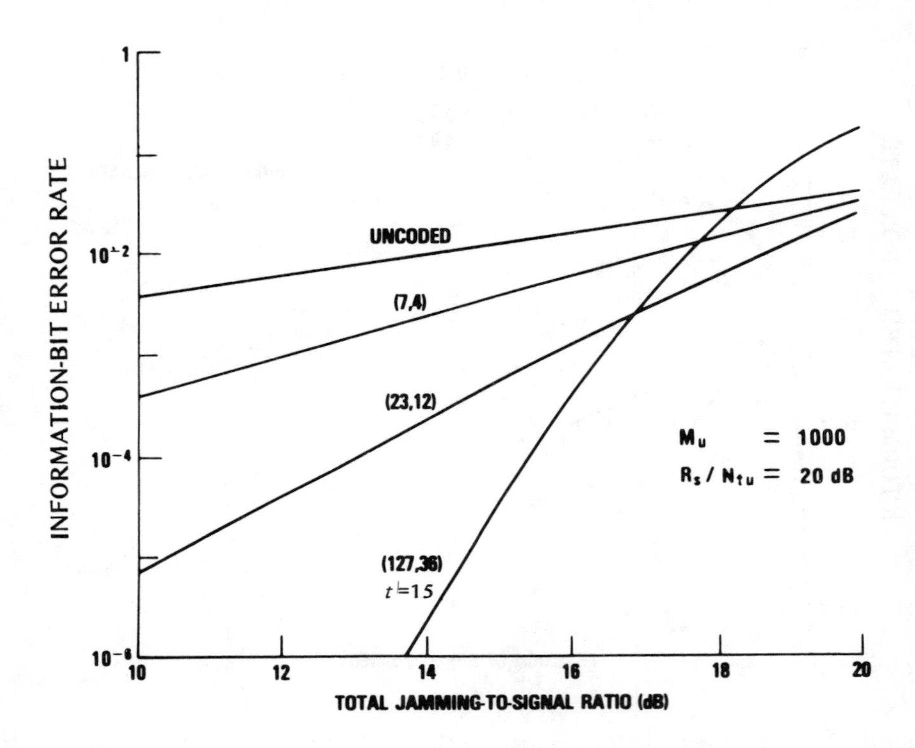

Figure 3.16 Worst-case performance for binary block (n, k) codes with hard decisions.

weight of the incorrect segment, and d_f is the minimum free distance. For hard-decision decoding and independent symbol errors, (1.7.37) gives

$$Q(\delta) = \begin{cases} \displaystyle\sum_{i=(\delta+1)/2}^{\delta} \binom{\delta}{i} P_s^i (1 - P_s)^{\delta-i}, & \delta \text{ is odd} \\ \displaystyle\sum_{i=\delta/2+1}^{\delta} \binom{\delta}{i} P_s^i (1 - P_s)^{\delta-i} + \frac{1}{2} \binom{\delta}{\delta/2} [P_s(1 - P_s)]^{\delta/2}, & \delta \text{ is even} \end{cases} \tag{3.3.12}$$

The values of $A(d_f + i)$, $0 \leqslant i \leqslant 7$, for rate-1/2, rate-1/3, and rate-1/4 binary convolutional codes with favorable distance properties are listed in Tables 1-2 to 1-4 of Section 1.7. In all the subsequent numerical examples, the series in (3.3.11) is truncated after the first eight terms.

Figures 3.19 and 3.20 depict the worst-case performance of an FH/MFSK system with rate-1/2 and rate-1/3 convolutional codes, respectively. Equations (3.3.6) to (3.3.9), (3.3.11) and (3.3.12) are used with $k = 1, M_u = 1000$, and $R_s/N_{tu} = 20$ dB. Codes of moderate constraint lengths are clearly preferable to the Reed-Solomon codes in terms of the worst-case performance that can be achieved for a given level of receiver complexity.

3.3.3 Concatenated Codes

An FH/MFSK system with two-level concatenated coding has the form shown in Figure 3.21. A fast frequency-hopping system does not require the inner interleaver and the inner deinterleaver. For slow frequency hopping, they ensure the random distribution of errors at the input of the inner decoder.

Compact expressions can sometimes be derived for P_{ib}, the information-bit error rate at the output of the outer decoder. If the inner code is a binary block

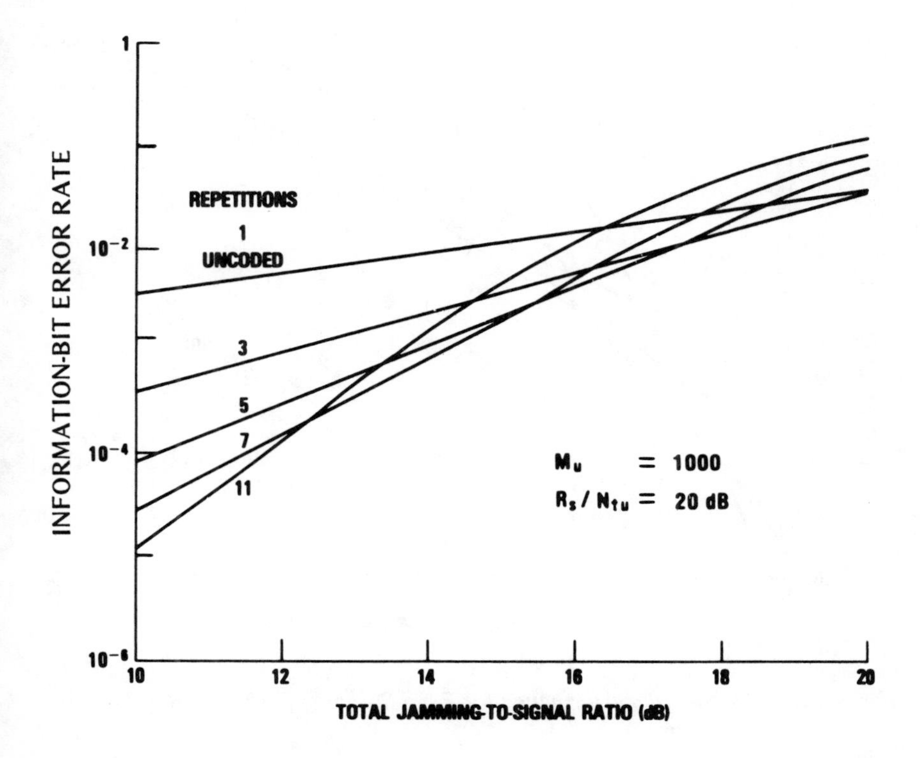

Figure 3.17 Worst-case performance for repetition codes with hard decisions.

code with minimum distance $d_1 = 2t_1 + 1$, the average bit error probability at the output of the inner decoder, P_{b1}, is determined for hard decisions from the right-hand side of (3.3.1); that is,

$$P_{b1} \approx \sum_{i=t_1+1}^{2t_1+1} \frac{2t_1+1}{n_1} \binom{n_1}{i} P_s^i (1-P_s)^{n_1-i} + \sum_{i=2t_1+2}^{n_1} \frac{i}{n_1} \binom{n_1}{i} P_s^i (1-P_s)^{n_1-i} \tag{3.3.13}$$

where P_s is the channel-symbol error probability given by (3.3.3) to (3.3.9). The ratio of information bits to channel symbols is

$$r = r_1 r_2 \tag{3.3.14}$$

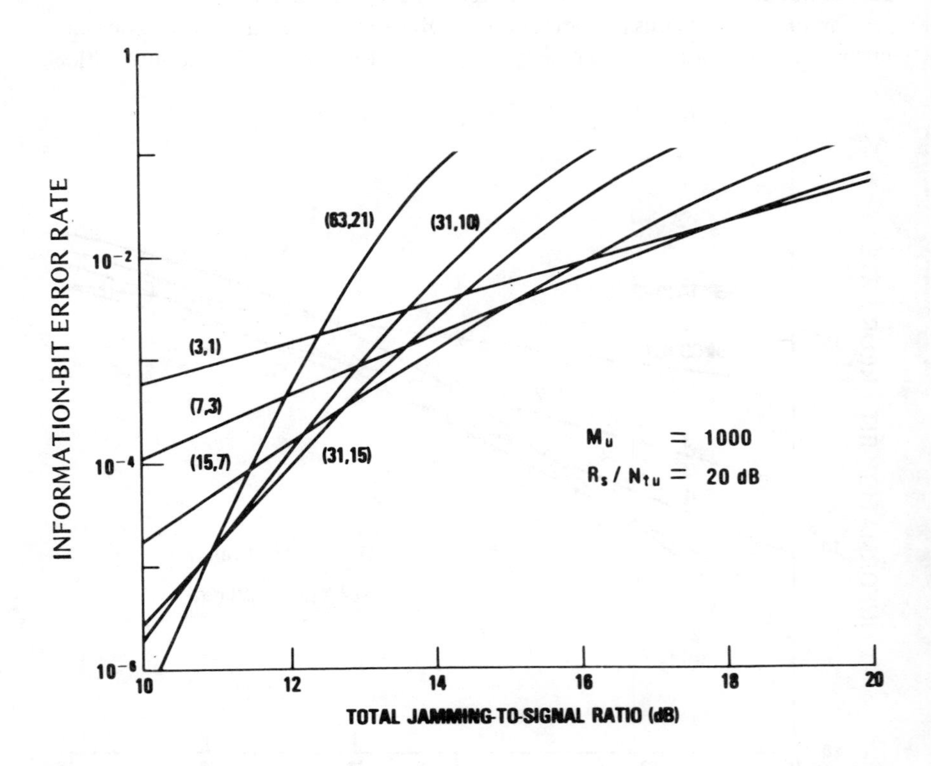

Figure 3.18 Worst-case performance for Reed-Solomon (n, k) codes with hard decisions.

where r_2 is the ratio of information bits to outer-code symbols and r_1 is the ratio of outer-code symbols to inner-code symbols. The outer deinterleaver provides independent errors at the input of the outer decoder. Thus, if the outer code is a binary block code, then P_{ib} is approximated by (3.3.1) with P_{b1} substituted in place of P_s and d and n referring to the outer code. If the outer code is a binary convolutional code, then an approximate upper bound on P_{ib} is determined from (3.3.11) and (3.3.12) with P_{b1} replacing P_s.

As specific examples, Figures 3.22 and 3.23 depict the worst-case performance against partial-band jamming of a system with a repetition code as the inner code, M_u = 1000, and R_s/N_{tu} = 20 dB. In Figure 3.22, the outer code is a Golay (23,12) code; in Figure 3.23, the outer code is a convolutional code of constraint length 7 and rate 1/2. The figures illustrate that the degree to which the repetition code improves or degrades the performance of the outer code

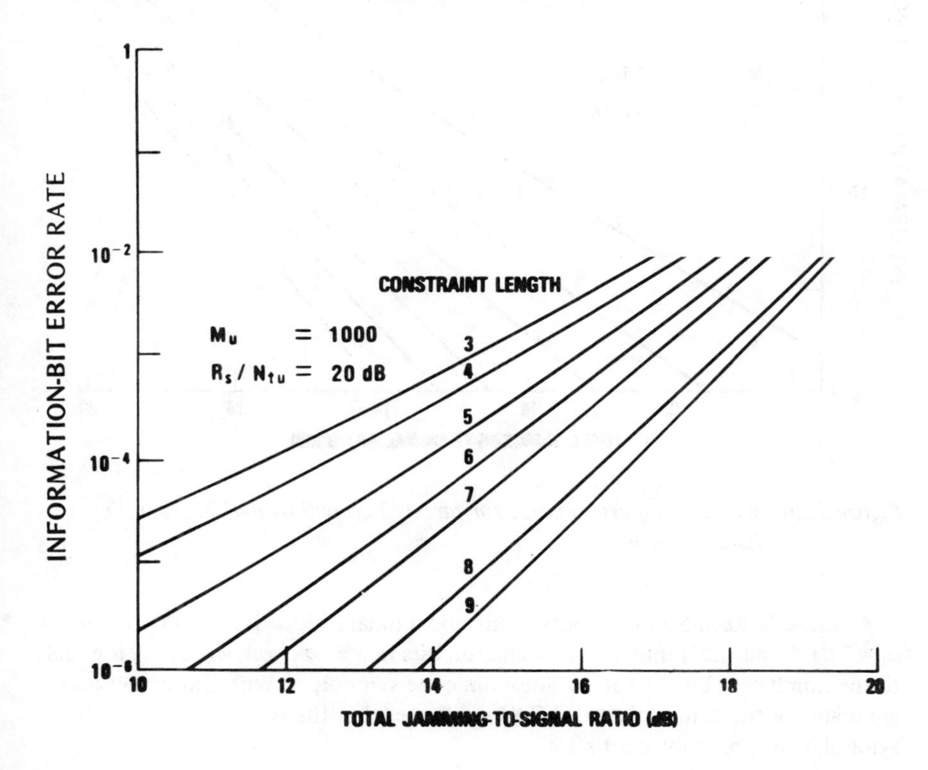

Figure 3.19 Worst-case performance for rate-1/2 convolutional codes with hard decisions.

alone varies with the required P_{ib}. In the system implementation, the outer interleaver and the outer deinterleaver are unnecessary. When a repetition code is used, the communication system is often regarded as a system with a single code (the outer code) and time or frequency diversity.

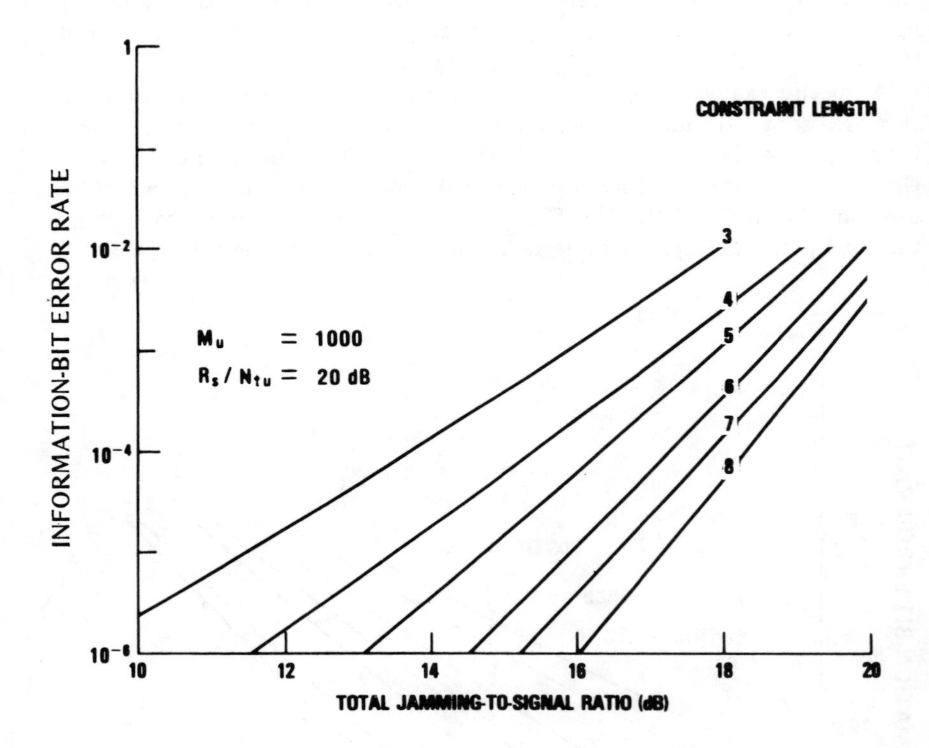

Figure 3.20 Worst-case performance for rate-1/3 convolutional codes with hard decisions.

Consider a Reed-Solomon outer code and a binary block inner code for which $d_1 = 2t_1 + 1$ and the number of information bits in a block codeword, k_1, is equal to the number of bits in a Reed-Solomon code symbol, m. When hard decisions are made by the demodulator, (1.7.84) indicates that the Reed-Solomon code-symbol error probability satisfies

$$P_{s1} \leqslant \sum_{i=t_1+1}^{n_1} \binom{n_1}{i} P_s^i (1 - P_s)^{n_1 - i} \tag{3.3.15}$$

where equality holds for bounded-distance decoding. Because hard decisions are presented to the outer decoder, an approximate upper bound on P_{ib} is determined from (3.3.2) with P_{s1} substituted in place of P_s. Figure 3.24 shows examples of the worst-case performance, assuming that M_u = 1000 and R_s/N_{tu} = 20 dB.

Consider a Reed-Solomon outer code and a binary convolutional inner code using the Viterbi algorithm. Equation (1.7.87) gives the error probability for a Reed-Solomon symbol:

$$P_{s1} \cong 2 \sum_{i=0}^{\infty} A(d_f + i)\,Q(d_f + i) \tag{3.3.16}$$

Hard-decision decoding and ideal symbol interleaving provide a P_{ib} that is approximately determined by (3.3.2) with P_{s1} substituted in place of P_s. Figure 3.25 depicts examples of the worst-case performance for M_u = 1000, R_s/N_{tu} = 20 dB, and an inner convolutional code with K = 7 and rate = 1/2.

Because only the inner code is directly affected by the jamming, soft-decision decoding in the outer decoder is a plausible strategy; it is possible if the inner decoder produces symbol metrics. For a practical example, consider a system with binary modulation, a repetition code of length n_1 as the inner code, and a convolutional code as the outer code. The demodulator produces a series of 1's and 0's. After the deinterleaving, the inner decoder counts n_2, the number of 1's in each group of n_1 inner-code symbols representing an outer-code symbol. The symbol metric associated with an outer code 1 is n_2, whereas $n_1 - n_2$ is associated with an outer-code 0. The outer decoder forms metrics that are sums of the symbol metrics. The selection of the largest of the metrics determines the decoded output.

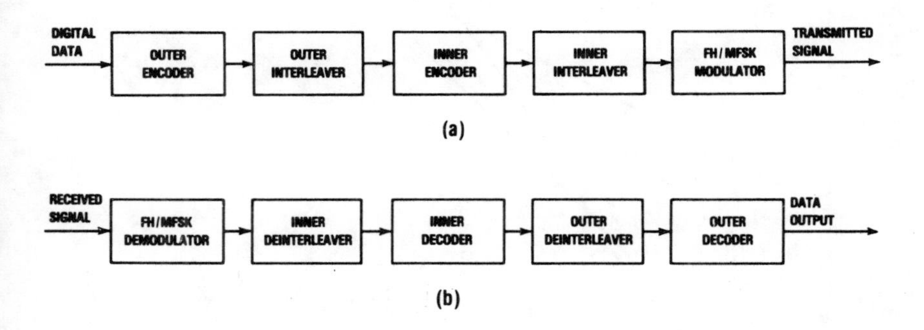

Figure 3.21 FH/MFSK system with concatenated coding: (a) transmitter and (b) receiver.

Let $m_1(\nu, i)$ denote a symbol metric determined from code symbol i. Let $\nu = 1$ label a correct sequence and $\nu = 2$ label an incorrect one. If there are j errors in the n_1 bits representing outer-code symbol i, then $m_1(2,i) = j$, $m_1(1,i) = n_1 - j$, and the difference between the symbol metric corresponding to the incorrect symbol and the one corresponding to the correct symbol is

$$m_1(2,i) - m_1(1,i) = 2j - n_1 \tag{3.3.17}$$

A direct application of (1.7.70) yields

$$\begin{aligned} Z &= \min_{0<s} \sum_{j=0}^{n_1} \binom{n_1}{j} P_s^j (1 - P_s)^{n_1 - j} \exp[s(2j - n_1)] \\ &= \min_{0<s} e^{-sn_1}(1 - P_s + P_s e^{2s})^{n_1} \\ &= [4P_s(1 - P_s)]^{n_1/2} \end{aligned} \tag{3.3.18}$$

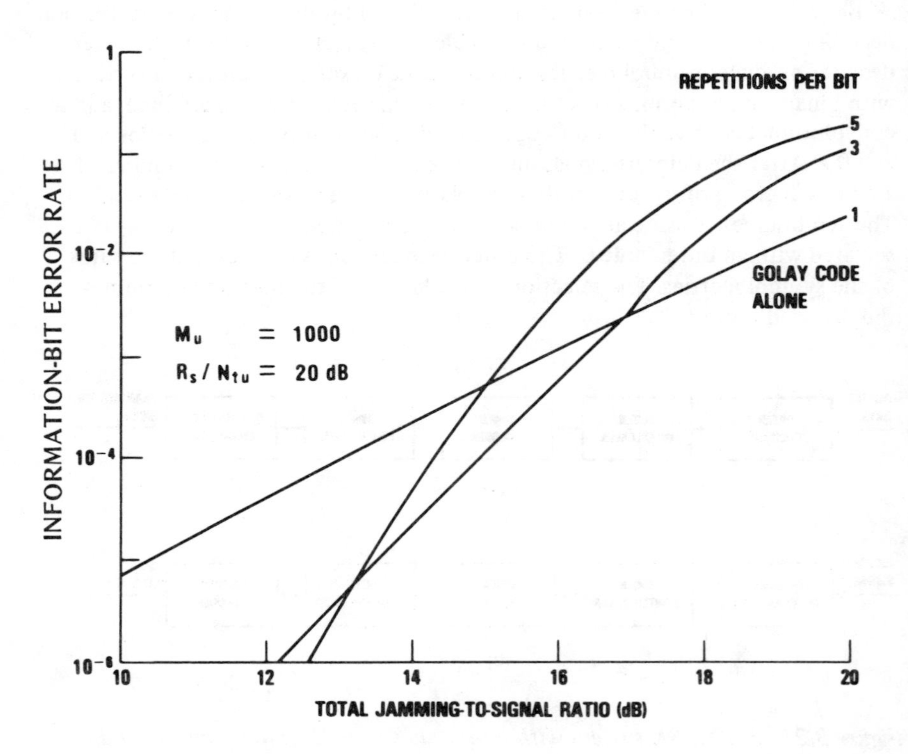

Figure 3.22 Worst-case performance for concatenated codes with outer Golay (23,12) code, inner repetition code, and hard decisions.

where $P_s = P_{s2}$ is determined from (3.3.4) to (3.3.9). If the outer code is a rate-r_2 binary convolutional code, (3.3.14) implies that $r = r_2/n_1$. Combining (3.3.11) and (1.7.71) and setting $k = 1$ and $\alpha = 1/2$ yields

$$P_{ib} \leqslant \frac{1}{2} \sum_{i=0}^{\infty} A(d_f + i) Z^{d_f + i} \tag{3.3.19}$$

Figure 3.26 illustrates the worst-case performance against partial-band jamming of an FH/MFSK system having an outer convolutional code with $K = 7$ and $r_2 = 1/2$, an inner repetition code with $n_1 = 2, 3, 4,$ or 5, $M_u = 1000$ and $R_s/N_{tu} = 20$ dB. For a required $P_{ib} = 10^{-5}$, the optimal choice for n_1 is 3. Comparing Figures 3.23 and 3.26 indicates that for $n_1 = 3$, a soft-decision outer decoder requires a signal power of approximately 1.3 dB less than a hard-decision one does to provide $P_{ib} = 10^{-5}$. Because of the error in the Chernoff bound, it is probable that the actual advantage of the soft-decision outer decoder is close to 2 dB.

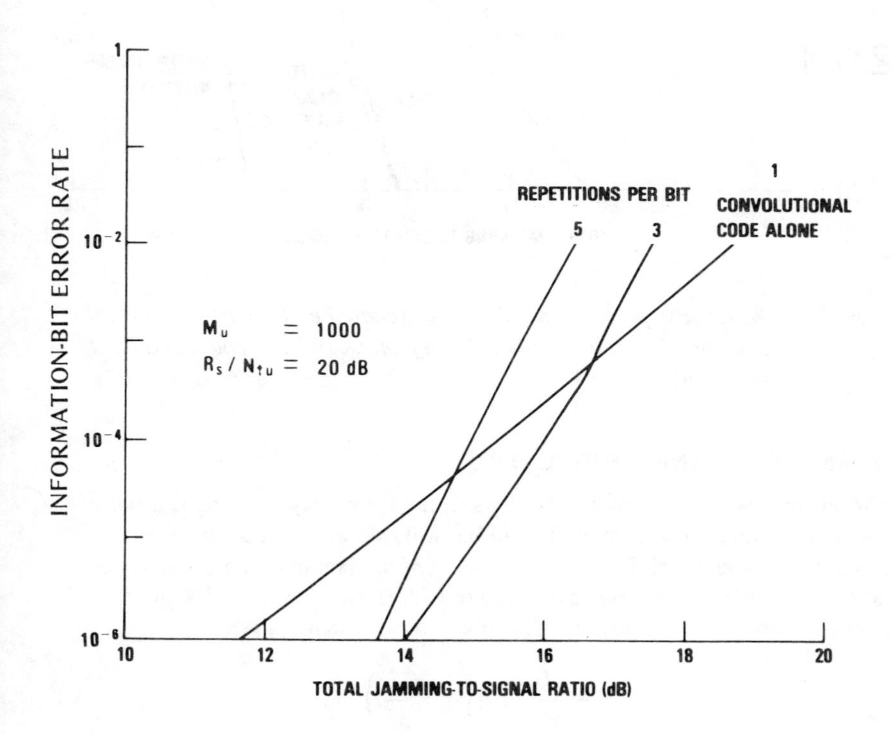

Figure 3.23 Worst-case performance for concatenated codes with outer convolutional code (K = 7, rate = 1/2), inner repetition code, and hard decisions.

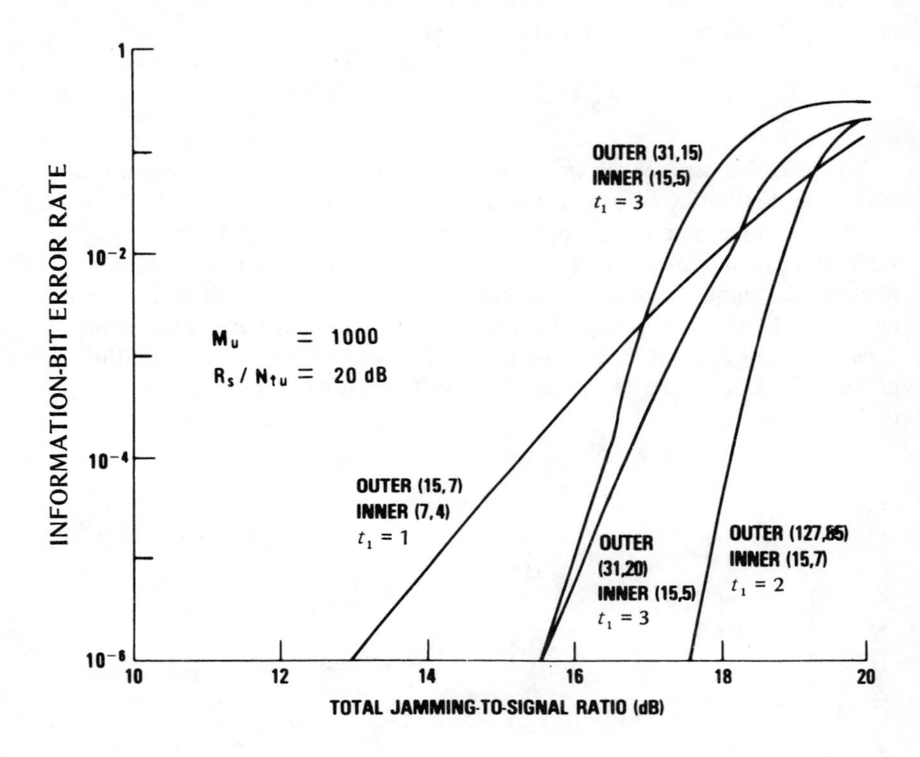

Figure 3.24 *Worst-case performance for concatenated codes with outer Reed-Solomon (n, k) code, inner binary block (n_1, k_1) code, and hard decisions.*

3.4 SINGLE-CHANNEL MODULATION

When single-channel modulation is used, the frequency hopping is usually slow. The symbol error probability can be easily obtained since there is no complementary channel. If J channels out of M are jammed, the probability that the transmission channel is jammed is J/M. If each jammed channel receives the same jamming power, the probability of a symbol error is

$$P_S = \frac{J}{M} S_1 + \left(1 - \frac{J}{M}\right) S_0 \tag{3.4.1}$$

where S_1 is the probability of a symbol error given that the transmission channel is jammed, and S_0 is the probability of a symbol error given that the channel is not jammed.

Consider a single-channel slow frequency-hopping system with interleaving and block coding. If a frequency channel is randomly selected for each transmitted symbol of a codeword of length n, then the word error probability satisfies

$$P_W \leqslant \sum_{i=t+1}^{n} \binom{n}{i} \left[\frac{J}{M} S_1 + \left(1 - \frac{J}{M}\right) S_0 \right]^m \left[1 - \frac{J}{M} S_1 - \left(1 - \frac{J}{M}\right) S_0 \right]^{n-m} \tag{3.4.2}$$

Equality holds for bounded-distance decoding.

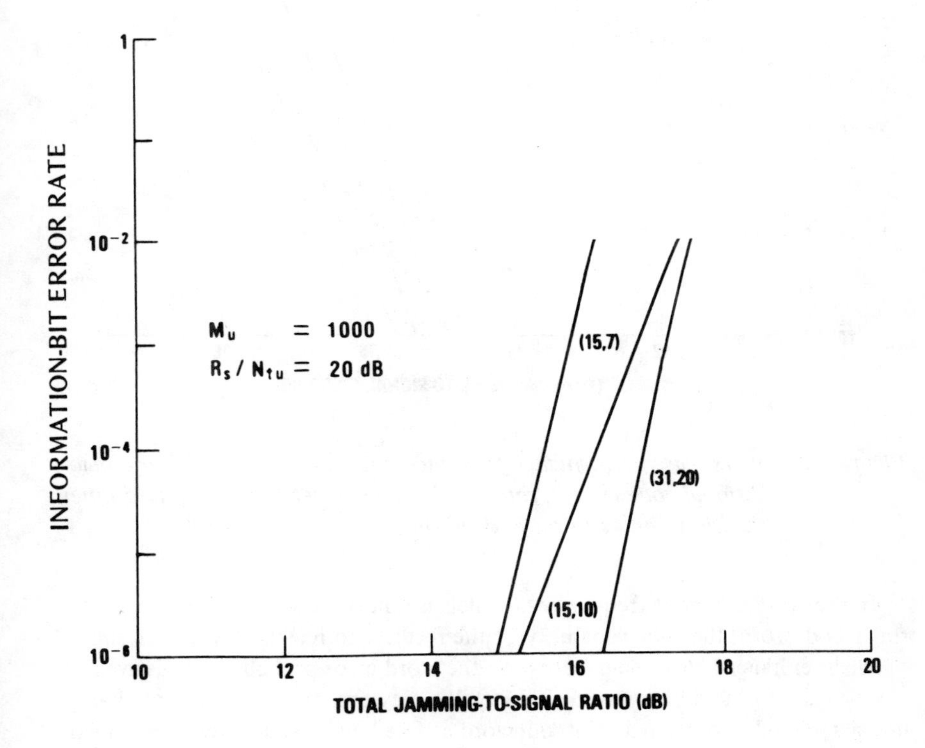

Figure 3.25 Worst-case performance for concatenated codes with outer Reed-Solomon (n, k) code, inner convolutional code (K = 7, rate = 1/2), and hard decisions.

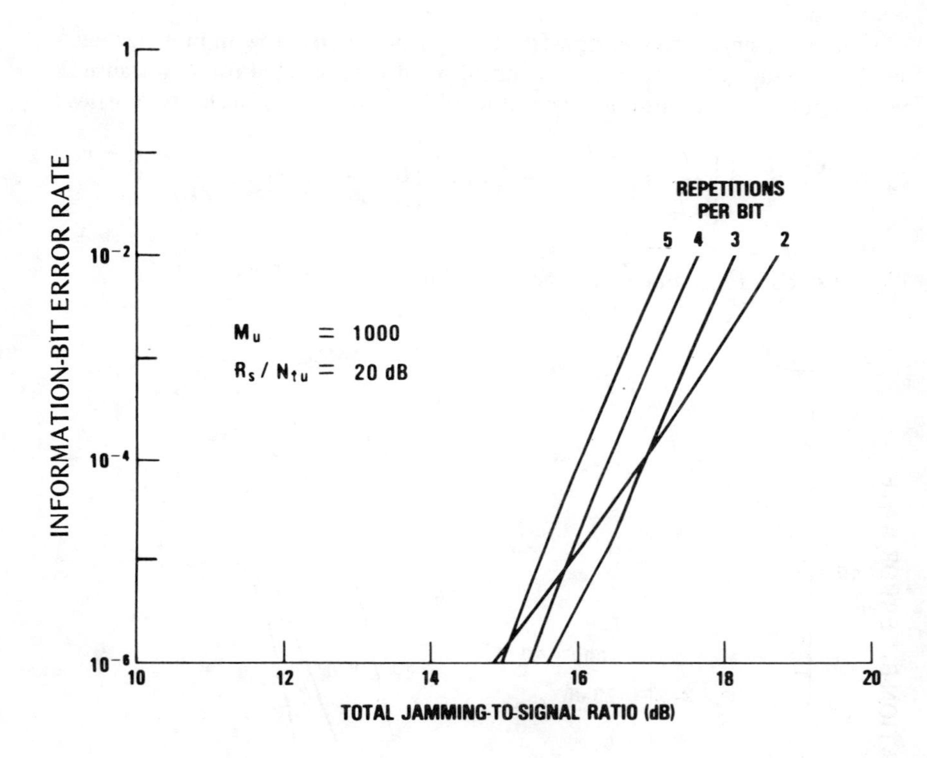

Figure 3.26 Worst-case performance for concatenated codes with outer convolutional code (K = 7, rate = 1/2), inner repetition code, hard inner decisions, and soft outer decisions.

If a set of n different frequency channels is randomly selected for transmitting a codeword, then the capability of the receiver to reject a few interfering signals is enhanced. An upper bound on the word error probability is derived analogously to the derivation of (3.2.29). We may regard the n different channels associated with a particular codeword as fixed and the interference as introduced into J randomly chosen channels. Alternatively, we may consider the J jammed channels as fixed and the set of n different transmission channels as randomly chosen for each codeword. In either case, we derive

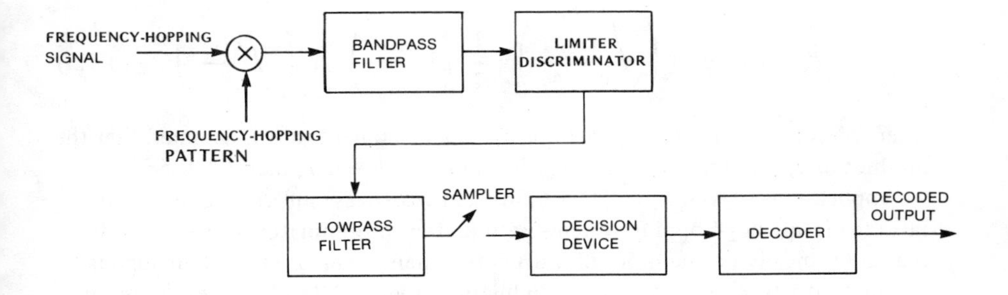

Figure 3.27 FH/CPFSK receiver.

$$P_W \leqslant \sum_{i=t+1}^{n} \sum_{j=j_0}^{j_1} \sum_{\alpha=\alpha_0}^{\alpha_1} \frac{\binom{J}{j}\binom{M-J}{n-j}\binom{n-j}{i-\alpha}\binom{j}{\alpha}}{\binom{M}{n}}$$

$$\times \left[S_0^{i-\alpha} S_1^{\alpha} (1-S_0)^{n-j-i+\alpha} (1-S_1)^{j-\alpha}\right] \tag{3.4.3}$$

where equality holds for bounded-distance decoding and

$$j_0 = \max(0, n+J-M), \quad \alpha_0 = \max(0, i-n+j),$$
$$j_1 = \min(n, J), \quad \alpha_1 = \min(i, J) \tag{3.4.4}$$

Slow frequency-hopping systems with *continuous-phase frequency-shift keying* (CPFSK) offer reduced spectral splatter (Section 3.6) and a potentially excellent performance. During the time between hops, a CPFSK signal changes frequency while maintaining a continuous phase function. We consider a binary CPFSK signal that shifts between two frequencies separated by h/T_s, where h is the deviation ratio and T_s is the channel-symbol duration. Because of the difficulties of coherent demodulation in a frequency-hopping system, the CPFSK waveform is generally demodulated by a limiter-discriminator, but differential demodulation is also possible. Although the frequency of a CPFSK signal varies, its compact spectrum allows it to be considered a single-channel modulation for which there is only a single carrier frequency between hops. The basic form of an FH/CPFSK receiver with hard-decision decoding is shown in Figure 3.27.

From (3.4.1) and (1.5.25), it follows that the channel-symbol error probability for an FH/CPFSK system in the presence of partial-band noise interference is

$$P_s \cong \frac{J}{2M} \operatorname{erfc}\left(\sqrt{\frac{\xi R_s}{N_t + N_i}}\right) + \frac{1}{2}\left(1 - \frac{J}{M}\right) \operatorname{erfc}\left(\sqrt{\frac{\xi R_s}{N_t}}\right) \qquad (3.4.5)$$

where N_i is the interference power per jammed channel and it is assumed that the product of T_s and the noise bandwidth is unity. If $h \cong 0.7$, then $\xi \cong 0.56$.

Applications of (3.4.5) with $\xi = 0.56$ yield worst-case information-bit error rates for frequency-hopping systems with limiter-discriminator demodulation that are typically on the order of 4 dB better than the error rates of analogous systems with envelope-detector demodulation. Figure 3.28 illustrates the potentially improved performance against worst-case partial-band jamming when the

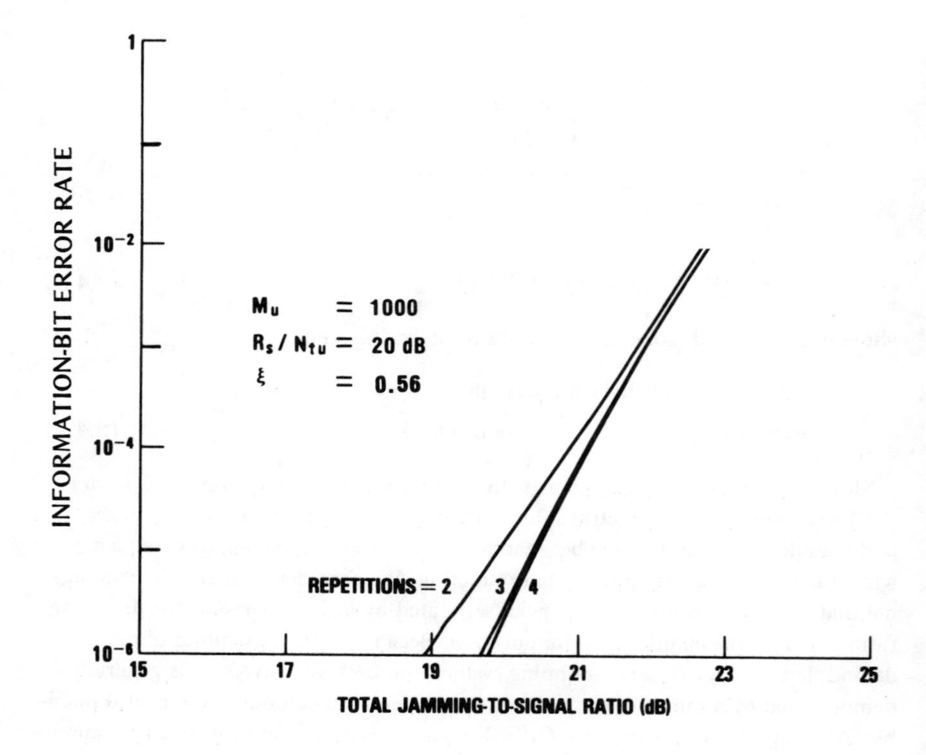

Figure 3.28 Worst-case performance for limiter-discriminator demodulation and concatenated codes with outer convolutional code (K = 7, rate = 1/2), inner repetition code, hard inner decisions, and soft outer decisions.

concatenated codes of Figure 3.26 are used with CPFSK and limiter-discriminator demodulation.

The potential performance of an FH/CPFSK system is significantly degraded in practice by the effects of the phase discontinuities that occur during the frequency-hopping transitions in the receiver. Because the bandwidth of the bandpass filter in Figure 3.27 is on the order of $1/T_s$, it is plausible that the effect of a phase discontinuity on the limiter-discriminator output will be significant for roughly T_s seconds. Let V denote the number of CPFSK symbols occurring during the dwell time. Suppose that the symbols do not occupy the first T_s seconds after a frequency-hopping transition, but occupy the fraction $V/(V+1)$ of the dwell time. Although the effects of the phase discontinuities are greatly mitigated, the loss of energy per channel symbol causes a performance degradation of $10 \log_{10}[V/(V+1)]$ in decibels. For $V \geqslant 2$, the loss is less than 1.8 dB; for $V \geqslant 7$, the loss is less than 0.6 dB. Thus, FH/CPFSK systems are expected to provide a net performance advantage of roughly 2 to 4 dB relative to analogous frequency-hopping systems with envelope detectors.

Because phase coherence between the transmitter and receiver can usually be maintained only at fixed frequencies, fast frequency hopping with *differential phase-shift keying* (DPSK) requires the grouping of received symbols that were transmitted at the same frequency; thus, it is seldom a useful method for secure communications. For slow frequency hopping with DPSK, the usual lack of phase coherence from hop to hop necessitates an extra phase-reference symbol every hop duration, which causes a performance degradation. Although slow frequency hopping with DPSK is practical, it is not as attractive as FH/CPFSK because of the spectral efficiency of the CPFSK modulation, which is valuable for multiple-access communications (Section 3.6).

3.5 SOFT-DECISION DECODING

The effect of partial-band jamming on soft-decision decoding in an FH/MFSK system is similar to the effect of pulsed jamming on soft-decision decoding in a fixed-frequency or a direct-sequence system. If perfect measurements of the jamming power level could be made, then soft-decision decoding would give considerable protection against worst-case partial-band jamming. At the other extreme, if no measurements of the jamming power or desired-signal power are made, the usual soft-decision metrics perform poorly. Metrics with clipping improve the performance, but a highly effective implementation requires an accurate measurement of the signal power, which is often impractical. Although a practical soft-decision metric using imperfect measurements may be able to provide a performance that is significantly superior to that attainable with hard-decision decoding, it will probably necessitate a large increase in the implementation complexity.

We consider an FH/MFSK system that uses a repetition code and noncoherent demodulation with envelope detectors (Figure 1.19). Each information symbol is transmitted as n code symbols. The jamming is modeled as noise jamming. Fast frequency hopping or slow hopping with ideal interleaving are assumed to ensure the independence of code-symbol errors.

To accommodate Rayleigh fading, the usual metric is

$$m_0(\nu, n) = \sum_{l=1}^{n} R_{\nu l}^2, \qquad \nu = 1, 2, \ldots, N \tag{3.5.1}$$

where $R_{\nu l}$ is the sample value of the envelope-detector output that is associated with symbol l of candidate sequence ν. This metric has the advantage that no power measurements are required for its implementation. A performance analysis of a frequency-hopping system with binary FSK, repetition coding, and soft-decision decoding with this *Rayleigh metric* indicates that it is a poor choice for use against worst-case partial-band jamming [4]. Furthermore, the repetition code is counterproductive.

If the thermal-noise and jamming powers are known, a better choice of metric is

$$m_0(\nu, n) = \sum_{l=1}^{n} \frac{R_{\nu l}^2}{N_t + N_{il}}, \qquad \nu = 1, 2, \ldots, N \tag{3.5.2}$$

where N_{il} is the interference (jamming) power in sample output l. To give a simplified demonstration of the effectiveness of this metric, it is assumed that $N_t = 0$ and that either all N of the MFSK frequency channels are jammed or none of them are [5]. When an MFSK symbol is jammed, $N_{il} = N_i$; when it is not, $N_{il} = 0$.

The union bound, given by (1.7.28), implies that the probability of an information-symbol error satisfies

$$P_{is} \leqslant (N - 1) P_w(2) \tag{3.5.3}$$

where $P_w(2)$ is the probability of an error in comparing the metrics associated with two n-symbol codewords. Using (1.7.21) and $N = 2^m$, we find that the probability of an information-bit error satisfies

$$P_{ib} \leqslant 2^{m-1} P_w(2) \tag{3.5.4}$$

where m is the number of bits represented by a symbol. As $N_t \rightarrow 0$, each term of (3.5.2) for which $N_{il} = 0$ tends to infinity. Therefore, an error can occur only if all n code symbols are jammed, an event that occurs with probability μ^n. When all code symbols are jammed, a symbol metric is $m_1(\nu, l) = R_{\nu l}^2/N_i$. The Chernoff bound of (1.7.71) with $\alpha = 1/2$ and the independence of symbol errors imply that

$$P_w(2) \leqslant \frac{\mu^n}{2} Z^n \qquad (3.5.5)$$

where Z is given by (1.7.70). Thus,

$$P_{ib} \leqslant 2^{m-2} \mu^n Z^n \qquad (3.5.6)$$

Following the derivation of (1.7.81) and (1.7.82), we obtain

$$Z = \frac{\exp\left[-\left(\frac{\lambda}{1+\lambda}\right)\left(\frac{R_s}{N_i}\right)\right]}{1 - \lambda^2} \qquad (3.5.7)$$

where

$$\lambda = -\left(\frac{1}{2} + \frac{R_s}{4N_i}\right) + \left[\left(\frac{1}{2} + \frac{R_s}{4N_i}\right)^2 + \frac{R_s}{2N_i}\right]^{1/2} \qquad (3.5.8)$$

The number of information bits per code symbol is

$$r = \frac{m}{n} \qquad (3.5.9)$$

The jamming power per frequency channel in the absence of the repetition coding is

$$N_{ju} = \frac{N_{jt}}{M_u} \qquad (3.5.10)$$

Equations (3.5.10), (3.5.9), (3.3.9), (3.3.8), and (3.3.6) imply that

$$\frac{R_s}{N_i} \cong \frac{\mu m R_s}{n N_{ju}} \qquad (3.5.11)$$

If $B_u T_{ib} \cong 1$, then

$$\frac{R_s}{N_{ju}} \cong \frac{E_b}{N_{0j}} \qquad (3.5.12)$$

as indicated by (3.3.10) and (3.5.10).

An upper bound on P_{ib} for worst-case partial-band jamming is obtained by maximizing the right-hand side of (3.5.6) with respect to μ, where $0 \leqslant \mu \leqslant 1$. Substituting (3.5.7), (3.5.8), and (3.5.11), we find that the value of μ that maximizes the upper bound is

$$\mu_0 = \begin{cases} \dfrac{3n}{m}\left(\dfrac{R_s}{N_{ju}}\right)^{-1}, & \dfrac{R_s}{N_{ju}} \geqslant \dfrac{3n}{m} \\ 1, & \dfrac{R_s}{N_{ju}} < \dfrac{3n}{m} \end{cases} \tag{3.5.13}$$

The upper bound on P_{ib} for worst-case partial-band jamming is given by

$$P_{ib} \leqslant \begin{cases} 2^{m-2}\left[\dfrac{4n}{me}\left(\dfrac{R_s}{N_{ju}}\right)^{-1}\right]^n, & \dfrac{R_s}{N_{ju}} \geqslant \dfrac{3n}{m} \\ 2^{m-2}(1-\lambda_0^2)^{-n}\exp\left[-\left(\dfrac{m\lambda_0}{1+\lambda_0}\right)\dfrac{R_s}{N_{ju}}\right], & \dfrac{R_s}{N_{ju}} < \dfrac{3n}{m} \end{cases} \tag{3.5.14}$$

where

$$\lambda_0 = -\left(\frac{1}{2}+\frac{\gamma}{4}\right) + \left[\left(\frac{1}{2}+\frac{\gamma}{4}\right)^2 + \frac{\gamma}{2}\right]^{1/2}, \quad \gamma = \frac{mR_s}{nN_{ju}} \tag{3.5.15}$$

Since μ_0 is obtained by maximizing a bound rather than an equality, it is not necessarily equal to the actual worst-case μ.

If R_s/N_{ju} is known, then n can be chosen to minimize the upper bound on P_{ib} for worst-case partial-band jamming, which occurs if the jammer is able to set $\mu = \mu_0$. We treat n as a continuous variable such that $n \geqslant 1$. Because the derivative with respect to n of the second line on the right-hand side of (3.5.14) is positive, the minimizing value of n is determined from the first line. The value of n that minimizes the upper bound on P_{ib} is found to be

$$n_0 = \begin{cases} \dfrac{mR_s}{4N_{ju}}, & \dfrac{R_s}{N_{ju}} \geqslant \dfrac{4}{m} \\ 1, & \dfrac{R_s}{N_{ju}} < \dfrac{4}{m} \end{cases} \tag{3.5.16}$$

The upper bound on P_{ib} for worst-case partial-band jamming when $n = n_0$ is given by

$$P_{ib} \leqslant \begin{cases} 2^{m-2} \exp\left(-\dfrac{mR_s}{4N_{ju}}\right), & \dfrac{R_s}{N_{ju}} \geqslant \dfrac{4}{m} \\ \\ \dfrac{2^m}{me}\left(\dfrac{R_s}{N_{ju}}\right)^{-1}, & \dfrac{R_s}{N_{ju}} < \dfrac{4}{m} \end{cases} \tag{3.5.17}$$

This result shows that P_{ib} decreases exponentially as R_s/N_{ju} increases if the appropriate number of repetitions is chosen and R_s/N_{ju} is large enough. Substituting (3.5.16) into (3.5.13), we obtain

$$\mu_0 = \begin{cases} \dfrac{3}{4}, & \dfrac{R_s}{N_{ju}} \geqslant \dfrac{4}{m} \\ \dfrac{3}{m}\left(\dfrac{R_s}{N_{ju}}\right)^{-1}, & \dfrac{3}{m} < \dfrac{R_s}{N_{ju}} < \dfrac{4}{m} \\ 1, & \dfrac{R_s}{N_{ju}} \leqslant \dfrac{3}{m} \end{cases} \tag{3.5.18}$$

This result shows that the appropriate choice of n forces the jammer to attempt to jam over approximately three-fourths or more of the total hopping band.

If the jamming power is spread uniformly over the total hopping band, then a repetition code is not useful. Consider an uncoded FH/MFSK system with $n = 1$. If $\mu = 1$, then $P_w(2) = S_2$, which is given by (3.2.17). Since we have assumed that $N_t = 0$, (3.5.4), (3.5.11), (3.2.17), and $\mu = n = 1$ imply that

$$P_{ib} \leqslant 2^{m-2} \exp\left(-\frac{mR_s}{2N_{ju}}\right) \tag{3.5.19}$$

A comparison of (3.5.19) and (3.5.17) for $R_s/N_{ju} \geqslant 4/m$ indicates that the performance degradation caused by worst-case partial-band jamming can be limited to 3 dB if the optimal repetition code is added to the FH/MFSK system. An extension of the derivation of (3.5.17) indicates that when concatenated codes with inner repetition codes are used, this degradation can be decreased or eliminated [6].

For frequency hopping with binary FSK and repetition coding, more precise calculations [7] that do not use the Chernoff bound and allow $N_t > 0$ confirm that the degradation caused by worst-case partial-band jamming is approximately 3 dB at $P_{ib} = 10^{-5}$ when N_t is small. Using the results of Figure 3.16, it is found that soft-decision decoding with the metric of (3.5.2) and an optimal number of repetitions provides more than a 3 dB advantage at $P_{ib} = 10^{-5}$ relative to hard-decision decoding.

The implementation of (3.5.2) requires the measurement of the jamming power. One might attempt to measure this power in frequency channels immediately before the hopping of the signal into those channels. However, this method will not be reliable if the jamming is frequency hopping or nonstationary. Another approach is to use an *automatic-gain-control* (AGC) *device* to measure the average power in the currently used frequency channels. The output of the AGC device is used to form the *AGC metric.* Because the signal power contaminates the power measurement, the AGC metric can be expressed as

$$m_0(\nu, n) = \sum_{l=1}^{n} \frac{R_{\nu l}^2}{N_t + N_{il} + R_s / N} \qquad (3.5.20)$$

The practical implementation problems for this metric are discussed in Section 1.7. It is expected that the AGC metric potentially provides a performance only slightly inferior to that provided by the metric of (3.5.2), but the implementation complexity and losses may be substantial.

A major advantage of frequency-hopping systems relative to direct-sequence systems is that it is possible to hop in frequency over a much greater band than can be occupied by a direct-sequence signal. This advantage more than compensates for the relatively inefficient noncoherent demodulation that is almost always required for frequency-hopping systems. Another advantage is the option of excluding frequency channels with steady or frequent interference. Repeater jamming is not a threat against a practical direct-sequence system because of the shortness of the chips. It is not a threat against a frequency-hopping system if the dwell time is kept sufficiently short.

Frequency-hopping systems reject interference by avoiding it whereas direct-sequence systems reject interference by spreading it. However, the effect of partial-band interference on a frequency-hopping system is similar to the effect of high-power pulsed interference on a direct-sequence system. The interleaving and error-correcting codes that are effective against one of these types of interference are effective against the other. Error-correcting codes are more essential for frequency-hopping systems than for direct-sequence systems because partial-band interference is a more pervasive threat than high-power pulsed interference.

3.6 FREQUENCY-HOPPING MULTIPLE-ACCESS NETWORKS

Frequency-hopping communicators do not often operate in isolation. Instead, they are usually elements of a network of frequency-hopping systems that cause mutual multiple-access interference. This network is called a *frequency-hopping multiple-access* (FHMA) *network.*

If the hoppers of an FHMA network all use the same M frequency channels, but coordinate their frequency transitions and their hopping sequences, then the multiple-access interference can be greatly reduced. In the absence of spectral splatter, as many as M frequency-hopping signals can be simultaneously accommodated by the network with no multiple-access interference at any of the active receivers. Network coordination is much simpler to implement than for a CDMA network of direct-sequence systems because the timing alignments must be within a fraction of a hop duration, rather than a fraction of a sequence chip. Errors in range estimates can be accommodated at some cost in the energy per information bit by reducing the dwell time of frequency-hopping pulses relative to the hop duration. However, since the operational complications of coordination are still significant, asynchronous FHMA networks are usually preferable. The effect of multiple-access interference on an asynchronous FHMA network is analyzed in this section.

Spectral splatter is the spectral overlap in extraneous frequency channels produced by a time-limited transmitted pulse. Whether or not spectral splatter is significant in causing errors in a network depends upon the deployment, the hopping rate, the frequency separation between channels, and the spectrum of the transmitted signals.

Since the transmitted symbol rate equals the hopping rate in fast frequency-hopping systems, the hopping rate strongly influences the transmitted spectrum and the number of available channels. In slow frequency-hopping systems, the hopping rate influences the spectrum through the *switching time* or *settling time,* which is defined as the duration of the part of the time interval between hops during which the frequency synthesizer is not operating plus any rise time or fall time not directly due to the data modulation. The nonzero switching time decreases the transmitted symbol duration, which in turn affects the transmitted spectrum. The switching time, T_{sw}, plus the dwell time, T_d, is equal to the hop duration, T_h.

If the total bandwidth over which hopping occurs is fixed, increasing the frequency separation between channels may reduce the effect of spectral splatter, but it also decreases the number of available channels. As a result, the rate at which network systems hop into the same channel increases, which may cancel the improvement due to the reduced splatter.

Assuming single-channel modulation, we derive equations for the channel-symbol error probability when the splatter is significant only in the two channels adjacent to the transmission channel. The generalization of the derivation to the case in which many channels are affected is straightforward but notationally complicated. A rough approximation of the symbol error probability for multiple-channel splatter is given subsequently.

We consider a network of independent frequency-hopping systems that have omnidirectional antennas, generate the same output power, share the same M contiguous frequency channels, are nearly stationary in location over a symbol duration, and use similar waveforms.

Consider the transmission of a symbol from a hopper at A to a receiver at B, as depicted in Figure 3.29. The distance between the two is D. The numbered light dots in the figure represent some of the N potentially interfering hoppers in the network of $N + 2$ total hoppers. In the subsequent analysis, we assume that each interferer uses at most one frequency channel during the symbol reception, which is true if T_{sw} exceeds the symbol duration, T_s. Results for networks with $T_{sw} \leqslant T_s$ can be derived analogously. The jammer in the figure represents any source of interference outside the network.

Let B_{jk} denote the event that j of the interferers use the transmission channel and each of k interferers uses one of the two channels adjacent to the transmission channel. The probability of B_{jk} is denoted by $P(B_{jk})$. The probability of a symbol error given B_{jk} is denoted by $P_s(j,k)$. Thus, the symbol error probability is

$$P_s = \sum_{j=0}^{N} \sum_{k=0}^{N-j} P(B_{jk}) P_s(j,k) \tag{3.6.1}$$

Let d represent the *duty factor,* which is defined as the probability that an interferer can potentially degrade the reception of a symbol. Thus, d is the product of the probability p_1 that an interferer is transmitting and the probability p_2 that a significant portion of the interferer's transmitted waveform occurs during the symbol interval. The probability p_2 is upper bounded by the probability that there is any overlap in time of the interference and the symbol interval. Because $T_{sw} > T_s$, it follows from elementary probability that $p_2 \leqslant (T_d + T_s)/T_h$. Let d_0 represent the hopper *duty cycle,* which is defined by $d_0 = T_d/T_h$. The preceding results imply that

$$d \leqslant p_1 \left(d_0 + \frac{T_s}{T_h} \right) \tag{3.6.2}$$

Because we assume that an interferer may transmit in any frequency channel with equal probability, the probability that power from an interferer enters the transmission channel is d/M. We assume that M is sufficiently large that we may neglect the fact that a channel at one of the ends of the total band has only one adjacent channel instead of two. Consequently, the probability that the power from an interferer enters one of the two adjacent channels is $2d/M$ and the probability that the power enters neither the transmission channel nor the adjacent channels is $(1 - 3d/M)$. Since there are $\binom{N}{j} \binom{N-j}{k}$ ways to select one set of j interferers and another set of k interferers when $j + k \leqslant N$,

$$P(B_{jk}) = \left(\frac{d}{M}\right)^j \left(\frac{2d}{M}\right)^k \left(1 - \frac{3d}{M}\right)^{N-j-k} \binom{N}{j} \binom{N-j}{k} \tag{3.6.3}$$

Equations (3.6.1) to (3.6.3) give P_s in terms of $P_s(j,k)$. To avoid evaluating multiple integrals of order greater than $L < N$, we determine lower and upper bounds on P_s. A truncation of (3.6.1) gives the lower bound

$$P_L = \sum_{\substack{j=0 \\ j+k \leq L}}^{N} \sum_{k=0}^{N-j} P(B_{jk})P_s(j,k) \tag{3.6.4}$$

If $N > L$, an upper bound on P_s is obtained from (3.6.1) by setting $P_s(j,k) = 1/2$ for $j + k > L$. The difference between the two bounds is one-half times $P(j + k > L)$, the probability that more than L interferers produce power in the transmission channel or the adjacent channels. The upper bound is given by

$$P_U = P_L + \frac{1}{2} P(j + k > L) \tag{3.6.5}$$

where

$$P(j + k > L) = \sum_{\substack{j=0 \\ j+k>L}}^{N} \sum_{k=0}^{N-j} P(B_{jk}) \tag{3.6.6}$$

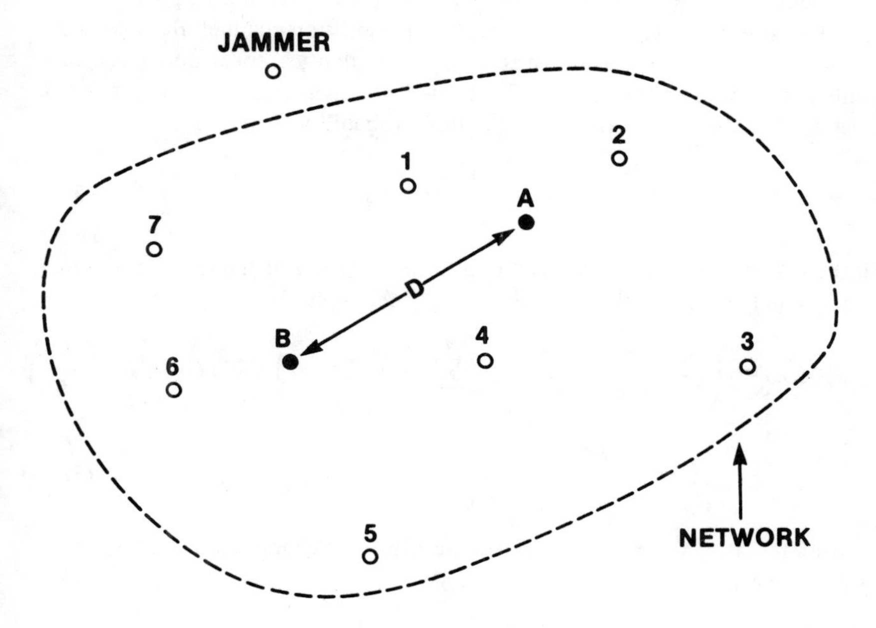

Figure 3.29 Frequency-hopping network and jammer.

As dN/M decreases, $P(j + k > L)$ decreases, and the upper and lower bounds become tighter.

If interferer i is using the transmission channel, the ratio of the power from interferer i to the power of the desired signal at the demodulator is denoted by x_i. If interferer i is using an adjacent channel, the ratio of the power from interferer i to the power of the desired signal at the demodulator is denoted by z_i. Let $P_1(x_1, \ldots, x_j, z_1, \ldots, z_k)$ denote the probability of a symbol error given that $x_1, x_2, \ldots, x_j, z_1, z_2, \ldots, z_k$ are the interference-to-signal ratios caused by j interferers using the transmission channel and k interferers using an adjacent channel. If the interfering signals are modeled as independent zero-mean processes, this probability is a function $P_0(\ )$ of the sum of the interference-to-signal ratios; therefore,

$$P_1(x_1, \ldots, x_j, z_1, \ldots, z_k) = P_0\left(\sum_{i=1}^{j} x_i + \sum_{i=1}^{k} z_i \right) \tag{3.6.7}$$

Let $f(u)$ denote the probability density function for an interference-to-signal ratio given that the interference enters the transmission channel. Let $f_1(u)$ denote the probability density function for an interference-to-signal ratio given that the interference enters an adjacent channel. We denote by K the ratio of the power due to an adjacent-channel interferer to the corresponding power that would exist if the interferer were using the transmission channel. Because each hopper in the network is assumed to produce an identical spectrum, K is a constant independent of the index i. Thus, $z_i = Kx_i'$, where x_i' has the density $f(u)$. Since $f_1(u)$ is the density for z_i, elementary probability theory gives

$$f_1(u) = \frac{1}{K} f\left(\frac{u}{K}\right) \tag{3.6.8}$$

Since each interferer is located and hops independently of the other interferers, (3.6.7) and (3.6.8) and the definition of $P_s(j,k)$ imply that

$$P_s(j,k) = \left(\frac{1}{K}\right)^k \int_0^\infty \cdots \int_0^\infty P_0\left(\sum_{i=1}^{j} x_i + \sum_{i=1}^{k} z_i \right) \prod_{i=1}^{j} f(x_i) \prod_{i=1}^{k} f\left(\frac{z_i}{K}\right)$$

$$\times\ dx_1 \ldots dx_j dz_1 \ldots dz_k \tag{3.6.9}$$

An alternative form of this equation results if we change variables to $y_i = z_i/K$. We obtain

$$P_s(j,k) = \int_0^\infty \cdots \int_0^\infty P_0\left(\sum_{i=1}^{j} x_i + K \sum_{i=1}^{k} y_i\right) \prod_{i=1}^{j} f(x_i) \prod_{i=1}^{k} f(y_i)$$

$$\times \; dx_1 \ldots dx_j \, dy_1 \ldots dy_k \tag{3.6.10}$$

To evaluate $P_s(j,k)$, we need expressions for $P_0(x)$ and $f(x)$. The former function depends upon the data modulation, and the latter upon the deployment statistics.

To simplify the evaluation of Equation (3.6.10), we model each interfering signal that enters a frequency-hopping receiver as independent noise interference. The total noise power in a demodulator is

$$N_1 = N_t + R_s x \tag{3.6.11}$$

where N_t is the thermal and background noise power, and $R_s x$ is the sum of the interference powers due to the other network hoppers. If repeater jamming is present, the jamming power is included in N_t. In the absence of partial-band jamming, we assume that $P_0(x)$ has the form

$$P_0(x) = \frac{1}{2} \exp\left(-\frac{\xi R_s}{N_t + R_s x}\right) \tag{3.6.12}$$

where ξ is determined by the modulation.

When partial-band jamming is present, let D_0 denote the event that the transmission channel is not jammed and D_1 denote the event that it is jammed. If J out of M frequency channels are jammed, then

$$P_0(x) = \left(1 - \frac{J}{M}\right) S_0(x) + \left(\frac{J}{M}\right) S_1(x) \tag{3.6.13}$$

where $S_j(x)$, $j = 1, 2$, is the probability of a symbol error given x and D_j. We assume that a jammed channel always produces jamming power N_i in the demodulator. Based upon the assumed form of (3.6.12), we infer that

$$S_j(x) = \frac{1}{2} \exp\left(-\frac{\xi R_s}{N_t + R_s x + jN_i}\right), \qquad j = 0, 1 \tag{3.6.14}$$

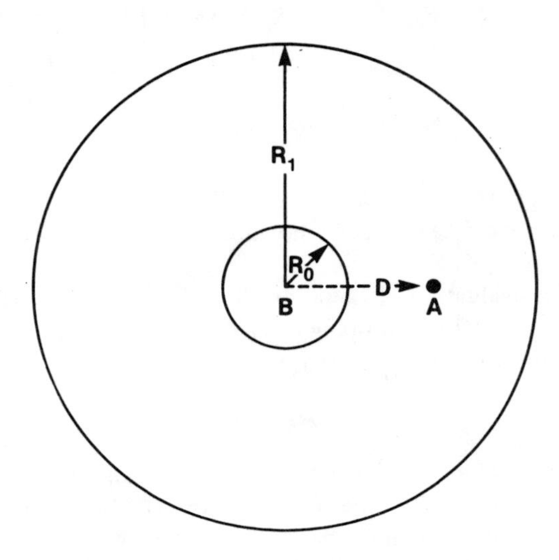

Figure 3.30 Geometry of uniform deployment.

3.6.1 Deployment Statistics

Let R represent the distance between an interferer and the receiver at B in Figure 3.29, and let U represent the potential interference-to-signal ratio at the receiver. If $g(r)$, the radial density function for R, and a propagation model are specified, then $f(u)$, the density function for U, can be determined. We assume that the received power varies inversely as the nth power of the distance to the source. Thus, if the hoppers have identical system parameters, the interference-to-signal ratio at the receiver is

$$U = \left(\frac{D}{R} \right)^n \tag{3.6.15}$$

where D is the distance between the receiver and the intended transmitter.

A plausible statistical deployment model is illustrated in Figure 3.30. The receiver is at B, which is the center of two circles with radii R_0 and R_1, such that $R_1 > R_0 > 0$. The intended transmitter is located at A, a distance D from the receiver. Each significant interferer is assumed to have a uniform location probability in the annular ring between R_0 and R_1. Radius R_0 is the minimum possible separation between an interferer and the receiver. Radius R_1 is the maximum possible separation between the receiver and a significant interferer.

In other words, if an interferer is at a distance greater than R_1 from the receiver, the interferer contributes negligibly to the symbol error probability. The radial density corresponding to a uniform location probability is

$$g(r) = \begin{cases} \left(\dfrac{2}{R_1^2 - R_0^2}\right) r, & R_0 \leqslant r \leqslant R_1 \\ 0, & \text{otherwise} \end{cases} \tag{3.6.16}$$

which is depicted in Figure 3.31. If for each interferer R has the radial density of (3.6.16), the deployment of the interferers is called a *uniform deployment.* Elementary probability theory and (3.6.15) and (3.6.16) imply that

$$f(u) = \begin{cases} \left[\dfrac{2D^2}{n(R_1^2 - R_0^2)}\right] u^{-(n+2)/n}, & \left(\dfrac{D}{R_1}\right)^n \leqslant u \leqslant \left(\dfrac{D}{R_0}\right)^n \\ 0, & \text{otherwise} \end{cases} \tag{3.6.17}$$

3.6.2 Examples

Figures 3.32 to 3.36 plot the upper and lower bounds on the symbol error probability as a function of the number of interferers, assuming that $n = 4$, $L = 2$, $R_0/D = 0.2$, and $R_1/D = 2$. The values of P_U and P_L are usually so close that only one curve appears on the graph. Thus, this single curve can be considered a plot of P_s. In the absence of partial-band jamming, it is convenient to

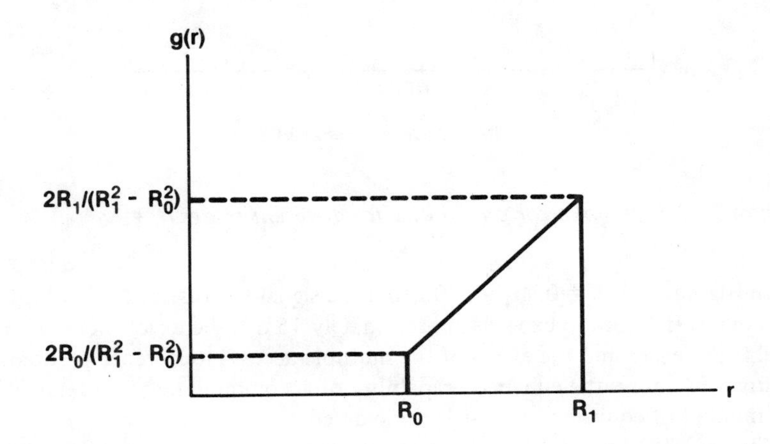

Figure 3.31 Radial density for interferer in uniform deployment of interferers.

use the parameter $M_1 = M/d$, which can be interpreted as an equivalent number of available frequency channels.

Figure 3.32 illustrates the effect of the parameter ξ in (3.6.12) when no jamming is present and the signal-to-noise ratio is 15 dB. When a required P_s is specified, an increase in ξ from 0.5 to 1.0 increases the number of interferers that can be accommodated by a factor of roughly 1.4. In the subsequent examples, we set $\xi = 1$, which corresponds to MSK modulation with ideal differential demodulation and no intersymbol interference (see (1.5.27)).

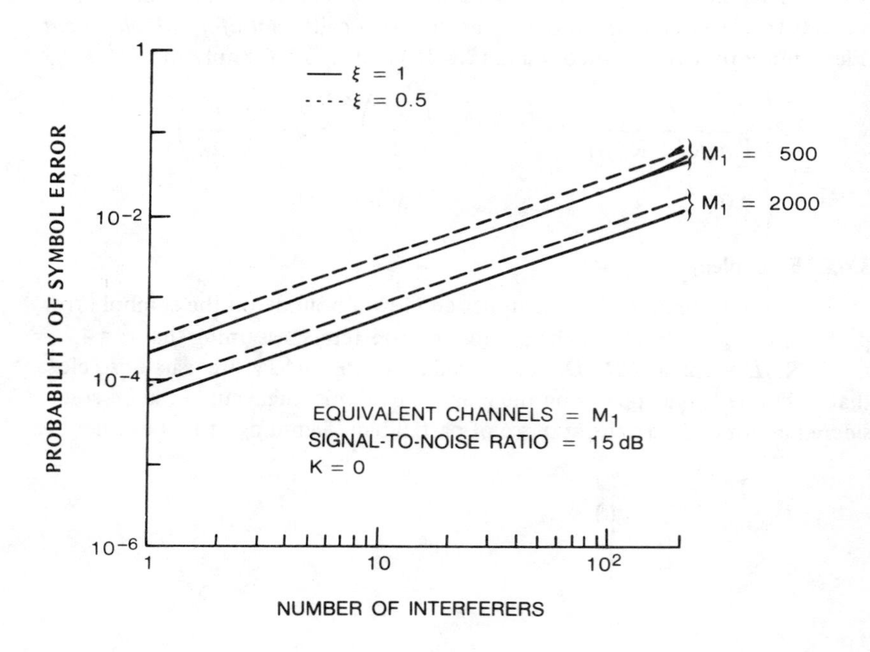

Figure 3.32 Symbol error probability for different values of ξ.

In Figure 3.33, $K = 0, M_1 = 2000$, and the signal-to-noise ratio, R_s/N_t, is a parameter. If this ratio exceeds approximately 15 dB, the exact thermal noise level is irrelevant and the effect of the multiple-access interference predominates. Figure 3.34 shows the enhanced capability of accommodating interferers when the number of equivalent channels is increased.

Figure 3.35 illustrates the effect of adjacent-channel spectral splatter when the parameter K, called the *adjacent-splatter ratio*, equals 0.05. The result of the splatter is to raise the curves by a small amount relative to the corresponding

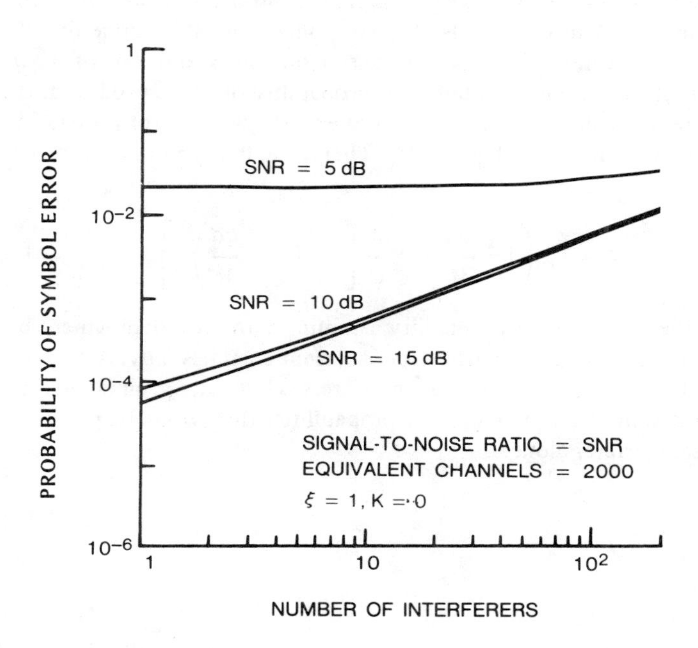

Figure 3.33 Symbol error probability for various signal-to-noise ratios.

curves for no splatter. The value of the adjacent splatter might arise in the following way. If the channels are designed to capture 90 percent of the desired-signal power, then less than five percent, or 0.05, of the power can fall into one of the adjacent channels. If the modulation is ordinary MSK, the channel bandwidth required is approximately $0.8/T_s$, where T_s is the symbol duration. With this bandwidth value, the effect of splatter on a received symbol from channels farther in frequency than the adjacent channels is negligible if $R_0/D \geqslant 0.2$ in a uniform deployment.

Figure 3.36 provides an example of the impact of combined partial-band jamming and multiple-access interference. The jamming-to-signal ratio per jammed channel, N_j/R_s, and the fraction of the band that contains jamming, μ, are parameters in this figure.

3.6.3 Close Interferers

In most practical deployments, only a few interferers are close enough to a receiver to cause significant spectral splatter when hopping in frequency channels beyond the adjacent channels. Suppose that there are ν of these *close interferers* and also N other interferers uniformly deployed beyond a minimum radius

R_0. To make a rough estimate of the symbol error probability, we ignore the effect of channels being near the ends of the hopping band and assume that if one or more of the ν interferers hops into a transmission channel or one of the $q - 1$ channels closest to it, then a symbol error probability of 1/2 is produced. If no close interferer hops into these q channels, then the symbol error probability is determined by the interferers beyond R_0. Therefore, the symbol error probability is

$$P_s \approx P_{s1}\left(1 - \frac{qd}{M}\right)^{\nu} + \frac{1}{2}\left[1 - \left(1 - \frac{qd}{M}\right)^{\nu}\right] \tag{3.6.18}$$

where P_{s1} is the symbol error probability assuming a uniform deployment beyond a minimum radius and splatter from adjacent channels only. Alternatively, if the exact deployment of the ν close interferers is known, P_s can be bounded by P_{s1} plus the sum of the symbol error probabilities that would be produced by each close interferer alone.

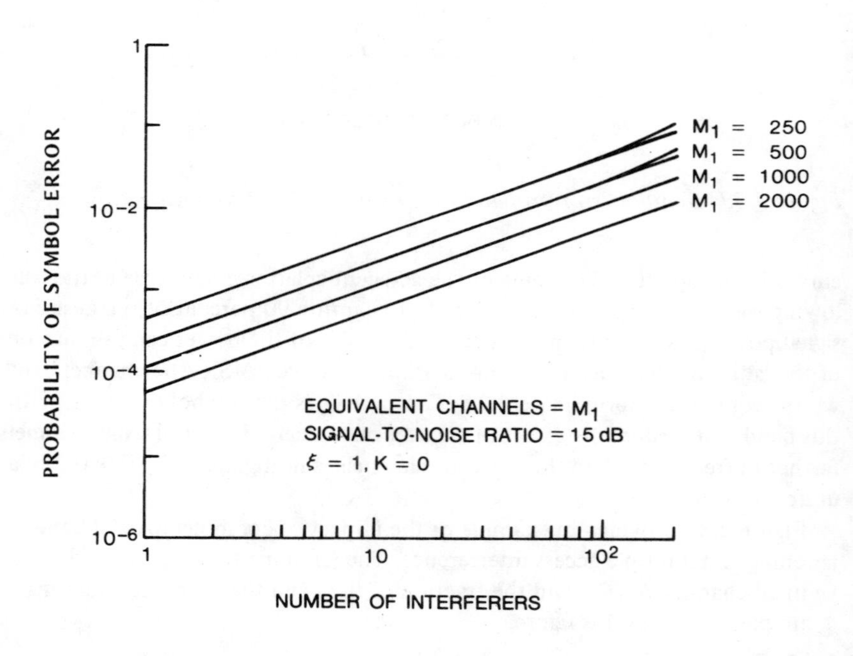

Figure 3.34 Symbol error probability for various numbers of equivalent channels.

Let $H(f)$ denote the transfer function of the bandpass filter in the receiver of Figure 3.1(b). Its center frequency is denoted by f_c. Let $S_0(f)$ denote the two-sided power spectral density of a dehopped interference signal that is modeled as a stationary stochastic process. Let $S(f)$ denote the power spectral density of its complex envelope. Suppose that the interference signal hops into the transmission channel of the desired signal, $S_0(f)$ occupies the band $f_c - W/2 \leqslant |f| \leqslant f_c + W/2$, and $f_c > W/2 \geqslant 0$. Equation (A.3.33) of Appendix A, the facts that $S_0(f)$ and $|H(f)|$ are even functions, and a change of variables indicate that the fraction of interference power that enters the demodulator is

$$K_0 = \frac{\int_{-\infty}^{\infty} S_0(f)|H(f)|^2 df}{\int_{-\infty}^{\infty} S_0(f)\, df} = \frac{\int_{0}^{\infty} S_0(f)|H(f)|^2 df}{\int_{0}^{\infty} S_0(f)\, df}$$

$$= \frac{\int_{-f_c}^{\infty} S_0(f+f_c)|H(f+f_c)|^2 df}{\int_{-f_c}^{\infty} S_0(f+f_c)\, df}$$

$$= \frac{\int_{-\infty}^{\infty} S(f)|H(f+f_c)|^2 df}{\int_{-\infty}^{\infty} S(f)\, df} \qquad (3.6.19)$$

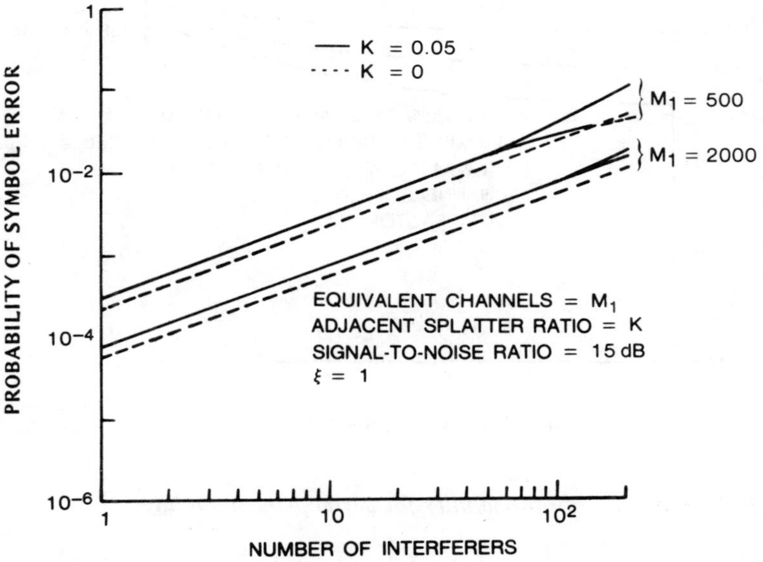

Figure 3.35 Symbol error probability for different adjacent-splatter ratios.

Let B denote the bandwidth of a frequency channel. For an ideal rectangular bandpass filter of bandwidth B,

$$K_0 = 1 - G\left(\frac{B}{2}\right) \tag{3.6.20}$$

where $G(f)$ is the fractional out-of-band power defined by (1.5.11). Suppose that an interference signal hops into a channel with a center frequency that is offset by iB, where i is a nonzero integer, from the center frequency of the transmission channel of the desired signal. If $S_0(f)$ occupies the band $f_c + iB - W/2 \leqslant |f| \leqslant f_c + iB + W/2$, and $f_c + iB > W/2 \geqslant 0$, then the fraction of the interference power that enters the demodulator is

$$K_i = \frac{\int_{-\infty}^{\infty} S(f)|H(f + f_c + iB)|^2 df}{\int_{-\infty}^{\infty} S(f) df}, \quad i = \pm 1, \pm 2, \ldots \tag{3.6.21}$$

If $S_0(f)$ and $|H(f)|$ are symmetric about their center frequencies, then $K_i = K_{-i}$ and the adjacent-splatter ratio is $K = K_1/K_0$. For an ideal rectangular bandpass filter of bandwidth B,

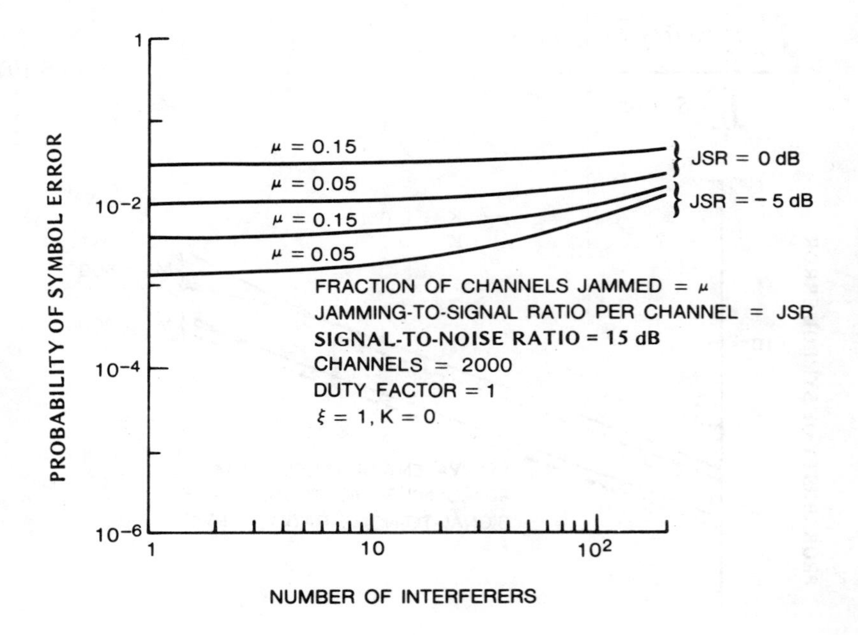

Figure 3.36 Symbol error probability for partial-band jamming.

$$K_i = \frac{1}{2}\left[G\left(iB - \frac{B}{2}\right) - G\left(iB + \frac{B}{2}\right)\right], \quad i = 1, 2, \ldots \tag{3.6.22}$$

$$K = \frac{K_1}{K_0} = \frac{G\left(\frac{B}{2}\right) - G\left(\frac{3B}{2}\right)}{2\left[1 - G\left(\frac{B}{2}\right)\right]} \tag{3.6.23}$$

The *near-far ratio, F,* is defined as the maximum interference-to-signal ratio due to a single interferer. Let Y denote a threshold power ratio below which the power due to an interferer is neglected. A reasonable analytical definition for q is $2i_0 + 1$, where i_0 is the largest index i for which $K_iF/K_0 \geq Y$. For example, if the set of frequency channels form a continuous band and $K_4F/K_0 < Y$, but $K_3F/K_0 \geq Y$, then $q = 7$, which includes the transmission channel plus three channels on each side of it.

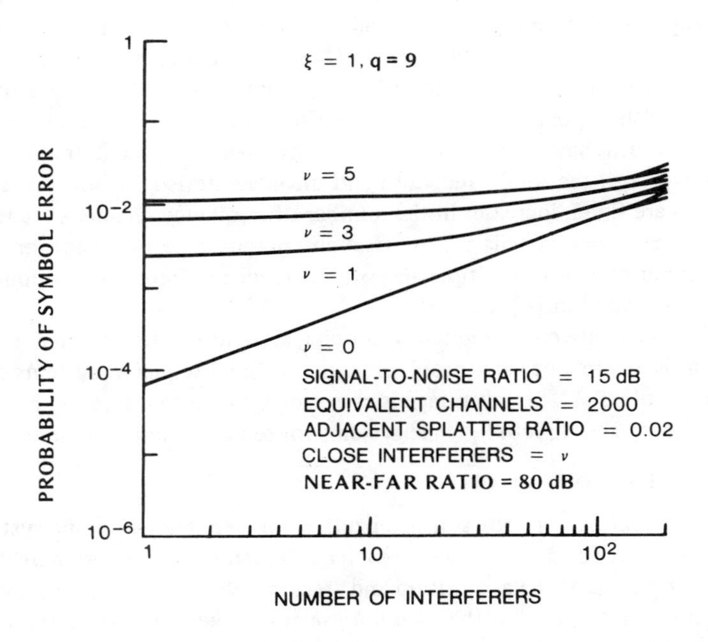

Figure 3.37 Symbol error probability for close interferers.

As an example, suppose that the channel bandwidth is $1.2/T_s$ and interferers can be located as close as $0.01D$. Then (3.6.15) with $n = 4$ implies a near-far ratio equal to 80 dB. Such a large near-far ratio causes significant splatter in many channels if ordinary MSK is used. Thus, we assume that SFSK is the modulation. If $Y = 1$ and the channels form a continuous band, Figure 1.10 indicates that $q = 9$ is appropriate and $K \approx 0.02$. Figure 3.37 shows the resulting symbol error probability for $\xi = 1, L = 2, R_0/D = 0.2, R_1/D = 2, M_1 = 2000$, no jamming, and various values of ν. If ordinary MSK is used instead of SFSK and $\nu \geqslant 1$, the large value of q increases the symbol error probability despite the fact that the adjacent-splatter ratio is reduced to $K \approx 0.005$. However, if there are no close interferers, the symbol error probability is lower for ordinary MSK than for SFSK.

Although the performance of a frequency-hopping multiple-access network depends upon a host of factors, a few general conclusions can be drawn. The effect of multiple-access interference on a system is a sensitive function of the number of interferers and the proximity of close interferers that contribute spectral splatter. If there are no close interferers and $R_0 \geqslant 0.2D$, then spectral splatter usually does not significantly degrade communications if the interferer uses a spectrally efficient modulation, such as MSK. To reduce the susceptibility of a frequency-hopping system to jamming, it is desirable to have as large a total hopping bandwidth as possible. To reduce the effects of multiple-access interference, this bandwidth may be divided into a large number of available frequency channels. However, if the total bandwidth and the characteristics of the transmitted waveforms are fixed, increases in the number of frequency channels eventually lead to sufficient spectral splatter to offset any potential performance improvement. Other models of multiple-access interference in frequency-hopping networks are presented in [8] and [9].

In asynchronous networks, frequency-hopping systems can be designed to be much more resistant to a near-far problem than can direct-sequence systems of comparable complexity. This advantage is sometimes decisive in deciding between frequency-hopping and direct-sequence systems for secure communications.

3.7 CODE SYNCHRONIZATION

The principal features of code synchronization for frequency-hopping systems are the same as those for direct-sequence systems, which are presented in Section 2.7. The configuration of the acquisition and tracking systems usually has the form of Figure 2.21 except that the pseudonoise code generator feeds a frequency synthesizer, the output of which is applied to the mixer. During acquisition, the synthesizer output is synchronized with the received frequency-hopping signal to within a fraction of a hop duration. The tracking system further reduces the synchronization error, or at least maintains it within certain bounds. For secure communication systems that require a strong capability to reject interference, matched-filter acquisition and serial-search acquisition are usually the most effective techniques. The use of a matched filter for the acquisition of a short hopping pattern while a serial-search system provides the acquisition of a long pattern is illustrated in Figure 2.23.

3.7.1 Acquisition

Figure 3.38 shows a matched-filter acquisition system that provides substantial protection against interference. It is assumed that unmodulated pulses are transmitted during acquisition. One or more frequency synthesizers produce tones at frequencies $f_1, f_2, \ldots, f_n$, which are offset by a constant frequency from the consecutive frequencies of the hopping pattern for code acquisition. Each tone multiplies the received frequency-hopping signal and the result is filtered so that most of the received energy is blocked except the energy in a frequency-hopping pulse at a specific frequency. The threshold detector produces the symbol 1 if its threshold is exceeded, which ideally occurs only if the received signal hops to a specific frequency. Otherwise, the threshold detector produces the symbol 0. The use of binary detector outputs prevents the system from being overwhelmed by a few strong interference signals. Input 1 of the comparator is the number of frequencies in the hopping pattern that were received in succession. Input 2 is the number of frequencies simultaneously being received, which is an indication of the amount of interference. The comparator output indicates acquisition if input 1 exceeds a function of input 2. Thus, the comparator usually does not give a false indication of acquisition even if most of the frequency channels contain interference. In an analog version of this matched-filter acquisition system, the threshold detectors are replaced by soft limiters, the digital adders by analog adders, and the envelope detectors by square-law devices. The inputs to the comparator are power levels.

A serial-search acquisition system for frequency-hopping signals is nearly the same as one for direct-sequence signals, as illustrated by Figure 3.39. Thus, the theoretical description of the statistics of acquisition given in Section 2.7 is applicable. The serial-search acquisition of frequency-hopping signals is faster than the acquisition of direct-sequence signals because the hop duration is much greater than a pseudonoise chip duration for practical systems. Given the same time uncertainty, fewer cells have to be searched to acquire frequency-hopping signals because the step sizes are longer in time.

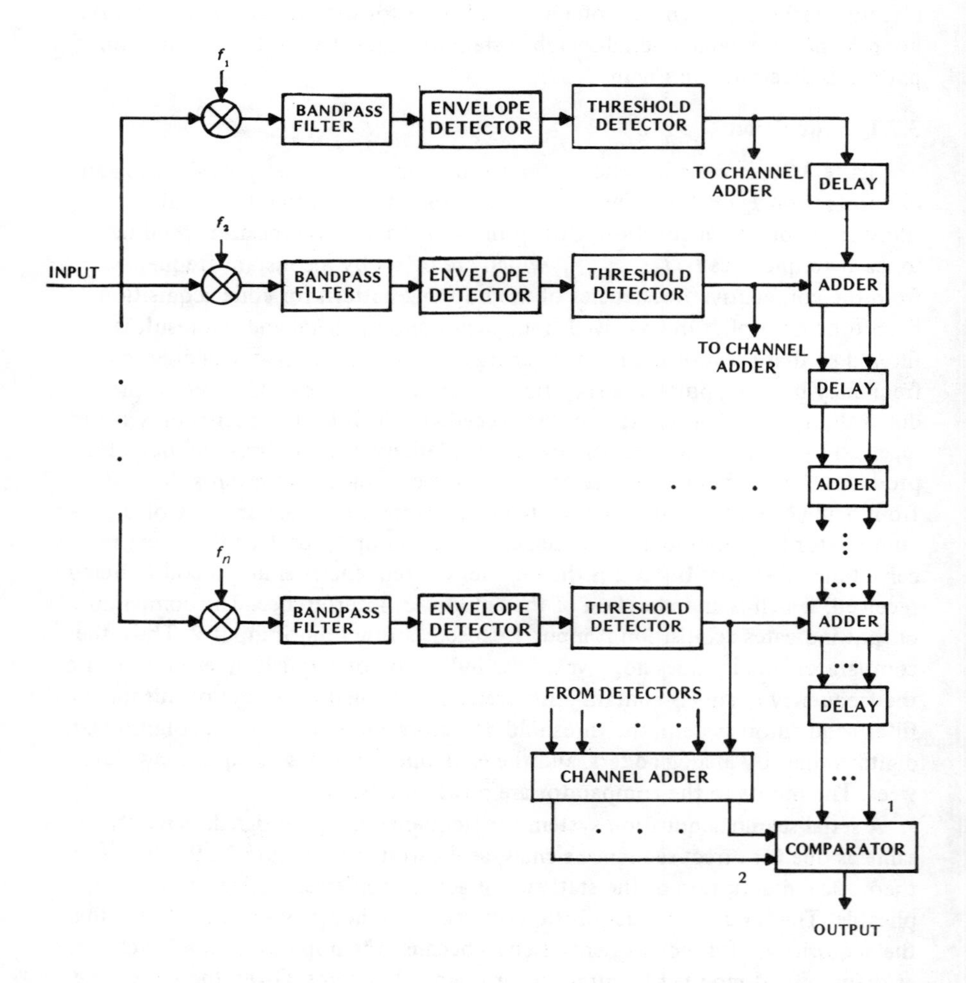

Figure 3.38 Matched-filter acquisition system with protection against interference.

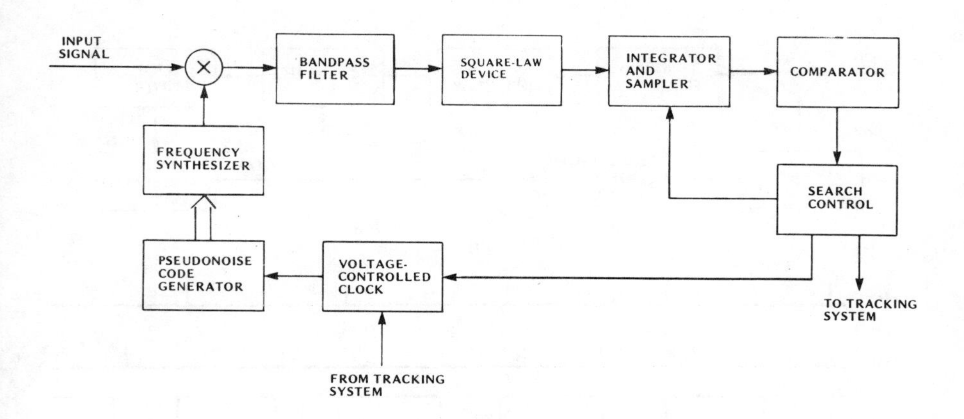

Figure 3.39 Serial-search acquisition system.

3.7.2 Tracking

The *early-late gate tracking loop* [10] for a frequency-hopping system with single-channel modulation is shown in Figure 3.40. The ideal associated waveforms for a typical example are also depicted. In the absence of noise, the envelope detector produces a positive output only when the received frequency-hopping signal, $r(t)$, and the receiver-generated frequency-hopping replica, $r_1(t)$, coincide in center frequency. The gating signal, $g(t)$, is a square wave with transitions from -1 to $+1$ that coincide with the frequency transitions of $r_1(t)$. The early-late gate output, $u(t)$, is the product of the gating signal and the envelope detector output, $v(t)$. The error signal, $e(\tau_e)$, is approximately equal to the time integral of $u(t)$ and is a function of τ_e, the delay of $r_1(t)$ relative to $r(t)$. The error signal can be expressed as a discriminator characteristic, $S(\epsilon)$, which is a function of $\epsilon = \tau_e/T_h$, the normalized delay error. For the typical waveforms shown, ϵ is positive. Hence, $S(\epsilon)$ is positive, the clock rate increases, and the alignment of $r_1(t)$ with $r(t)$ increases.

If the data modulation is MFSK, then the outputs of parallel branches, each with a bandpass filter and envelope detector, can be combined and applied to the early-late gate. If the frequency channels of an MFSK set are not contiguous, additional frequency-synthesizer outputs and mixers are also required.

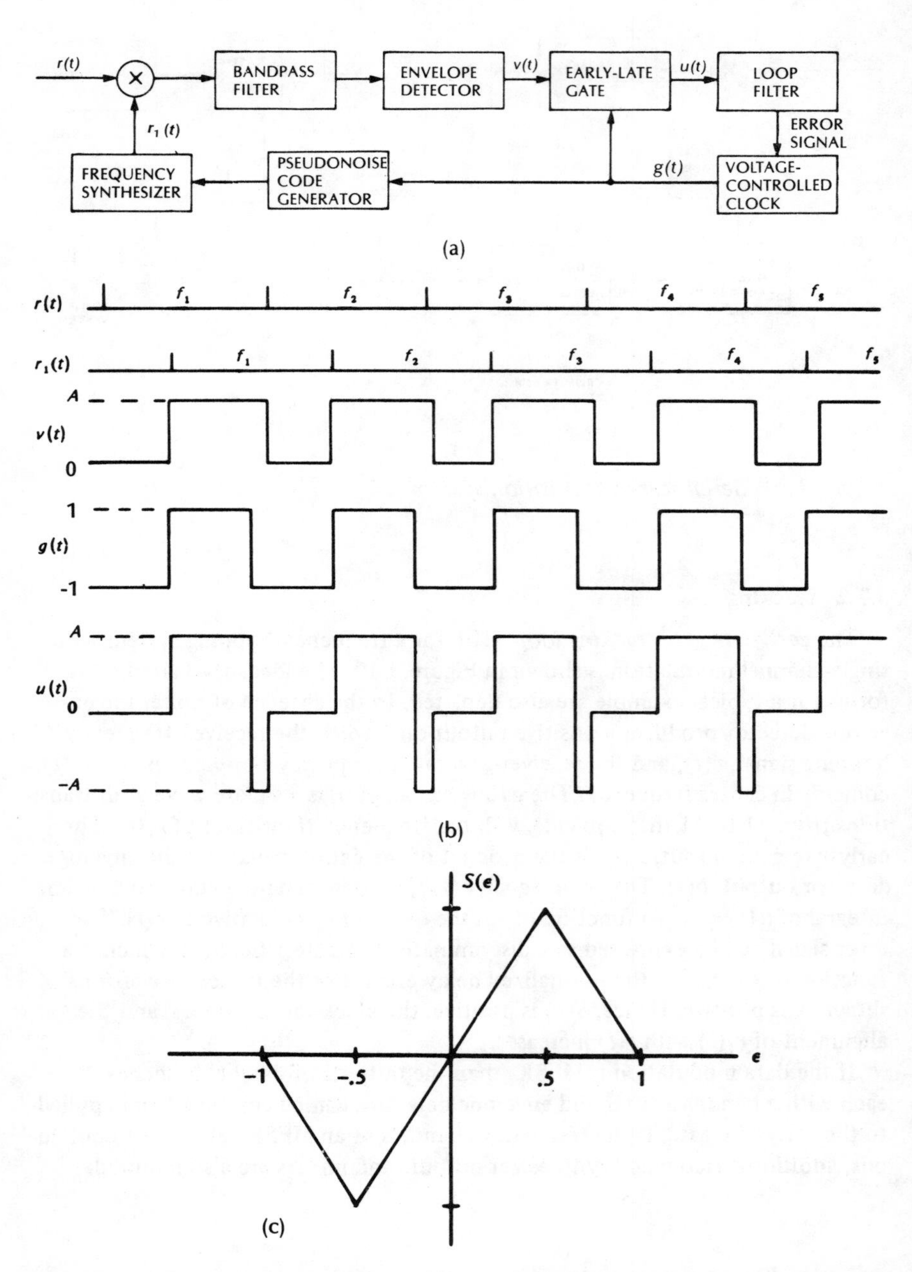

***Figure 3.40** Early-late gate tracking: (a) loop, (b) signals, and (c) discriminator characteristic.*

3.8 HYBRID SYSTEMS

A *hybrid frequency-hopping direct-sequence system* is a direct-sequence system in which the carrier frequency changes periodically. The hybrid system takes advantage of the fact that it is possible to hop in frequency over a much greater band than the bandwidth of a practical direct-sequence signal. Compared with

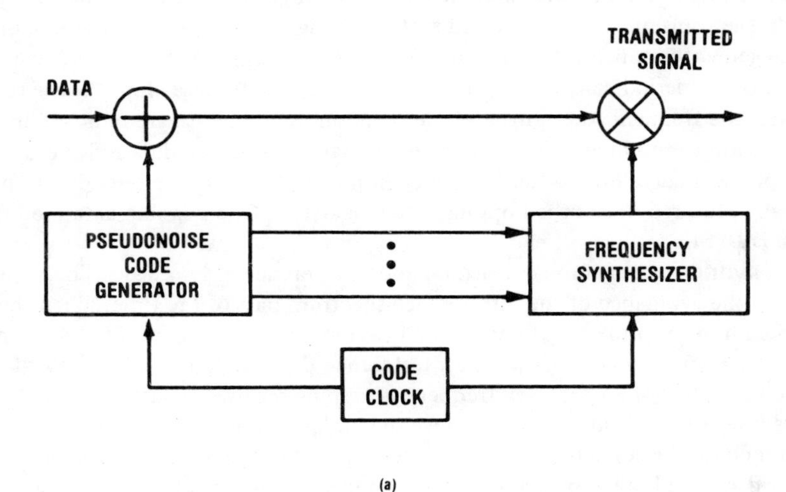

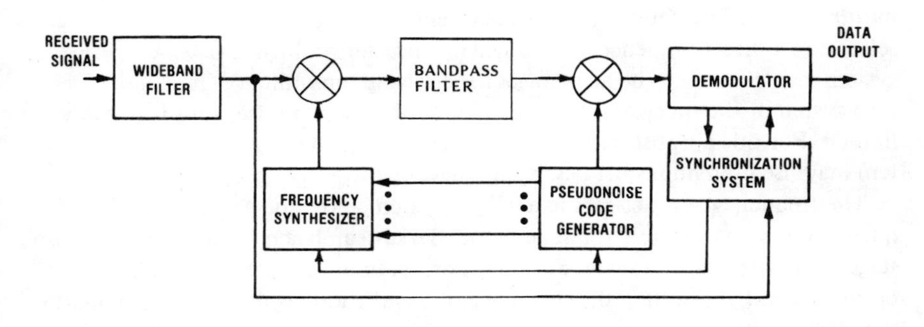

Figure 3.41 Hybrid frequency-hopping direct-sequence system: (a) transmitter and (b) receiver.

ordinary frequency-hopping systems, hybrid systems offer the improved multipath rejection capability of direct-sequence systems and a reduction in the number of frequency channels needed to use the entire available bandwidth. If it is not feasible to hop at a fast enough rate to eliminate a repeater jamming threat, then a hybrid system may solve the problem since the despreading by the pseudonoise sequence impairs the jamming even if the hopping is easily followed.

In the transmitter of the hybrid system of Figure 3.41, a single pseudonoise code generator controls the spreading and the choice of hopping frequencies. Hops occur periodically after a fixed number of pseudonoise chips. In the receiver, the frequency hopping and the pseudonoise sequence are removed in succession to produce a carrier with the biphase message modulation. Because of the phase changes due to the frequency hopping, noncoherent demodulation is usually required unless the hopping rate is very low. One possible data modulation is DPSK.

A hybrid system combats partial-band interference in two ways. The hopping allows the avoidance of the interference spectrum part of the time. When the system hops into the interference, the interference is spread and filtered as in a direct-sequence system. However, interference that would be blocked by the bandpass filter of an ordinary frequency-hopping receiver is passed by the bandpass filter of a hybrid receiver because the bandwidth must be increased to accommodate the spreading waveform. Consequently, the usual performance of a hybrid system against partial-band interference is not expected to differ greatly from the performance of a similar ordinary frequency-hopping system. Against multiple-access interference, a hybrid system provides a performance advantage relative to a direct-sequence system when a near-far problem exists. A hybrid system may sometimes be advantageous relative to an ordinary frequency-hopping system if the spectral splatter caused by the pseudonoise chips is suitably limited. For this purpose, the underlying direct-sequence signal of a hybrid system may use minimum-shift keying and have the form given by (2.1.5).

The frequency synthesizers in a hybrid system allow the generation of a frequency-hopping preamble for acquisition. Serial-search acquisition occurs in two stages. The first stage provides alignment of the hopping patterns. If a hop occurs every n chips, then the first stage of acquisition results in the alignment of the receiver-generated pseudonoise sequence to within fewer than n chips of the received sequence. A second stage of acquisition over the remaining uncertainty in the phase of the sequence finishes acquisition relatively rapidly if n is much less than the number of chips in the original time uncertainty region. Thus, the serial-search acquisition for a hybrid system is slower than for a frequency-hopping system with the same hopping rate, but is often much faster than for a direct-sequence system with the same chip rate and time uncertainty region.

REFERENCES

1. U. Rohde, *Digital PLL Frequency Synthesizers.* Englewood Cliffs, N.J.: Prentice-Hall, 1983.
2. R.M. Gagliardi, *Satellite Communications.* Belmont, California: Lifetime Learning, 1984.
3. D.J. Torrieri, "Frequency-Hopping with Multiple Frequency-Shift Keying and Hard Decisions," *IEEE Trans. Commun.* COM-32, 574, May 1984.
4. J.S. Lee, R.H. French, and L.E. Miller, "Probability of Error Analyses of a BFSK Frequency-Hopping System with Diversity under Partial-Band Jamming Interference – Part I: Performance of Square-Law Linear Combining Soft Decision Receiver," *IEEE Trans. Commun.* COM-32, 645, June 1984.
5. A.J. Viterbi and I.M. Jacobs, "Advances in Coding and Modulation for Noncoherent Channels Affected by Fading, Partial Band, and Multiple-Access Interference," in A.J. Viterbi, ed., *Advances in Communication Systems,* vol. 4. New York: Academic Press, 1975.
6. B.K. Levitt and J.K. Omura, "Coding Tradeoffs for Improved Performance of FH/MFSK Systems in Partial Band Noise," *Nat. Telecommun. Conf.,* vol. 2, D.9.1, 1981.
7. J.S. Lee, L.E. Miller, and Y.K. Kim, "Probability of Error Analyses of a BFSK Frequency-Hopping System with Diversity under Partial-Band Jamming Interference – Part II: Performance of Square-Law Nonlinear Combining Soft Decision Receivers," *IEEE Trans. Commun.* COM-32, 645, 1243, December 1984.
8. D.J. Torrieri, "Simultaneous Mutual Interference and Jamming in a Frequency-Hopping Network," *U.S. Army DARCOM Report* CM/CCM-80-3, Nat. Tech. Inform. Serv. AD-A087 598, 1980.
9. S.A. Musa and W. Wasylkiwskyj, "Co-Channel Interference of Spread-Spectrum Systems in a Multiple User Environment," *IEEE Trans. Commun.* COM-26, 1405, October 1978.
10. C.A. Putman, S.S. Rappaport, and D.L. Schilling, "Tracking of Frequency-Hopped Spread-Spectrum Signals in Adverse Environments," *IEEE Trans. Commun.* COM-31, 955, August 1983.
11. R.C. Dixon, *Spread Spectrum Systems,* 2nd ed. New York: John Wiley and Sons, 1984.
12. C.R. Cahn *et al., Spread Spectrum Communication,* NATO Advisory Group AGARD-LS-58, Nat. Tech. Inform. Serv. AD-766-914, 1973.

Chapter 4
Interception

4.1 INTRODUCTION

Interception of hostile communications is attempted for many diverse reasons, such as reconnaissance, surveillance, position fixing, identification, or a prelude to jamming. Different purposes require different systems, but whatever the purpose, an interception system nearly always must achieve the three basic functions of *detection, frequency estimation,* and *direction finding.* Although these three elements of interception are usually integrated in a practical system, they are discussed separately in this chapter for clarity of presentation. The basic concepts and issues of the three elements are presented at the systems level, assuming that little is known about the signals to be intercepted. Although this chapter is concerned with the interception of communications, only slight modifications of the results are required to apply them to the interception of radar.

The potential interceptor has at least one major advantage over the communicators. The accuracies of detection, frequency estimation, and direction finding are determined by the energy of the entire received message, which may include many symbols. In contrast, the intended receiver makes decisions with accuracies determined by the energy of a small number of symbols. From another point of view, the intended receiver generally must make many separate decisions, whereas the interception receiver must make only a few decisions.

4.2 DETECTION

If the form and the parameters of the signal to be intercepted, $s(t)$, were known, optimum *detection* in white Gaussian noise, $n(t)$, could be accomplished by a matched filter or an ideal correlator. Figure 4.1 depicts a correlator for the received signal, $r(t) = s(t) + n(t)$, and an observation interval of duration T. The comparator input is compared with a fixed threshold level, V_T, to determine the presence of an intercepted signal. It is a standard result that the probability of false alarm P_F, and the probability of detection, P_D, are given by [1]

$$P_F = \frac{1}{2} \operatorname{erfc} \left[\frac{V_T}{(N_0 E)^{1/2}} \right] \tag{4.2.1}$$

$$P_D = \frac{1}{2}\,\text{erfc}\left[\frac{V_T}{(N_0 E)^{1/2}} - \left(\frac{E}{N_0}\right)^{1/2}\right] \tag{4.2.2}$$

where E is the signal energy, $N_0/2$ is the two-sided noise power spectral density, and the complementary error function is defined as

$$\text{erfc}(x) = \frac{2}{\sqrt{\pi}} \int_x^\infty \exp(-y^2)dy \tag{4.2.3}$$

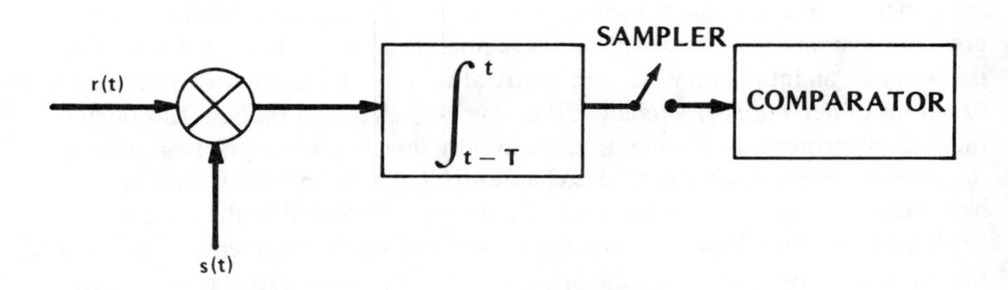

Figure 4.1 Correlator.

Denoting the inverse complementary error function by erfc^{-1}, we define

$$\beta = \text{erfc}^{-1}(2P_F) \tag{4.2.4}$$

$$\xi = \text{erfc}^{-1}(2P_D) \tag{4.2.5}$$

From (4.2.1) and (4.2.2), we can calculate the value of E/N_0 necessary to ensure specified values of P_F and P_D. The result is

$$\frac{E}{N_0} = (\beta - \xi)^2 \tag{4.2.6}$$

Although the ideal correlator cannot be used when $s(t)$ is unknown, (4.2.6) provides a basis of comparison for more realistic interception receivers.

Suppose that the intercepted signal has random phase and frequency and an unknown constant amplitude. The signal frequency is assumed to be one of M possible values; that is, the band to be searched is divided into M channels with center frequencies ω_1, ω_2, *et cetera.* Each discrete frequency, ω_i, is associated with a hypothesis, H_i. Thus, the multiple alternative hypotheses over an observation interval are

$$H_0: \ r(t) = n(t), \ \ 0 \leqslant t \leqslant T$$

$$H_1: \ r(t) = A \sin(\omega_1 t + \theta_1) + n(t), \ \ 0 \leqslant t \leqslant T$$

$$\vdots$$

$$H_M: \ r(t) = A \sin(\omega_M t + \theta_M) + n(t), \ \ 0 \leqslant t \leqslant T$$

where the θ_i are phase angles. If the phase angles are uniformly distributed and each frequency is equally likely to occur, detection theory [1] yields the receiver depicted in Figure 4.2. The decision rule is the following: choose H_i, $i = 1, \ldots, M$, if x_i is the largest detector output and x_i exceeds the threshold, and choose H_0 otherwise. If a signal is detected, this receiver automatically identifies the frequency as the center frequency of the filter with the largest output.

The matched filters of Figure 4.2 are matched to intercepted signals that are pulsed sinusoids of known duration and arrival times. To accommodate more general, unknown signals, the matched filters could be replaced by bandpass filters, but the receiver would no longer be optimum.

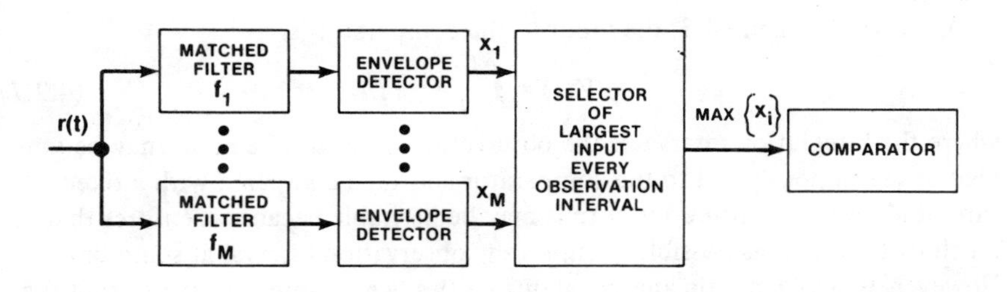

Figure 4.2 Optimum receiver for pulsed sinusoid of unknown frequency.

There remain other problems with this receiver. It is doubtful that the envelope detectors can function efficiently against some signal forms. Because the receiver is designed to operate on a narrowband signal, the detection of direct-sequence communications requires additional hardware.

4.2.1 Radiometer

Another approach is to model the signal as a stationary Gaussian process with a flat power spectral density. Assuming that the noise present is white and Gaussian, detection theory yields the optimum receiver depicted in Figure 4.3, which is called an *energy detector* or a *radiometer* [2]. This receiver has the major advantages that it requires relatively little hardware, and no additional hardware is needed for the detection of spread-spectrum communications.

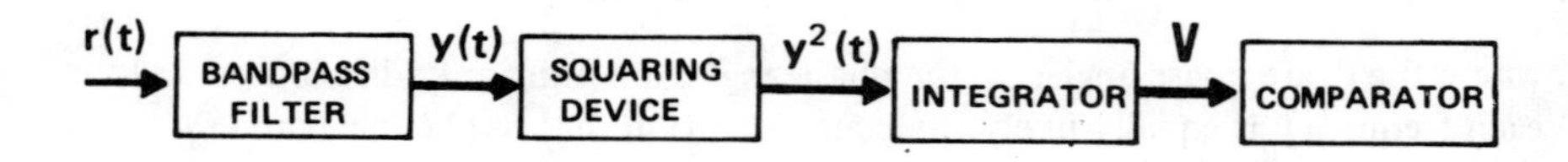

Figure 4.3 Radiometer.

In addition to being optimum if the signal is modeled as a stationary Gaussian process, the radiometer is a reasonable configuration for determining the presence of unknown deterministic signals. We give a performance analysis of the radiometer, assuming a deterministic signal and the presence of white Gaussian noise [3].

As shown in Figure 4.3, the input to the comparator is

$$V(t) = \int_{t-T}^{t} y^2(\tau)\, d\tau \tag{4.2.7}$$

where the integration interval is the observation interval. The input may be sampled or continuously fed to the comparator and then compared with a fixed threshold level. A continuous output may be desirable because it ensures that as much of the signal as possible occupies the observation interval at some time. However, to avoid certain analytical difficulties, we assume henceforth that the integrator output is sampled periodically. To determine the probabilities of false alarm and detection, it is convenient to normalize the test statistic to

$$V = \int_{0}^{T} y^2(t)\, dt \tag{4.2.8}$$

The bandpass filter is assumed to be an ideal rectangular filter with center frequency f_c and bandwidth W. Its output is

$$y(t) = s(t) + n(t) \tag{4.2.9}$$

where $s(t)$ is a deterministic signal, and $n(t)$ is bandlimited white Gaussian noise with a two-sided power spectral density equal to $N_o/2$. Substituting (4.2.9) into (4.2.8), taking the expected value, and observing that $n(t)$ is a zero-mean process, we obtain

$$E[V] = \int_0^T s^2(t)\,dt + \int_0^T E[n^2(t)]\,dt$$

$$= E + N_0 TW \tag{4.2.10}$$

where $E[x]$ denotes the expected value of x and E denotes the signal energy. This equation indicates that the radiometer output is an unbiased estimate of the total energy in $y(t)$.

According to the results of Appendix A, the deterministic signal can be represented as

$$s(t) = s_c(t) \cos \omega_c t - s_s(t) \sin \omega_c t \tag{4.2.11}$$

where $\omega_c = 2\pi f_c$. Since the spectrum of $s(t)$ is confined within the filter passband, $s_c(t)$ and $s_s(t)$ have frequency components confined to the band $|f| \leqslant W/2$. The Gaussian noise emerging from the bandpass filter can be represented in terms of quadrature components as (Appendix A)

$$n(t) = n_c(t) \cos \omega_c t - n_s(t) \sin \omega_c t \tag{4.2.12}$$

where $n_c(t)$ and $n_s(t)$ have flat power spectral densities, each equal to N_0 over $|f| \leqslant W/2$. Substituting (4.2.9), (4.2.11), and (4.2.12) into (4.2.8) and assuming that $f_c \gg W$ and $f_c \gg 1/T$, we obtain

$$V \cong \frac{1}{2}\int_0^T [s_c(t) + n_c(t)]^2\,dt + \frac{1}{2}\int_0^T [s_s(t) + n_s(t)]^2\,dt. \tag{4.2.13}$$

The *sampling theorems* for deterministic and stochastic processes (Appendix A) provide expansions of $s_c(t)$, $s_s(t)$, $n_c(t)$, and $n_s(t)$ that facilitate a statistical performance analysis. For example,

$$s_c(t) = \sum_{i=-\infty}^{\infty} s_c\left(\frac{i}{W}\right) \text{sinc}(Wt - i) \tag{4.2.14}$$

where sinc $x = (\sin \pi x)/\pi x$. After substituting the appropriate expansions into (4.2.11), we make the following approximations:

$$\int_0^T \text{sinc}^2(Wt - i)\,dt \cong 0, \quad i \leqslant 0 \text{ or } i > TW \tag{4.2.15}$$

$$\int_0^T \text{sinc}^2(Wt - i)\,dt \cong \int_{-\infty}^{\infty} \text{sinc}^2(Wt - i)\,dt = \frac{1}{W}, \quad 0 < i \leqslant TW \tag{4.2.16}$$

$$\int_0^T \text{sinc}(Wt - i)\,\text{sinc}(Wt - j)\,dt \cong \int_{-\infty}^{\infty} \text{sinc}(Wt - i)\,\text{sinc}(Wt - j)\,dt = 0, \quad i \neq j \tag{4.2.17}$$

As a result, we obtain

$$V \cong \frac{1}{2W} \sum_{i=1}^{\gamma} \left[s_c\left(\frac{i}{W}\right) + n_c\left(\frac{i}{W}\right) \right]^2 + \frac{1}{2W} \sum_{i=1}^{\gamma} \left[s_s\left(\frac{i}{W}\right) + n_s\left(\frac{i}{W}\right) \right]^2 \tag{4.2.18}$$

where γ is the largest integer less than or equal to TW. Equation (4.2.15) and the first equalities in (4.2.16) and (4.2.17) are implied by plots of the integrand. The second equalities in (4.2.16) and (4.2.17) are proved by observing that the sinc function is the Fourier transform of the rectangle function, interpreting the integrals as convolutions, using the convolution theorem, and evaluating the results.

Assuming that $TW \geqslant 1$, the error introduced by (4.2.15) at $i = 0$ is nearly $1/2W$. For other values of i, except possibly $i = TW$, the errors caused by the approximations are much less than $1/2W$ and decrease as TW increases. Equation (4.2.18) becomes an increasingly accurate approximation of (4.2.13) as γ increases. It is always assumed that $\gamma \geqslant 1$.

Because $n(t)$ has a power spectral density that is symmetrical about f_c, $n_c(t)$ and $n_s(t)$ are independent Gaussian processes (Appendix A). Thus, $n_c(i/W)$ and $n_s(j/W)$ are independent Gaussian random variables. The power spectral densities of $n_c(t)$ and $n_s(t)$ are

$$S_c(f) = S_s(f) = \begin{cases} N_0, & |f| \leqslant W/2 \\ 0, & |f| > W/2 \end{cases} \tag{4.2.19}$$

The associated autocorrelation functions are

$$R_c(\tau) = R_s(\tau) = N_0 W \,\text{sinc}\, W\tau \tag{4.2.20}$$

This expression indicates that $n_c(i/W)$ is statistically independent of $n_c(j/W)$, $i \neq j$, and similarly for $n_s(i/W)$ and $n_s(j/W)$. Because $n(t)$ is assumed to be zero mean, so are $n_c(i/W)$ and $n_s(i/W)$. Using these facts, we rewrite (4.2.18) as

$$V = \frac{N_0}{2}\left\{\sum_{i=1}^{\gamma} A_i^2 + \sum_{i=1}^{\gamma} B_i^2\right\} \tag{4.2.21}$$

where the A_i and the B_i are statistically independent Gaussian random variables with unit variances and means

$$m_{1i} = E[A_i] = \frac{1}{\sqrt{N_0 W}} s_c\left(\frac{i}{W}\right) \tag{4.2.22}$$

$$m_{2i} = E[B_i] = \frac{1}{\sqrt{N_0 W}} s_s\left(\frac{i}{W}\right) \tag{4.2.23}$$

Before continuing the analysis, we digress to establish the statistical properties of the sum

$$Z = \sum_{i=1}^{N} A_i^2 \tag{4.2.24}$$

where the A_i are independent Gaussian random variables with means m_i and unit variances. The random variable Z is said to have a *noncentral chi-squared* $(\chi)^2$ *distribution* with N degrees of freedom and a *noncentral parameter*

$$\lambda = \sum_{i=1}^{N} m_i^2 \tag{4.2.25}$$

To derive the density function of Z, we first note that each A_i has the density function

$$p_{A_i}(x) = \frac{1}{\sqrt{2\pi}} \exp\left[-\frac{(x - m_i)^2}{2}\right] \tag{4.2.26}$$

From elementary probability, the density of $Y_i = A_i^2$ is

$$p_{Y_i}(x) = \frac{1}{2\sqrt{x}} [p_{A_i}(\sqrt{x}) + p_{A_i}(-\sqrt{x})] u(x) \tag{4.2.27}$$

where $u(x) = 1, x \geqslant 0$, and $u(x) = 0, x < 0$. Substituting (4.2.26) into (4.2.27), expanding the exponentials, and simplifying, we obtain the density

$$p_{Y_i}(x) = \frac{1}{\sqrt{2\pi x}} \exp\left(-\frac{x + m_i^2}{2}\right) \cosh(m_i\sqrt{x})\, u(x) \tag{4.2.28}$$

The characteristic function of a random variable X is defined as

$$M_X(jv) = E[e^{jvX}] = \int_{-\infty}^{\infty} p_X(x)\exp(jvx)\,dx \tag{4.2.29}$$

where $j = \sqrt{-1}$, and $p_X(x)$ is the density of X. Because $M_X(jv)$ is the conjugate Fourier transform of $p_X(x)$,

$$p_X(x) = \frac{1}{2\pi}\int_{-\infty}^{\infty} M_X(jv)\exp(-jvx)\,dv \tag{4.2.30}$$

From Laplace or Fourier transform tables, it is found that the characteristic function of $p_{Y_i}(x)$ is

$$M_{Y_i}(jv) = \exp\left(-\frac{m_i^2}{2}\right)(1 - j2v)^{-1/2}\exp\left[\frac{m_i^2}{2(1 - j2v)}\right] \tag{4.2.31}$$

The characteristic function of a sum of independent random variables is equal to the product of the individual characteristic functions. Because Z is the sum of the Y_i, the characteristic function of Z is

$$M_Z(jv) = \exp\left(-\frac{\lambda}{2}\right)(1 - j2v)^{-N/2}\exp\left[\frac{\lambda}{2(1 - j2v)}\right] \tag{4.2.32}$$

where we have used (4.2.25). From (4.2.30) and Laplace or Fourier transform tables, we obtain

$$p_Z(x) = \frac{1}{2}\left(\frac{x}{\lambda}\right)^{(N-2)/4}\exp\left(-\frac{x+\lambda}{2}\right)I_{N/2-1}(\sqrt{x\lambda})\,u(x) \tag{4.2.33}$$

where $I_n(\ )$ is the modified Bessel function of the first kind and order n. Equation (4.2.33) is the density function of a noncentral χ^2 random variable with N degrees of freedom and a noncentral parameter λ. Using the series expansion in λ of the Bessel function and setting $\lambda = 0$, we obtain

$$p_Z(x) = \frac{1}{2^{N/2}\Gamma(N/2)}\,x^{N/2-1}\exp\left(-\frac{x}{2}\right)u(x), \qquad \lambda = 0 \tag{4.2.34}$$

where $\Gamma(x)$ is the gamma function. Alternatively, we can obtain (4.2.34) by setting $\lambda = 0$ in (4.2.32) and using transform tables. Equation (4.2.34) is the density function for a central χ^2 random variable with N degrees of freedom.

From (4.2.32), it follows that the sum of two independent noncentral χ^2 random variables with N_1 and N_2 degrees of freedom and noncentral parameters λ_1 and λ_2, respectively, is a noncentral χ^2 random variable with $N_1 + N_2$ degrees of freedom and noncentral parameter $\lambda_1 + \lambda_2$.

From these results and (4.2.21) to (4.2.23), it follows that $2V/N_o$ has a χ^2 distribution with 2γ degrees of freedom and noncentral parameter

$$\lambda = \frac{1}{N_0 W} \sum_{i=1}^{\gamma} s_c^2\left(\frac{i}{W}\right) + \frac{1}{N_0 W} \sum_{i=1}^{\gamma} s_s^2\left(\frac{i}{W}\right)$$

$$\cong \frac{1}{N_0} \int_0^T [s_c^2(t) + s_s^2(t)]\, dt$$

$$\cong \frac{2}{N_0} \int_0^T s^2(t)\, dt \tag{4.2.35}$$

In terms of the signal energy, E,

$$\lambda \cong \frac{2E}{N_0} \tag{4.2.36}$$

By straightforward calculations using the statistics of Gaussian variables, (4.2.21) and the subsequent results yield

$$E[V] \cong E + N_0\gamma \tag{4.2.37}$$

$$\mathrm{VAR}(V) \cong 2N_0 E + N_0^2\gamma \tag{4.2.38}$$

Equation (4.2.37) approaches the exact result of (4.2.10) as TW increases.

In the absence of the signal, the probability density function of $2\,V/N_0$ has the form of (4.2.34) with $N = 2\gamma$. Let V_T denote the threshold level to which V is compared. A false alarm occurs if $V > V_T$ when the signal is absent. Thus, the probability of a false alarm is

$$P_F = \int_{2V_T/N_0}^{\infty} \frac{1}{2^\gamma \Gamma(\gamma)}\, v^{\gamma-1} e^{-v/2}\, dv \tag{4.2.39}$$

Integrating by parts $\gamma - 1$ times yields

$$P_F = \exp\left(-\frac{V_T}{N_0}\right) \sum_{i=0}^{\gamma-1} \frac{V_T^i}{N_0^i\, i!} \tag{4.2.40}$$

Rewriting this finite series and then extending it into an infinite series, we obtain

$$P_F = \frac{V_T^{\gamma-1} \exp(-V_T/N_0)}{N_0^{\gamma-1}(\gamma-1)!}\left[1 + \frac{N_0(\gamma-1)}{V_T} + \frac{N_0^2(\gamma-1)(\gamma-2)}{V_T^2} + \ldots + \frac{N_0^{\gamma-1}(\gamma-1)!}{V_T^{\gamma-1}}\right]$$

$$\leqslant \frac{V_T^{\gamma-1} \exp(-V_T/N_0)}{N_0^{\gamma-1}(\gamma-1)!}\left[1 + \frac{N_0(\gamma-1)}{V_T} + \frac{N_0^2(\gamma-1)^2}{V_T^2} + \ldots\right] \tag{4.2.41}$$

Summing the infinite series yields

$$P_F \leqslant \frac{V_T^{\gamma-1}\exp(-V_T/N_0)}{N_0^{\gamma-1}(\gamma-1)![1-N_0(\gamma-1)/V_T]}, \qquad V_T > N_0(\gamma-1) \tag{4.2.42}$$

For small P_F, V_T is large and this bound is tight. Thus, it can be used as a simple approximation for P_F.

If the signal is present, the χ^2 probability density function of V has the form of (4.2.33) with $N = 2\gamma$. If $V > V_T$, correct detection occurs. Thus, the probability of detection is

$$P_D = \int_{2V_T/N_0}^{\infty} \frac{1}{2}\left(\frac{v}{\lambda}\right)^{(\gamma-1)/2} \exp\left(-\frac{v+\lambda}{2}\right) I_{\gamma-1}(\sqrt{v\lambda})\, dv \tag{4.2.43}$$

A change of variables yields

$$P_D = Q_\gamma(\sqrt{\lambda}, \sqrt{2V_T/N_0}) \tag{4.2.44}$$

where

$$Q_M(\alpha,\beta) = \int_\beta^\infty x\left(\frac{x}{\alpha}\right)^{M-1} \exp\left(-\frac{x^2+\alpha^2}{2}\right) I_{M-1}(\alpha x)\, dx \tag{4.2.45}$$

The function $Q_M(\alpha,\beta)$ is called the *generalized Q-function.* Many algorithms for its approximate evaluation can be found in the literature [1].

Large values of TW are particularly interesting because this case includes the interception of spread-spectrum communications. When TW is large, $\gamma \cong TW$, and the central limit theorem indicates that the distribution of V is approximately Gaussian. Using (4.2.37) and (4.2.38) with $E = 0$ and the Gaussian distribution, we obtain

$$\begin{aligned} P_F &= \frac{1}{(2\pi N_0^2 TW)^{1/2}} \int_{V_T}^{\infty} \exp\left[-\frac{(v - N_0 TW)^2}{2N_0^2 TW}\right] dv \\ &= \frac{1}{2}\operatorname{erfc}\left[\frac{V_T - N_0 TW}{(2N_0^2 TW)^{1/2}}\right], \qquad TW \gg 1 \end{aligned} \tag{4.2.46}$$

Similarly, we obtain

$$P_D = \frac{1}{2}\operatorname{erfc}\left[\frac{V_T - N_0 TW - E}{(2N_0^2 TW + 4N_0 E)^{1/2}}\right], \qquad TW \gg 1 \tag{4.2.47}$$

A suitable performance measure of an interception system is the received signal power required to achieve specified values of P_F and P_D when the signal is optimally aligned with the observation interval. Alternatively, one might specify

P_D and the *false alarm rate*, F, which is the expected number of false alarms per unit time. If successive observation intervals do not overlap each other except possibly at end points, then

$$F = \frac{P_F}{T} \tag{4.2.48}$$

Combining (4.2.4), (4.2.5), (4.2.46), and (4.2.47) gives

$$(2N_0^2 TW + 4N_0 E)^{1/2} \xi = (2N_0^2 TW)^{1/2} \beta - E \tag{4.2.49}$$

Solving for the value of E/N_0 necessary to achieve specified values of P_D and either P_F or F, we obtain

$$\frac{E}{N_0} = 2\xi^2 + \beta(2TW)^{1/2} - \xi\left[2TW + 4\xi^2 + 4\beta(2TW)^{1/2}\right]^{1/2} \tag{4.2.50}$$

We expand this equation as a Taylor series in $\beta/\sqrt{TW}$ and $\xi/\sqrt{TW}$ and retain only the lowest order terms. The result is

$$\frac{E}{N_0} = (2TW)^{1/2}(\beta - \xi) , \qquad TW \gg \max(\beta^2, \xi^2) \tag{4.2.51}$$

A comparison of (4.2.51) and (4.2.6) shows that the disparity in performance between the radiometer and the matched filter increases with TW. Equation (4.2.51) indicates that detection difficulties increase as spectrum spreading forces a larger value of W.

Denoting the intercepted signal power by R_s and the signal duration by T_1, we find that the intercepted power necessary to achieve specified values of P_D and either P_F or F is

$$R_s = \begin{cases} N_0 \dfrac{(2TW)^{1/2}}{T_1} (\beta - \xi), & T_1 < T, \; TW \gg \max(\beta^2, \xi^2) \\[2ex] N_0 \left(\dfrac{2W}{T}\right)^{1/2} (\beta - \xi), & T_1 \geq T, \; TW \gg \max(\beta^2, \xi^2) \end{cases} \tag{4.2.52}$$

As long as $T_1 \geq T$, this equation indicates that increasing the observation interval decreases the required power. However, if $T_1 < T$, an increase in the observation interval increases the required power. Equation (4.2.52) can be combined with (1.2.6) to determine the median range to the transmitter that can be accommodated by the interception receiver. If the outputs of ν independent radiometers are averaged, a straightforward calculation shows that the required R_s can be reduced by a factor of $\nu^{-1/2}$.

4.2.2 Channelized Radiometer

A *channelized radiometer* results when M radiometers are inserted into the branches of Figure 4.2, as depicted in Figure 4.4(a). Each block labeled radiometer contains a bandpass filter of bandwidth W/M, a squaring device, and an integrator, but no comparator. The channelized radiometer is potentially effective against narrowband, multiple-frequency-shift-keying (MFSK), and frequency-hopping signals. It is useful against direct-sequence signals if preliminary processing produces a signal with a narrow bandwidth, as in (2.3.12). If the presence of more than one signal is to be verified, it is preferable to employ an array of radiometers of the form of Figure 4.3 with the comparator outputs feeding into a processor that analyzes the activity of individual channels, as shown in Figure 4.4(b).

To operate efficiently against frequency-hopping or MFSK signals, the channelized radiometer requires approximate knowledge of both the frequency channels and the frequency-transition times. With this information, the processor of Figure 4.4(a) examines N consecutive comparator output samples corresponding to N signal pulses and determines that a signal is present if the comparator threshold has been exceeded r or more times. Ideally, the number of parallel branches, M, is set equal to the number of frequency channels, but many fewer branches may be a practical or economic necessity. Additional circuitry is necessary to prevent interference in a single frequency channel from causing a false alarm.

Let T_s denote the duration of the sampling interval, which is the observation interval of the constituent radiometers. The effective observation time of the channelized radiometer, given by $T = NT_s$, should be less than the minimum expected message duration to avoid processing extraneous noise. If the signals to be intercepted have fixed center frequencies, then $N = 1$.

To simplify the analysis of the interception of a single signal, we assume that the N sets of radiometer outputs are statistically independent. If P_{F1} is the probability that a particular radiometer output exceeds the threshold when no signal is present, then the probability that none of the radiometer outputs exceeds the threshold is $(1 - P_{F1})^M$, assuming that the channel noises are statistically independent. The probability that exactly i out of N comparator inputs exceed the threshold is

$$P(i,N) = \begin{cases} = \binom{N}{i}\left[1 - (1 - P_{F1})^M\right]^i (1 - P_{F1})^{M(N-i)}, & i \leq N \\ = 0, & i > N \end{cases} \tag{4.2.53}$$

It follows that the probability of false alarm associated with the observation interval is

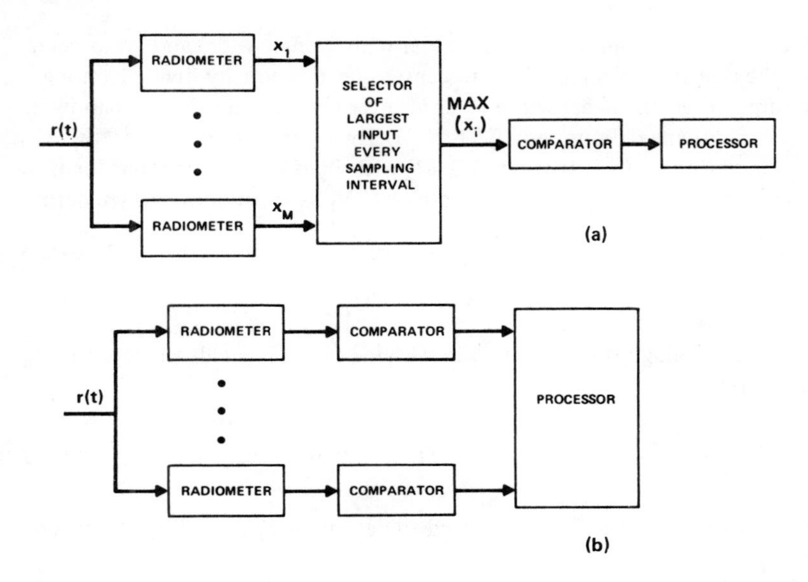

Figure 4.4 Channelized radiometers: (a) for detection of presence of hostile communications and (b) for simultaneous detection of multiple signals.

$$P_F = \sum_{i=r}^{N} P(i,N) \tag{4.2.54}$$

If the intercepted signal duration, T_1, is less than the observation time, T, we assume for simplicity that $N_1 = T_1/T_s$, the number of sampling intervals during which the signal is present, is an integer. Furthermore, we assume that a single radiometer contains the intercepted signal during each sampling interval. Let P_{D2} denote the probability that the threshold is exceeded at the end of a sampling interval when a signal is present. Let P_{D1} denote the probability that a particular radiometer output exceeds the threshold when a signal is present in that radiometer. From these definitions, it follows that

$$P_{D2} = 1 - (1 - P_{D1})(1 - P_{F1})^{M-1} \tag{4.2.55}$$

The probability of detection associated with the observation interval is determined by reasoning similar to that which led to (4.2.53) and (4.2.54). The result is

$$P_D = \sum_{i=r}^{N} \sum_{j=0}^{i} \binom{N_1}{j} (P_{D2})^j (1 - P_{D2})^{N_1-j} P(i-j, N-N_1) \tag{4.2.56}$$

where the summand is equal to the probability that the threshold is exceeded i times and the signal is actually present during j of these times.

To compare the channelized radiometer with a single wideband radiometer, we assume that the energy of the intercepted signal is equally divided among N_1 sampling intervals. Because the total receiver bandwidth is W, the bandwidth of each constituent radiometer is $W_s = W/M$. Thus, for large values of $T_s W_s$, P_{F1} and P_{D1} are given by (4.2.46) and (4.2.47) with $W_s = W/M$ substituted for W, $T_s = T_1/N_1 = T/N$ substituted for T, and $E_s = E/N_1$ substituted for E. We define

$$\beta_1 = \text{erfc}^{-1}(2P_{F1}) \tag{4.2.57}$$

$$\xi_1 = \text{erfc}^{-1}(2P_{D1}) \tag{4.2.58}$$

A calculation analogous to that used in deriving (4.2.52) yields the required R_s for detection:

$$R_s = N_0 \left(\frac{2WN}{MT} \right)^{1/2} (\beta_1 - \xi_1), \qquad TW \gg MN \max(\beta_1^2, \xi_1^2) \tag{4.2.59}$$

Equations (4.53) to (4.56) can be solved to determine P_{F1} and P_{D1} in terms of P_F and P_D.

The required power has been expressed in terms of a fixed observation interval of duration T. Successive observation intervals may be related to each other in a variety of ways. If the successive observation intervals do not overlap, except possibly at end points, then the false alarm rate defined in (4.2.48) is an appropriate design parameter. Another possibility, called *binary moving-window detection,* is for the observation interval to be constructed by dropping the first sampling interval of the preceding observation interval and adding a new sampling interval. Let P_r denote the probability that an observation interval results in a false alarm, which occurs only if the comparator input for an added sampling interval exceeds the threshold, but the comparator input for the discarded sampling interval did not. It follows that

$$P_r = P(0,1)P(r-1, N-1)P(1,1) \tag{4.2.60}$$

Because the false alarm rate is $F = P_r/T_s$,

$$F = \frac{N}{T} P(0,1)P(r-1, N-1)P(1,1) \tag{4.2.61}$$

Because $\beta_1 - \xi_1$ is a slowly varying function of P_{F1} and P_{D1}, a rough comparison of (4.2.59) and (4.2.52) indicates that if $M > N$, then a channelized receiver requires approximately the same power as or less power than a wideband radiometer with the same values of T_1, T, W, P_D, and F. Numerical computations using the chi-squared densities verify this result [4]. They also indicate that with binary moving-window detection and an optimal value of r, the channelized receiver sometimes requires less power even when $M < N$. However, if the number of constituent radiometers, M, is less than the number of frequency channels for a frequency-hopping signal, then T_1 for the channelized radiometer may be less than T_1 for the wideband radiometer. As a result, the wideband radiometer may provide a better performance even if $M > N$.

To process frequency-hopping signals efficiently, the sampling interval duration, $T_s = T/N$, should be approximately equal to the hop duration. Because each radiometer bandwidth should equal or exceed the bandwidth of a frequency channel, $W_s = W/M$ remains the same or increases with the hopping rate. Thus, (4.2.59) indicates that the required power usually increases with the hopping rate.

Consider a narrowband signal with a fixed center frequency. If nearly all the signal energy enters a single radiometer and $N = 1$, then

$$P_F = 1 - (1 - P_{F1})^M \tag{4.2.62}$$

$$P_D = 1 - (1 - P_{D1})(1 - P_{F1})^{M-1} \tag{4.2.63}$$

The required value of R_s for detection is determined by the usual method to be

$$R_s = \begin{cases} N_0 \left(\dfrac{2TW}{MT_1^2}\right)^{1/2} (\beta_1 - \xi_1), & T_1 < T, \quad TW \gg M \max(\beta_1^2, \xi_1^2) \\[2ex] N_0 \left(\dfrac{2W}{MT}\right)^{1/2} (\beta_1 - \xi_1), & T_1 \geq T, \quad TW \gg M \max(\beta_1^2, \xi_1^2) \end{cases} \tag{4.2.64}$$

where

$$\beta_1 = \text{erfc}^{-1} [2 - 2(1 - P_F)^{1/M}] \tag{4.2.65}$$

$$\xi_1 = \text{erfc}^{-1} \left[2 - \frac{2(1 - P_D)}{(1 - P_F)^{1-1/M}}\right] \tag{4.2.66}$$

Thus, the required power falls approximately as the inverse of the square root of the number of constituent radiometers.

4.2.3 Cross Correlator

The ideal correlator of Figure 4.1 can be approximated if the signal is intercepted at two spatially separated antennas. The *cross correlation* of the two antenna outputs is computed for various relative arrival times, and the peak value of this function is applied to a comparator. Figure 4.5 shows an analytically tractable, but not necessarily practical, realization using the discrete Fourier transform (DFT). One way to implement the DFT is to use a digital filter and the fast Fourier transform algorithm. An alternative implementation is to use the chirp z-transform algorithm and charge-coupled devices. An analog realization of the cross correlation that is similar to the configuration of Figure 4.5

can be implemented with chirp filters providing Fourier transforms (Section 4.3). Elegant realizations are possible with acousto-optical devices [5].

Figure 4.5(a) depicts the initial processing of each antenna output. After passage through the rectangular bandpass filter of bandwidth W, the intercepted waveform, $r(t) = s(t) + n(t)$, can be represented as

$$r(t) = r_c(t) \cos \omega_c t - r_s(t) \sin \omega_c t \tag{4.2.67}$$

where the quadrature components, $r_c(t)$ and $r_s(t)$,are confined to the band $|f| \leqslant W/2$. If the two rectangular lowpass filters have bandwidths $W/2$ and $f_c = \omega_c/2\pi > W/2$, then $r_c(t)$ and $r_s(t)$ are extracted by the operations shown in Figure 4.5(a). Analog-to-digital converters produce the discrete sequences $r_c(i/W_1)$ and $r_s(i/W_1)$, where the sampling rate is $W_1 > W$.

We regard one of the antennas as a reference and denote its output by the subscript 1. The output of the other antenna is denoted by the subscript 2. In terms of the signal and the noise, (4.2.11) and (4.2.12) imply that

$$r_1(t) = s(t) + n_1(t) \tag{4.2.68}$$

$$r_{1c}(t) = s_c(t) + n_{1c}(t) \tag{4.2.69}$$

$$r_{1s}(t) = s_s(t) + n_{1s}(t) \tag{4.2.70}$$

Let T_r denote the arrival time of the intercepted signal at the second antenna output relative to the arrival time at the reference antenna output. By inserting a sufficiently long delay, we ensure that $T_r \geqslant 0$. By making the separation of the antennas small enough, we ensure that $\omega_c T_r \ll 1$. Consequently, $T_r W \ll 1/\pi$ and

$$r_2(t) = s(t - T_r) + n_2(t) \tag{4.2.71}$$

$$r_{2c}(t) \cong s_c(t - T_r) + n_{2c}(t)\,, \quad \omega_c T_r \ll 1 \tag{4.2.72}$$

$$r_{2s}(t) \cong s_s(t - T_r) + n_{2s}(t)\,, \quad \omega_c T_r \ll 1 \tag{4.2.73}$$

The observation interval has duration $T \gg T_r$. Let

$$a_i = r_{1c}\left(\frac{i+1}{W_1}\right), \quad i = 0, 1, \ldots, \gamma_1 - 1 \tag{4.2.74}$$

$$b_i = r_{2c}\left(\frac{i+1}{W_1}\right), \quad i = 0, 1, \ldots, \gamma_1 - 1 \tag{4.2.75}$$

where γ_1 is the largest integer less than or equal to TW_1. Sequences with $K = \gamma_1 + \nu - 1$ points are formed by augmenting the a_i and b_i with $\nu - 1$ zeros, where $T_r W_1 < \nu \ll \gamma_1$. As indicated in Figure 4.5(b), the conjugate DFT of the a_i is calculated, giving

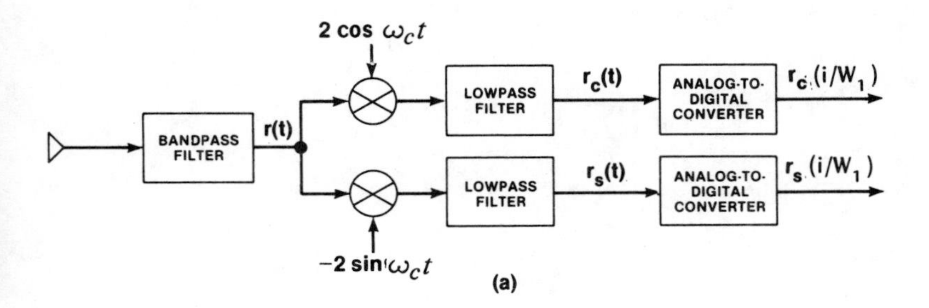

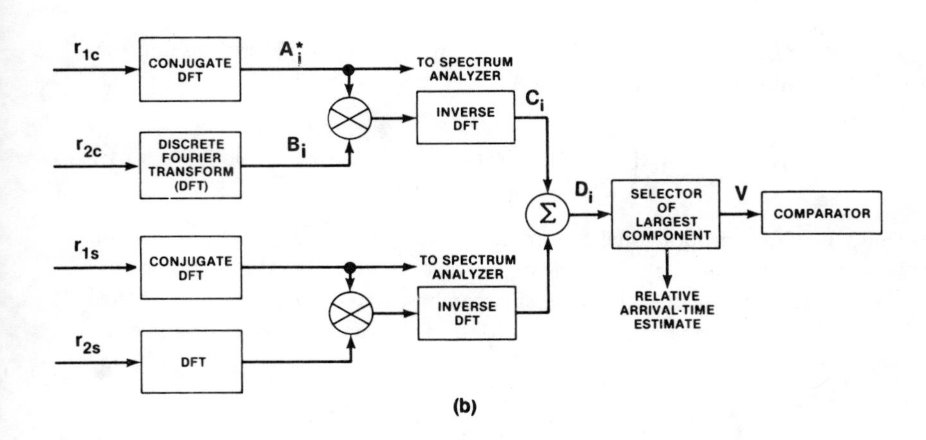

Figure 4.5 Cross correlator: (a) initial processing of each antenna output and (b) final processing.

$$A_i^* = \sum_{n=0}^{K-1} a_n \Omega_K^{-in} \quad i = 0, 1, \ldots, K-1 \tag{4.2.76}$$

where

$$\Omega_K = \exp\left(-\frac{j2\pi}{K}\right) \tag{4.2.77}$$

and $j = \sqrt{-1}$. Similarly, the DFT of the b_i is

$$B_i = \sum_{n=0}^{K-1} b_n \Omega_K^{in}, \qquad i = 0, 1, \ldots, K-1 \tag{4.2.78}$$

The inverse DFT of the product $A_i^* B_i$ for $0 \leqslant i \leqslant \nu - 1$ is

$$\begin{aligned} C_i &= \frac{1}{K} \sum_{n=0}^{K-1} A_n^* B_n \Omega_K^{-in} \\ &= \frac{1}{K} \sum_{m=0}^{K-1} \sum_{k=0}^{K-1} a_m b_k \sum_{n=0}^{K-1} \Omega_K^{n(-m+k-i)} \\ &= \sum_{k=0}^{\gamma_1-1} \sum_{m=0}^{\gamma_1-1} a_m b_k \delta_{m,k-i} \\ &= \sum_{k=i}^{\gamma_1-1} a_{k-i} b_k, \qquad i = 0, 1, \ldots, \nu - 1 \end{aligned} \tag{4.2.79}$$

where $\delta_{ik} = 0, i \neq k$, and $\delta_{ik} = 1, i = k$. From the original definitions and the sampling theorem, we obtain

$$C_i \cong W_1 \int_{i/W_1}^{T} r_{1c}\left(t - \frac{i}{W_1}\right) r_{2c}(t)\,dt, \qquad i = 0, 1, \ldots, \nu - 1 \tag{4.2.80}$$

Thus, the C_i are proportional to sample values of a cross-correlation function. This sequence is the output of one of the inverse DFT operations shown in Figure 4.5(b). An analogous expression can be written for the output of the other inverse DFT operation. The addition of the two sequences produces a sequence

$$D_i = W_1 \int_{i/W_1}^{T} \left[r_{1c}\left(t - \frac{i}{W_1}\right) r_{2c}(t) + r_{1s}\left(t - \frac{i}{W_1}\right) r_{2s}(t) \right] dt \tag{4.2.81}$$

Let q denote the index value that corresponds to the largest of the D_i. If the signal-to-noise ratio is large and the autocorrelations of $s_c(t)$ and $s_s(t)$ have distinct peaks, then q is the index value closest to $T_r W_1$ with high probability. The ratio q/W_1 provides an estimate of T_r. Observe that if the sampling rate were $W_1 = W$, then $q = 0$ would usually occur regardless of the value of T_r since we must have $T_r W \ll 1/\pi$. Thus, we require that $W_1 > W$.

Assuming that the largest D_i has $q = T_r W_1$, it is convenient to use the normalized test statistic $V = D_q/2W_1$ as the input to the comparator in Figure 4.5(b). Equation (4.2.81) and a change of variable give

$$\begin{aligned} V &= \frac{1}{2} \int_{T_r}^{T} [r_{1c}(t - T_r) r_{2c}(t) + r_{1s}(t - T_r) r_{2s}(t)]\,dt \\ &= \frac{1}{2} \int_0^{T_a} [r_{1c}(t') r_{2c}(t' + T_r) + r_{1s}(t') r_{2s}(t' + T_r)]\,dt' \end{aligned} \tag{4.2.82}$$

where $T_a = T - T_r$. In a manner similar to the derivation of (4.2.18), we obtain the series expansion

$$V = \frac{1}{2W} \sum_{i=1}^{\gamma_a} \left[s_c\left(\frac{i}{W}\right) + n_{1c}\left(\frac{i}{W}\right) \right] \left[s_c\left(\frac{i}{W}\right) + n'_{2c}\left(\frac{i}{W}\right) \right]$$
$$+ \frac{1}{2W} \sum_{i=1}^{\gamma_a} \left[s_s\left(\frac{i}{W}\right) + n_{1s}\left(\frac{i}{W}\right) \right] \left[s_s\left(\frac{i}{W}\right) + n'_{2s}\left(\frac{i}{W}\right) \right] \quad (4.2.83)$$

where $n'_{2c}(t) = n_{2c}(t + T_r)$, $n'_{2s}(t) = n_{2s}(t + T_r)$, and γ_a is the largest integer less than or equal to $T_a W$. Assuming that $n_1(t)$ and $n_2(t)$ are statistically independent, zero-mean, Gaussian processes, a straightforward calculation yields

$$E[V] = E_a \quad (4.2.84)$$

$$\mathrm{VAR}(V) = N_0 E_a + N_0^2 \gamma_a / 2 \quad (4.2.85)$$

where E_a is the signal energy in interval T_a.

For large values of $T_a W$, the test statistic is approximately normally distributed. It follows that

$$P_D = \frac{1}{2} \operatorname{erfc} \left[\frac{V_T - E_a}{(N_0^2 T_a W + 2N_0 E_a)^{1/2}} \right], \quad T_a W \gg 1 \quad (4.2.86)$$

The maximum possible value of T_r may be such that only a few sample values of D_i need to be computed to obtain an appropriate test statistic. When no signal is present, a false alarm occurs if any of the ν estimated autocorrelation values exceeds the threshold. If ν is sufficiently small, it is reasonable to assume that each estimated value has approximately the same probability, denoted by P_{F1}, of exceeding the threshold. This assumption implies the approximation

$$P_F = 1 - (1 - P_{F1})^{\nu} \quad (4.2.87)$$

In the absence of the signal, the mean and variance of V are given by (4.2.84) and (4.2.85) with $E_a = 0$. For large values of $T_a W$, we obtain

$$P_{F1} = \frac{1}{2} \operatorname{erfc} \left[\frac{V_T}{(N_0^2 T_a W)^{1/2}} \right], \quad T_a W \gg 1 \quad (4.2.88)$$

Equations (4.2.57) and (4.2.87) yield

$$\beta_1 = \operatorname{erfc}^{-1} [2 - 2(1 - P_F)^{1/\nu}] \quad (4.2.89)$$

We obtain in the usual manner the required R_s to detect a signal with specified values of P_D and either P_F or F. The result is

$$R_s = \begin{cases} N_0 \dfrac{(T_a W)^{1/2}}{T_1} (\beta_1 - \xi), & T_1 < T_a, \quad T_a W \gg \max(\beta_1^2, \xi^2) \\ N_0 \left(\dfrac{W}{T_a}\right)^{1/2} (\beta_1 - \xi), & T_1 \geq T_a, \quad T_a W \gg \max(\beta_1^2, \xi^2) \end{cases} \tag{4.2.90}$$

where T_1 is the signal duration, and ξ is given by (4.2.5).

Comparison with (4.2.52) indicates that the cross correlator can give a theoretical improvement of approximately 1.5 dB over a single wideband radiometer. Taking into account the approximations made to derive (4.2.90), it is possible that in practice the cross correlator provides no improvement at all. A comparison of Figures 4.3 and 4.5 indicates that the implementation of the cross correlator entails considerably more hardware than the implementation of a wideband radiometer. However, as discussed in subsequent sections, the cross correlator requires little additional hardware to provide frequency estimation and direction finding. Furthermore, interference or jamming can often be reduced by inserting filters between the DFT blocks and the multipliers in Figure 4.5.

The *channelized cross correlator* is an array of M cross correlators, each of which has a bandwidth of W/M. The outputs of the array are applied to a processor. Analogously to the channelized radiometer, the channelized cross correlator may be preferable to a single wideband cross correlator when the hostile communications are narrowband or when two or more simultaneous signals are to be intercepted.

Equations (4.2.90) and (4.2.52) indicate that increasing the bandwidth of a frequency-hopping system degrades the performance of both the wideband cross correlator and the wideband radiometer, but neither of these receivers is sensitive to the hopping rate. Increasing the hopping rate makes the practical design of a channelized receiver more difficult and degrades its performance. If the rate is sufficiently high, the channelized receiver may have to be abandoned in favor of a wideband receiver.

If the form of the signal to be intercepted is known, or partially known, then a detector can sometimes be designed to yield better performance than the wideband radiometer, channelized radiometer, or cross correlator. However, the theoretically superior detectors are very complex and their implementation losses and costs probably negate their advantages in practical applications.

4.3 FREQUENCY ESTIMATION

The purpose of a *frequency-estimation system* is to determine the center frequency and possibly the spectral shape of an intercepted signal. If a frequency-hopping signal is intercepted, the purpose is to determine each hopping frequency or at least the frequency range over which the hopping occurs. As described in Section 2.3, nonlinear processing of direct-sequence signals is desirable before frequency estimation is attempted.

The most important performance measures of a frequency-estimation system are the estimation accuracy and the frequency resolution. Let $\text{VAR}(\hat{f})$ denote the variance of a frequency estimate $\hat{f}$ of a center frequency f. Consider a sinusoidal signal of constant amplitude, unknown initial phase, and duration T in white Gaussian noise. The Cramer-Rao bound for an unbiased frequency estimate gives (Appendix B)

$$\text{VAR}(\hat{f}) \geqslant \left(\frac{2\pi^2 T^2 E}{3N_0} \right)^{-1} \tag{4.3.1}$$

Although it is difficult to accurately determine $VAR(\hat{f})$ for a particular frequency-estimation system, this bound provides at least a rough measure of what might be achieved in a well-designed system.

The *frequency resolution* of a frequency-estimation system is the minimum difference in center frequencies between two signals of equal power that is necessary if the presence of two distinct signals is to be recognized when the noise is weak. If the observation interval has duration T, then the achievable resolution for unknown signals is usually found to be

$$\Delta \approx \frac{1}{T} \tag{4.3.2}$$

A lower resolution is possible if *a priori* information is available, for example, if it is known that the signals to be observed are always unmodulated tones.

4.3.1 Channelized Receiver

Estimation theory leads to the receiver of Figure 4.2 for frequency estimation, assuming that the frequency belongs to a discrete set and that the arrival time and the signal waveform, except for a uniformly distributed phase angle, are known [1]. The unknown frequency is estimated as the center frequency of the filter producing the largest output. The channelized radiometers of Figure 4.4 are practical approximations to the ideal frequency estimator.

Consider an array of parallel bandpass filters. Each filter is followed by a detector which may be a radiometer. Suppose that we desire a frequency resolution of Δ for two ideal tones, one of which has its carrier frequency equal to the center frequency of one of the filters. If the filter passbands are rectangular and disjoint, then each filter must have a bandwidth equal to 2Δ. If the entire range of monitored frequencies has a bandwidth W, then the number of required filters is $M = W/2\Delta$. For intercepted signals with significant bandwidths and more practical filters, the outputs of adjacent filters must be analyzed and compared to determine the number of different received signals and their center frequencies.

If $M \geqslant 6$, the number of required filters can be reduced by arranging them in successive stages, as shown in Figure 4.6 for the case in which each stage has the same number of filters. The resolution provided by the first filter bank is $\Delta_1 = W/2m$, where m is the number of filters in each stage and $2\Delta_1$ is the bandwidth of each of the first stage filters. A bank of mixers ensures that the filter outputs are shifted in frequency so that the input to the second stage has a frequence between $f_{11} - f_{c1} - \Delta_1$ and $f_{11} - f_{c1} + \Delta_1$, where f_{11} is the center frequency of the top filter in the first bank and f_{c1} is the frequency of a local oscillator. The resolution provided by the first and second filter banks together is $\Delta_2 = \Delta_1/m$. If n stages of m filters each are used, then a resolution of Δ is attained if

$$m^n = \frac{W}{2\Delta} \tag{4.3.3}$$

The total number of filters required is mn. Disadvantages of the channelized receiver of Figure 4.6, relative to that of Figure 4.4, are the increased processing time required for frequency estimation, the reduced amount of noise and interference filtering, and the ambiguities that may arise when more than one signal is intercepted.

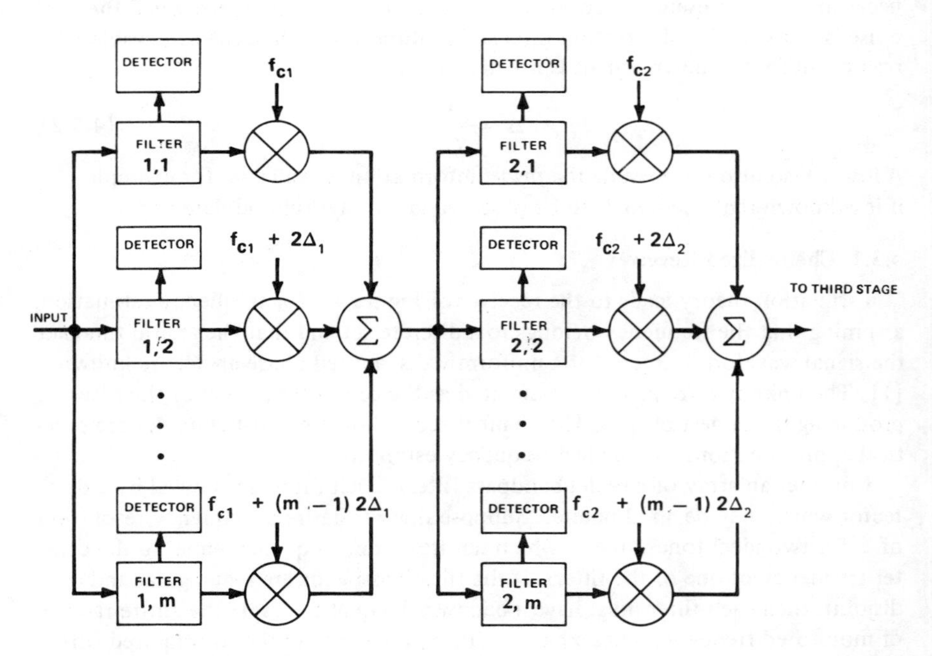

Figure 4.6 Channelized receiver with filters arranged in successive stages.

It is not necessary that each stage have the same number of filters. However, if n and W/Δ are fixed, it can easily be shown, by using Lagrange multipliers, that the total number of filters is minimized if each stage has approximately the same number of filters (exactly the same number if an integer m exists that satisfies (4.3.3)). If each stage has the same number of filters and W/Δ is fixed, it can be shown that the total number of filters is minimized if each stage has three filters (Lagrange multipliers yield $m = e$, but m must be an integer) provided that W/Δ is chosen so that (4.3.3) is satisfied when $m = 3$ and n is some positive integer.

Even when minimized, the number of filters and detectors required in a channelized receiver may make this method of frequency estimation expensive. The detectors of a stage can be eliminated by replacing the summer with a commutator and detector, but then a signal may be missed. Alternatively, a limited number of filters can be used to reduce the total bandwidth examined by other frequency estimation devices.

4.3.2 Acousto-Optical Spectrum Analyzer

Acousto-optical diffraction provides a means of implementing a channelized radiometer without filter banks. The principal components of an *acousto-optical spectrum analyzer* are shown in Figure 4.7. The diffraction geometry associated with the *Bragg cell* is illustrated in Figure 4.8. The Bragg cell converts an electronic input at frequency f_0 into a traveling acoustic wave with velocity v and wavelength $\Lambda_a = v/f_0$. The laser light has wavelength Λ_l in free space. Under certain conditions, the acoustic wave interacts with the light beam to produce a principal diffracted beam that is offset from the incident beam by an angle

$$\theta = 2 \sin^{-1} \left(\frac{\Lambda_l f_0}{2v} \right) \tag{4.3.4}$$

outside the cell [5]. This equation is valid provided that the acoustic wave has a single wavelength across the cell. For small values of the argument, (4.3.4) becomes

$$\theta \cong \frac{\Lambda_l f_0}{v}, \quad \Lambda_l f_0 \ll 2v \tag{4.3.5}$$

The lens produces a Fourier transform on its focal plane at the photodetector array. The center of the diffracted beam converges to a position at a distance approximately

$$F\theta = \frac{F \Lambda_l f_0}{v} \tag{4.3.6}$$

from the center of the corresponding undiffracted beam, where F is the focal length of the lens. Thus, the frequency f_0 can be estimated by selecting the largest of the outputs of the photodetector array elements. Because an element responds to the intensity of the light incident upon it, an element serves as a radiometer.

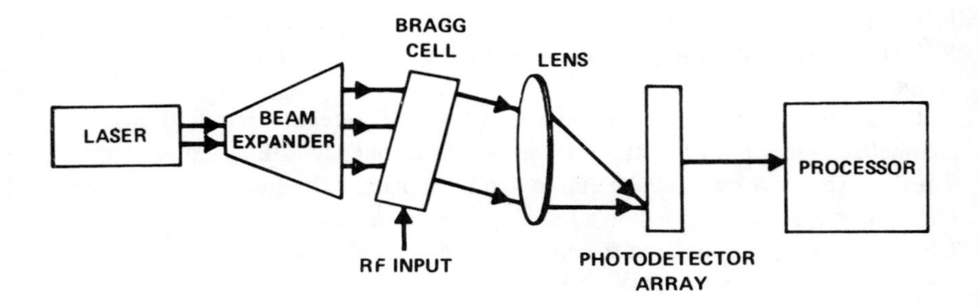

Figure 4.7 Acousto-optical spectrum analyzer.

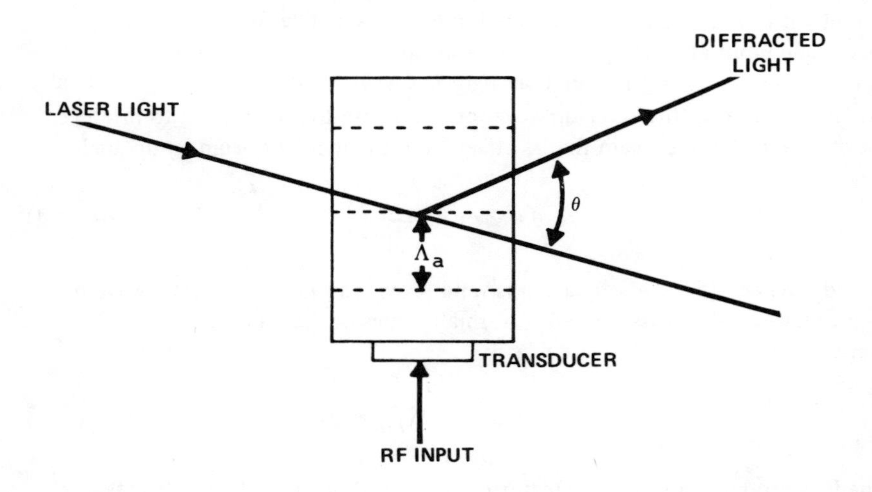

Figure 4.8 Acousto-optical diffraction geometry for input at single frequency.

The diffracted beam has an angular half-width on the order of Λ_l/D, where D is the effective aperture of the Bragg cell, which is approximately equal to the width of the incident optical beam. Consequently, the diffracted beam forms a spot of half-width $F\Lambda_l/D$ in the focal plane. The frequency resolution is defined to be the difference in frequency between two intercepted signals such that the corresponding center positions in the focal plane differ by the half-width of a spot in the focal plane. From this definition and (4.3.6), the resolution is

$$\Delta \cong \frac{v}{D} = \frac{1}{T_c} \tag{4.3.7}$$

where T_c is the time that it takes an acoustic wave to cross the cell aperture. To achieve this resolution in practice, the photodetector element spacing must not exceed the beam half-width and the acoustic wave must be sufficiently narrowband. To achieve this resolution against a frequency-hopping signal, T_c must be less than the dwell time of the pulses. Thus, the resolution is no better than the hopping rate.

4.3.3 Spectrum Analysis from the Discrete Fourier Transform

The discrete Fourier transform (DFT) can be used to estimate the spectrum of a truncated version of an intercepted signal. The truncation results because only the part of the signal over an observation interval of duration T is used in the computation of the DFT. Thus, it is convenient to assume that $s(t)$, $s_c(t)$, and $s_s(t)$, which are defined in Section 4.2, vanish outside the interval $0 \leqslant t \leqslant T$. The Fourier transforms of these signals are denoted by $S(f)$, $S_c(f)$, and $S_s(f)$, respectively. From (4.2.11), it follows that

$$S(f) = \frac{1}{2} S_c(f - f_c) + \frac{1}{2} S_c(f + f_c) - \frac{1}{2j} S_s(f - f_c) + \frac{1}{2j} S_s(f + f_c) \tag{4.3.8}$$

It is usually of interest to determine $S(f)$ for $f \geqslant 0$. If the center frequency f_c exceeds half the receiver bandwidth, W, then

$$S(f) = \frac{1}{2} S_c(f - f_c) - \frac{1}{2j} S_s(f - f_c), \quad f \geqslant 0, \ f_c \geqslant W/2 \tag{4.3.9}$$

Thus, $S(f)$ can be estimated by first estimating $S_c(f)$ and $S_s(f)$. For spectrum analysis, it is not necessary for the sampling rate, W_1, to exceed the bandwidth. Hence, we set $W_1 = W$ in the subsequent analysis.

To show how $S_c(f)$ is estimated, we first establish the relation between the discrete and continuous Fourier transforms. The continuous Fourier transform of an absolutely integrable function $g(t)$ is defined as

$$G(f) = \int_{-\infty}^{\infty} g(t)\exp(-j2\pi ft)\,dt \tag{4.3.10}$$

The periodic extension of $G(f)$ is defined as

$$\overline{G}(f) = \frac{1}{W} \sum_{i=-\infty}^{\infty} G(f+iW) \tag{4.3.11}$$

where the period of $\overline{G}(f)$ is equal to W and it is assumed that the series converges uniformly. Let K denote a positive integer. *Poisson's sum formula,* which is given by (A.4.7) of Appendix A and is valid under mild conditions, implies that

$$\overline{G}\left(\frac{iW}{K}\right) = \frac{1}{W} \sum_{k=-\infty}^{\infty} g\left(\frac{k}{W}\right)\Omega_K^{ik},\ i=0, 1, \ldots, K-1 \tag{4.3.12}$$

where Ω_K is defined by (4.2.77).

Any integer k can be expressed as $k = mK + n$, where m is some integer and n is an integer such that $0 \leqslant n \leqslant K-1$. Therefore, (4.3.12) can be written as

$$\overline{G}\left(\frac{iW}{K}\right) = \frac{1}{W} \sum_{n=0}^{K-1} \sum_{m=-\infty}^{\infty} g\left(\frac{n+mk}{W}\right)\Omega_K^{i(n+mK)} \tag{4.3.13}$$

We define the periodic extension of the signal,

$$\overline{g}(t) = \sum_{m=-\infty}^{\infty} g\left(t + m\frac{K}{W}\right) \tag{4.3.14}$$

where the series is assumed to converge for all t. Observing that

$$\Omega_K^{imK} = 1 \tag{4.3.15}$$

we obtain

$$\overline{G}\left(\frac{iW}{K}\right) = \frac{1}{W} \sum_{n=0}^{K-1} \overline{g}\left(\frac{n}{W}\right)\Omega_K^{in}, \quad i=0, 1, \ldots, K-1 \tag{4.3.16}$$

This equation relates a periodic extension of the signal spectrum to the discrete Fourier transform of a periodic extension of the signal.

We apply (4.3.16) to the estimation of $S_c(f)$; the estimation of $S_s(f)$ is similar. This equation implies that

$$\overline{S}_c\left(\frac{iW}{K}\right) = \frac{1}{W} \sum_{n=0}^{K-1} \overline{s}_c\left(\frac{n}{W}\right)\Omega_K^{in}, \quad i=0, 1, \ldots, K-1 \tag{4.3.17}$$

where

$$\bar{s}_c\left(\frac{n}{W}\right) = \sum_{m=-\infty}^{\infty} s_c\left(\frac{n}{W} + m\frac{K}{W}\right) \tag{4.3.18}$$

$$\bar{S}_c\left(\frac{iW}{K}\right) = \sum_{m=-\infty}^{\infty} S_c\left(\frac{iW}{K} + mW\right) \tag{4.3.19}$$

Because $s_c(t) = 0$ unless $0 \leqslant t \leqslant T$, then if $K > TW$, (4.3.18) implies that

$$\bar{s}_c\left(\frac{n}{W}\right) = s_c\left(\frac{n}{W}\right), \quad 0 \leqslant n \leqslant K-1 \tag{4.3.20}$$

We assume that $S(f) \cong 0$ unless $|f - f_c| < W/2$ or $|f + f_c| < W/2$. Consequently, $S_c(f) \cong 0$ for $|f| \geqslant W/2$, and (4.3.19) yields the approximate result

$$\bar{S}_c\left(\frac{iW}{K}\right) = \begin{cases} S_c\left(\dfrac{iW}{K}\right), & 0 \leqslant i \leqslant K/2 \\[2ex] S_c\left(\dfrac{iW}{K} - W\right), & K/2 \leqslant i \leqslant K \end{cases} \tag{4.3.21}$$

Equations (4.3.17), (4.3.20), and (4.3.21) imply that

$$S_c\left(\frac{iW}{K}\right) = \frac{1}{W} \sum_{n=0}^{K-1} s_c\left(\frac{n}{W}\right) \Omega_K^{in}, \quad |i| \leqslant K/2 \tag{4.3.22}$$

We conclude that, in the presence of noise, a reasonable estimate of the sample values of $S_c(f)$ is given by

$$\hat{S}_c\left(\frac{iW}{K}\right) = \begin{cases} \dfrac{1}{W} A_i, & 0 \leqslant i \leqslant K/2 \\[2ex] \dfrac{1}{W} A_{i+K}, & -K/2 \leqslant i < 0 \end{cases} \tag{4.3.23}$$

where

$$A_i = \sum_{n=0}^{K-1} r_c\left(\frac{n}{W}\right) \Omega_K^{in}, \qquad i = 0, 1, \ldots, K-1 \tag{4.3.24}$$

and $r_c(t)$ is defined by (4.2.67). Thus, simple operations on the DFT give the estimates $\hat{S}_c(f)$ and $\hat{S}_s(f)$ and, hence, $\hat{S}(f)$ at sampled frequency values. An estimate of the power spectral density of $s(t)$ is proportional to $|\hat{S}(f)|^2$.

By sampling the input every $1/W$ seconds and using a DFT of size $K > TW$, accurate sampled values of the Fourier transform can be obtained. However, as W increases, the attainable speed of the logic circuitry becomes a limiting factor. Faster rates are possible if many parallel devices are used, but then the power dissipation and system size become problems. Thus, for a large bandwidth, analog processing may be necessary.

If the bandwidth is not too large, digital processing with the DFT has major advantages. For example, an almost arbitrary linear dynamic range is obtainable by increasing the number of bits per sample and the system complexity.

The main lobes of the spectral responses of weak signals can be masked by higher sidelobes from the spectral responses of stronger signals. To reduce the sidelobes, the data samples can be weighted, a process known as *data windowing.* However, the reduction of sidelobes invariably leads to a loss of resolution relative to (4.3.2).

The duration T of the observation interval is limited by the duration of the measured signal and by the time dependency of the signal spectrum. For a fixed T, an improvement in resolution beyond that predicted by (4.3.2) depends upon fitting the measured data to an assumed model that approximates certain signal characteristics. Thus, nonzero values can be assumed for the received signal outside the observation interval. For example, the *maximum entropy method* assumes that the sample autocorrelation function extrapolates beyond the measured values in a manner that maximizes the randomness of the corresponding time series. The improvement provided by the model depends upon the accuracy of its assumptions, especially when the signal-to-noise ratio is low. However, even if the assumptions are justified, the computational requirements of spectral estimation based on a model may make real-time implementation infeasible [6].

4.3.4 Instantaneous Frequency Measurement

The *instantaneous frequency measurement* (IFM) *receiver* is primarily useful as a supplement to other frequency estimators. Its operation is based on the relationship among carrier frequency, path length, and phase shift of a signal.

One version of the IFM receiver is illustrated in Figure 4.9. Suppose that, after passage through the bandpass filter of bandwidth W, an intercepted signal has the form

$$s(t) = A(t)\cos[2\pi f_0 t + \phi(t)] \tag{4.3.25}$$

where $A(t)$ is the amplitude modulation and $\phi(t)$ is the angle modulation. This signal is delayed by time δ in one branch relative to the other branch. If δ is sufficiently small, then

$$A(t-\delta) \cong A(t), \quad \phi(t-\delta) \cong \phi(t) \tag{4.3.26}$$

for most of the time. It follows that

$$s_1(t) \cong A(t)\cos[2\pi f_0 t + \phi(t) - 2\pi f_0 \delta] \tag{4.3.27}$$

By trigonometric identities, the outputs of the sum and difference operations in the figure are found to be

$$s_2(t) = 2A(t)\cos(\pi f_0 \delta)\cos[2\pi f_0 t + \phi(t) - \pi f_0 \delta] \tag{4.3.28}$$

$$s_3(t) = 2A(t)\sin(\pi f_0 \delta)\sin[2\pi f_0 t + \phi(t) - \pi f_0 \delta] \tag{4.3.29}$$

for most of the time. The envelope detectors produce the magnitudes of the first factors in these equations. The detector outputs pass through logarithmic amplifiers, and the difference is taken. The processor input is

$$s_4(t) = \log|\tan(\pi f_0 \delta)| \tag{4.3.30}$$

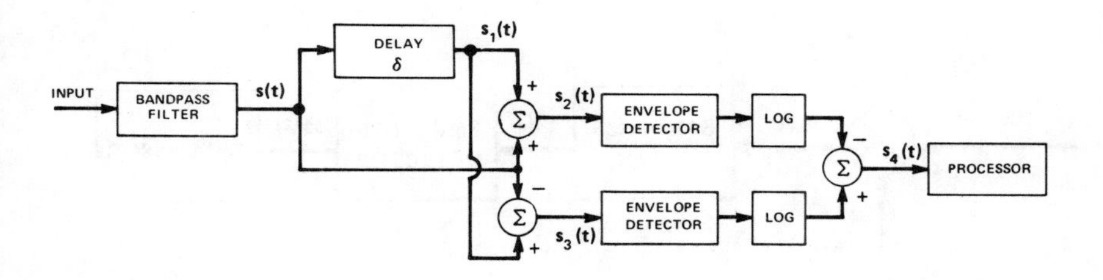

Figure 4.9 Instantaneous frequency measurement receiver.

Because the modulation effects have been removed and δ is known, the processor can calculate an estimate of f_0. The function on the right-hand side of (4.3.30) has an unambiguous inverse only over a range of $\pi/2$ radians. Thus, for unambiguous operation over the frequency range of W hertz, it is necessary that $\delta \leqslant 1/2W$, which is also usually adequate for the validity of (4.3.26). However, the estimation accuracy improves with increases in δ. To achieve unambiguous and accurate frequency estimation, two or more different values of δ may be used. The smaller values of δ primarily ensure an unambiguous estimate, while the larger values of δ primarily determine the accuracy [7].

The major problem of the IFM receiver is its poor response to two or more simultaneously intercepted signals of comparable magnitudes. The IFM output may be regarded as a seriously degraded estimate of the frequency of the larger signal. Consequently, the frequency range may be limited, and a device to detect the presence of more than one signal may be necessary.

4.3.5 Scanning Superheterodyne Receiver

Figure 4.10 shows a block diagram of a realization of a *scanning superheterodyne receiver* for frequency estimation. To explain the operation, we consider the system response to one scan of the generator and an input that has constant amplitude and frequency over the scan period, T. The input is represented by

$$s(t) = A\cos(2\pi f_0 t + \theta_0)\,, \quad 0 \leqslant t \leqslant T \tag{4.3.31}$$

where f_0 is the carrier frequency and θ_0 is the phase angle at $t = 0$, which defines the beginning of the scan. The *scanning waveform* is a periodic function. Over one scan period, it is represented by

$$y(t) = \cos(2\pi f_s t + \pi\mu t^2 + \theta_s)\,, \; 0 \leqslant t \leqslant T \tag{4.3.32}$$

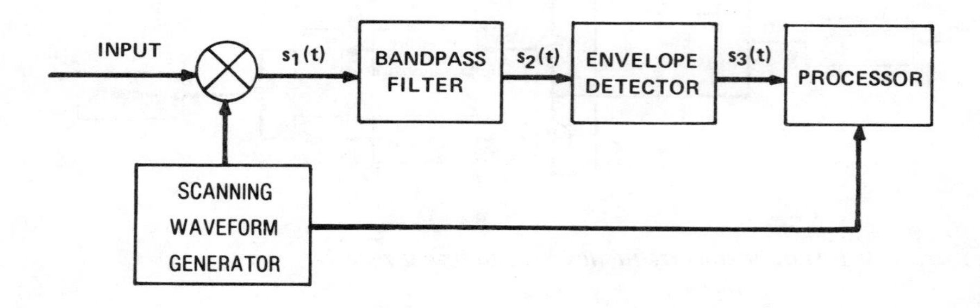

Figure 4.10 Scanning superheterodyne receiver.

where f_s is the frequency at $t = 0$, μ is the scan rate, and θ_s is the phase angle at $t = 0$. For simplicity, it is assumed that $\mu > 0$. The output of the mixer, $s_1(t) = s(t)y(t)$, passes through a bandpass filter with impulse response $h(t)$ and bandwidth $2B$. It is assumed that a high-frequency term is suppressed by the bandpass filter so that the relevant component of the mixer output is

$$s_1(t) = \frac{1}{2} A \cos(2\pi f_1 t + \pi\mu t^2 + \theta)\,, \quad 0 \leqslant t \leqslant T \tag{4.3.33}$$

where $f_1 = f_s - f_0$ and $\theta = \theta_s - \theta_0$. The symmetrical bandpass filter has transfer function $H(f)$ that can be written as

$$H(f) = H_1(f - f_c) + H_1(f + f_c) \tag{4.3.34}$$

where $H_1(f)$ is the transfer function of a lowpass filter and f_c is the center frequency of the bandpass filter. The first term on the right-hand side has significant values only for positive frequencies, and the second term has significant values only for negative frequencies. If $h_1(t)$ is the impulse response of the lowpass filter, then

$$h(t) = 2h_1(t)\cos 2\pi f_c t \tag{4.3.35}$$

The output of the bandpass filter is

$$s_2(t) = \int_{-\infty}^{\infty} s_1(\tau)\, h(t-\tau)\, d\tau \tag{4.3.36}$$

The substitution of (4.3.33) and (4.3.35) and the pertinent trigonometric relations into (4.3.36) yields

$$s_2(t) = \frac{A}{2}\int_0^T h_1(t-\tau)\cos[2\pi(f_1 - f_c)\tau + \pi\mu\tau^2 + \theta + 2\pi f_c t]\, d\tau$$

$$+ \frac{A}{2}\int_0^T h_1(t-\tau)\cos[2\pi(f_1 + f_c)\tau + \pi\mu\tau^2 + \theta - 2\pi f_c t]\, d\tau \tag{4.3.37}$$

It is assumed that $H_1(\omega)$ has a sufficiently narrow bandwidth relative to $f_1 + f_c$ that the second integral on the right-hand side of this equation is negligible. The time-frequency diagram of Figure 4.11 illustrates the effect of the filter. The filter output, $s_2(t)$, is significant only over a portion of the scan period. Thus, the limits of the first integral can be extended to $\pm\infty$ with negligible error if $f_1 + \mu T \gg f_c + B/2$ and $f_1 \ll f_c - B/2$, for which it is necessary that

$$\mu T \gg B \tag{4.3.38}$$

Under these assumptions, it follows that

$$s_2(t) \cong \frac{A}{2}\int_{-\infty}^{\infty} h_1(t-\tau)\cos(2\pi f_2\tau + \pi\mu\tau^2 + \theta_1)d\tau \tag{4.3.39}$$

where $f_2 = f_s - f_0 - f_c$ and $\theta_1 = \theta + 2\pi f_c t$. To further simplify (4.3.39), we assume a Gaussian bandpass filter with

$$H_1(f) = \exp\left[-\frac{\pi^2 f^2}{a^2} - j2\pi f\delta\right] \tag{4.3.40}$$

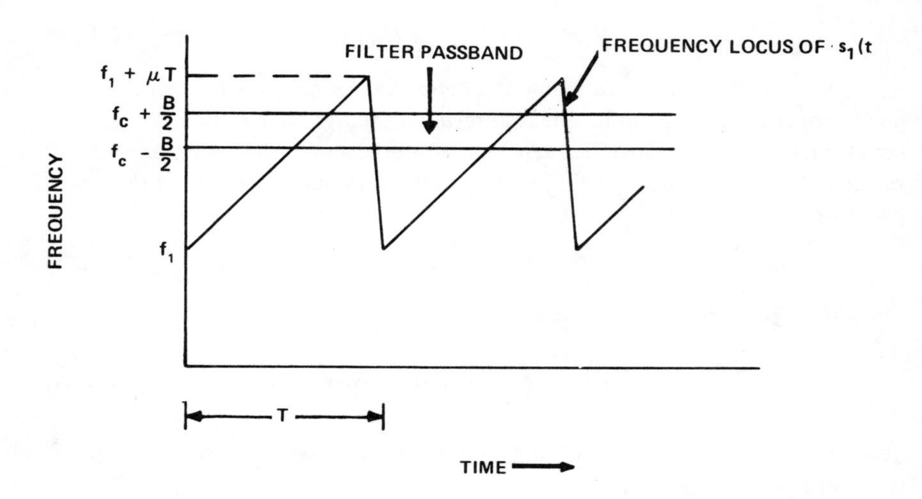

Figure 4.11 Time-frequency diagram for scanning superheterodyne receiver.

where parameter a is proportional to the bandwidth and δ is the filter delay. If δ is sufficiently large, (4.3.40) approximates a realizable filter. The corresponding impulse response is

$$h_1(t) = \int_{-\infty}^{\infty} H_1(f) e^{j2\pi ft} df = \frac{a}{\sqrt{\pi}} \exp\left[-a^2(t-\delta)^2\right] \tag{4.3.41}$$

Substituting (4.3.41) into (4.3.39), expressing the cosine in terms of complex exponentials, and simplifying the result, we obtain

$$s_2(t) = \text{Re}\left\{ \frac{Aa}{2\sqrt{\pi}} \exp\left[-a^2(t-\delta)^2 + j\theta_1 + sc^2\right] \int_{-\infty}^{\infty} \exp\left[-s(\tau + c)^2\right] d\tau \right\} \tag{4.3.42}$$

where $Re(x)$ denotes the real part of x and

$$s = a^2 - j\pi\mu \tag{4.3.43}$$

$$c = \frac{-2a^2(t-\delta) - j2\pi f_2}{2s} \tag{4.3.44}$$

The integral in Equation (4.3.42) is [8]

$$\int_{-\infty}^{\infty} \exp[-s(\tau + c)^2]\, d\tau = \left(\frac{\pi}{s}\right)^{1/2}, \quad Re(s) > 0, \; Re(\sqrt{s}) > 0 \tag{4.3.45}$$

Therefore,

$$s_2(t) = Re \left\{ \exp[-a^2(t - \delta)^2 + j\theta_1 + sc^2] \, \frac{Aa}{2\sqrt{s}} \right\}$$

$$= s_3(t)\cos[\phi(t)] \tag{4.3.46}$$

where

$$\phi(t) = \frac{\pi a^4 \mu (t - \delta)^2 + 2\pi a^4 f_2 (t - \delta) - \pi^3 \mu f_2^2}{a^4 + \pi^2 \mu^2} + \theta_1 + \frac{1}{2}\tan^{-1}\left(\frac{\pi\mu}{a^2}\right) \tag{4.3.47}$$

$$s_3(t) = \frac{A}{2}\left(1 + \frac{\pi^2 \mu^2}{a^4}\right)^{-1/4} \exp\left[-\frac{\pi^2 a^2 (\mu t - \mu\delta + f_2)^2}{a^4 + \pi^2 \mu^2}\right] \tag{4.3.48}$$

As indicated in Figure 4.10, $s_2(t)$ is applied to an envelope detector that extracts the envelope, $s_3(t)$, from the input. The peak value of $s_3(t)$ is attained when $t = \delta - f_2/\mu = \delta + (f_0 + f_c - f_s)/\mu$. Thus, the input frequency f_0 can easily be estimated from the time location of the peak value. The normalized peak value, α, which is defined as the peak value relative to $A/2$, the peak value for small μ and f_2, is

$$\alpha = \left(1 + \frac{\pi^2 \mu^2}{a^4}\right)^{-1/4} \tag{4.3.49}$$

The half-power points of $s_3(t)$ are determined by setting the exponential factor in (4.3.48) equal to $1/\sqrt{2}$. The pulse duration of $s_3(t)$ between half-power points is found to be

$$T_p = \frac{a(2 \ln 2)^{1/2}}{\pi\mu}\left(1 + \frac{\pi^2 \mu^2}{a^4}\right)^{1/2} \tag{4.3.50}$$

When $t = \delta - f_2/\mu \pm T_p$, $s_3(t)$ falls to one-fourth of its peak value. Thus, it is reasonable to assume that the frequency resolution, Δ, is approximately equal to μT_p, the frequency range scanned during an interval of duration T_p. Therefore,

$$\Delta = \frac{a(2\ ln\ 2)^{1/2}}{\pi} \left(1 + \frac{\pi^2\mu^2}{a^4}\right)^{1/2} \tag{4.3.51}$$

From Equation (4.3.40), the 3-dB power-spectrum bandwidth in hertz is related to parameter a by

$$B = \frac{(2\ ln\ 2)^{1/2}}{\pi}\, a \tag{4.3.52}$$

In terms of B,

$$\alpha = \left(1 + 0.195\,\frac{\mu^2}{B^4}\right)^{-1/4} \tag{4.3.53}$$

$$\Delta = B\left(1 + 0.195\,\frac{\mu^2}{B^4}\right)^{1/2} \tag{4.3.54}$$

These equations have long been used to explain the operational characteristics of spectrum analyzers.

If the scan rate is high,

$$\alpha \cong 1.5\,\frac{B}{\sqrt{\mu}}, \qquad \Delta \cong 0.44\,\frac{\mu}{B}, \qquad \mu \gg B^2 \tag{4.3.55}$$

which clearly show the effects of increasing μ.

Using elementary calculus, we determine the optimal filter bandwidth, B_0, to minimize Δ. Substituting B_0 into (4.3.54), we obtain Δ_0, the minimum resolution as a function of μ. The results are

$$B_0 = 0.66\sqrt{\mu} \tag{4.3.56}$$

$$\Delta_0 = \sqrt{2}\,B_0 = 0.94\sqrt{\mu} \tag{4.3.57}$$

The corresponding normalized peak value is

$$\alpha_0 = 0.84 \tag{4.3.58}$$

which is no longer a function of μ. If the optimal bandwidth is used, these equations indicate that the achievable resolution becomes worse as μ increases, but the peak value does not change. Relations (4.3.56) and (4.3.38) indicate that

$$\sqrt{\mu} \gg \frac{0.66}{T} \tag{4.3.59}$$

is a necessary condition for the validity of (4.3.56) to (4.3.58). Relations (4.3.57) and (4.3.59) imply that the scanning superheterodyne receiver at best might barely achieve the intrinsic resolution of (4.3.2).

4.3.6 Compressive Receiver

Improved resolution at a high scan rate is provided by the *compressive receiver,* which is sometimes called the *microscan receiver.* A compressive receiver contains chirp filters or possibly other filters with dispersive transfer functions. A *chirp filter* has an impulse response with a linearly varying frequency modulation. The version of the compressive receiver shown in Figure 4.12 is similar to the scanning superheterodyne receiver except that a chirp filter is used instead of a bandpass filter.

For carrier frequency estimation, the impulse response of the chirp filter is

$$h(t) = \cos(2\pi f_c t - \pi\mu t^2 + \theta_c), \quad 0 \leqslant t \leqslant T \tag{4.3.60}$$

where the amplitude has been normalized to unity, $\mu > 0$, and the duration is equal to the scan period. It is assumed that the chirp filter suppresses the high-frequency component of $s_1(t)$ so that the relevant part of $s_1(t)$ is represented by (4.3.33). Substituting (4.3.60) and (4.3.33) into (4.3.36) yields the chirp filter response to the receiver input over one scan period. Using $2 \cos u \cos v = \cos(u - v) + \cos(u + v)$, we may write the result as the sum of two integrals. Assuming that $|f_1 + f_c - 2\mu T|$ is much larger than $|f_1 - f_c + 2\mu T|$ and $1/T$, we can usually neglect one of the integrals, thereby obtaining the approximation

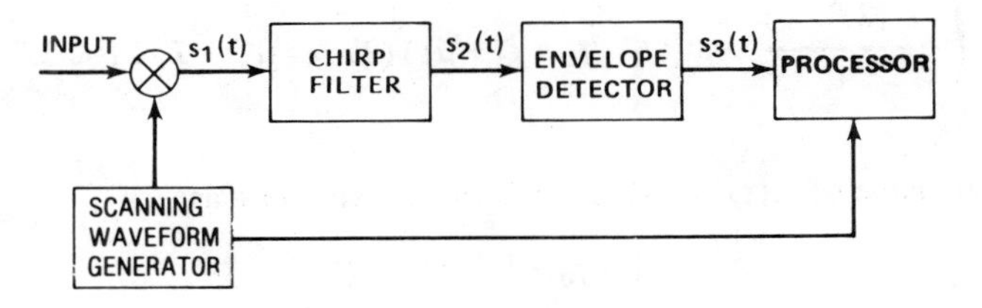

Figure 4.12 Compressive receiver for estimating carrier frequency or magnitude of Fourier transform.

$$s_2(t) = \begin{cases} \dfrac{A}{4} \displaystyle\int_{T_1}^{T_2} \cos(2\pi f_3 \tau + \theta_2)\, d\tau, & 0 \leqslant t \leqslant 2T \\ 0, & t < 0 \text{ or } t > 2T \end{cases} \tag{4.3.61}$$

where

$$f_3 = f_s - f_0 - f_c + \mu t \tag{4.3.62}$$

$$\theta_2 = \theta_s - \theta_0 + \theta_c + 2\pi f_c t - \pi \mu t^2 \tag{4.3.63}$$

$$T_1 = \max(t - T, 0) \tag{4.3.64}$$

$$T_2 = \min(t, T) \tag{4.3.65}$$

If $f_3 \neq 0$, (4.3.61) yields

$$\begin{aligned} s_2(t) &= \frac{A}{8\pi f_3}\,[\sin(2\pi f_3 T_2 + \theta_2) - \sin(2\pi f_3 T_1 + \theta_2)] \\ &= \frac{A}{4\pi f_3}\,\sin[\pi f_3(T_2 - T_1)]\,\cos[\pi f_3(T_1 + T_2) + \theta_2], 0 \leqslant t \leqslant 2T \end{aligned} \tag{4.3.66}$$

The final form of (4.3.66) is valid even if $f_3 = 0$ because $\text{sinc}(0) = 1$. For practical values of the parameters, the cosine factor varies much more rapidly with time than the other factors. Consequently, after using (4.3.62) to (4.3.65), the output of the envelope detector is

$$s_3(t) = \begin{cases} \dfrac{At}{4}\,\text{sinc}[(f_s - f_0 - f_c)t + \mu t^2], & 0 \leqslant t \leqslant T \\ \dfrac{A(2T - t)}{4}\,\text{sinc}[(f_s - f_0 - f_c + \mu t)(2T - t)], & T \leqslant t \leqslant 2T \end{cases} \tag{4.3.67}$$

The peak value of $s_3(t)$ is $AT/4$. We define parameter ϵ as

$$\epsilon = f_0 + f_c - f_s - \mu T \tag{4.3.68}$$

If the frequency of the intercepted signal is such that $\epsilon = 0$, then

$$s_3(T) = \frac{AT}{4}, \quad \epsilon = 0 \tag{4.3.69}$$

Furthermore, (4.3.67) and (4.3.68) yield

$$s_3\left(T \pm \frac{1}{2\mu T}\right) \cong \frac{AT}{4}\,(0.64), \quad \epsilon = 0, \quad T \gg \frac{1}{\mu T} \tag{4.3.70}$$

$$s_3\left(T \pm \frac{1}{\mu T}\right) \cong 0, \quad \epsilon = 0, \quad T \gg \frac{1}{\mu T} \tag{4.3.71}$$

These equations indicate that $1/\mu T$ is an approximate measure of the width of the compressed output pulse. Satisfying the inequality ensures that the response of the compressive receiver due to one scan does not interact significantly with the response due to the next scan.

If the frequency of the intercepted signal shifts slightly, then $\epsilon \neq 0$. A small shift yields

$$s_3\left(T + \frac{\epsilon}{\mu}\right) \cong \frac{AT}{4}, \quad \left|\frac{\epsilon}{\mu}\right| \ll T \tag{4.3.72}$$

where the right-hand side of the equation is the peak value of $s_3(t)$. Thus, the peak value occurs at time $t = T + \epsilon/\mu$. Using (4.3.68), we can estimate the input frequency, f_0, from the time location of the peak value.

The resolution of the compressive receiver, Δ, is approximately equal to the frequency shift, ϵ, that produces a change in the time of the peak output equal to the interval between the peak and the first null when $\epsilon = 0$. It follows from (4.3.71) and (4.3.72) that

$$\Delta \cong \frac{1}{T} \tag{4.3.73}$$

Thus, the resolution of the compressive receiver is independent of the scan rate, in sharp contrast to the resolution of the scanning superheterodyne receiver.

The total bandwidth scanned is

$$W = \mu T \tag{4.3.74}$$

The ratio of the input pulse width, T, to the compressed output pulse width, $1/\mu T = 1/W$, is TW, which is called the *compression ratio* of the compressive receiver. When this ratio is large, (4.3.73), (4.3.74), and (4.3.57) indicate that there is a substantial improvement in resolution of the compressive receiver over the scanning superheterodyne receiver. A comparison of the receiver outputs for typical inputs of the form of (4.3.31) is depicted in Figure 4.13.

Equation (4.3.67) exhibits sidelobes in addition to the main peak. The sidelobes generated by a strong signal may mask main peaks due to weaker signals or may be mistaken for main peaks. To alleviate this problem, one may multiply the received waveform by a weighting function, use a scanning waveform with amplitude modulation, follow the chirp filter with a shaping filter, or modify the impulse response of the chirp filter [9]. However, any of these methods inevitably at least slightly degrades the resolution and the signal-to-noise ratio at the main peak.

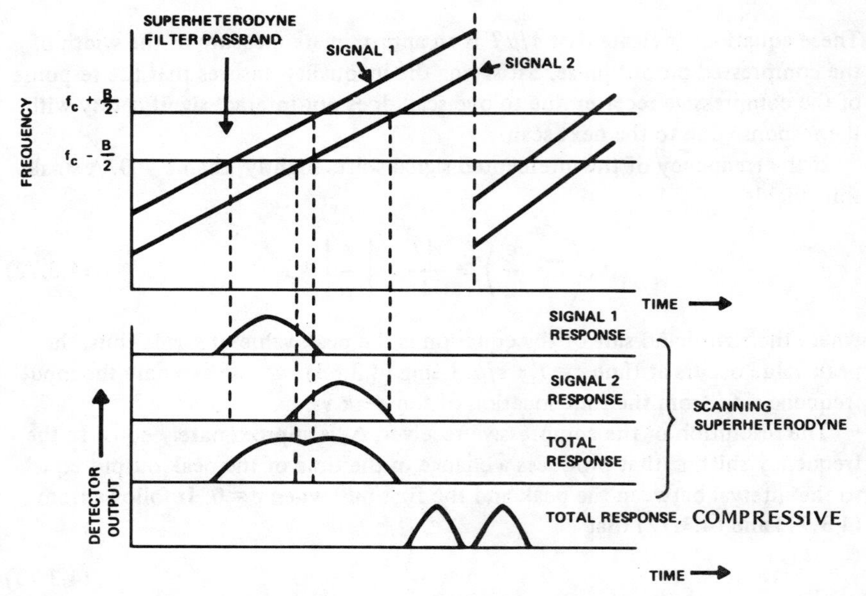

Figure 4.13 Response of scanning superheterodyne and compressive receivers to simultaneous signals.

The *Fresnel integrals* are tabulated functions defined as

$$C(x) = \int_0^x \cos\left(\frac{\pi t^2}{2}\right) dt \tag{4.3.75}$$

and

$$S(x) = \int_0^x \sin\left(\frac{\pi t^2}{2}\right) dt \tag{4.3.76}$$

Taking the Fourier transform of (4.3.60), we find that the transfer function of the chirp filter is

$$H(f) = \frac{1}{2} H_1(f - f_c) + \frac{1}{2} H_1^*(-f - f_c) \tag{4.3.77}$$

where the asterisk denotes the complex conjugate, and

$$H_1(f) = \frac{1}{\sqrt{2\mu}} \exp\left(\frac{j\pi f^2}{\mu} + j\theta_c\right) \left\{ \left[C\left(\frac{\sqrt{2}f}{\sqrt{\mu}} + \sqrt{2\mu}\,T\right) - C\left(\frac{\sqrt{2}f}{\sqrt{\mu}}\right) \right] \right.$$

$$\left. -j\left[S\left(\frac{\sqrt{2}f}{\sqrt{\mu}} + \sqrt{2\mu}\,T\right) - S\left(\frac{\sqrt{2}f}{\sqrt{\mu}}\right) \right] \right\} \tag{4.3.78}$$

If $\sqrt{2\mu}\, T \gg 1$, (4.3.77) and (4.3.78) indicate that the transfer function has a nearly rectangular amplitude response for $f_c - \mu T < |f| < f_c$, a quadratic phase response, and a bandwidth that is well approximated by (4.3.74).

The preceding analysis is valid if the modulation period of the input is large compared with the scan period. If a more rapidly modulated input is present, the compressive receiver can be designed to produce an output that is an approximation of the Fourier transform of the input. In this case, the system is often called a *chirp transform processor.*

In the most common version, the input $s(t)$ is mixed with a scanning waveform that is given over one period by

$$y(t) = \begin{cases} \cos(2\pi f_s t - \pi\mu t^2 + \theta_s)\,, & 0 \leqslant t \leqslant T \\ 0\,, & T < t \leqslant T_c \end{cases} \tag{4.3.79}$$

where T_c is the scan period and $\mu > 0$. The input to the chirp filter is

$$s_1(t) = \begin{cases} s(t)\cos(2\pi f_s t - \pi\mu t^2 + \theta_s)\,, & 0 \leqslant t \leqslant T \\ 0\,, & T < t \leqslant T_c \end{cases} \tag{4.3.80}$$

The impulse response of the chirp filter is

$$h(t) = \cos(2\pi f_c t + \pi\mu t^2 + \theta_c)\,, \quad 0 \leqslant t \leqslant T_c \tag{4.3.81}$$

where T_c is the duration of the response. We derive the system output in the time interval $T \leqslant t \leqslant T_c$. Substituting (4.3.80) and (4.3.81) into (4.3.36), using trigonometry, dropping an integral that is usually negligible, and substituting a complex exponential, we obtain

$$\begin{aligned} s_2(t) &= \mathrm{Re}\left\{\frac{1}{2}\exp[j(2\pi f_c t + \pi\mu t^2 + \theta)]\int_{-\infty}^{\infty} s(\tau)q(\tau)\exp[-j2\pi\tau(f_c - f_s + \mu t)]\,d\tau\right\} \\ &= \mathrm{Re}\left\{\frac{1}{2}\exp[j(2\pi f_c t + \pi\mu t^2 + \theta)]S(f_c - f_s + \mu t)\right\} \\ &= \frac{1}{2}|S(f_c - f_s + \mu t)|\cos[2\pi f_c t + \pi\mu t^2 + \theta + \phi(f_c - f_s + \mu t)]\,, T \leqslant t \leqslant T_c \end{aligned} \tag{4.3.82}$$

where $\theta = \theta_c + \theta_s$, $q(t) = 1$ for $0 \leqslant t \leqslant T$ and $q(t) = 0$ otherwise, $S(f)$ is the Fourier transform of $s(t)q(t)$, and $\phi(f)$ is the phase angle of the Fourier transform. The output of the envelope detector is proportional to

$$s_3(t) = |S(f_c - f_s + \mu t)|, \quad T \leqslant t \leqslant T_c \tag{4.3.83}$$

Thus, the magnitude of the Fourier transform of the input modulation has been produced as a time signal. The power spectrum can be obtained by squaring.

During the interval $T \leqslant t \leqslant T_c$, $s_3(t)$ exhibits all the values of $|S(f)|$ for which

$$f_c - f_s + \mu T \leqslant f \leqslant f_c - f_s + \mu T_c \tag{4.3.84}$$

Thus, the bandwidth exhibited is

$$B_e = \mu(T_c - T) \tag{4.3.85}$$

The bandwidth of $s(t)$ plus any uncertainty in its center frequency must not exceed B_e if all of the significant power spectrum of $s(t)$ is to be available. Because the power spectrum of a real function is even, usually only positive frequencies are of interest. Inequality (4.3.84) indicates that only positive frequencies are exhibited if

$$f_s < f_c + \mu T \tag{4.3.86}$$

The scanning waveform is a periodic sequence of signals of the form of (4.3.79). Because of this form, the output due to a scan does not interfere with the Fourier transform generated by the next scan. If T_c is specified, a large value of T/T_c limits the fraction of the receiver input signal that cannot be processed. However, B_e decreases as T increases. As a compromise, we may choose T to maximize the product B_eT. Equation (4.3.85) and calculus yield $T = T_c/2$ and a maximum product

$$B_eT = \frac{T_cW_c}{4}, \qquad T_c = 2T \tag{4.3.87}$$

where $W_c = \mu T_c$ is the approximate bandwidth of the chirp filter. Larger values of B_eT can be obtained by using other versions of the chirp transform processor [10].

If $s_2(t)$ is multiplied by $h(t)$, then after elimination of a double frequency term by filtering, the phase-shifted real part of $S(f)$ is produced as a time signal. As shown in Figure 4.14, the waveform $h(t) = \cos(2\pi f_c t + \pi\mu t^2)$, $T \leqslant t \leqslant T_c$, can be produced by applying an impulse at time T to a chirp filter with impulse response $h_1(t) = h(t + T)$, $0 \leqslant t \leqslant T_c - T$. The phase-shifted imaginary part of $S(f)$ can be produced as a time signal by multiplying $s_2(t)$ by $\sin(2\pi f_c t + \pi\mu t^2)$. In Figure 4.14, the scanning waveform is produced by applying impulses to a chirp filter.

The chirp transform processor can be used as the basic building block of an analog version of the cross correlator of Figure 4.5. The output of the processor can be time-gated or amplitude-limited to provide notch filtering of narrowband interference.

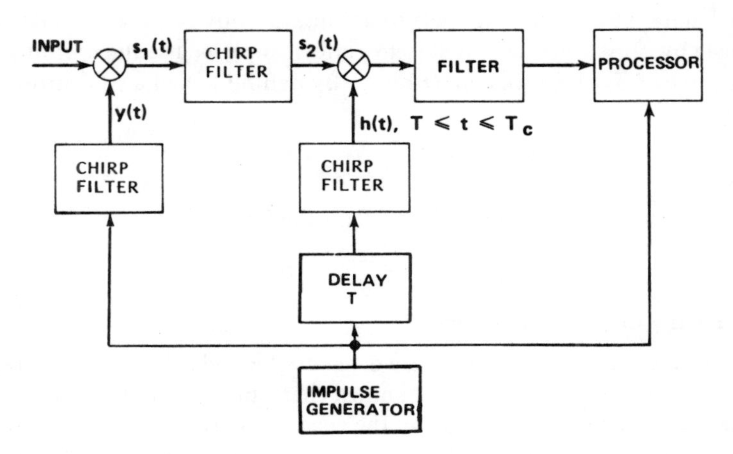

Figure 4.14 Compressive receiver for estimating real part of Fourier transform.

If the scanning period is less than the period between frequency changes of the intercepted signal, the compressive and scanning superheterodyne receivers can estimate each frequency of a frequency-hopping or MFSK signal. If not, some frequencies may be missed.

4.4 DIRECTION FINDING

A signal must be detected, and sometimes its frequency must be estimated, if the direction to its source is to be found. Conversely, *direction finding* enhances signal sorting, which restricts the number of signals that the detection and frequency-estimation systems must process simultaneously.

For simplicity, the estimation of a single bearing angle is usually assumed subsequently. In a ground-based interception system, a single angle may be all that is needed. However, airborne systems may require estimates of both the azimuth and the elevation angles to the intercepted transmitter.

The attainable direction-finding accuracy of an antenna system is limited by physical and electrical design imperfections. For example, component variations cause gain mismatches in multibeam systems. Receiver noise generated internally or externally causes an additional loss of accuracy. Here, we assess the effect of white Gaussian noise, keeping in mind that a residual angle-estimation error remains even in the absence of noise.

We denote the true bearing angle to a transmitter by ϕ and an estimate of this angle by $\hat{\phi}$. As a measure of system performance, we use the *root-mean-square error* of $\hat{\phi}$, which we denote by ζ_ϕ. By definition and a straightforward expansion,

$$\zeta_\phi^2 = E[(\hat{\phi} - \phi)^2] = \text{VAR}(\hat{\phi}) + B_\phi^2 \tag{4.4.1}$$

where $\text{VAR}(\hat{\phi})$ is the variance of $\hat{\phi}$ and B_ϕ is the *bias,*

$$B_\phi = E[\hat{\phi}] - \phi \tag{4.4.2}$$

4.4.1 Energy Comparison Systems

Amplitude comparison systems are often used in radar [10, 11, 12]. However, when communications with unknown characteristics are to be intercepted, it is logical to base comparisons upon the energy. Thus, we analyze *energy comparison systems* rather than the analogous amplitude comparison systems, which are sometimes equally viable. Although Gaussian antenna radiation patterns are assumed for mathematical convenience, the analytical methods can be used to obtain analogous results for non-Gaussian patterns.

A stationary multibeam system for direction finding is shown in Figure 4.15. The beam-forming network is used if the antennas are elements of a phased array. The N beam outputs are applied to N receivers simultaneously. Alternatively, the beam outputs could be sequentially switched to a single receiver, or the beams could be created by a single moving antenna. However, these alternatives cause a performance degradation when the intercepted signal has a short duration or a rapidly changing power level.

For the system of Figure 4.15, the largest of the receiver outputs is selected. Then, the larger of the two receiver outputs corresponding to beams adjacent to the beam that produced the largest output is selected. The two selected outputs are denoted by L_1 and L_2. The processor compares the outputs with thresholds for detection. The signal direction is estimated from the logarithm of the ratio of L_1 to L_2. The radiation patterns of the adjacent beams are illustrated in Figure 4.16, where ϕ represents the angle of arrival of an intercepted signal, and beam pattern $F_i(\theta)$ produces L_i. The origin of the coordinate system is defined so that $+\psi$ and $-\psi$ indicate the peak responses of the two beams, respectively. Suppose that the beam patterns are approximately Gaussian; that is, they are described by

$$F_1(\theta) = G \exp\left[-\frac{2(\theta - \psi)^2}{b^2}\right] \tag{4.4.3}$$

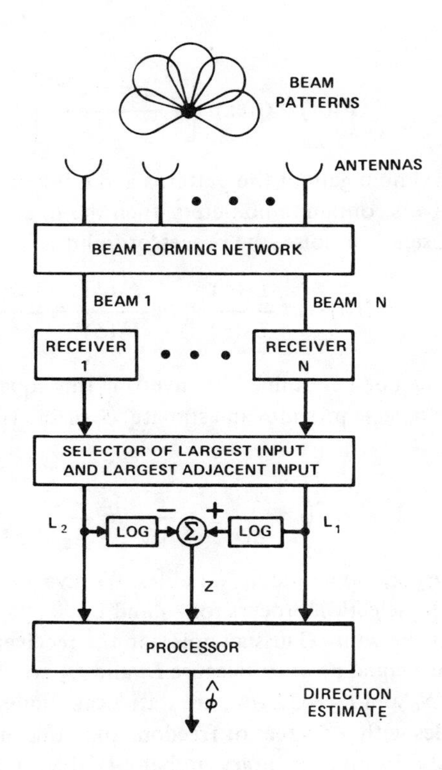

Figure 4.15 Stationary multibeam system.

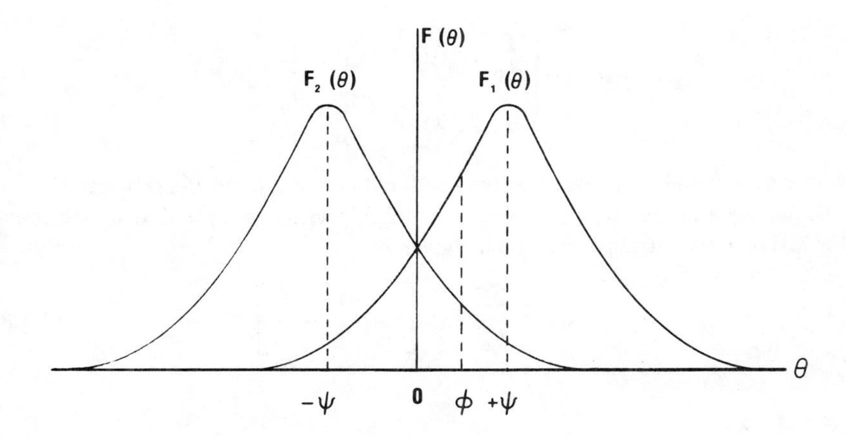

Figure 4.16 Adjacent antenna radiation patterns.

$$F_2(\theta) = G \exp\left[-\frac{2(\theta + \psi)^2}{b^2}\right] \tag{4.4.4}$$

where G is the maximum gain of the patterns and b is a measure of the beamwidth. If the receivers contain radiometers, then the L_i are proportional to the F_i. Thus, in the absence of noise, the processor input is

$$Z(\phi) = ln \frac{L_1(\phi)}{L_2(\phi)} = ln \frac{F_1(\phi)}{F_2(\phi)} = \frac{8\psi\phi}{b^2} \tag{4.4.5}$$

The bearing angle can be determined by inverting this equation. In the presence of noise, the same inverse provides an estimate, $\hat{\phi}$, of the true angle ϕ. The estimate is

$$\hat{\phi} = \frac{b^2}{8\psi} Z = \frac{b^2}{8\psi} ln \frac{L_1}{L_2} \tag{4.4.6}$$

where Z, L_1, and L_2 are now random variables. We give an error analysis neglecting the effect of the selection process for L_1 and L_2.

We assume that the white Gaussian noises in the receivers are statistically independent and have equal powers. Because L_1 and L_2 are the outputs of radiometers, $L_1' = 2L_1/N_0$ and $L_2' = 2L_2/N_0$ are statistically independent, noncentral χ^2 random variables with γ degrees of freedom and noncentral parameters λ_1 and λ_2, respectively. From elementary probability theory, the probability density function of $Y = L_1/L_2 = L_1'/L_2'$, the quotient of two non-negative random variables, is

$$p_2(y) = \begin{cases} \int_0^{\infty} x p_{11}(yx) p_{12}(x)\, dx\,, & y \geqslant 0 \\ 0\,, & y < 0 \end{cases} \tag{4.4.7}$$

where $p_{11}(\ )$ and $p_{12}(\ )$ are the density functions of L_1' and L_2', respectively. The density function $p_{1i}(\ )$ is given by (4.2.33) with $N = 2\gamma$ and λ_i substituted for λ. The noncentral parameters are given by

$$\lambda_1 = \frac{2E}{N_0} \exp\left[-\frac{2(\phi - \psi)^2}{b^2}\right] \tag{4.4.8}$$

$$\lambda_2 = \frac{2E}{N_0} \exp\left[-\frac{2(\phi+\psi)^2}{b^2}\right] \tag{4.4.9}$$

where E is the energy received when the intercepted signal enters the center of one of the beams.

When $\phi = 0$, (4.4.8) and (4.4.9) give $\lambda_1 = \lambda_2$ so that L'_1 and L'_2 are identically distributed random variables. Because $Z = \ln L'_1 - \ln L'_2$, it follows that

$$E[Z] = 0, \quad \phi = 0 \tag{4.4.10}$$

At other values of ϕ, exact expressions for the mean and other moments are difficult to obtain.

To obtain an approximate expression for $E[Z]$, we can bypass the evaluation of (4.4.7) by observing that since $Z = \ln(L'_1/L'_2)$,

$$E[Z] = \int_0^\infty \int_0^\infty \ln\left(\frac{y}{x}\right) p_{11}(y) p_{12}(x)\, dx\, dy \tag{4.4.11}$$

The logarithm is approximated by the first few terms of its two-dimensional Taylor-series expansion about the point $y = m_1, x = m_2$, where m_1 and m_2 are the mean values of L'_1 and L'_2, respectively. Thus,

$$\ln\left(\frac{y}{x}\right) \cong \ln\left(\frac{m_1}{m_2}\right) + \frac{y - m_1}{m_1} - \frac{x - m_2}{m_2} - \frac{(y - m_1)^2}{2m_1^2} + \frac{(x - m_2)^2}{2m_2^2} \tag{4.4.12}$$

This approximation is accurate over some range of y and x about $y = m_1, x = m_2$. If $p_{11}(y)$ and $p_{12}(x)$ are negligible outside this range, then the substitution of (4.4.12) into (4.4.11) yields an accurate approximation of $E[Z]$. The ranges of significant values of $p_{11}(y)$ and $p_{12}(x)$ are approximately limited by $|y - m_1| < 3\sigma_1$ and $|x - m_2| < 3\sigma_2$, where σ_1 and σ_2 are the standard deviations of L'_1 and L'_2, respectively. Consequently, the remainder of the Taylor series indicates that sufficient conditions for the validity of using (4.4.12) are that $\sigma_1 \ll m_1$ and $\sigma_2 \ll m_2$.

Making the substitution and using the properties of density functions, we obtain

$$E[Z] \cong \ln\left(\frac{m_1}{m_2}\right) - \frac{\sigma_1^2}{2m_1^2} + \frac{\sigma_2^2}{2m_2^2}, \qquad \sigma_1 \ll m_1, \quad \sigma_2 \ll m_2 \tag{4.4.13}$$

Since $L_i' = 2L_i/N_0$, (4.2.37) and (4.2.38) imply that

$$m_i = \lambda_i + 2\gamma, \qquad i = 1, 2 \tag{4.4.14}$$

$$\sigma_i^2 = 4\lambda_i + 4\gamma, \qquad i = 1, 2 \tag{4.4.15}$$

An approximate expression for $E[Z^2]$ is obtained in an analogous manner. Combining this expression with (4.4.13) and dropping terms higher than second order in σ_1/m_1 and σ_2/m_2, we obtain

$$\mathrm{VAR}(Z) \cong \frac{\sigma_1^2}{m_1^2} + \frac{\sigma_2^2}{m_2^2}, \qquad \sigma_1 \ll m_1, \qquad \sigma_2 \ll m_2 \tag{4.4.16}$$

By using (4.4.1), (4.4.6), (4.4.13), and (4.4.16), an equation for ζ_ϕ can be derived. Dropping terms higher than second order gives

$$\zeta_\phi = \frac{b^2}{8\psi}\left\{\left[\ln\left(\frac{m_1}{m_2}\right) - \frac{8\psi\phi}{b^2}\right]^2 - \left[\ln\left(\frac{m_1}{m_2}\right) - \frac{8\psi\phi}{b^2}\right]\right.$$
$$\left.\times\left(\frac{\sigma_1^2}{m_1^2} - \frac{\sigma_2^2}{m_2^2}\right) + \frac{\sigma_1^2}{m_1^2} + \frac{\sigma_2^2}{m_2^2}\right\}^{1/2}, \qquad \sigma_1 \ll m_1, \quad \sigma_2 \ll m_2 \tag{4.4.17}$$

This equation and (4.4.14), (4.4.15), (4.4.8), and (4.4.9) indicate that ζ_ϕ is symmetric about $\phi = 0$. At $\phi = 0$, $m_1 = m_2 = m$ and $\sigma_1 = \sigma_2 = \sigma$ so that

$$\zeta_\phi = \frac{\sqrt{2}\, b^2 \sigma}{8\psi m}, \qquad \phi = 0, \quad \sigma \ll m \tag{4.4.18}$$

To assess the implications of (4.4.17) and (4.4.18), it is convenient to define the parameter

$$\rho = \frac{E}{N_0 \gamma} \tag{4.4.19}$$

which is approximately equal to the signal-to-noise ratio in a receiver when the bearing angle is aligned with the center of the corresponding beam. For large values of ρ, (4.4.18) becomes

$$\zeta_\phi = \frac{b^2}{4\psi} (\rho\gamma)^{-1/2} \exp\left(\frac{\psi^2}{b^2}\right), \qquad \phi = 0\,, \quad \rho \gg \exp\left(\frac{2\psi^2}{b^2}\right) \tag{4.4.20}$$

This equation indicates that we can achieve $\zeta_\phi \ll b$ if ρ is sufficiently large. Thus, the potential direction-finding accuracy is much better than a beamwidth.

For a fixed value of ψ and $\phi = 0$, the beamwidth $b = \psi$ minimizes ζ_ϕ. However, this choice of beamwidth is not optimal for all values of ϕ. Figure 4.17 illustrates the effect of the normalized beamwidth, b/ψ, on ζ_ϕ. For $\rho = 10$, $\gamma = 100$, and $\phi = \psi$, the normalized beamwidth that minimizes ζ_ϕ is approximately 3.6.

Figure 4.18 illustrates the effect of increasing the duration of the observation interval, T. Assuming that an increase in T causes proportionate increases in E and γ, we set $\rho = 10$ for each value of γ. A nearly optimal normalized beamwidth for each value of γ and $\phi = \psi$ is chosen. The curves show a steady decrease in ζ_ϕ as γ increases.

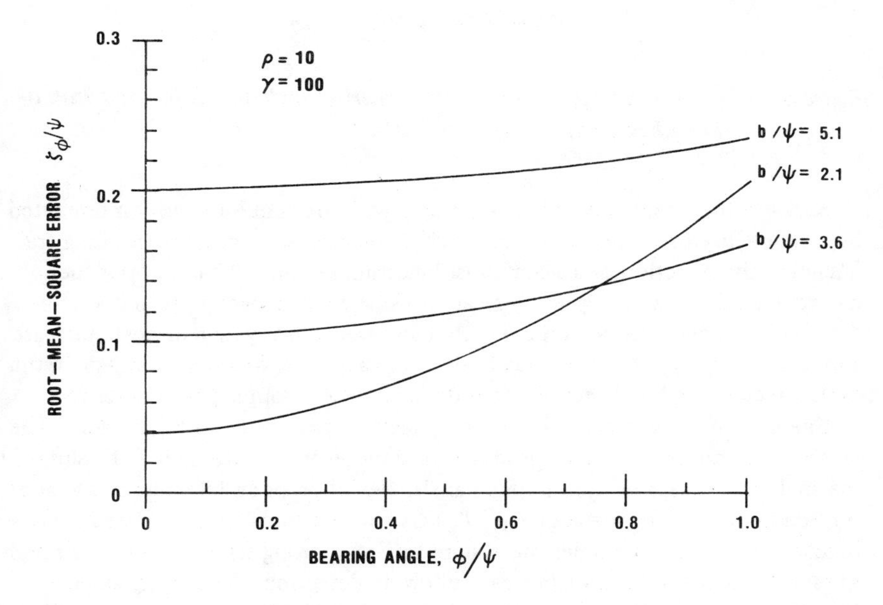

Figure 4.17 Root-mean-square error versus bearing angle for different beamwidths.

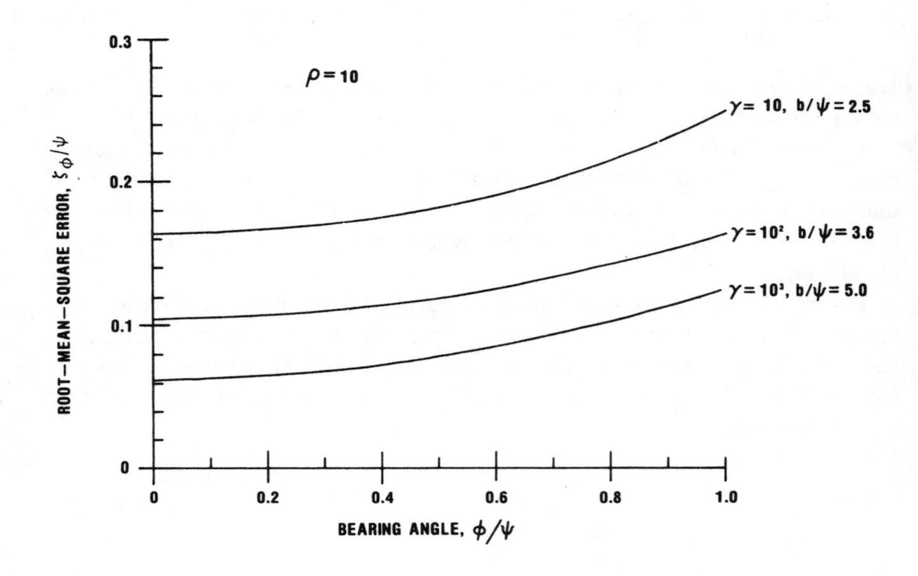

Figure 4.18 *Root-mean-square error versus bearing angle for different values of γ and beamwidths optimized at $\phi = \psi$.*

Suppose that many frequency-hopping signals are simultaneously intercepted by a channelized receiver. If it is impossible to correlate successive hopping frequencies, the direction of a signal must be estimated on the basis of the energy due to a single pulse. Let γ and ρ refer to a single pulse dwell time and a single frequency channel. As the hopping rate increases, either γ or ρ or both decrease and, consequently, (4.4.20) indicates that ζ_ϕ increases. We conclude that, in this case, direction finding becomes more difficult as the hopping rate increases.

Energy comparison with *two rotating beams* is illustrated in Figure 4.19. The receivers are radiometers that produce continuous or discrete signals. To simplify the analysis, we assume continuous signals. Detection of an intercepted signal is verified by comparing $Z = \log[(L_1 + L_2)/L_3]$ with a threshold. Because $Z(t)$ is a function of a ratio, it enables the system to reject strong signals entering through the sidelobes of the rotating beams. Following detection, the bearing angle is identified as the angle corresponding to the time when the energies received by the two displaced beams are equal, which is the time when $L_1(t) - L_2(t)$ crosses zero.

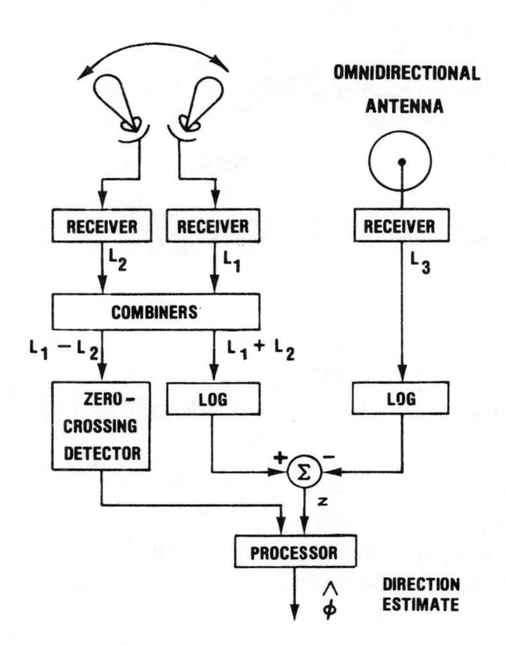

Figure 4.19 Energy comparison with two rotating beams.

Figure 4.16 illustrates the displaced beam patterns at the time of the zero crossing if we define the angular coordinates so that the origin corresponds to boresight. Thus, at the zero-crossing time, the bearing angle is estimated as $\hat{\phi} = 0$. Let ϕ denote the true bearing angle relative to boresight at the zero-crossing time. Because the zero-crossing time is a random variable, ϕ is a random variable.

We assume that the beam rotation is negligible during an observation interval and that the signal energy produced by the source per observation interval does not vary significantly during a scan. Then at the zero-crossing time t_0, we make the decomposition

$$L_i'(t_0) = m_i(t_0) + n_i(t_0)\,, \quad i = 1, 2 \tag{4.4.21}$$

where $m_i(t_0)$, the expected value of $L_i'(t_0) = 2L_i(t_0)/N_0$, is given by (4.4.14).

The zero-mean random variable $n_i(t_0)$ has a variance given by (4.4.15). Because $n_1(t_0)$ and $n_2(t_0)$ are produced by different receivers, it is reasonable to assume that they are statistically independent. We further assume that ρ is sufficiently large that $\phi \ll \psi$ with high probability and

$$\rho = \frac{E}{N_0 \gamma} \gg \exp\left(\frac{2\psi^2}{b^2}\right), \quad \gamma \geq 1 \tag{4.4.22}$$

It follows that $n_1(t_0)$ and $n_2(t_0)$ have nearly the same variance given by

$$\sigma^2 \cong \frac{8E}{N_0} \exp\left(-\frac{2\psi^2}{b^2}\right) + 4\gamma$$

$$\cong \frac{8E}{N_0} \exp\left(-\frac{2\psi^2}{b^2}\right) \tag{4.4.23}$$

Because $L_1'(t_0) = L_2'(t_0)$, (4.4.21), (4.4.14), (4.4.8), and (4.4.9) imply that

$$n_2 - n_1 = m_1 - m_2 = \frac{4E}{N_0} \exp\left[-\frac{2(\phi^2 + \psi^2)}{b^2}\right] \sinh\left(\frac{4\phi\psi}{b^2}\right) \tag{4.4.24}$$

where the time dependencies are suppressed for notational convenience. Because $\phi \ll \psi$ with high probability, (4.4.24) yields the approximate solution

$$\phi \cong \frac{b^2}{4\psi} \sinh^{-1}\left[\frac{N_0}{4E} \exp\left(\frac{2\psi^2}{b^2}\right)(n_2 - n_1)\right] \tag{4.4.25}$$

The random variable $n_2 - n_1$ has a standard deviation of $\sqrt{2}\,\sigma$. Thus, (4.4.23) and (4.4.22) imply that the factor in brackets in (4.4.25) is small with high probability. Since $\sinh^{-1}(x) \cong x$ for small x,

$$\phi \cong \frac{b^2 N_0}{16\psi E} \exp\left(\frac{2\psi^2}{b^2}\right)(n_2 - n_1) \tag{4.4.26}$$

Because $\hat{\phi} = 0$, (4.4.1) gives $\zeta_\phi^2 = E[\phi^2]$. Thus,

$$\zeta_\phi = \frac{\sqrt{2}\, b^2 N_0 \sigma}{16\psi E} \exp\left(\frac{2\psi^2}{b^2}\right)$$

$$= \frac{b^2}{4\psi} (\rho\gamma)^{-1/2} \exp\left(\frac{\psi^2}{b^2}\right), \quad \rho \gg \exp\left(\frac{2\psi^2}{b^2}\right) \tag{4.4.27}$$

Because (4.4.27) is identical to (4.4.20), the stationary multibeam array potentially performs as well as the rotating-beam system when $\phi = 0$. However, if $\phi \neq 0$, ζ_ϕ increases in the multibeam case. The main disadvantage with rotating-beam systems is the narrow instantaneous field of view, which may cause a signal to be missed or may decrease the possible observation time. On the other hand, when many hostile communications are present, the narrow field of view provides a valuable signal-sorting capability.

Referring to Figures 4.19 and 4.16, the threshold for $Z(t)$ is set so that, with high probability, once $Z(t)$ crosses the threshold, it stays above it until $L_1(t) - L_2(t)$ crosses zero. To achieve this high probability, it is intuitively reasonable to require that, in the absence of noise, $L_1(t) + L_2(t)$ have a local maximum near the zero-crossing time. Thus, we might require $F_1(\theta) + F_2(\theta)$ to have a maximum at $\theta = 0$. Using calculus and (4.4.3) and (4.4.4), we obtain $b \geqslant 2\psi$ as a necessary condition for the local maximum. Figure 4.20 depicts (4.4.27) with $b = 2\psi$.

An alternative system uses a *single rotating beam,* as shown in Figure 4.21. Detection of a signal is verified by comparing L_1 and $Z = \log (L_1/L_2)$ with thresholds. Following detection, the direction to the signal source may be estimated as the angle corresponding to the peak value of $L_1(t)$ or $Z(t)$. Alternatively, the system may measure the angle at which L_1 or $Z(t)$ first crosses a threshold and the angle at which $L_1(t)$ or $Z(t)$ subsequently drops below this threshold; the direction estimate is the average of the two angles.

To avoid processing the noise in $L_2(t)$, it is preferable to use the threshold crossings of $L_1(t)$ rather than $Z(t)$ if the signal energy produced by the source per observation interval does not vary significantly during a scan. The threshold may be set at a fixed level to accommodate signals with a predetermined minimum energy, or adaptive thresholding may be used. *Adaptive thresholding systems* [13] can be designed so that the threshold crossings occur at the times at which the slope of $L_1(t)$ has its maximum magnitude, regardless of the signal energy. As a result, adaptive thresholding often provides more accurate measurements of threshold crossings than thresholding with a fixed level, thereby allowing an improved estimation of the bearing angle.

We derive ζ_ϕ for the rotating-beam system with adaptive thresholding. We again assume that the beam rotation is negligible during an observation interval and that the signal energy produced by the source per observation interval does not vary significantly during a scan. Let α denote a uniform scan rate. The rotating beam has the radiation pattern

$$F(\theta) = G \exp\left(-\frac{2\theta^2}{b^2}\right) \tag{4.4.28}$$

In the absence of noise, $L_1'(t) = 2L_1(t)/N_0$ is proportional to $F(\alpha t)$, where $t = 0$ is the time of the peak value of $L_1(t)$. Let $\hat{t}_1$ and $\hat{t}_2$ denote the times at which $L_1'(t)$ crosses the threshold in the presence of noise. The bearing-angle estimate is

$$\hat{\phi} = \alpha\left(\frac{\hat{t}_1 + \hat{t}_2}{2}\right) \tag{4.4.29}$$

We assume that the time interval between the threshold crossings is long compared to both the reciprocal of the receiver bandwidth and the duration of an observation interval. Thus, the random variables $\hat{t}_1$ and $\hat{t}_2$ may be considered statistically independent. We assume that the sources of noise do not change significantly between threshold crossings so that $\hat{t}_1$ and $\hat{t}_2$ have equal variances. Consequently, (4.4.29) implies that

$$\mathrm{VAR}(\hat{\phi}) = \frac{\alpha^2}{2}\ \mathrm{VAR}(\hat{t}_1) \tag{4.4.30}$$

To evaluate the variance of $\hat{t}_1$, we make the decomposition

$$L_1'(t) = m_1(t) + n_1(t) \tag{4.4.31}$$

where $m_1(t)$ is the expected value of $L_1'(t)$. Let E denote the signal energy received when $t = 0$. Equations (4.4.14) and (4.4.28) imply that

$$m_1(t) = \frac{2E}{N_0} \exp\left[-\frac{2(\alpha t)^2}{b^2}\right] + 2\gamma \tag{4.4.32}$$

The zero-mean stochastic process $n_1(t)$ is nonstationary and nearly Gaussian in distribution for each value of t and large values of γ. Using (4.4.15) and (4.4.28), we obtain the mean power of $n_1(t)$:

$$\sigma^2(t) = \frac{8E}{N_0} \exp\left[-\frac{2(\alpha t)^2}{b^2}\right] + 4\gamma \tag{4.4.33}$$

We interpret $m_1(t)$ as the "signal" in $L_1'(t)$ and $n_1(t)$ as additive "noise."

The maximum magnitude of the slope of $m_1(t)$ occurs at times

$$t_1 = -\frac{b}{2\alpha}, \quad t_2 = \frac{b}{2\alpha} \tag{4.4.34}$$

At these times, the slope magnitude, $|dm_1(t)/dt|$, is

$$S = \frac{4\alpha E}{\sqrt{e}\, bN_0} \tag{4.4.35}$$

If $\rho \gg 1$, $\hat{t}_1$ and $\hat{t}_2$ are near t_1 and t_2, respectively.

At times near t_1 or t_2, $n_1(t)$ can be approximated by a zero-mean stationary process with variance

$$\sigma^2 = \frac{8E}{\sqrt{e}\, N_0} + 4\gamma \tag{4.4.36}$$

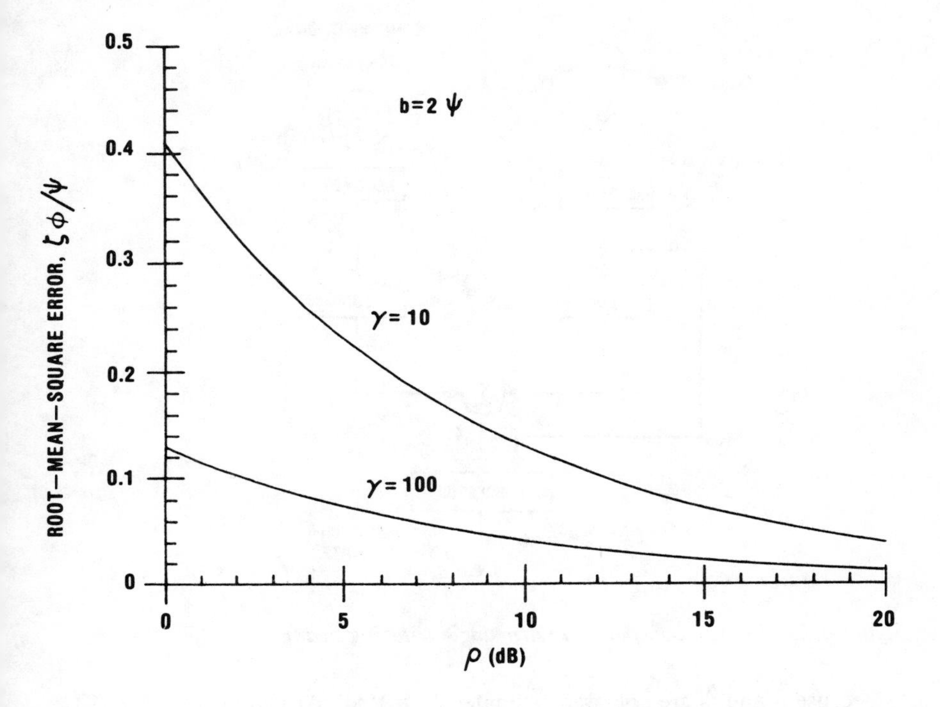

Figure 4.20 Root-mean-square error versus ρ for different values of γ.

From arrival-time estimation theory, it follows that for $\rho \gg 1$, $\hat{t}_1$ is an unbiased estimate of t_1 with variance

$$\mathrm{VAR}(\hat{t}_1) \cong \kappa^2 \frac{\sigma^2}{S^2} \tag{4.4.37}$$

where κ^2 is a constant on the order of unity [13]. Using (4.4.35) to (4.4.37), we obtain

$$\mathrm{VAR}(\hat{t}_1) \cong \kappa^2 \left(\frac{8E}{\sqrt{e}\, N_0} + 4\gamma \right) \frac{e b^2 N_0^2}{16 \alpha^2 E^2}$$

$$\cong \frac{\sqrt{e}\, \kappa^2 b^2 N_0}{2\alpha^2 E}, \quad \rho \gg 1 \tag{4.4.38}$$

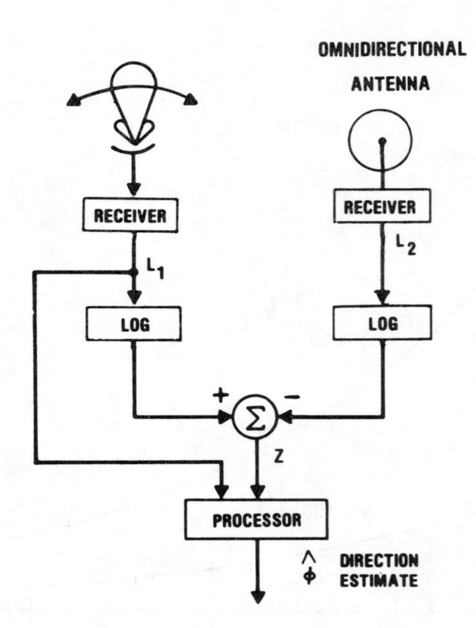

Figure 4.21 Energy comparison with single rotating beam.

Because $\hat{t}_1$ and $\hat{t}_2$ are unbiased estimates, $\zeta_\phi^2 = \text{VAR}(\hat{\phi})$. Combining (4.4.19), (4.4.30), and (4.4.38), yields

$$\zeta_\phi = \frac{\sqrt[4]{e}\,\kappa b}{2}\,(\rho\gamma)^{-1/2}, \qquad \rho \gg 1 \tag{4.4.39}$$

Because $\kappa \approx 1$ in (4.4.39) and typically $\psi \approx b/2$ in (4.4.27), the two types of rotating-beam systems have approximately the same potential performance.

4.4.2 Interferometers

An *interferometer* consists of two or more antennas or groups of elements of a phased array that use phase or arrival-time information to estimate direction. Consider a plane wave arriving at two antennas separated by distance d. The bearing angle ϕ to the source is defined as the angle between the normal to the line connecting the antennas and the normal to the wavefront, as illustrated in Figure 4.22, where the plane wave arrives at antenna 1 later than it arrives at antenna 2. It is assumed that the modulation bandwidth of the signals excited in the antennas is much less than the carrier frequency. Because phase angles are modulo 2π numbers, the phase of the signal at antenna 1 lags the phase of the signal at antenna 2 by

$$\theta = \frac{2\pi d \sin\phi}{\Lambda} - 2\pi n, \quad |\theta| \leqslant \pi \tag{4.4.40}$$

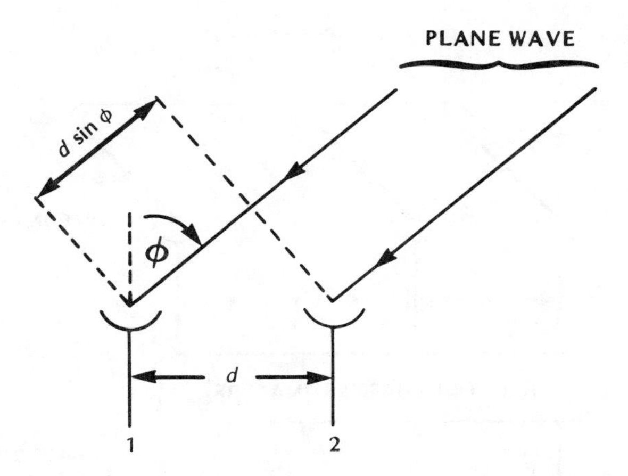

Figure 4.22 Plane wave arriving at two antennas.

where Λ is the wavelength corresponding to the carrier frequency and n is an integer that ensures satisfaction of the inequality. For a unique correspondence between the values of ϕ and the values of θ, it is necessary to assume that $|\phi| \leqslant \pi/2$. If $d \leqslant \Lambda/2$, then $n = 0$. If $d > \Lambda/2$, the n varies with ϕ, taking positive, negative, and zero values.

Let $\hat{\theta}$, $\hat{d}$, $\hat{\Lambda}$, and $\hat{n}$ denote the estimated or measured values of θ, d, Λ, and n. A *phase interferometer* estimates the bearing by inverting (4.4.40), which yields

$$\hat{\phi} = \sin^{-1}\left[\frac{\hat{\Lambda}}{2\pi\hat{d}}(\hat{\theta} + 2\pi\hat{n})\right] \tag{4.4.41}$$

Suppose that the interferometer suppresses or does not receive signals with $|\phi| > \phi_m$, where $\phi_m \leqslant \pi/2$. To ensure that $|\hat{\phi}| \leqslant \phi_m$, the estimates must satisfy

$$|\hat{\theta} + 2\pi\hat{n}| \leqslant \frac{2\pi\hat{d}\sin\phi_m}{\hat{\Lambda}} \tag{4.4.42}$$

The device that produces the estimate of the phase difference, $\hat{\theta}$, may be similar to the part of the IFM receiver in Figure 4.9 that is fed by $s(t)$ and $s_1(t)$.

The accuracy of the bearing estimate increases with d. However, if $\hat{d}/\hat{\Lambda}$ is sufficiently large, more than one value of $\hat{n}$ may satisfy (4.4.42). For each value of $\hat{n}$, (4.4.41) gives a corresponding value of $\hat{\phi}$. Thus, the bearing estimate obtained from two antennas is ambiguous unless it is known that $d \leqslant \Lambda/2$.

For a bearing estimate that is both accurate and unambiguous, three or more antennas are used, as illustrated in Figure 4.23. A simple approach is to place the

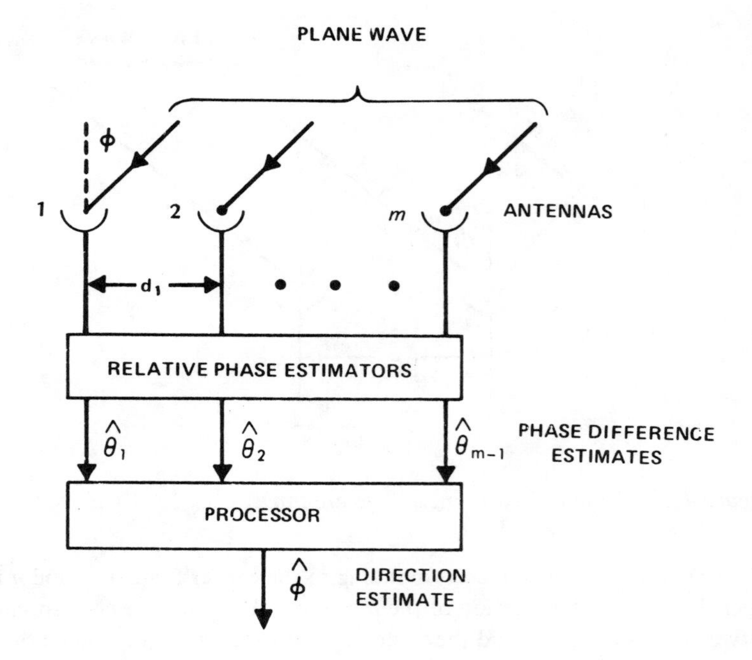

Figure 4.23 Phase interferometer.

antennas along a straight line and to use antenna 1 as a reference. Antenna 2 is separated from antenna 1 by $d_1 \leqslant \Lambda/2$ so that this antenna pair provides an unambiguous estimate of ϕ. The other antennas are separated by increasingly greater distances from the reference. Antenna 3 and antenna 1 provide a more accurate estimate of ϕ, but an ambiguous one. However, this ambiguity can be resolved by using the first estimate, as illustrated in Figure 4.24. Subsequent antennas allow increasingly accurate estimates of ϕ, provided that the ambiguities can be resolved by using the less accurate estimates.

If physical constraints make it necessary for all separations of antenna pairs to exceed $\Lambda/2$, ambiguities can still be resolved. For three antennas lying along a straight line, the two independent phase differences are

$$\theta_i = \frac{2\pi d_i \sin \phi}{\Lambda} - 2\pi n_i \,, \quad i = 1, 2 \tag{4.4.43}$$

where θ_1, d_1, and n_1 refer to antennas 2 and 1, while θ_2, d_2, and n_2 refer to antennas 3 and 1. Therefore, the phase differences must satisfy

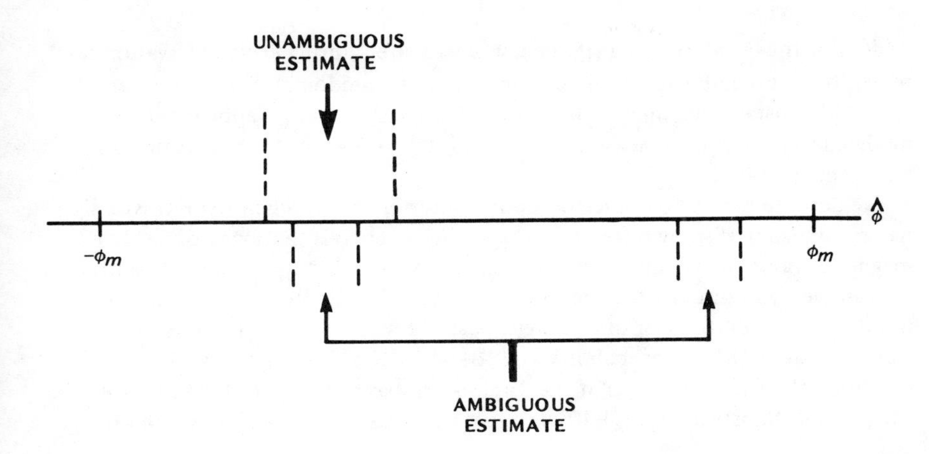

Figure 4.24 Relation between unambiguous inaccurate estimate and ambiguous accurate estimate.

$$\theta_1 - \frac{d_1}{d_2}\,\theta_2 = -2\pi n_1 + \frac{d_1}{d_2}\,2\pi n_2 \tag{4.4.44}$$

Since $|\theta_i| \leqslant \pi$, $i = 1, 2$, (4.4.43) implies that estimates $\hat{n}_1$ and $\hat{n}_2$ should satisfy

$$|\hat{n}_i| \leqslant \frac{\hat{d}_i \sin \phi_m}{\hat{\Lambda}} + \frac{1}{2}, \qquad i = 1, 2 \tag{4.4.45}$$

We define

$$\epsilon = \hat{\theta}_1 - \frac{\hat{d}_1}{\hat{d}_2}\,\hat{\theta}_2 + 2\pi\hat{n}_1 - \frac{\hat{d}_1}{\hat{d}_2}\,2\pi\hat{n}_2 \tag{4.4.46}$$

If exact estimates were available, then (4.4.44) and (4.4.46) indicate that ϵ would be zero. Thus, a logical way to determine $\hat{n}_1$ and $\hat{n}_2$ is to substitute all integer pairs satisfying (4.4.45) into (4.4.46) and then select the pair that minimizes $|\epsilon|$. Subsequently, $\hat{\theta}_1$ and $\hat{\theta}_2$ may be adjusted to make $\epsilon = 0$ while satisfying

$$|\hat{\theta}_i + 2\pi\hat{n}_i| \leqslant \frac{2\pi\hat{d}_i \sin \phi_m}{\hat{\Lambda}}, \qquad i = 1, 2 \tag{4.4.47}$$

The bearing estimate is then determined from (4.4.41) with $\hat{\theta}$, $\hat{d}$, and $\hat{n}$ referring to the antenna pair with the largest separation.

If Λ or the corresponding frequency is estimated during each observation interval, then it may be appropriate to model $\hat{\Lambda}$ as a random variable. However, if $\hat{\Lambda}$ is held constant over many observation intervals, it is more appropriately modeled as a constant that causes a bias in $\hat{\phi}$ when $\hat{\Lambda} \neq \Lambda$. Another source of bias is an error in $\hat{d}$.

We analyze the performance of a phase interferometer when the received signal is a sinusoid of known frequency $f_0 = c/\Lambda$, where c is the speed of electromagnetic waves. It is assumed that $\hat{d} = d$, $\hat{\Lambda} = \Lambda$, and that the phase ambiguities are resolved without error. In general, $\hat{\phi}$ is a biased estimator of ϕ, whether or not $\hat{\theta}$ is a biased estimator of θ, except possibly when $\phi = 0$. If $\hat{\theta} - \theta$ is sufficiently small with high probability and $|\phi| \leqslant \phi_m < \pi/2$, then $\hat{\phi}$ is well approximated by the first two terms of the Taylor-series expansion of (4.4.41) about the point θ. Substituting (4.4.40) and $\Lambda = c/f_0$ into this expansion, we obtain

$$\hat{\phi} \cong \phi + \frac{c}{2\pi f_0 d \cos\phi} (\hat{\theta} - \theta) \tag{4.4.48}$$

Taking the expected value of this expression, we find that the bias of $\hat{\phi}$ is

$$B_\phi = \frac{c}{2\pi f_0 d \cos\phi} B_\theta \tag{4.4.49}$$

where B_θ is the bias of $\hat{\theta}$. Thus, if the initial assumptions are valid, B_ϕ is proportional to B_θ. Equation (4.4.48) implies that

$$\zeta_\phi = \left(\frac{c}{2\pi f_0 d \cos\phi}\right) \zeta_\theta \tag{4.4.50}$$

where ζ_θ, the root-mean-square error of θ, is defined by

$$\zeta_\theta^2 = E[(\hat{\theta} - \theta)^2] = \mathrm{VAR}(\hat{\theta}) + B_\theta^2 \tag{4.4.51}$$

If the antenna outputs contain independent white Gaussian noises, then $\mathrm{VAR}(\hat{\theta})$ is bounded by (B.5.12) of Appendix B. The relation $\beta(\theta) = E[\hat{\theta}] = \theta + B_\theta$ implies that

$$\mathrm{VAR}(\hat{\theta}) \geqslant \left(1 + \frac{\partial B_\theta}{\partial \theta}\right)^2 \left[\frac{1}{2E_1/N_{01}} + \frac{1}{2E_2/N_{02}}\right], \tag{4.4.52}$$

where E_1 and E_2 are the signal energies at the two different antenna outputs and $N_{01}/2$ and $N_{02}/2$ are the two-sided noise power spectral densities. Relations (4.4.50) to (4.4.52) provide a lower bound for ζ_ϕ. If $E_1 = E_2 = E$, $N_{01} = N_{02} = N_0$, and $B_\theta = 0$, we obtain

$$\zeta_\phi \geqslant \left(\frac{c}{2\pi f_0 d \cos\phi}\right) \left(\frac{E}{N_0}\right)^{-1/2} \tag{4.4.53}$$

Because this bound is minimized when $\phi = 0$, it may be advantageous to rotate a sufficiently small interferometer until the measured phase differences are approximately zero and then set $\hat{\phi} = 0$.

To estimate a bearing angle such that $|\phi| \leqslant \pi$ rather than $|\phi| \leqslant \pi/2$, or to estimate both an azimuthal and an elevation angle, a nonlinear deployment of the antennas is necessary. A planar configuration of five antennas [14], which is often called an *Adcock array,* is depicted in Figure 4.25(a). The array may be regarded as two linear, mutually orthogonal arrays with a common antenna in the center. The central antenna is used to resolve ambiguities, while the direction of a signal is determined by measuring the phase differences between other antenna outputs. Let θ_{ij} denote the phase lag of the signal at antenna i relative to the corresponding signal at antenna j, $\mathbf{p}_{ij}$ denote the vector from antenna i to antenna j, and $\mathbf{r}$ denote the unit vector in the direction of arrival. It follows that

$$\theta_{12} = \frac{2\pi}{\Lambda} \mathbf{p}_{12} \cdot \mathbf{r} - 2\pi n_{12} \tag{4.4.54}$$

$$\theta_{34} = \frac{2\pi}{\Lambda} \mathbf{p}_{34} \cdot \mathbf{r} - 2\pi n_{34} \tag{4.4.55}$$

where n_{12} and n_{34} are integers. Using the spherical coordinates indicated in Figure 4.25(b), we obtain

$$\theta_{12} = \frac{2\pi d_{12}}{\Lambda} \sin\psi \sin\phi - 2\pi n_{12} \tag{4.4.56}$$

$$\theta_{34} = \frac{2\pi d_{34}}{\Lambda} \sin\psi \cos\phi - 2\pi n_{34} \tag{4.4.57}$$

where d_{ij} is the distance between antennas i and j. To estimate n_{12}, we can use the measured phase differences $\hat{\theta}_{12}$ and $\hat{\theta}_{15}$ in the manner described previously for a linear array of three antennas. Similarly, n_{34} can be estimated by using $\hat{\theta}_{34}$ and $\hat{\theta}_{35}$. Thus, the output of the central antenna is used to determine $\hat{n}_{12}$ and $\hat{n}_{34}$. If $0 \leqslant \psi \leqslant \pi/2$, (4.4.56) and (4.4.57) imply that the elevation angle ψ and the azimuth angle ϕ may be estimated by

$$\hat{\psi} = \sin^{-1}\left\{([\hat{\Lambda}(\hat{\theta}_{12} + 2\pi\hat{n}_{12})/2\pi\hat{d}_{12}]^2 + [\hat{\Lambda}(\hat{\theta}_{34} + 2\pi\hat{n}_{34})/2\pi\hat{d}_{34}]^2)^{1/2}\right\} \tag{4.4.58}$$

$$\hat{\phi} = \tan^{-1}[\hat{d}_{34}(\hat{\theta}_{12} + 2\pi\hat{n}_{12})/\hat{d}_{12}(\hat{\theta}_{34} + 2\pi\hat{n}_{34})] \tag{4.4.59}$$

where the quadrant of $\hat{\phi}$ is determined from the signs of $\hat{\theta}_{12} + 2\pi\hat{n}_{12}$ and $\hat{\theta}_{34} + 2\pi\hat{n}_{34}$.

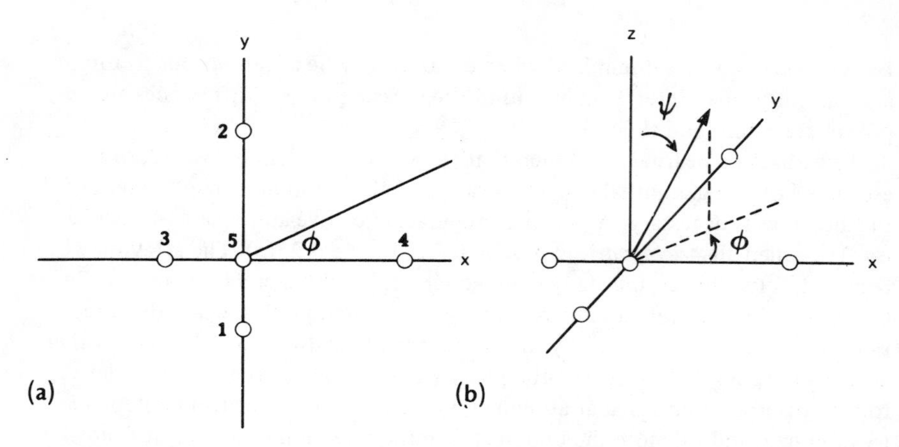

Figure 4.25 Adcock array: (a) planar configuration and (b) spherical coordinates.

The interferometer depicted in Figure 4.26, which eliminates the needs to estimate Λ and to resolve phase ambiguities, is based on the direct measurement of the relative arrival times of a plane wave at two antennas. If $|\phi| \leqslant \phi_m \leqslant \pi/2$ and the received signal is continuous, the bearing estimate is unambiguous. The delay in the arrival time at antenna 1 relative to the arrival time at antenna 2 is

$$T_r = \frac{d \sin \phi}{c} \tag{4.4.60}$$

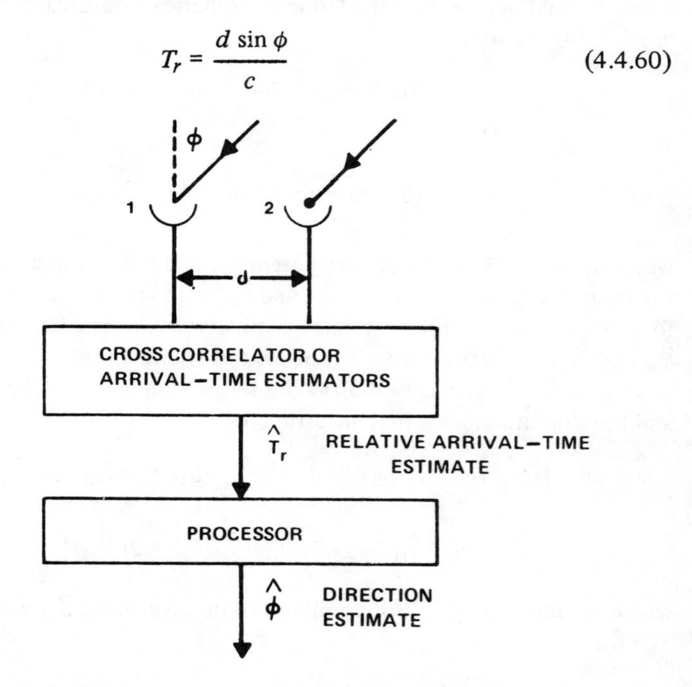

Figure 4.26 Interferometer using arrival-time information.

Because $|\sin\phi| \leqslant 1$, it is prudent to require that the relative-arrival-time estimator satisfy

$$|\hat{T}_r| \leqslant \frac{\hat{d}\sin\phi_m}{c} \tag{4.4.61}$$

Equation (4.4.56) implies that a suitable bearing estimator is

$$\hat{\phi} = \sin^{-1}\left(\frac{c\hat{T}_r}{\hat{d}}\right), \quad |\phi| \leqslant \phi_m \leqslant \pi/2 \tag{4.4.62}$$

Inequality (4.4.61) ensures that the arcsine is defined.

In the subsequent analysis of the *arrival-time interferometer*, we assume that $\hat{d} = d$. In general, $\hat{\phi}$ is a biased estimator of ϕ, whether or not $\hat{T}_r$ is a biased estimator of T_r, except possibly at $\phi = 0$. If $\hat{T}_r - T_r$ is sufficiently small with high probability and $|\phi| \leqslant \phi_m < \pi/2$, then $\hat{\phi}$ is well approximated by the first two terms of the Taylor-series expansion of (4.4.62) about the point T_r. Substituting (4.4.60) into this expansion, we obtain

$$\hat{\phi} \cong \phi + \frac{c}{d\cos\phi}(\hat{T}_r - T_r) \tag{4.4.63}$$

From this equation, it follows that

$$B_\phi = \frac{c}{d\cos\phi} B_T \tag{4.4.64}$$

$$\zeta_\phi = \frac{c}{d\cos\phi} \zeta_T \tag{4.4.65}$$

where B_T is the bias of T_r and ζ_T is the root-mean-square error of $\hat{T}_r$, which is defined by

$$\zeta_T^2 = E[(\hat{T}_r - T_r)^2] = \mathrm{VAR}(\hat{T}_r) + B_T^2 \tag{4.4.66}$$

If the antenna outputs contain independent white Gaussian noises, then $\mathrm{VAR}(\hat{T}_r)$ is bounded by (B.5.22) of Appendix B. The relation $\beta(T_r) = E[\hat{T}_r] = T_r + B_T$ implies that

$$\mathrm{VAR}(\hat{T}_r) \geqslant \left(1 + \frac{\partial B_T}{\partial T_r}\right)^2 \left(\frac{1}{2\beta_r^2}\right)\left[\frac{1}{E_1/N_{01}} + \frac{1}{E_2/N_{02}}\right] \tag{4.4.67}$$

The parameter β_r^2, which is a normalized function of the spectral characteristics of the signal and is assumed to be the same at both antennas, is defined by

$$\beta_r^2 = \frac{(2\pi)^2 \int_{-\infty}^{\infty} f^2 |S(f)|^2 df}{\int_{-\infty}^{\infty} |S(f)|^2 df} \tag{4.4.68}$$

where $S(f)$ is the Fourier transform of the desired signal. Relations (4.4.65) to (4.4.68) provide a lower bound for ζ_ϕ. If $E_1 = E_2 = E, N_{01} = N_{02} = N_0$, and $B_T = 0$, then

$$\zeta_\phi \geqslant \frac{c}{\beta_r d \cos\phi} \left(\frac{E}{N_0}\right)^{-1/2} \tag{4.4.69}$$

Consider the signal that results from passing a truncated sinusoid with an ideal rectangular envelope of duration T through an ideal rectangular bandpass filter of bandwidth B centered at the sinusoidal frequency. Equation (4.4.68) yields

$$\beta_r^2 \approx \frac{2B}{T}, \qquad BT \gg 1 \tag{4.4.70}$$

In contrast, for a signal with a uniform Fourier transform over a bandwidth B, (4.4.68) gives

$$\beta_r^2 = \frac{\pi^2 B^2}{3} \tag{4.4.71}$$

One way to estimate the relative arrival time is to subtract the estimated arrival time at antenna 2, $\hat{T}_2$, from the estimated arrival time at antenna 1, $\hat{T}_1$; that is, we use

$$\hat{T}_r = \hat{T}_1 - \hat{T}_2 \tag{4.4.72}$$

If the arrival-time estimates are uncorrelated, then

$$\mathrm{VAR}(\hat{T}_r) = \mathrm{VAR}(\hat{T}_1) + \mathrm{VAR}(\hat{T}_2) \tag{4.4.73}$$

In the absence of a replica in the receiver to compare to the received signal, the arrival time may be measured by adaptive thresholding or by various other methods [13]. Suppose that a received signal at antenna 1 is applied to an envelope detector. Let $\hat{t}_1$ denote the time at which the leading edge of the detector output pulse exceeds a threshold. Let $\hat{t}_2$ denote the time at which the trailing edge of the pulse drops below the threshold. The arrival time of the signal may be estimated as

$$\hat{T}_1 = \frac{\hat{t}_1 + \hat{t}_2}{2} \tag{4.4.74}$$

If $\hat{t}_1$ and $\hat{t}_2$ are uncorrelated and have equal variances, then

$$\mathrm{VAR}(\hat{T}_1) = \frac{1}{2}\,\mathrm{VAR}(\hat{t}_1) \tag{4.4.75}$$

If adaptive thresholding is used and the signal-to-noise ratio is large, (4.4.37) applies. Thus,

$$\mathrm{VAR}(\hat{T}_1) = \frac{\kappa^2\sigma_1^2}{2S_1^2} \tag{4.4.76}$$

where σ_1^2 is the noise power in the detector output, S_1 is the maximum slope magnitude of the pulse edge, and κ^2 is a constant on the order of unity. A similar expression can be written for $\mathrm{VAR}(\hat{T}_2)$. Using (4.4.73), we obtain

$$\mathrm{VAR}(\hat{T}_r) = \frac{\kappa^2}{2}\left(\frac{\sigma_1^2}{S_1^2} + \frac{\sigma_2^2}{S_2^2}\right) \tag{4.4.77}$$

This equation can be used in place of (4.4.67) when (4.4.72) is applicable.

The relative arrival time can be estimated from the cross correlation of two antenna outputs. The sequence D_i of Figure 4.5 provides discrete-time estimates of the cross correlation. However, to obtain an accurate $\hat{T}_r$, interpolation between the sequence values is required. If the received signals can be modeled as stationary stochastic processes with known spectra, then generalizations of the cross correlator are useful [15].

The cross-spectral density of received signals $s(t)$ and $s(t - T_r)$ has a phase that is a linear function of frequency. The slope of the phase function is proportional to T_r. Thus, if the phase function is measured, regression analysis can be used to obtain $\hat{T}_r$ [16]. The product $A_i^* B_i$ in Figure 4.5 provides a discrete version of the cross-spectral density.

To compare the performance of the phase interferometer with those of the arrival-time interferometer and the energy comparison systems, we assume that the intercepted signal is narrowband with a small uncertainty in the carrier frequency and that the phase ambiguities can be resolved by the phase interferometer with a high probability. Comparing (4.4.69) and (4.4.53) for equal antenna separations, we find that the phase interferometer potentially produces a more accurate bearing estimate than the arrival-time interferometer if

$$f_0 > \frac{\beta_r}{2\pi}. \tag{4.4.78}$$

This inequality is satisfied if β_r is given by (4.4.71) since $f_0 \geqslant B/2$ for a well-defined carrier frequency and $f_0 \gg B$ when the intercepted signal is narrowband. To compare the phase interferometer with the energy comparison systems, we consider $\phi = 0$ and assume that (4.4.69) is nearly an equality. From (4.4.19),

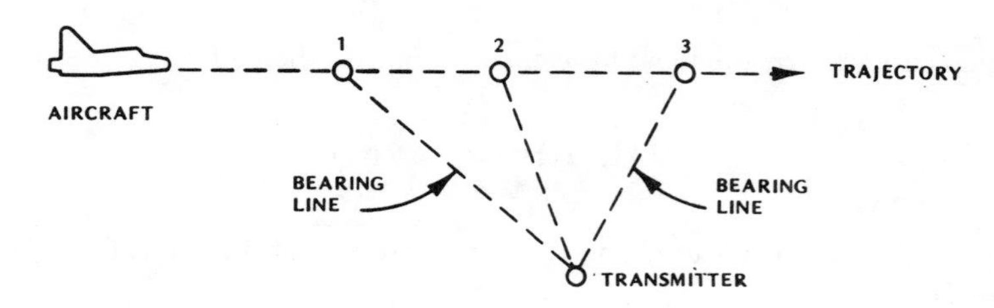

Figure 4.27 Bearing lines from aircraft positions.

(4.4.20), and (4.4.27), we find that the phase interferometer is potentially more accurate than the energy comparison systems for $\phi = 0$ if

$$f_0 > \frac{2c\psi}{\pi d b^2} \exp\left(-\frac{\psi^2}{b^2}\right) \tag{4.4.79}$$

When $b = \psi$, the phase interferometer is potentially more accurate for $\phi = 0$ if

$$d > \frac{2\Lambda}{\pi e \psi} \tag{4.4.80}$$

4.4.3 Passive Location Systems

The output of a direction-finding system can be used in reconnaissance, directional jamming, and other applications. Bearing measurements at two or more stations can be combined to estimate the location of a transmitter [17]. However, accurate bearing measurements may require large antennas or the wide separation of interferometer elements. Consequently, location estimates are sometimes based upon signal arrival-time measurements at three or more stations [17]. Even in this case, bearing measurements are usually necessary at the stations for signal sorting when many signals are intercepted simultaneously.

Figure 4.27 depicts an aircraft with a *direction-finding location system* that makes bearing measurements at three different points in its trajectory. The intersection of two bearing lines provides an estimate of the location of the transmitter, which may be on the surface of the earth or airborne. In the presence of noise, more than two bearing lines will not intersect at a single point. However, the appropriate processing allows an improved estimate of the transmitter position.

A *hyperbolic location system,* which uses arrival-time measurements to estimate location, is a generalization of the interferometer concept. Measurements at two stations are combined to produce a relative arrival time that, in the absence of noise and interference, restricts the possible transmitter location to a hyperboloid with the corresponding pair of stations as foci. Transmitter location is estimated from the intersections of hyperboloids. If the transmitter and the stations lie in the same plane, location is estimated from the intersections of two or more hyperbolas determined from three or more stations. In contrast, the

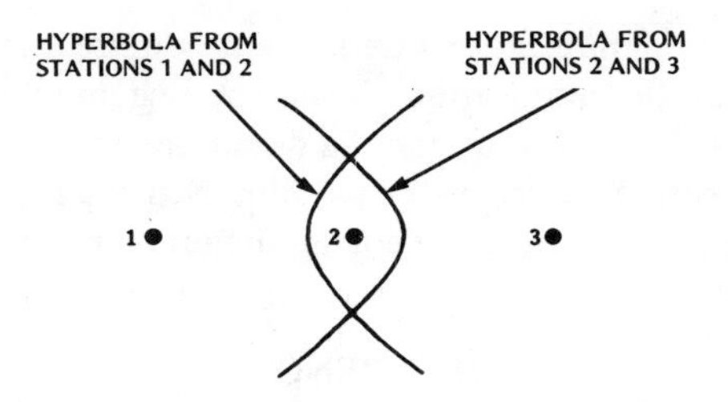

Figure 4.28 Intersecting hyperbolas from three stations.

plane-wave assumption of interferometer bearing estimation is equivalent to the approximation of a hyperbola by one of its asymptotes.

Figure 4.28 illustrates two hyperbolas and three stations. The measurements at two stations predict that the transmitter is located on a hyperbola with those stations as its foci. The two hyperbolas have two points of intersection. The resulting location ambiguity may be resolved by using *a priori* information about the location, bearing measurements at one or more of the stations, or a fourth station to generate an additional hyperbola.

In this chapter, it has usually been assumed that little is known about the signals to be intercepted. As *a priori* information about the signals increases, the receiver designs can be improved accordingly. Detection systems may resemble the receivers being used by the intended communicators. Cross correlators may process the outputs of demodulators rather than the modulated waveforms. The methods of spectrum analysis and linear prediction may be useful in frequency estimation or direction finding.

4.5 COUNTERMEASURES TO INTERCEPTION

By what electronic countermeasures can communicators thwart interception? The signal durations and the transmission powers can be kept to a minimum. Cables and optical-fiber links are effective whenever feasible. Propagation losses can sometimes be exploited. Directional antennas help to conceal the existence of communications from the opponent. However, there are constraints on the degree of directionality that can be designed into an antenna to be used in the battlefield. An important constraint is the need to keep the antenna small to hide it from sight.

Because the antenna beam angle can be decreased by the use of a smaller wavelength as well as by a larger antenna, millimeter-wave or even higher frequencies are sometimes viable alternatives to radio frequencies. The decision to use smaller wavelengths is tempered by such things as cost, available power, and propagation properties. The shorter wavelengths generally are attenuated more than the longer wavelengths and are more easily blocked by obstructions in their path. Furthermore, if the beamwidth is exceedingly narrow, it is difficult to keep it centered on another station of a communication network.

Spread-spectrum communications are inherently more difficult to intercept than are conventional communications. Time hopping, in which transmissions are increased in total duration, but contain pseudorandom time gaps, is another general countermeasure. Because of the pseudorandom gaps, an interception receiver must either process more noise energy than it would otherwise or decrease its observation interval. In either case, performance degrades.

REFERENCES

1. A. Whalen, *Detection of Signals in Noise.* New York: Academic Press, 1971.
2. H.L. Van Trees, *Detection, Estimation, and Modulation Theory*, Vol. III. New York: John Wiley and Sons, 1971.
3. H. Urkowitz, "Energy Detection of Unknown Deterministic Signals," *Proc. IEEE* 55, 523, April 1967.
4. R.A. Dillard, "Detectability of Spread-Spectrum Signals," *IEEE Trans. Aerosp. Electron. Syst.* AES-15, 526, July 1979.
5. N.J. Berg and J.N. Lee, eds., *Acousto-Optic Signal Processing.* New York: Marcel Dekker, 1983.
6. S.L. Kay and S.L. Marple, "Spectrum Analysis – A Modern Perspective," *Proc. IEEE* 69, 1380, November 1981.
7. J.B.G. Roberts, G.L. Moule, and G. Parry, "Design and Application of Real-Time Spectrum Analyzer Systems," *IEE Proc.* 127, Pt. F, 76, April 1980.
8. A. Papoulis, *Signal Analysis.* New York: McGraw-Hill, 1977.
9. M.A. Jack, P.M. Grant and J.H. Collins, "The Theory, Design, and Applications of Surface Acoustic Wave Fourier-Transform Processors," *Proc. IEEE* 68, 450, April 1980.
10. M.I. Skolnik, *Introduction to Radar Systems,* 2nd ed. New York: McGraw-Hill, 1980.
11. B.O. Steinberg, *Principles of Aperture and Array System Design.* New York: John Wiley and Sons, 1976.
12. D.C. Cooper, "Errors in Directional Measurements Using the Relative Amplitudes of Signals Received by Two Aerials," *IEE Proc.* 114, 1834, December 1967.
13. D.J. Torrieri, "Adaptive Thresholding Systems," *IEEE Trans. Aerosp. Electron. Syst.* AES-13, 273, May 1977.
14. E. Jacobs and E.W. Ralston, "Ambiguity Resolution in Interferometry," *IEEE Trans. Aerosp. Electron. Syst.* AES-17, 766, November 1981.
15. *Special Issue on Time Delay Estimation, IEEE Trans. Acoust., Speech, Signal Proc.* ASSP-29, June 1981.
16. Y.T. Chan, R.V. Hattin, and J.B. Plant, "The Least Squares Estimation of Time Delay and Its Use in Signal Detection," *IEEE Trans. Acoust., Speech, Signal Proc.* ASSP-26, 217, June 1978.
17. D.J. Torrieri, "Statistical Theory of Passive Location Systems," *IEEE Trans. Aerosp. Electron. Syst.* AES-20, 183, March 1984.

Chapter 5
Adaptive Antenna Systems

5.1 INTRODUCTION

Nulls in antenna radiation patterns can often be formed in nearly arbitrary directions outside the main beam. Thus, if the arrival angles of a desired signal and interference are known and adequately separated, it is possible in principle to enhance the desired signal while nulling the interference. However, a change in the geometry due to the movement of the intended transmitter, the receiver, or the interference source may cause the interference initially in a null to leave the null. An *adaptive antenna system* can change the direction of a null to accommodate geometrical changes.

An adaptive antenna system automatically monitors its output and adjusts its parameters accordingly. It does so to reduce the impact of interference that enters through the sidelobes, or possibly the mainlobe, of its antenna radiation pattern, while still allowing reception of an intended transmission. The design of an adaptive antenna system requires little *a priori* knowledge of the interference characteristics.

Although any type of antenna can be used in an adaptive antenna system, *phased-array antennas* are the most common. A phased-array antenna consists of a number of antenna elements, the outputs of which are processed and combined to produce the desired antenna radiation pattern. In an adaptive array, usually each antenna-element output is applied to a tunable multiplier. The other input to the multiplier effectively weights the element output. The outputs of the multipliers are summed to produce the array output.

5.2 SIDELOBE CANCELLER

The *sidelobe canceller* is a classic example of an adaptive antenna system. It is not only of practical importance, but also provides an introduction to the fundamental concepts of adaptive antenna systems. The heuristic mathematical analysis in this section shows that the adaptation in a sidelobe canceller can be interpreted not only as noise cancellation, but also as adaptive beam forming and null steering.

Figure 5.1 shows a version of a sidelobe canceller. The *primary* and *auxiliary* antennas are two separate antennas or two different groups of elements in a single phased-array antenna. It is intended that the auxiliary signal, $X_1(t)$, provide an estimate of the interference in the primary signal, $X_0(t)$. After suitable processing, this estimate is subtracted from the primary signal. As a result, the interference, which may have entered through the sidelobes of the primary antenna, is reduced or eliminated by cancellation.

Ideally, $X_1(t)$ has a large interference component and a small desired-signal component, whereas $X_0(t)$ may have a much larger desired-signal component. The quarter-wavelength delay shown in Figure 5.1 produces a signal, $X_2(t)$, that is in phase quadrature with $X_1(t)$. (A quadrature hybrid could be used instead of the delay.) The *weights*, $W_1(t)$ and $W_2(t)$, regulate the amounts of $X_1(t)$ and $X_2(t)$ that are subtracted from $X_0(t)$. The relative magnitudes of the weights determine the magnitude and the phase of the total waveform that is subtracted from $X_0(t)$. If the magnitudes and the phases of the interference component of $X_0(t)$ and the total subtracted waveform are equal, the interference is cancelled and does not appear in the output, $y(t)$. If the cancellation is nearly complete, the weights are nearly constants; if it is not, the weights vary in such a way that the total subtracted waveform gradually becomes a facsimile of the interference component of $X_0(t)$.

In the following analysis, no assumptions are made about the directionality of the primary antenna beam or the reference antenna beam. However, it is highly desirable to have the primary beam point in the direction of the desired signal. The *beam-steering network* forms a beam in the appropriate direction by using various types of *a priori* information. If the antenna is part of a radar system, the information may be radar return characteristics. If the antenna is part of a communication system, the information may be the characteristics of a pilot signal transmitted along with the message signal. The antenna system then locks onto the pilot to form a beam in the direction of the transmitter.

5.2.1 Steady-State Operation

In general, the desired signal received by the primary antenna has the form

$$s(t) = A_s(t)\cos[\omega_0 t + \phi_s(t)] \tag{5.2.1}$$

where ω_0 is the carrier frequency, and $A_s(t)$ and $\phi_s(t)$ are modulation functions. We assume that the interference emerging from the bandpass filter of the primary branch has the form

$$I(t) = A_i(t)\cos[\omega_1 t + \phi_i(t)] \tag{5.2.2}$$

where ω_1 is the carrier frequency, and $A_i(t)$ and $\phi_i(t)$ are modulation functions. If we neglect the thermal noise, the output of the bandpass filter of the primary branch is

$$X_0(t) = A_s(t)\cos[\omega_0 t + \phi_s(t)] + A_i(t)\cos[\omega_1 t + \phi_i(t)] \tag{5.2.3}$$

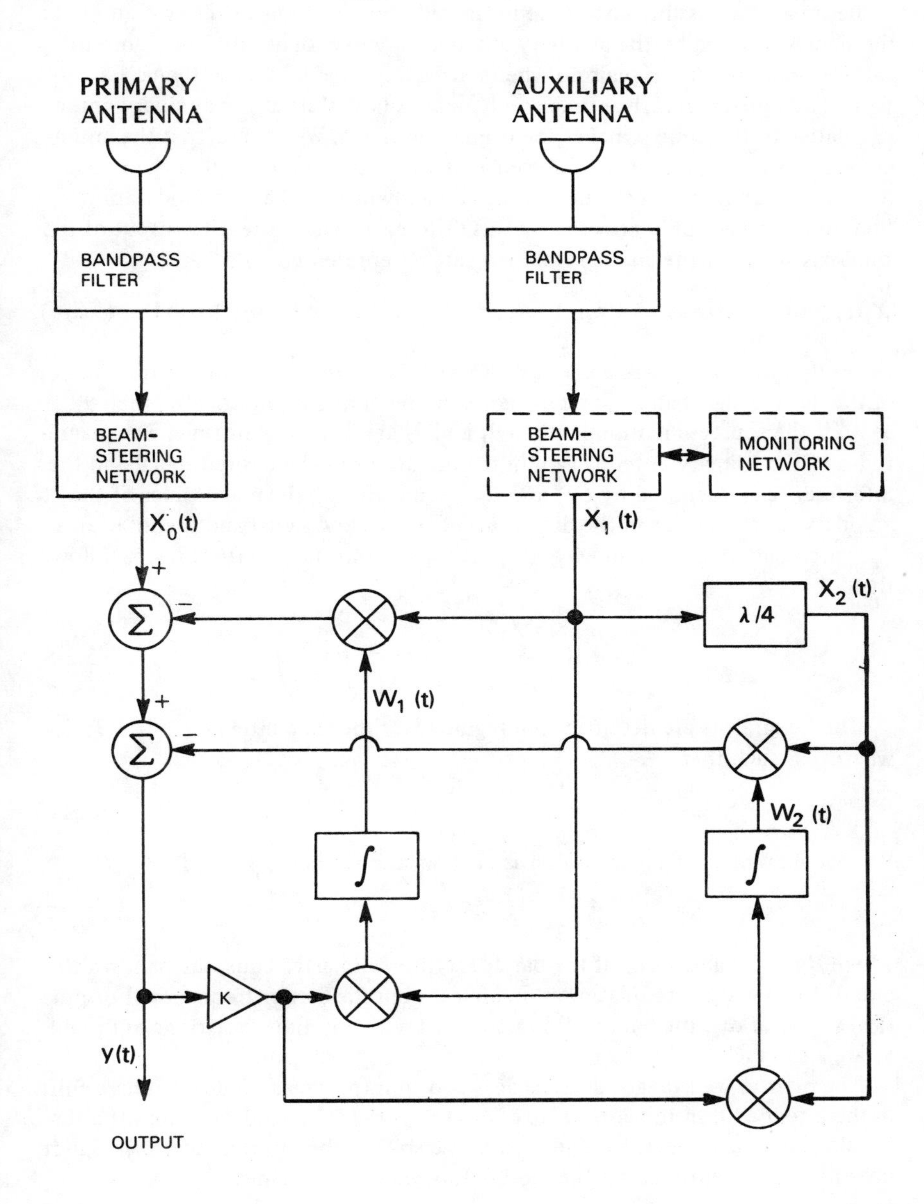

Figure 5.1 Sidelobe canceller.

Because of a possible difference in the radiation patterns of the two antennas, the signals received by the auxiliary antenna may experience different amplifications than the same signals do when received by the primary antenna. Each signal has a different arrival time and hence, a phase shift at the auxiliary antenna relative to the same signal at the primary antenna. We assume that the antennas are close enough that the difference in arrival time of a signal at the two antennas is much less than the inverse signal bandwidth so that the modulation functions are negligibly affected by this difference. Therefore, the output of the band-pass filter of the auxiliary branch can be represented by

$$X_1(t) = C_1 A_s(t)\cos[\omega_0 t + \phi_s(t) + \theta_1] + C_2 A_i(t)\cos[\omega_1 t + \phi_i(t) + \theta_2] \quad (5.2.4)$$

where θ_1 and θ_2 are phase angles, and C_1 and C_2 are real constants. If the sources of the desired signal and the interference are separated geometrically, then $\theta_1 \neq \theta_2$. If the sources are mobile, then θ_1 and θ_2 are functions of time. The magnitudes of the primary antenna gain in the directions of the desired signal and the interference are denoted by G_{ps} and G_{pi}, respectively. The magnitudes of the auxiliary antenna gains in the directions of the desired signal and the interference are denoted by G_{as} and G_{ai}, respectively. From these definitions, it follows that

$$C_1 = \left(\frac{G_{as}}{G_{ps}}\right)^{1/2}, \quad C_2 = \left(\frac{G_{ai}}{G_{pi}}\right)^{1/2} \quad (5.2.5)$$

The integrators are designed to integrate over the time interval $I = [t - T, t]$, where T is such that

$$\omega_0 T \gg 1 \quad (5.2.6)$$

We assume that the bandpass filter is narrowband so that

$$\omega_0 \gg 2\pi B \quad (5.2.7)$$

where B is the bandwidth of the bandpass filters in hertz. Thus, the bandwidth due to the message modulation is much less than the carrier frequency. For practical values of ω_0, the bandwidths associated with any time variations of θ_1 and θ_2 are also much less than ω_0.

The quarter-wavelength delay is designed to introduce a 90-degree phase shift in the intended transmission. If $|\omega_0 - \omega_1| < 2\pi B$, (5.2.7) indicates that this delay also introduces nearly a 90-degree phase shift in the interference, although it does not significantly affect the modulation waveforms. Thus,

$$X_2(t) = C_1 A_s(t)\sin[\omega_0 t + \phi_s(t) + \theta_1] + C_2 A_i(t)\sin[\omega_1 t + \phi_i(t) + \theta_2] \quad (5.2.8)$$

The *signal-to-interference ratio at the primary input* is defined by

$$\rho_i = \frac{\dfrac{1}{T}\displaystyle\int_I s^2(u)\,du}{\dfrac{1}{T}\displaystyle\int_I I^2(u)\,du}$$

$$= \frac{\displaystyle\int_I A_s^2\,du + \int_I A_s^2 \cos(2\omega_0 u + 2\phi_s)\,du}{\displaystyle\int_I A_i^2\,du + \int_I A_i^2 \cos(2\omega_i u + 2\phi_i)\,du} \tag{5.2.9}$$

In general, ρ_i is a function of time. The steady state is defined to exist when the first terms in the numerator and the denominator are nearly constant.

During steady state, (5.2.6) and (5.2.7) imply that the second terms in the numerator and the denominator are negligible compared with the first terms. Thus,

$$\rho_i \cong \frac{\displaystyle\int_I A_s^2(u)\,du}{\displaystyle\int_I A_i^2(u)\,du} = \frac{G_{ps}}{G_{pi}}\,\rho_n \tag{5.2.10}$$

where ρ_n is the ratio that would exist if the primary pattern were omnidirectional. Similarly, *the signal-to-interference ratio at the auxiliary input* is found to be

$$\rho_a \cong \left(\frac{C_1}{C_2}\right)^2 \rho_i$$

$$= \frac{G_{as}}{G_{ai}}\,\rho_n \tag{5.2.11}$$

Although ρ_i, ρ_a, and ρ_n are functions of time in general, they are nearly constant during the steady state.

From Figure 5.1, the output and the weights are given by

$$y(t) = X_0(t) - W_1(t)X_1(t) - W_2(t)X_2(t) \tag{5.2.12}$$

$$W_1(t) = K\int_I y(u)X_1(u)\,du \tag{5.2.13}$$

$$W_2(t) = K\int_I y(u)X_2(u)\,du \tag{5.2.14}$$

where the constant K is the gain of a linear amplifier. Substituting (5.2.12) into (5.2.13), we obtain

$$W_1(t) = K\int_I X_0(u)X_1(u)\,du - K\int_I W_1(u)X_1^2(u)\,du$$
$$- K\int_I W_2(u)X_1(u)X_2(u)\,du \tag{5.2.15}$$

It is plausible that $W_1(t)$ and $W_2(t)$ are nearly constant if ρ_a, θ_1, and θ_2 are nearly constant. In this case, $W_1(t)$ and $W_2(t)$ can be removed outside the integrals with negligible error so that

$$W_1(t) \cong K\int_I X_0(u)X_1(u)\,du - KW_1(t)\int_I X_1^2(u)\,du - KW_2(t)\int_I X_1(u)X_2(u)\,du \tag{5.2.16}$$

In a similar manner, we obtain

$$W_2(t) \cong K\int_I X_0(u)X_2(u)\,du - KW_2(t)\int_I X_2^2(u)\,du - KW_1(t)\int_I X_1(u)X_2(u)\,du \tag{5.2.17}$$

Relations (5.2.4) and (5.2.6) to (5.2.8) and simple trigonometry show that

$$\int_I X_2^2(u)\,du \cong \int_I X_1^2(u)\,du \tag{5.2.18}$$

during the steady state. Using this approximation to solve (5.2.16) and (5.2.17) simultaneously yields

$$W_1(t) = \frac{V_1V_3 - V_2V_4}{V_1^2 - V_2^2} \tag{5.2.19}$$

$$W_2(t) = \frac{V_1V_4 - V_2V_3}{V_1^2 - V_2^2} \tag{5.2.20}$$

where

$$V_1 = \frac{1}{K} + \int_I X_1^2(u)\,du \tag{5.2.21}$$

$$V_2 = \int_I X_1(u)X_2(u)\,du \tag{5.2.22}$$

$$V_3 = \int_I X_0(u)X_1(u)\,du \tag{5.2.23}$$

$$V_4 = \int_I X_0(u)X_2(u)\,du \tag{5.2.24}$$

Evaluating the integrals with (5.2.3), (5.2.4), and (5.2.8) and substituting (5.2.19) and (5.2.20) into (5.2.12) given an expression for the output in terms of the desired signal and interference.

In most cases of interest, some simplification is possible. If T and K are sufficiently large, the first term on the right-hand side of (5.2.21) can be ignored. Substituting (5.2.4) and (5.2.8) into (5.2.22) shows that V_2 is very small during steady-state operation. Consequently, it is reasonable to neglect V_2 in (5.2.19) and (5.2.20). As a result of these approximations, we obtain

$$W_1(t) \cong \frac{\int_I X_0(u) X_1(u)\, du}{\int_I X_1^2(u)\, du} \tag{5.2.25}$$

$$W_2(t) \cong \frac{\int_I X_0(u) X_2(u)\, du}{\int_I X_1^2(u)\, du} \tag{5.2.26}$$

Thus, the weights are essentially normalized cross-correlation functions. From (5.2.3), (5.2.4), (5.2.6), (5.2.7), (5.2.8), we obtain

$$\int_I X_1^2(u)\, du = \int_I \Bigg\{ \frac{C_1^2}{2} A_s^2(u) + \frac{C_2^2}{2} A_i^2(u) + C_1 C_2 A_s(u) A_i(u) \times \cos[(\omega_0 - \omega_1)u + \phi_s(u) - \phi_i(u) + \theta_1 - \theta_2] \Bigg\} du \tag{5.2.27}$$

$$\begin{aligned}\int_I X_0(u) X_1(u)\, du = \int_I \Bigg\{ & \frac{C_1}{2} A_s^2(u) \cos \theta_1 + \frac{C_2}{2} A_i^2(u) \cos \theta_2 \\ & + \frac{C_1}{2} A_s(u) A_i(u) \cos[(\omega_0 - \omega_1)u + \phi_s(u) - \phi_i(u) + \theta_1] \\ & + \frac{C_2}{2} A_s(u) A_i(u) \cos[(\omega_0 - \omega_1)u + \phi_s(u) - \phi_i(u) - \theta_2] \Bigg\}\end{aligned} \tag{5.2.28}$$

$$\begin{aligned}\int_I X_0(u) X_2(u)\, du = \int_I \Bigg\{ & \frac{C_1}{2} A_s^2(u) \sin \theta_1 + \frac{C_2}{2} A_i^2(u) \sin \theta_2 \\ & + \frac{C_1}{2} A_s(u) A_i(u) \sin[(\omega_0 - \omega_1)u + \phi_s(u) - \phi_i(u) + \theta_1] \\ & - \frac{C_2}{2} A_s(u) A_i(u) \sin[(\omega_0 - \omega_1)u + \phi_s(u) - \phi_i(u) - \theta_2] \Bigg\}\end{aligned} \tag{5.2.29}$$

In general, because the desired signal is unsynchronized with the interference, we expect that during steady-state operation

$$\int_I s_1(u, \phi_1) I_1(u, \phi_2)\, du \cong 0 \tag{5.2.30}$$

where $s_1(u, \phi_1)$ is the desired signal with an arbitrary phase shift of ϕ_1, and $I_1(u, \phi_2)$ is the interference signal with an arbitrary phase shift of ϕ_2. If (5.2.30) is valid, the desired signal and interference are said to be uncorrelated in the time-average sense.

Equation (5.2.30) implies that only the first two terms of (5.2.27) to (5.2.29) need to be retained. Using (5.2.10), (5.2.11), and (5.2.25) to (5.2.30), we obtain

$$W_1 \cong \frac{C_1 \cos\theta_2 + C_2 \rho_a \cos\theta_1}{C_1 C_2 (1 + \rho_a)} \tag{5.2.31}$$

$$W_2 \cong \frac{C_1 \sin\theta_2 + C_2 \rho_a \sin\theta_1}{C_1 C_2 (1 + \rho_a)} \tag{5.2.32}$$

These equations indicate that W_1 and W_2 are approximately constant if ρ_a, θ_1, and θ_2 are approximately constant. Thus, the initial assumption that the weights are constant has been verified as consistent with the other approximations made in deriving these equations.

The substitution of (5.2.31) and (5.2.32) into (5.2.12) and use of trigonometric identities and (5.2.5) yield

$$y(t) = \beta A_s(t) \cos\left\{\omega_0 t + \phi_s(t) + \tan^{-1}\left[\frac{\sin(\theta_2 - \theta_1)}{\alpha - \cos(\theta_2 - \theta_1)}\right]\right\}$$
$$+ \beta\alpha\rho_a A_i(t) \cos\left\{\omega_1 t + \phi_i(t) + \tan^{-1}\left[\frac{\alpha \sin(\theta_2 - \theta_1)}{1 - \alpha\cos(\theta_2 - \theta_1)}\right]\right\} \tag{5.2.33}$$

where

$$\alpha = \left(\frac{G_{ps} G_{ai}}{G_{pi} G_{as}}\right)^{1/2} \tag{5.2.34}$$

$$\beta = \frac{\left[1 - \frac{2\cos(\theta_2 - \theta_1)}{\alpha} + \frac{1}{\alpha^2}\right]^{1/2}}{1 + \rho_a} \tag{5.2.35}$$

The *signal-to-interference ratio at the output* of the sidelobe canceller, ρ_0, is calculated as the ratio of the average power over interval I in the first term to the average power over interval I in the second term of the right-hand side of (5.2.33). If $y(t) \neq 0$, the result is the remarkably simple formula:

$$\rho_0 = \frac{1}{\rho_a} \tag{5.2.36}$$

Thus, the interference component of the primary signal has been nearly cancelled if $\rho_a \ll 1$. We conclude that the output signal distortion is small when the signal power at the auxiliary antenna is relatively low. An important approximation made in deriving (5.2.33) and (5.2.36) is that the thermal noise is negligible.

If the interference source is almost directly behind the source of the desired signal, then $\alpha \cong 1$ and $\theta_1 \cong \theta_2$. Because (5.2.33) indicates that $y(t) \cong 0$, the output signal is buried in the thermal noise. We conclude that the sidelobe canceller is ineffective in this case.

A strong desired signal in the auxiliary branch can induce cancellation of the desired signal at the output. Specifically, if $\rho_i \rho_a \gg 1$, (5.2.36) gives $\rho_0 \ll \rho_i$. Consequently, the adaptive system causes a performance degradation relative to the performance of the primary antenna operating alone. Equation (5.2.11) indicates that designing the system so that $G_{as} \ll G_{ai}$ is helpful in ensuring small values of ρ_a. Thus, an auxiliary antenna with a beam that points in the direction of the interference is highly desirable. One possible implementation is to use a beam-steering network that enables an auxiliary beam to search for interfering signals of high power or special characteristics.

Even if $G_{as} \ll G_{ai}$, serious cancellation of the desired signal can occur if the interference is so small that $\rho_a \gg 1$. To prevent this disaster, the auxiliary input may be monitored and its propagation may be blocked unless its amplitude or power exceeds a fixed threshold. Thus, the adaptive mechanism is disabled unless the interference is strong enough to warrant its use.

If (5.2.30) is not satisfied, the output signal distortion can be significant even if $\rho_a = 0$. To demonstrate this phenomenon, we set $C_1 = 0$ and consider tone interference at frequency $\omega_1 = \omega_0$. Assuming that $A_i(t)$ and $\phi_i(t)$ are constants, (5.2.3), (5.2.4), (5.2.8), (5.2.12), and (5.2.25) to (5.2.29) yield

$$y(t) = A_s(t)\cos[\omega_0 t + \phi_s(t)] - \frac{1}{T}\int_I A_s(u)\cos[\omega_0 t + \phi_s(u)]du \qquad (5.2.37)$$

If the desired signal has frequency or phase modulation only, then $A_s(t)$ is a constant. Therefore, if T is large, the second term in (5.2.37) becomes small, indicating that the distortion may be negligible. However, if the desired signal has amplitude modulation and $\phi_s(t)$ is a constant, then (5.2.37) becomes

$$y(t) = \left[A_s(t) - \frac{1}{T}\int_I A_s(u)\,du\right]\cos(\omega_0 t + \phi_s) \qquad (5.2.38)$$

which indicates that a desired-signal spectral component at the same frequency as the tone interference is cancelled.

5.2.2 Adaptive Null Steering

The operation of the sidelobe canceller can be interpreted in terms of *adaptive null steering.* Equation (5.2.33) can be expressed in the forms

$$\begin{aligned} y(t) &= \beta s_1(t) + \beta\alpha\rho_a J_1(t) \\ &= [s_1(t) + J_1(t)] - [(1-\beta)s_1(t) + (1-\beta\alpha\rho_a)J_1(t)] \end{aligned} \qquad (5.2.39)$$

where $s_1(t)$ and $J_1(t)$ are phase-shifted versions of $s(t)$ and $I(t)$. The first bracketed term on the right-hand side of (5.2.39) can be interpreted as the response due to the primary antenna with gains G_{ps} and G_{pi} in the directions of the desired signal and interference, respectively. The second bracketed term can be interpreted as the response due to an equivalent pattern, the output of which is subtracted from the primary output to give $y(t)$.

With this interpretation, the equivalent cancellation pattern or adaptive beam has gains

$$G_s' = (1 - \beta)^2 G_{ps} \tag{5.2.40}$$

$$G_i' = (1 - \beta\alpha\rho_a)^2 G_{pi} \tag{5.2.41}$$

in the directions of the desired signal and interference, respectively. The equivalent overall pattern of the sidelobe canceller provides the gains

$$G_s'' = (\sqrt{G_{ps}} - \sqrt{G_s'})^2 = \beta^2 G_{ps} \tag{5.2.42}$$

$$G_i'' = (\sqrt{G_{pi}} - \sqrt{G_i'})^2 = (\beta\alpha\rho_a)^2 G_{pi} \tag{5.2.43}$$

in the directions of the desired signal and interference, respectively. If ρ_a is small and α is large, then $G_s'' \cong G_{ps}$. If ρ_a is so small that $\rho_a \alpha \ll 1$, then $G_i'' \ll G_{pi}$. Thus, an approximate null can be created in the direction of the interference. For this reason, the action of an adaptive antenna system is sometimes called null steering. A pictorial representation is shown in Figure 5.2.

The sidelobe canceller of Figure 5.1 exhibits the main features of all adaptive antenna systems, but has at least two important limitations. It is designed for narrowband signals only and cannot reject interference from two or more sources in different directions. Thus, more general and alternative configurations are considered in the remainder of the chapter.

5.3 POTENTIAL PERFORMANCE OF TWO-ELEMENT ADAPTIVE ARRAY

An adaptive antenna system consists of an antenna array, fixed processing components, and adaptive components. Its purpose is to remove externally generated interference from a received signal. Figure 5.3 depicts an array with two antenna elements separated by distance D. Behind the elements are *adaptive filters* with parameters that are adjusted by adaptive control signals. The outputs of the adaptive filters are summed to produce the system output. Let ϕ denote the direction of an incoming plane wave; ϕ is positive in the clockwise direction from the perpendicular to the line connecting the antennas. The received signal at element 1 is taken as the reference with zero phase. The received signal at element 2 arrives before the reference by the amount

$$T_r = \frac{D \sin \phi}{c} \tag{5.3.1}$$

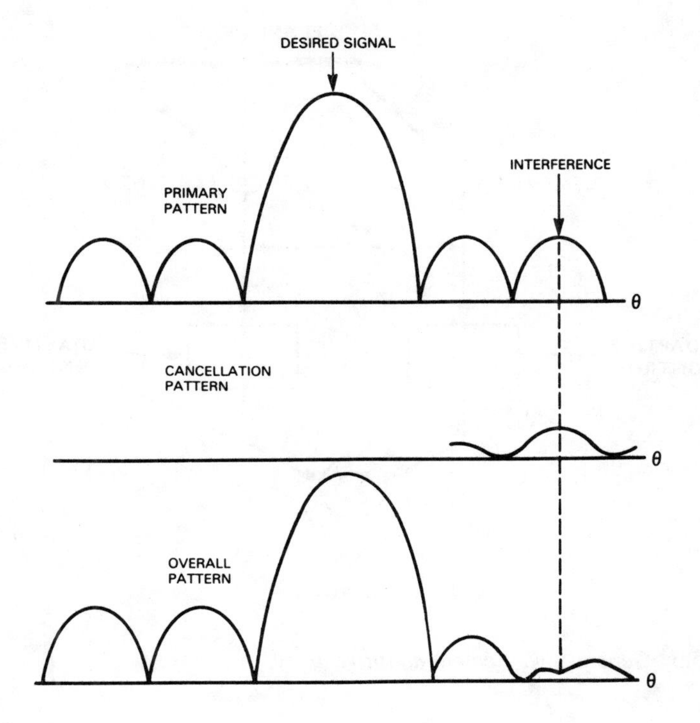

Figure 5.2 Adaptive beam forming and null steering.

where c is the velocity of an electromagnetic wave. Let $h(\phi,f)$ represent the voltage response of each element to a unit-amplitude signal arriving from angle ϕ at frequency f. The mutual coupling between the antenna elements is neglected. The transfer function between the received signal at the reference and the system output is

$$H_s(\phi,f) = h(\phi,f)H_1(f) + h(\phi,f)H_2(f)\exp(j2\pi fT_r) \quad (5.3.2)$$

where $j = \sqrt{-1}$ and $H_1(f)$ and $H_2(f)$ are the transfer functions of the adaptive filters.

Suppose that interference arrives from angle $\phi = \phi_0$ so that $T_r = T_0$. Let Ω denote the domain of frequencies over which the adaptive system must operate. If the adaptive controls adjust the filters so that

$$H_2(f) = -H_1(f)\exp(-j2\pi fT_0) \quad (5.3.3)$$

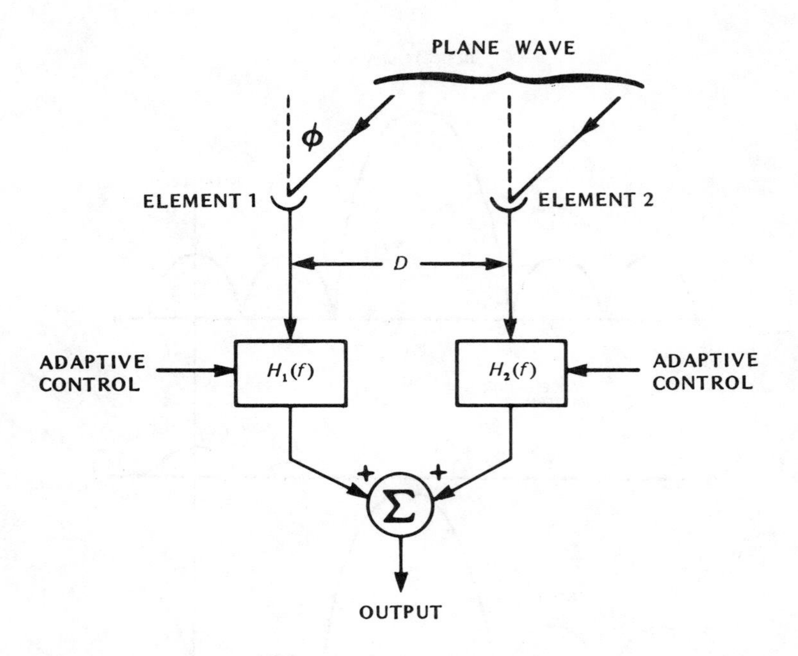

Figure 5.3 General two-element adaptive array.

for f in Ω, then $H_s(\phi_0,f) = 0$ and the interference is completely eliminated at the system output. Thus, the approximate nulling of one interferer is possible if the adaptive filters have enough adjustable parameters that (5.3.3) can be approximately satisfied for all possible ϕ_0 and all f in Ω. If (5.3.3) is satisfied, it follows from (5.3.2) and (5.3.1) that the transfer function for a signal from an arbitrary direction ϕ is

$$H_s(\phi,f) = h(\phi,f)H_1(f)\left\{1 - \exp\left[\frac{j2\pi fD}{c}(\sin\phi - \sin\phi_0)\right]\right\} \tag{5.3.4}$$

In creating a null response to the interference from one direction, the adaptive array may cause additional nulls, called *grating nulls,* at other angles for some f in Ω. Equation (5.3.4) indicates that $H_s(\phi,f) = 0$ at $\phi = \phi_n$ if there is an integer n such that

$$\frac{fD}{c}(\sin\phi_n - \sin\phi_0) = n \tag{5.3.5}$$

Among the nulls determined by this equation, the null at ϕ_0 is the desired one; the others are grating nulls. Antenna-element nulls, which occur at angles for which $h(\phi,f) = 0$, are also possible. Because $f/c = 1/\lambda$, where λ is the wavelength, the desired null and grating nulls exist at angles

$$\phi_n = \sin^{-1}\left(\sin\phi_0 + \frac{n\lambda}{D}\right), \qquad N_1 \leqslant n \leqslant N_2 \tag{5.3.6}$$

where the limits N_1 and N_2 are determined by the requirement that the arcsine exist. There are no grating nulls when the frequency is f if

$$D < \frac{\lambda}{2} \tag{5.3.7}$$

Equation (5.3.4) indicates that it is not possible to completely null a second interferer if $H_1(f) \neq 0$ unless the interference is a tone that fortuitously arrives in a grating null or an antenna-element null. If (5.3.3) is exactly satisfied, the adaptation of $H_1(f)$ serves only to regulate the spectral characteristics of a desired signal that arrives from a direction that is not nulled. In general, an adaptive array of N elements can null at most $N - 1$ interferers.

For ϕ sufficientlv close to ϕ_n, a Taylor-series expansion of (5.3.4) gives

$$H_s(\phi,f) \cong h(\phi_n,f)H_1(f)\left(-\frac{j2\pi D}{\lambda}\cos\phi_n\right)(\phi - \phi_n) \tag{5.3.8}$$

The *angular extent* of a null is defined to be the range of values of ϕ near ϕ_n for which $H_s(\phi,f)$ is small. Equation (5.3.8) indicates that for omnidirectional antenna elements, the angular extent of a null increases with ϕ_n, which is the angular offset from boresight. As D/λ increases, the angular extent of each null decreases, but according to (5.3.6), the number of grating nulls increases. Consequently, there is a rough conservation of the total angular extent of all the nulls for $D/\lambda > 1/2$.

It is often necessary to have $D > \lambda/2$ for all frequencies of interest to prevent significant mutual coupling between the antenna elements. From the results of section 5.5, it can be shown that when the element separations exceed $\lambda/2$, grating-null problems can be mitigated by the use of three or more antenna elements, the use of elements that do not all have the same patterns and polarization responses, and the deployment of the elements in a nonlinear or nonplanar configuration [1].

The potential spectral extent of a desired null decreases with the number of adjustable parameters in the adaptive filters. As an example, it is assumed that for f in Ω the adaptive filters have the forms

$$H_1(f) = 1, \quad H_2(f) = a + b \exp\left(\frac{-j\pi f}{2f_0}\right) \tag{5.3.9}$$

where a and b are adjustable parameters and f_0 is fixed. The transfer function $H_2(f)$ can be realized with a delay line of length $\lambda_0/4 = c/4f_0$ and two attenuators, as shown in Figure 5.4. It is also assumed for simplicity that the antenna element patterns are omnidirectional and independent of frequency; that is, $h(\phi,f) = 1$.

Suppose that the adaptation creates a perfect null at angle ϕ_0 and frequency f_0 so that $H_s(\phi_0,f_0) = 0$. Equations (5.3.3) and (5.3.9) with $f = f_0$ allow the determination of a and b. Substitution into (5.3.2) then yields

$$H_s(\phi,f) = 1 - \left[\cos 2\pi f_0 T_0 + \sin 2\pi f_0 T_0 \exp\left(\frac{-j\pi f}{2f_0}\right)\right] \exp(j2\pi f T_r) \tag{5.3.10}$$

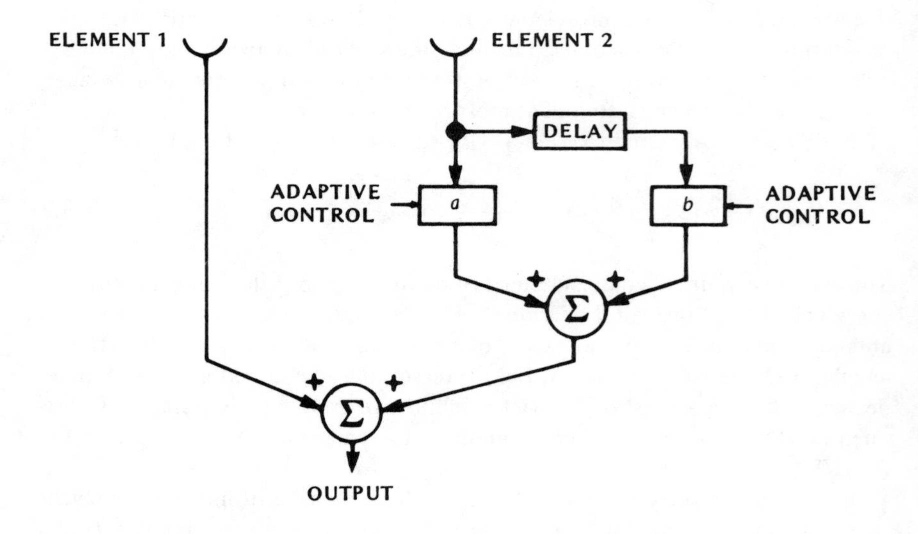

Figure 5.4 Simple two-element adaptive array.

If $\phi = \phi_0$ and f is sufficiently close to f_0, then a Taylor-series expansion gives

$$H_s(\phi_0, f) \cong \left[\frac{\pi}{2f_0} \sin 2\pi f_0 T_0 \exp(j2\pi f_0 T_0) - j\, 2\pi T_0 \right] (f - f_0) \qquad (5.3.11)$$

which shows that for most values of ϕ_0, if the interference has power at frequencies other than f_0, it cannot be completely cancelled.

Suppose that the interference has a uniform two-sided power spectral density of $J_0/2$ at the array inputs. The power spectral density of the interference at the system output is $J_0 |H_s(\phi_0, f)|^2/2$. The spectral width of the null is defined as the range of positive values of f such that $|H_s(\phi_0, f)|^2 \leqslant G^2$, where G is a constant that specifies an acceptable or required reduction in the power spectral density. For this inequality to be satisfied, (5.3.11) indicates that an approximate necessary condition is

$$|f - f_0| \leqslant G \left| \frac{\pi}{2f_0} \sin 2\pi f_0 T_0 \exp(j2\pi f_0 T_0) - j2\pi T_0 \right|^{-1} \qquad (5.3.12)$$

Thus, the spectral width of the null is approximated by twice the right-hand side of (5.3.12), provided that it is small compared with f_0. The parameter G is determined by the desired null depth. For example, for a null of at least 0.01 (−20 dB) in the power spectral density at the output, $G = 0.1$. The *fractional bandwidth* of the null, F, is defined as the spectral width divided by f_0. Thus,

$$F \approx 2G \left| \frac{\pi}{2} \sin 2\pi f_0 T_0 \exp(j2\pi f_0 T_0) - j2\pi f_0 T_0 \right|^{-1} \qquad (5.3.13)$$

provided that the right-hand side of this equation is much less than unity. For extreme values of $f_0 T_0 = D \sin \phi_0/\lambda_0$, we have

$$F \approx 2G \left(\frac{2\pi D}{\lambda_0} \sin \phi_0 \right)^{-1}, \qquad D \sin \phi_0 \gg \lambda_0 \qquad (5.3.14)$$

$$F \approx 1.1G \left(\frac{2\pi D}{\lambda_0} \sin \phi_0 \right)^{-1}, \quad D \sin \phi_0 \ll \lambda_0 \qquad (5.3.15)$$

Both of these equations indicate that F is inversely proportional to D/λ_0. However, decreasing D to increase F eventually leads to undesirable interactions between the antenna elements or to an unacceptably wide angular extent of the null, as indicated by (5.3.8).

If $f_0 T_0 = 1/2$, then (5.3.13) gives $F \approx 2G/\pi$. For $G = 0.1$, the fractional bandwidth is roughly six percent. If a fractional bandwidth of ten percent must be accommodated, then we must accept $G = 0.157$, which corresponds to a null of −16 dB. Thus, the simple adaptive system of Figure 5.4 is limited in the bandwidth over which it can suppress interference. Because perfect nulling at a fixed frequency is assumed and the spectral characteristics of the desired signal at the output are ignored, the foregoing results are not conclusive. Nevertheless, they indicate the spectral limitations of adaptive antenna systems with only a few adjustable parameters. In general, enough adjustable parameters are needed so that (5.3.3) can be approximately satisfied for all frequencies within the bandwidth of the desired signal and for all possible directions of the interference source.

A more realistic model of a two-element adaptive array with two-tap delay lines or quadrature hybrids in the branches emanating from the antenna elements gives similar results [2]. Significant performance degradation occurs if the adaptive system must process signals with rectangular spectra and fractional bandwidths exceeding roughly four percent. However, a two-element array with three-tap delay lines appears capable of accommodating far larger fractional bandwidths. In a practical system with two or more elements, the number of delay-line taps required in each branch is significantly influenced by the *frequency-dependent mismatches* among the antenna elements and among the other devices used in the implementation.

Suppose that a third antenna element with the same voltage response is added to the right of the elements in Figure 5.3. If the elements lie along a straight line and are uniformly spaced, then the perfect nulling of two interferers requires that

$$H_1(f) + H_2(f)\exp(j2\pi f T_0) + H_3(f)\exp(j4\pi f T_0) = 0 \tag{5.3.16}$$

$$H_1(f) + H_2(f)\exp(j2\pi f T_1) + H_3(f)\exp(j4\pi f T_1) = 0 \tag{5.3.17}$$

where T_0 and T_1 are the arrival-time shifts of the two interference waveforms and the $H_i(f)$, $i = 1, 2, 3$, are the filter transfer functions. Even without solving these equations, it is clear that if T_1 is arbitrary, then (5.3.3), which represents the solution to the two-element and one-interferer problem, is inconsistent with (5.3.16) and (5.3.17). To null the second interferer, $H_2(f)$ must be changed. Thus, a particular adaptive filter does not null a particular interferer. Instead, the adaptive filters act cooperatively to null all the interferers.

5.4 ADAPTIVE FILTERS

The adaptive filters used in adaptive antenna systems have one of the forms shown in Figure 5.5. In Figure 5.5(a), the *weight-adjustment mechanism* responds to the inputs, the output, or both. In Figure 5.5(b), the mechanism responds to the inputs and to the *error*, which is the difference between the reference and the output. Ideally, the *reference signal,* which is sometimes called the *desired response,* would be the received waveform minus the interference and the noise. In practice, the reference signal has the general characteristics, but not the detailed structure, of the desired signal that the antenna system is attempting

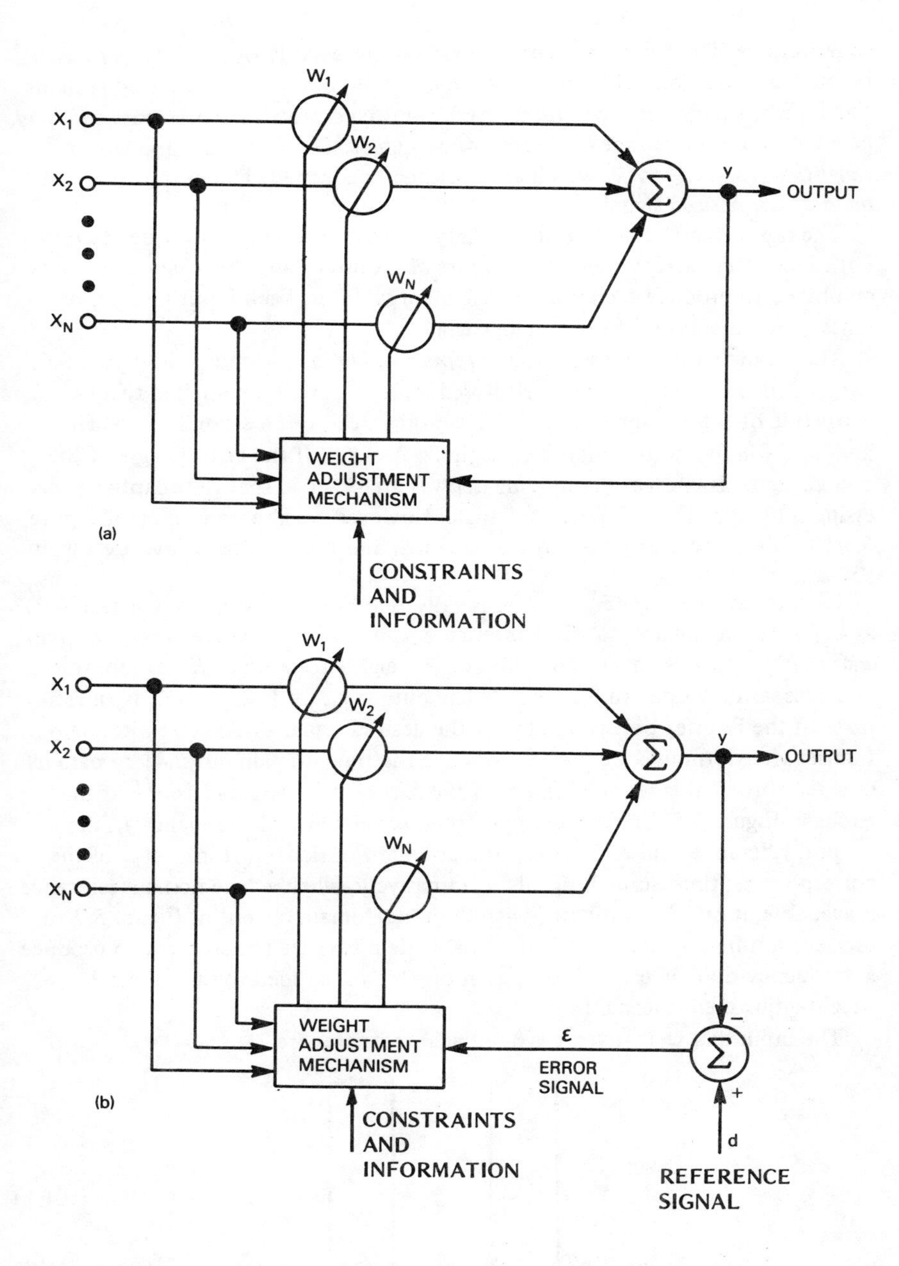

Figure 5.5 Basic adaptive filters with weight adjustments controlled by (a) inputs and/or output, or (b) inputs and error.

to reproduce. The forms in Figure 5.5 are not necessarily optimal, but are used because of their simplicity and compatibility with computer-controlled systems. The inputs, $x_1, x_2, \ldots, x_N$, are derived from fixed processing elements and may be either continuous-time or discrete-time signals. The inputs are applied to *weights* $w_1, w_2, \ldots, w_N$, which are continually adjusted. The output is an estimate of the desired signal.

The inputs may be derived in a variety of ways. Each input may be the output of an antenna. The weighted outputs of groups of antenna elements may be combined to produce each input to an adaptive filter. Each input then represents a separate beam of the antenna array.

The inputs may be derived from a *tapped delay line,* as shown in Figure 5.6. Each of the K antenna outputs is filtered, delayed, and then applied to a line consisting of L tap points and $L - 1$ ideal time delays of δ seconds each. The number of inputs to the adaptive weights is $N = KL$. If the arrival angle of the desired signal is known, the *steering delays* can be set so that the adaptive processing is independent of the arrival angle. However, for the processing of Figure 5.5(b), this knowledge is usually not assumed and the steering delays are usually absent.

Each input of Figure 5.5 may be the discrete Fourier transform (section 4.2) at a specific frequency, f_i, of an antenna output. A complete *frequency-domain adaptive system* has an adaptive filter of N weights for each of K discrete frequencies, as illustrated in Figure 5.7. The output y_i, $i = 1, 2, \ldots, K$, is an estimate of the Fourier transform at f_i of the desired signal. An inverse discrete Fourier transform may be used to produce the time-domain output y in parallel or serial form if it is required. Each of the K sets of N weights has one of the forms in Figure 5.5. For the adaptive filter associated with frequency f_i, the output, reference, and error are all the discrete Fourier transforms at f_i of the corresponding time-domain signals. Alternatively, when a time-domain reference is available, it may be combined with the time-domain output of Figure 5.7 to produce a time-domain error. This signal is then Fourier transformed to produce K frequency-domain error signals, each of which is applied to one of the K weight-adjustment mechanisms.

The input and weight vectors of an adaptive filter are

$$\mathbf{X} = \begin{bmatrix} x_1 \\ x_2 \\ \cdot \\ \cdot \\ \cdot \\ x_N \end{bmatrix}, \quad \mathbf{W} = \begin{bmatrix} w_1 \\ w_2 \\ \cdot \\ \cdot \\ \cdot \\ w_N \end{bmatrix} \tag{5.4.1}$$

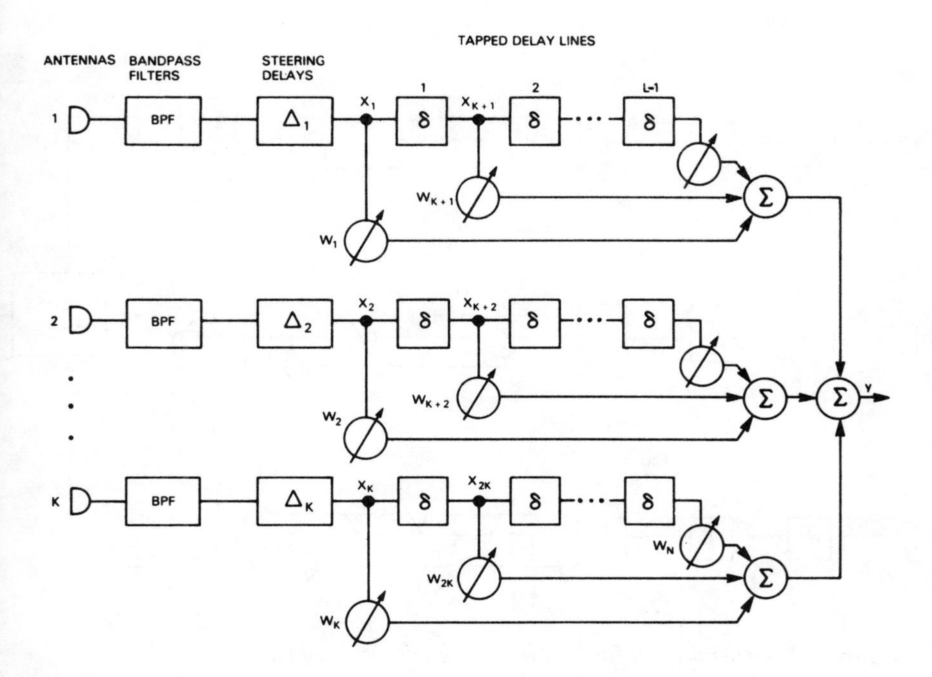

Figure 5.6 General form for tapped-delay-line processing.

The scalar output is

$$y = \mathbf{W}^T \mathbf{X} \tag{5.4.2}$$

where the superscript T denotes the transpose. The elements of **X** and **W** may be real or complex.

The *complex-variable representation* is useful in describing certain types of processing. In the system of Figure 5.7, the components of **X** are discrete Fourier transforms and, hence, complex variables. As another example, we consider two real inputs processed by four real weights that are related as shown in Figure 5.8(a). The two real outputs are

$$y_1 = w_1 x_1 - w_2 x_2 \tag{5.4.3}$$

$$y_2 = w_2 x_1 + w_1 x_2 \tag{5.4.4}$$

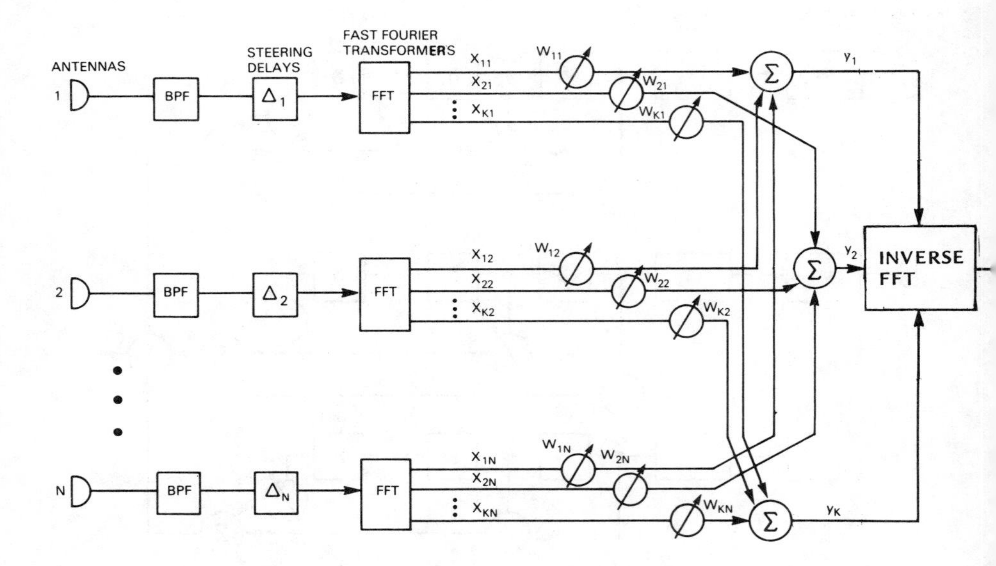

Figure 5.7 General form for frequency-domain processing.

If we define $y = y_1 + jy_2$, $x = x_1 + jx_2$, and $w = w_1 + jw_2$, then $y = wx$, which gives a compact complex-variable representation of the two real-variable equations.

If a real-valued signal is represented by an analytic signal or a complex envelope (Appendix A), then multiplication by a complex weight represents an amplitude change and a phase shift. As an example, a common configuration in adaptive antenna systems is depicted in Figure 5.8(b). Let $\breve{f}(t)$ denote the Hilbert transform of $f(t)$. The ideal *quadrature hybrid* produces the input $x(t)$ at one of its outputs and $\breve{x}(t)$ at the other. Alternatively, if $x(t) = g(t)\cos 2\pi f_0 t$ and $g(t)$ is sufficiently narrowband, then it follows from (A.1.11) of Appendix A that $\breve{x}(t)$ can be approximately produced by the output of a quarter-wavelength delay line. The output of Figure 5.8(b) is

$$y(t) = w_1 x(t) + w_2 \breve{x}(t) \tag{5.4.5}$$

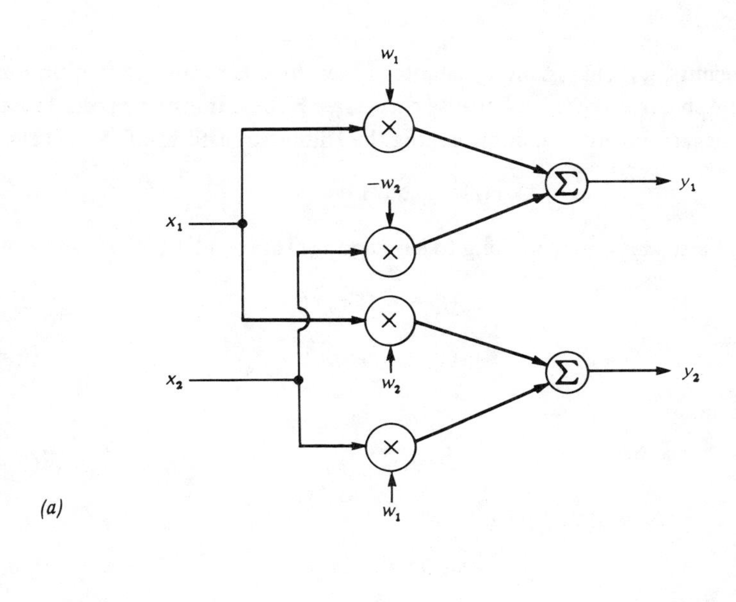

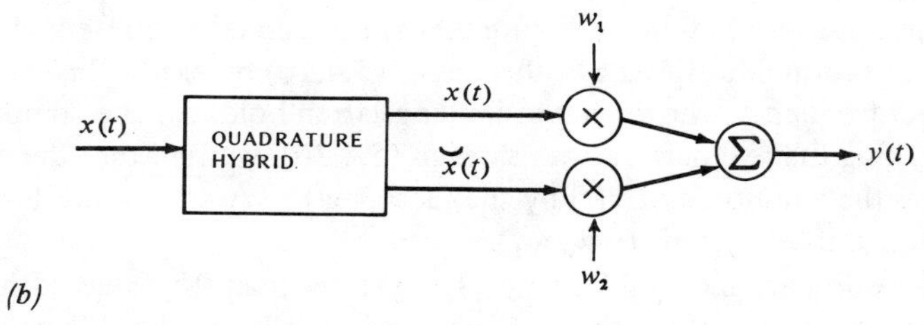

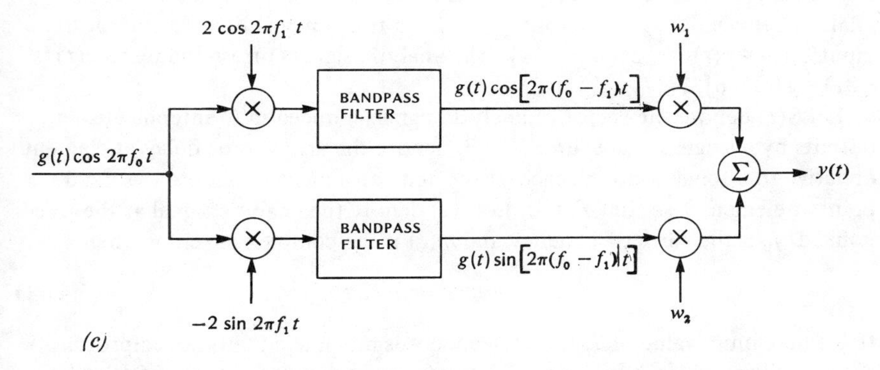

***Figure 5.8** Processing conveniently represented by complex notation.*

The weights, w_1 and w_2, are constants or have baseband spectra that do not overlap the spectrum of $x(t)$. Because successive Hilbert transforms of a function yield the negative of the function, taking the Hilbert transform of (5.4.5) gives

$$\breve{y}(t) = w_1 \breve{x}(t) - w_2 x(t) \tag{5.4.6}$$

The analytic signals corresponding to $x(t)$ and $y(t)$ are

$$x_a(t) = x(t) + j\breve{x}(t) \tag{5.4.7}$$

$$y_a(t) = y(t) + j\breve{y}(t) \tag{5.4.8}$$

If we define

$$w = w_1 - jw_2 \tag{5.4.9}$$

then (5.4.5) to (5.4.8) imply that

$$y_a(t) = wx_a(t) \tag{5.4.10}$$

Thus, the input-output relation for Figure 5.8(b) is succinctly represented by the complex notation. Multiplying both sides of (5.4.10) by $\exp(-j2\pi f_0 t)$, where f_0 is the center frequency, shows that a similar relation holds for the complex envelopes. Taking the real part of both sides of (5.4.10), we find that the processing changes the amplitude of $x(t)$ by the factor $|w| = \sqrt{w_1^2 + w_2^2}$, while the phase of $x(t)$ is shifted by $\tan^{-1}(-w_2/w_1)$.

According to Equation (A.1.11), if $f_0 - f_1$ is greater than the bandwidth of $g(t)$ in Figure 5.8(c), then the output of one bandpass filter is the Hilbert transform of the output of the other bandpass filter. Thus, (5.4.10) describes the relation between the final output, $y(t)$, and a frequency-shifted version of the input, $x(t) = g(t)\cos[2\pi(f_0 - f_1)t]$. The analytic signal corresponding to $x(t)$ is $x_a(t) = g(t)\exp[j2\pi(f_0 - f_1)t]$.

Let $\mathbf{S}(t)$ denote the vector of analytic signals induced in N antenna-element outputs by a single signal source. Let T_i denote the arrival-time delay at element i relative to a fixed point in space. It is often convenient to choose the fixed point at element 1 so that $T_1 = 0$. Let $s(t)$ denote the analytic signal at the fixed point. If f_0 is the carrier frequency and $m(t)$ is the complex envelope, then

$$s(t) = m(t)\exp(j2\pi f_0 t) \tag{5.4.11}$$

If the maximum value of T_i for any element is much less than the reciprocal of the signal bandwidth, then $m(t - T_i) \cong m(t)$ and the components of $\mathbf{S}(t)$ are proportional to

$$s(t - T_i) \cong s(t)\exp(-j2\pi f_0 T_i), \qquad i = 1, 2, \ldots, N \tag{5.4.12}$$

Therefore, for a sufficiently narrowband signal,

$$\mathbf{S}(t) = s(t)\,\mathbf{S}_0 \tag{5.4.13}$$

where

$$\mathbf{S}_0 = \begin{bmatrix} h_1(\theta)\exp(-j2\pi f_0 T_1) \\ h_2(\theta)\exp(-j2\pi f_0 T_2) \\ \cdot \\ \cdot \\ \cdot \\ h_N(\theta)\exp(-j2\pi f_0 T_N) \end{bmatrix} \tag{5.4.14}$$

and the $h_i(\theta)$, $i = 1, 2, \ldots, N$, are amplitudes that depend upon the element patterns and the propagation losses. If the element patterns are identical and the elements are close enough relative to the distances to the signal source that the propagation losses to all the elements are nearly the same, then the $h_i(\theta)$ are all equal. If the element patterns are isotropic and their common value is included in $s(t)$, then we can set $h_i(\theta) = 1$, $i = 1, 2, \ldots, N$.

5.5 OPTIMAL WEIGHTS

To develop adaptive algorithms, first the optimal fixed weights are derived in this section. In subsequent sections, we determine adaptive algorithms that yield weights converging to the optimal fixed weights. Ideally, the steady-state performance of an adaptive antenna system closely approximates the performance of an antenna system with optimal fixed weights.

The derivation of the optimal fixed weights depends upon the specification of a performance criterion or estimation procedure. A number of different estimators of the desired signal can be implemented by linear filters of the form of (5.4.2). Unconstrained estimators that depend only upon the second-order moments of $\mathbf{X}$ can be derived by using performance criteria based upon the mean square error or the signal-to-noise ratio of the filter output. Similar estimators result from using the maximum-*a-posteriori* or the maximum-likelihood criteria, but the standard application of these criteria entails the unwarranted assumption that any interference in $\mathbf{X}$ has a Gaussian distribution.

5.5.1 Mean-Square-Error Criterion

The most widely used method of estimation is based on the minimization of the *mean square error* and the assumption that the reference signal is equal to the desired response of the antenna system [3]. It is assumed in the subsequent analysis that $\mathbf{X}$ and d, the reference signal, are derived from *jointly stationary stochastic processes.* Similar results can be obtained by modeling $\mathbf{X}$ and d as

continuous-time or ***discrete-time deterministic variables*** and minimizing the ***time-average*** or ***sample-average square error.***

Because it is convenient and because the results are usually applied to systems of the form of Figure 5.6, we initially assume that all scalars and matrix components are real. Referring to Figure 5.5(b), the difference between the *reference* and the *output* is the *error signal.*

$$\epsilon = d - \mathbf{W}^T\mathbf{X} \tag{5.5.1}$$

We square both sides of this equation and take the expected value, regarding $\mathbf{W}$ as fixed. Because $\mathbf{W}^T\mathbf{X} = \mathbf{X}^T\mathbf{W}$, the mean square error is

$$\begin{aligned} E[\epsilon^2] &= E[(d - \mathbf{W}^T\mathbf{X})^2] \\ &= E[d^2] + \mathbf{W}^T\mathbf{R}_{xx}\mathbf{W} - 2\mathbf{W}^T\mathbf{R}_{xd} \end{aligned} \tag{5.5.2}$$

where $E[x]$ denotes the expected value of x,

$$\mathbf{R}_{xx} = E[\mathbf{X}\mathbf{X}^T] = E\begin{bmatrix} x_1x_1 & x_1x_2 \ldots x_1x_N \\ x_2x_1 & x_2x_2 \ldots x_2x_N \\ \cdot & \cdot \qquad \cdot \\ \cdot & \cdot \qquad \cdot \\ \cdot & \cdot \qquad \cdot \\ x_Nx_1 & x_Nx_2 \ldots x_Nx_N \end{bmatrix} \tag{5.5.3}$$

is the symmetric *correlation matrix* of $\mathbf{X}$, and

$$\mathbf{R}_{xd} = E[\mathbf{X}d] = E[x_1d \quad x_2d \quad \ldots \quad x_Nd]^T \tag{5.5.4}$$

is a *vector of cross correlations* between the input signals and the reference signal. Taking the gradient of (5.5.2) with respect to $\mathbf{W}$ and applying (C.3.5) of Appendix C, we obtain the column vector

$$\nabla_w E[\epsilon^2] = 2\mathbf{R}_{xx}\mathbf{W} - 2\mathbf{R}_{xd} \tag{5.5.5}$$

Setting the gradient equal to zero, we obtain a necessary condition for the *optimal weight vector*, $\mathbf{W}_0$. If we assume that the system output is always nonzero, then $\mathbf{R}_{xx}$ must be positive definite (Appendix C). Thus, it is nonsingular and we obtain

$$\mathbf{W}_0 = \mathbf{R}_{xx}^{-1}\mathbf{R}_{xd} \tag{5.5.6}$$

This equation is the *Wiener-Hopf equation* for the optimal weight vector. The associated linear filter is called the *Wiener filter.* To show that $\mathbf{W}_0$ produces the minimum mean square error, we first set $\mathbf{W} = \mathbf{W}_0$ in (5.5.2) and use (5.5.6), the symmetry of $\mathbf{R}_{xx}$, and (C.1.8) of Appendix C to obtain E_R, the mean square error corresponding to $\mathbf{W}_0$:

$$E_R = E[d^2] - \mathbf{W}_0^T \mathbf{R}_{xx} \mathbf{W}_0$$
$$= E[d^2] - \mathbf{R}_{xd}^T \mathbf{R}_{xx}^{-1} \mathbf{R}_{xd} \tag{5.5.7}$$

Substituting (5.5.7) into (5.5.2) to eliminate $E[d^2]$ and then using (5.5.6) to eliminate $\mathbf{R}_{xd}$, we can express the mean square error in the form

$$E[\epsilon^2] = E_R + (\mathbf{W} - \mathbf{W}_0)^T \mathbf{R}_{xx} (\mathbf{W} - \mathbf{W}_0) \tag{5.5.8}$$

Because $\mathbf{R}_{xx}$ is positive definite, this expression shows that $\mathbf{W}_0$ is a unique optimal weight vector and E_R is the minimum mean square error. Equation (5.5.8) describes a multidimensional quadratic function of the weights that can be visualized in two dimensions as a bowl-shaped surface. The purpose of adaptation is to continually seek the bottom of the bowl.

For complex signals and weights, the *mean-square-error criterion* entails the minimization of the expected value of the squared error magnitude, which is proportional to the mean power in the error signal. Let H denote the conjugate transpose and an asterisk denote the conjugate. We obtain

$$E[|\epsilon|^2] = E[\epsilon^* \epsilon] = E[(d^* - \mathbf{W}^H \mathbf{X}^*)(d - \mathbf{X}^T \mathbf{W})]$$
$$= E[|d|^2] - \mathbf{W}^H \mathbf{R}_{xd} - \mathbf{R}_{xd}^H \mathbf{W} + \mathbf{W}^H \mathbf{R}_{xx} \mathbf{W} \tag{5.5.9}$$

where

$$\mathbf{R}_{xx} = E[\mathbf{X}^* \mathbf{X}^T] \tag{5.5.10}$$

is the Hermitian (Appendix C) correlation matrix of $\mathbf{X}$ and

$$\mathbf{R}_{xd} = E[\mathbf{X}^* d] \tag{5.5.11}$$

is a cross-correlation vector.

In terms of its real part, $\mathbf{W}_R$, and its imaginary part, $\mathbf{W}_I$, a complex weight vector is defined as

$$\mathbf{W} = \mathbf{W}_R - j\mathbf{W}_I \tag{5.5.12}$$

where the minus sign is consistent with the convention established in section 5.4 to conveniently describe weights associated with quadrature hybrids or their equivalents. Let ∇_{WR} and ∇_{WI} denote the gradients with respect to $\mathbf{W}_R$ and $\mathbf{W}_I$, respectively. The complex gradient with respect to $\mathbf{W}$ is defined as

$$\overline{\nabla}_w = \nabla_{WR} - j\nabla_{WI} \tag{5.5.13}$$

Let W_i, W_{Ri}, and W_{Ii}, $i = 1, 2, \ldots, N$, denote the components of $\mathbf{W}$, $\mathbf{W}_R$, and $\mathbf{W}_I$, respectively. Let $g(\mathbf{W}, \mathbf{W}^*)$ denote a real-valued function of $\mathbf{W}$ and $\mathbf{W}^*$. Regarding $\mathbf{W}$ and $\mathbf{W}^*$ as independent variables, we assume that g is an analytic function of each W_i when $\mathbf{W}^*$ is held constant and an analytic function of each W_i^*

when $\mathbf{W}$ is held constant [4]. We define ∇_{w*} as the gradient with respect to $\mathbf{W}^*$. Because $W_i = W_{Ri} - jW_{Ii}$,

$$\frac{\partial W_i}{\partial W_{Ri}} = 1\,, \quad \frac{\partial W_i}{\partial W_{Ii}} = -j\,, \quad \frac{\partial W_i^*}{\partial W_{Ri}} = 1\,, \quad \frac{\partial W_i^*}{\partial W_{Ii}} = j$$

The chain rule of calculus then implies that

$$\begin{aligned}\bar{\nabla}_w g(\mathbf{W},\mathbf{W}^*) &= \nabla_{WR}\, g - j\nabla_{WI}\, g \\ &= \nabla_W g + \nabla_{W*} g - j(-j\nabla_W g + j\nabla_{W*} g) \\ &= 2\nabla_{w*} g(\mathbf{W},\mathbf{W}^*) \end{aligned} \tag{5.5.14}$$

Because $\mathbf{W}^H = \mathbf{W}^{*T}$, (C.3.2) of Appendix C, (5.5.9), and (5.5.14) yield

$$\bar{\nabla}_w E[|\epsilon|^2] = 2\mathbf{R}_{xx}\mathbf{W} - 2\mathbf{R}_{xd} \tag{5.5.15}$$

Since $\nabla_{WR}\, g = 0$ and $\nabla_{WI} g = 0$ imply that $\bar{\nabla}_W g = 0$, a necessary condition for the optimal weight is obtained from setting $\bar{\nabla}_W E[|\epsilon|^2] = 0$. Thus, if $\mathbf{R}_{xx}$ is positive definite and hence nonsingular, the Wiener-Hopf equation for the optimal weight vector is

$$\mathbf{W}_0 = \mathbf{R}_{xx}^{-1}\mathbf{R}_{xd} \tag{5.5.16}$$

This equation is formally identical to (5.5.6), but the definitions of $\mathbf{R}_{xx}$ and $\mathbf{R}_{xd}$ are different in the complex case. The mean square error corresponding to $\mathbf{W}_0$ is derived by substituting (5.5.16) into (5.5.9). Using the Hermitian character of $\mathbf{R}_{xx}$, we obtain

$$E_R = E[|d|^2] - \mathbf{R}_{xd}^H \mathbf{R}_{xx}^{-1}\mathbf{R}_{xd} \tag{5.5.17}$$

Equations (5.5.9), (5.5.16), and (5.5.17) imply that

$$E[|\epsilon|^2] = E_R + (\mathbf{W} - \mathbf{W}_0)^H \mathbf{R}_{xx}(\mathbf{W} - \mathbf{W}_0) \tag{5.5.18}$$

Because $\mathbf{R}_{xx}$ is positive definite, this equation shows that the Wiener-Hopf equation provides a unique optimal weight vector and that (5.5.17) gives the minimum mean square error.

The input vector can usually be decomposed into the sum of two components:

$$\mathbf{X}(t) = \mathbf{S}(t) + \mathbf{n}(t) \tag{5.5.19}$$

where $\mathbf{S}(t)$ is a vector containing the desired-signal components and $\mathbf{n}(t)$ is a vector containing the interference and thermal-noise components. If $\mathbf{X}(t)$ is the analytic signal at the output of an antenna array and the narrowband representation of (5.4.13) is valid, then

$$\mathbf{X}(t) = s(t)\mathbf{S}_0 + \mathbf{n}(t) \tag{5.5.20}$$

where $s(t)$ and the components of $\mathbf{n}(t)$ are analytic signals.

Let R_x denote the power in a stationary stochastic process $x(t)$. If $x_a(t)$ is the associated analytic signal, then (Appendix A)

$$R_x = E[x^2(t)] = \tfrac{1}{2}E[|x_a(t)|^2] \tag{5.5.21}$$

For an ideal reference signal, $d(t) = \alpha s(t)$, where α is a constant. If it is assumed that all signals are stationary processes and that the desired signal is uncorrelated with the interference and noise, (5.5.11), (5.5.20), and (5.5.21) give

$$\mathbf{R}_{xd} = 2\alpha R_s \mathbf{S}_0^* \tag{5.5.22}$$

where R_s is the desired-signal power at element 1. Thus, for narrowband signals and an ideal reference, knowledge of $\mathbf{R}_{xd}$ is equivalent to knowledge of $\mathbf{S}_0$. Equations (5.5.10) and (5.5.20) yield

$$\mathbf{R}_{xx} = 2R_s\mathbf{S}_0^*\mathbf{S}_0^T + \mathbf{R}_{nn} \tag{5.5.23}$$

where the correlation matrix of the interference plus noise is

$$\mathbf{R}_{nn} = E[\mathbf{n}^*\mathbf{n}^T] \tag{5.5.24}$$

The noise power is assumed to be always positive. Thus, $\mathbf{R}_{nn}$ is positive definite, and since it is Hermitian, it is nonsingular. Therefore, we subsequently always assume that $\mathbf{R}_{nn}^{-1}$ and $\mathbf{R}_{xx}^{-1}$ exist.

If $\mathbf{A}$ is an $N \times N$ non-singular matrix and $\mathbf{Y}$ is an N-dimensional complex column vector, then by direct matrix multiplication and exploiting the fact that $\mathbf{Y}^H\mathbf{A}^{-1}\mathbf{Y}$ is a scalar, it can be verified that

$$(\mathbf{A} + \mathbf{Y}\mathbf{Y}^H)^{-1} = \mathbf{A}^{-1} - \frac{\mathbf{A}^{-1}\mathbf{Y}\mathbf{Y}^H\mathbf{A}^{-1}}{1 + \mathbf{Y}^H\mathbf{A}^{-1}\mathbf{Y}} \tag{5.5.25}$$

Using this *matrix inversion identity* with (5.5.16), (5.5.22), and (5.5.23) and then simplifying the result, we obtain

$$\mathbf{W}_0 = \left[\frac{2\alpha R_s}{1 + 2R_s\mathbf{S}_0^T\mathbf{R}_{nn}^{-1}\mathbf{S}_0^*}\right]\mathbf{R}_{nn}^{-1}\mathbf{S}_0^* \tag{5.5.26}$$

where the factor within the brackets is a scalar.

5.5.2 SINR Criterion

A reasonable design criterion for adaptive radar systems is the maximization of the probability of detection given a fixed probability of false alarm. This maximization can be shown to be equivalent to the maximization of the *signal-to-interference-plus-noise ratio (SINR),* which is itself an intuitively appealing design criterion for both radar and communications [5].

Consider the adaptive system of Figure 5.5(a) with an input vector given by (5.5.20). All vector components and scalars of the system are assumed to be complex variables. The output component due to the desired signal is

$$y_s(t) = \mathbf{W}^T \mathbf{S}(t) = s(t)\mathbf{W}^T \mathbf{S}_0 \tag{5.5.27}$$

The output component due to the interference plus noise is

$$y_n(t) = \mathbf{W}^T \mathbf{n}(t) \tag{5.5.28}$$

If the desired signal is modeled as a stationary stochastic process, then (5.5.27) implies that the signal power at the output is

$$P_s = \frac{1}{2} E[|y_s(t)|^2] = R_s |\mathbf{W}^T \mathbf{S}_0|^2 \tag{5.5.29}$$

where R_s is the mean signal power at the input to element 1. If the desired signal is assumed to be deterministic, the time-average signal power at the output is again given by the far right-hand side of (5.5.29), but R_s then represents the time-average signal power at the input to element 1. The interference power plus the noise power at the output is

$$P_n = \frac{1}{2} E[(\mathbf{n}^T \mathbf{W})^H (\mathbf{n}^T \mathbf{W})] = \frac{1}{2} \mathbf{W}^H \mathbf{R}_{nn} \mathbf{W} \tag{5.5.30}$$

where $\mathbf{R}_{nn}$ is defined by (5.5.24). The SINR at the system output is

$$\rho_0 = \frac{P_s}{P_n} = \frac{2R_s |\mathbf{W}^T \mathbf{S}_0|^2}{\mathbf{W}^H \mathbf{R}_{nn} \mathbf{W}} \tag{5.5.31}$$

A positive-definite Hermitian matrix has positive eigenvalues. Thus, it can be diagonalized to the identity matrix. We define the nonsingular $N \times N$ matrix,

$$\mathbf{D} = \left[\frac{\mathbf{e}_1}{\sqrt{\lambda_1}} \quad \frac{\mathbf{e}_2}{\sqrt{\lambda_2}} \quad \cdots \quad \frac{\mathbf{e}_N}{\sqrt{\lambda_N}} \right] \tag{5.5.32}$$

where the λ_i are the eigenvalues and the $\mathbf{e}_i$ are the corresponding orthonormal eigenvectors of $\mathbf{R}_{nn}$. Straightforward matrix algebra using partitioned matrices yields

$$\mathbf{I} = \mathbf{D}^H \mathbf{R}_{nn} \mathbf{D} \tag{5.5.33}$$

where $\mathbf{I}$ is the identity matrix.

Combining this result with (5.5.30) gives

$$2P_n = \mathbf{W}^H(\mathbf{D}^H)^{-1}\mathbf{D}^H\mathbf{R}_{nn}\mathbf{D}\mathbf{D}^{-1}\mathbf{W}$$
$$= (\mathbf{D}^{-1}\mathbf{W})^H(\mathbf{D}^{-1}\mathbf{W}) = \|\mathbf{D}^{-1}\mathbf{W}\|^2 \qquad (5.5.34)$$

where $\|\mathbf{x}\|^2 = \mathbf{x}^H\mathbf{x}$ is the squared Euclidean norm of the vector $\mathbf{x}$. Using the Cauchy-Schwarz inequality (Appendix C), (5.5.29) yields

$$P_s/R_s = |\mathbf{W}^T(\mathbf{D}^T)^{-1}\mathbf{D}^T\mathbf{S}_0|^2$$
$$= |(\mathbf{D}^{-1}\mathbf{W})^T\mathbf{D}^T\mathbf{S}_0|^2$$
$$= |(\mathbf{D}^{-1}\mathbf{W})^H\mathbf{D}^H\mathbf{S}_0^*|^2$$
$$\leqslant \|\mathbf{D}^{-1}\mathbf{W}\|^2\,\|\mathbf{D}^H\mathbf{S}_0^*\|^2 \qquad (5.5.35)$$

Equality in this equation is attained if

$$\mathbf{D}^{-1}\mathbf{W} = \eta\mathbf{D}^H\mathbf{S}_0^* \qquad (5.5.36)$$

where η is an arbitrary constant. Substituting (5.5.34) and (5.5.35) into (5.5.31), it follows that

$$\rho_0 \leqslant 2R_s\,\|\mathbf{D}^H\mathbf{S}_0^*\|^2 \qquad (5.5.37)$$

The maximum value of ρ_0 is attained if (5.5.36) is satisfied. Thus, the optimal choice for $\mathbf{W}$ is

$$\mathbf{W}_0 = \eta\mathbf{D}\mathbf{D}^H\mathbf{S}_0^* \qquad (5.5.38)$$

By using (5.5.33), this relation becomes

$$\mathbf{W}_0 = \eta\mathbf{R}_{nn}^{-1}\mathbf{S}_0^* \qquad (5.5.39)$$

The value of ρ_0 corresponding to the optimal weights, obtained by the substitution of (5.5.39) into (5.5.31), is

$$\rho_0 = 2R_s\mathbf{S}_0^{\mathrm{T}}\mathbf{R}_{nn}^{-1}\mathbf{S}_0^* \qquad (5.5.40)$$

This equation gives the ideal steady-state *SINR* for an adaptive system with weights that converge to their optimal values. To implement (5.5.39), we must know the values of the components of $\mathbf{S}_0$. For arrays with more than one element, this knowledge implies that the direction of arrival of the desired signal is known. If the desired signal is not narrowband so that $\mathbf{S}(t)$ cannot be represented by $s(t)\,\mathbf{S}_0$, then (5.5.39) and (5.5.40) are not valid, and numerical analysis may be necessary to maximize the *SINR*.

In most applications, the antenna elements are separated by approximately half a wavelength or more and the cross-coupling among the elements is negligible. Under these conditions, $n_i(t)$, the thermal and background noise at the output of element i, is usually nearly uncorrelated with $n_k(t)$, $k \neq i$. Equation (A.3.10) of Appendix A indicates that the outputs of ideal quadrature hybrids are uncorrelated with each other. It is reasonable to assume that all the noise powers are equal to

$$\sigma^2 = E[n_i^2(t)] = \tfrac{1}{2}E[|n_{ia}(t)|^2] \tag{5.5.41}$$

where $n_{ia}(t)$ represents the analytic signal corresponding to $n_i(t)$. Thus, it is usually assumed subsequently that when interference is absent, $\mathbf{R}_{nn} = \sigma^2 \mathbf{I}$.

If the interference is narrowband, the interference vector due to a single source is

$$\mathbf{I}(t) = i(t)\mathbf{J}_0 \tag{5.5.42}$$

where $i(t)$ is the analytic signal associated with the interference at the output of element 1 and $\mathbf{J}_0$ is a spatial vector having a form similar to the right-hand side of (5.4.14). It follows that

$$\mathbf{R}_{nn} = 2\sigma^2\mathbf{I} + 2R_i\mathbf{J}_0^*\mathbf{J}_0^T \tag{5.5.43}$$

where R_i is the interference power at the output of element 1. Substituting (5.5.43) into (5.5.40), applying (5.5.25), and observing that $\|x\|^2 = \mathbf{x}^T\mathbf{x}^*$, we obtain

$$\rho_0 = \gamma_s \left[\|\mathbf{S}_0\|^2 - \frac{\gamma_i |\mathbf{S}_0^T\mathbf{J}_0^*|^2}{1 + \gamma_i \|\mathbf{J}_0\|^2} \right] \tag{5.5.44}$$

where $\gamma_s = R_s/\sigma^2$ is the signal-to-noise ratio at the output of element 1, and $\gamma_i = R_i/\sigma^2$ is the interference-to-noise ratio at the output of element 1.

Because (5.5.26) differs from (5.5.39) by a scalar, ρ_0 for the mean-square-error criterion and narrowband signals is also given by (5.5.44). For narrowband signals, therefore, the minimization of the mean square error is equivalent to the maximization of the *SINR*.

An upper bound on the *SINR* is obtained by equating to zero the second term within the brackets in (5.5.44). A lower bound follows upon application of the Cauchy-Schwarz inequality. Therefore,

$$\frac{\gamma_s \|\mathbf{S}_0\|^2}{1 + \gamma_i \|\mathbf{J}_0\|^2} \leqslant \rho_0 \leqslant \gamma_s \|\mathbf{S}_0\|^2 \tag{5.5.45}$$

The upper bound is achieved if $\gamma_i = 0$ or if the spatial vector $\mathbf{J}_0$ is orthogonal to $\mathbf{S}_0$; that is, $|\mathbf{S}_0^T\mathbf{J}_0^*| = 0$. The lower bound is realized if

$$|\mathbf{S}_0^T\mathbf{J}_0^*| = \|\mathbf{S}_0\| \, \|\mathbf{J}_0\| \tag{5.5.46}$$

which occurs if $\mathbf{S}_0 = \xi \mathbf{J}_0$ for some complex scalar ξ. This condition is satisfied if the interference and desired signal arrive from the same direction. Other cases in which it is satisfied correspond to the grating-null phenomenon, in which a null in the direction of the interference causes a null in the direction of the desired signal [1]. To prevent the grating nulls, the element patterns and the array configuration must be carefully chosen.

A *linear array* consists of elements that lie along a straight line. Consider an ideal linear array of N elements with identical isotropic element patterns, identical polarization responses, and no propagation loss differentials. The position of element 1 serves as a fixed reference point for the received signals. The spatial vectors of narrowband signals can be represented by

$$\mathbf{S}_0 = \begin{bmatrix} 1 \\ \exp(-j\beta_s) \\ \cdot \\ \cdot \\ \cdot \\ \exp[-j(N-1)\beta_s] \end{bmatrix}, \quad \mathbf{J}_0 = \begin{bmatrix} 1 \\ \exp(-j\beta_i) \\ \cdot \\ \cdot \\ \cdot \\ \exp[-j(N-1)\beta_i] \end{bmatrix} \tag{5.5.47}$$

where

$$\beta_s = \frac{2\pi D}{\lambda} \sin \phi_s \, , \quad \beta_i = \frac{2\pi D}{\lambda} \sin \phi_i \tag{5.5.48}$$

D is the separation between elements, λ is the wavelength corresponding to the common carrier frequency of the received signals, and θ_s and θ_i refer to the arrival angles of the desired signal and interference, respectively. It follows that

$$\begin{aligned} \mathbf{S}_0^T \mathbf{J}_0^* &= \sum_{k=0}^{N-1} \exp[jk(\beta_i - \beta_s)] \\ &= N L_N(\beta_i - \beta_s) \exp\left[\frac{j(N-1)(\beta_i - \beta_s)}{2}\right] \end{aligned} \tag{5.5.49}$$

where

$$L_N(x) = \frac{\sin(Nx/2)}{N \sin(x/2)} \tag{5.5.50}$$

Therefore, (5.5.44) becomes

$$\rho_0 = N\gamma_s \frac{1 + N\gamma_i - N\gamma_i L_N^2(\beta_i - \beta_s)}{1 + N\gamma_i} \tag{5.5.51}$$

Combining (5.5.51), (5.5.50), and (5.5.48) and using a Taylor-series expansion in ϕ_i about ϕ_s, we obtain

$$\rho_0 \cong \frac{N\gamma_s}{1+N\gamma_i}\left[1+N\gamma_i \frac{N^2-1}{12}\left(\frac{2\pi D}{\lambda}\cos\phi_s\right)^2(\phi_i-\phi_s)^2\right], \quad |\phi_i-\phi_s| \ll 1 \tag{5.5.52}$$

If $\gamma_i \gg 1/N$, the first factor on the right-hand side is approximately equal to the signal-to-interference ratio at an antenna element output. This equation indicates that for $0 \leqslant |\phi_i - \phi_s| \ll 1$, ρ_0 increases with D/λ and N, but decreases as ϕ_s increases.

Figure 5.9 depicts the *SINR* for a five-element ideal linear array when $\phi_s = 0$, $\gamma_s = 10$ dB, $\gamma_i = 30$ dB, and $D = 1.5\lambda$. The nearly exact expression, (5.5.51), is plotted along with the approximation of (5.5.52), which rapidly loses accuracy for ϕ_i beyond 0.05 radians. The precipitous drop in *SINR* when $\phi_i \approx 0.72$ is due to a grating null. If D is reduced to λ or less, there are no grating null effects for $|\phi_i| \leqslant 1.0$ radians.

In the derivation of (5.5.44) and (5.5.52), the effect of *mutual coupling* among the antenna elements was ignored. A computation for half-wavelength center-fed dipoles indicates that the mutual coupling can cause a serious performance degradation if the interelement spacing is less than half a wavelength [6]. The effect of the mutual coupling can be significant even when the spacing exceeds half a wavelength.

5.5.3 Constrained Minimum-Power Criterion

A performance criterion that inherently limits the inadvertent cancellation of the desired signal is the *constrained minimum-power criterion* [7], which is applied to systems with adaptive filters of the form of Figure 5.5(a). We assume that the components of $\mathbf{X}$ are derived from stationary stochastic processes. If the signals and weights are real variables, the criterion requires the minimization of the *mean output power*

$$E[y^2] = \mathbf{W}^T\mathbf{R}_{xx}\mathbf{W} \tag{5.5.53}$$

subject to the constraint

$$\mathbf{C}^T\mathbf{W} = \mathbf{F} \tag{5.5.54}$$

where $\mathbf{F}$ is an L-dimensional constraint vector with $L \leqslant N$, $\mathbf{W}$ is an N-dimensional weight vector, and $\mathbf{C}$ is an $N \times L$ matrix associated with the constraint. If $L = 1$, (5.5.54) represents a single linear constraint for the weights, whereas if $L \geqslant 2$, it represents multiple linear constraints. Quadratic and other non-linear constraints are sometimes plausible, but are not considered here. Each constraint removes one of the degrees of freedom in $\mathbf{W}$. Thus, L constraints leave $N - L$ degrees of freedom to be used in rejecting the interference.

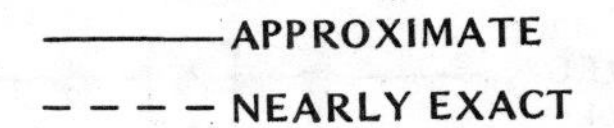

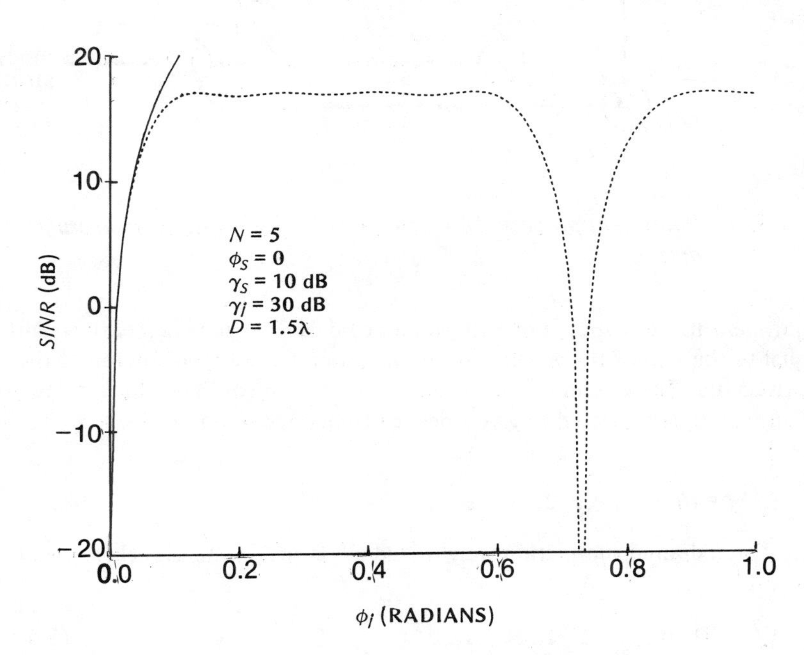

Figure 5.9 Performance of ideal linear array with identical isotropic element patterns, identical polarization responses, no propagation loss differentials, and narrowband received signals.

Suppose that the inputs of Figure 5.5(a) are the outputs of ideal steering delays, so that they differ from each other only because of the interference and noise. If we require (5.5.54) with

$$\mathbf{C}^T = [1 \quad 1 \dots \quad 1], \quad \mathbf{F} = F_1 \tag{5.5.55}$$

where F_1 is a scalar, this constraint forces the desired-signal component of the system output to be a conventional coherent sum of the desired-signal components of the inputs. Minimization of the output power then minimizes the interference power obtained from directions other than the main beam direction.

For the system of Figure 5.6 with $L > 1$, the constraint is somewhat different. We assume that the steering delays provide identical desired-signal components at the first K taps. Consequently, the filter response to the desired signal

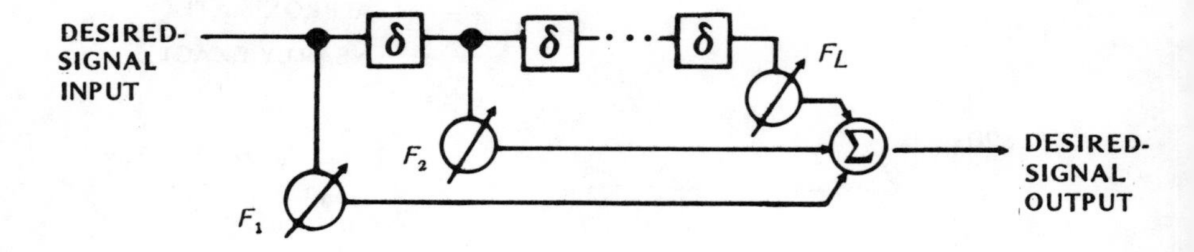

Figure 5.10 Equivalent processor for desired signal, assuming ideal steering delays.

is equivalent to the response of a single tapped delay line in which each weight is equal to the sum of the weights in the corresponding vertical column of the adaptive filter. These summation weights in the equivalent line, which is depicted in Figure 5.10, are selected to give a desired frequency response. Thus, we require that

$$\mathbf{C}_i^T \mathbf{W} = F_i, \qquad i = 1, 2, \ldots, L \tag{5.5.56}$$

where F_i is a desired summation weight, and $\mathbf{C}_i^T$ is an N-dimensional vector of the form

$$\mathbf{C}_i^T = [0 \;\; 0 \ldots \;\; 1 \;\; 1 \ldots \;\; 1 \;\; 0 \;\; 0 \ldots \;\; 0] \tag{5.5.57}$$

The ones correspond to the ith vertical column of K weights. To put (5.5.56) in matrix form, the constraint matrix is defined as

$$\mathbf{C} = [\mathbf{C}_1 \;\; \mathbf{C}_2 \;\; \ldots \;\; \mathbf{C}_L] \tag{5.5.58}$$

which has the dimensions $N \times L$. The L-dimensional vector of weights of the equivalent line is

$$\mathbf{F} = [F_1 \;\; F_2 \;\; \ldots \;\; F_L]^T \tag{5.5.59}$$

With these definitions, the constraints can be combined into a single matrix equation of the form of (5.5.54). Once again, the constrained minimization ensures that the output power from all directions except that of the desired signal is minimized.

Having shown how the constraint equation is constructed, we proceed to a derivation of the optimal weight vector, assuming that the weights are fixed constants. By the method of *Lagrange multipliers.* we minimize

$$H = \mathbf{W}^T \mathbf{R}_{xx} \mathbf{W} + \boldsymbol{\gamma}^T (\mathbf{C}^T \mathbf{W} - \mathbf{F}) \tag{5.5.60}$$

where Υ is a column vector of Lagrange multipliers. Taking the gradient of (5.5.60) with respect to the vector **W** yields the column vector

$$\nabla_w H = 2\mathbf{R}_{xx}\mathbf{W} + \mathbf{C}\Upsilon \tag{5.5.61}$$

A necessary condition for the minimum is determined by setting $\nabla_w H$ equal to zero. Thus,

$$\mathbf{W}_0 = -\frac{1}{2}\mathbf{R}_{xx}^{-1}\mathbf{C}\Upsilon \tag{5.5.62}$$

Because $\mathbf{W}_0$ must satisfy the constraint, we substitute (5.5.62) into (5.5.54) with the result that

$$-\frac{1}{2}\mathbf{C}^T\mathbf{R}_{xx}^{-1}\mathbf{C}\Upsilon = \mathbf{F} \tag{5.5.63}$$

This equation can be solved for Υ. Substituting the solution into (5.5.62) gives

$$\mathbf{W}_0 = \mathbf{R}_{xx}^{-1}\mathbf{C}(\mathbf{C}^T\mathbf{R}_{xx}^{-1}\mathbf{C})^{-1}\mathbf{F} \tag{5.5.64}$$

where we assume that the indicated inverse exists. Combining (5.5.53), (5.5.54), and (5.5.64), we find that

$$E[y^2] = (\mathbf{W} - \mathbf{W}_0)^T\mathbf{R}_{xx}(\mathbf{W} - \mathbf{W}_0) + \mathbf{F}^T(\mathbf{C}^T\mathbf{R}_{xx}^{-1}\mathbf{C})^{-1}\mathbf{F} \tag{5.5.65}$$

Because $\mathbf{R}_{xx}$ is positive definite, this equation shows that $\mathbf{W}_0$ is the unique weight vector that minimizes $E[y^2]$.

If the signals and weights are complex variables, the mean output power to be minimized is given by

$$\frac{1}{2}E[|y|^2] = \frac{1}{2}\mathbf{W}^H\mathbf{R}_{xx}\mathbf{W} \tag{5.5.66}$$

where $\mathbf{R}_{xx}$ is defined by (5.5.10). The constraint is again assumed to have the form of (5.5.54), but **C**, **W**, and **F** are allowed to be complex. Thus, this equation can be written as two equations in real variables. Consequently, to apply the method of Lagrange multipliers, we minimize the real scalar

$$H = \frac{1}{2}\mathbf{W}^H\mathbf{R}_{xx}\mathbf{W} + \Upsilon_1^T[\mathrm{Re}(\mathbf{C}^T\mathbf{W} - \mathbf{F})] + \Upsilon_2^T[\mathrm{Im}(\mathbf{C}^T\mathbf{W} - \mathbf{F})] \tag{5.5.67}$$

where Υ_1 and Υ_2 are column vectors of real-valued Lagrange multipliers. Defining $\Upsilon = \Upsilon_1 - j\Upsilon_2$, we obtain

$$H = \frac{1}{2} \mathbf{W}^H \mathbf{R}_{xx} \mathbf{W} + \text{Re}[\boldsymbol{\gamma}^T (\mathbf{C}^T \mathbf{W} - \mathbf{F})]$$

$$= \frac{1}{2} \mathbf{W}^H \mathbf{R}_{xx} \mathbf{W} + \frac{1}{2} \boldsymbol{\gamma}^T (\mathbf{C}^T \mathbf{W} - \mathbf{F}) + \frac{1}{2} \boldsymbol{\gamma}^H (\mathbf{C}^H \mathbf{W}^* - \mathbf{F}^*) \qquad (5.5.68)$$

Using (5.5.14), equating the complex gradient of H to zero, and applying (5.5.54) gives the optimal weight vector

$$\mathbf{W}_0 = \mathbf{R}_{xx}^{-1} \mathbf{C}^* (\mathbf{C}^T \mathbf{R}_{xx}^{-1} \mathbf{C}^*)^{-1} \mathbf{F} \qquad (5.5.69)$$

where the indicated inverse is assumed to exist. That $\mathbf{W}_0$ uniquely minimizes the output power can be proven by the same method used for real variables.

When a single linear constraint is imposed and $F_1 = 1$, (5.5.69) reduces to

$$\mathbf{W}_0 = \frac{\mathbf{R}_{xx}^{-1} \mathbf{C}^*}{\mathbf{C}^T \mathbf{R}_{xx}^{-1} \mathbf{C}^*} \qquad (5.5.70)$$

where the denominator is a scalar. If each input of Figure 5.5(a) is a narrowband signal derived from an array element, then the gain in the direction of the desired signal can be fixed by requiring

$$\mathbf{S}_0^T \mathbf{W} = 1 \qquad (5.5.71)$$

as the constraint, which implies that $\mathbf{C} = \mathbf{S}_0$. Using (5.5.23) and (5.5.25), (5.5.70) becomes

$$\mathbf{W}_0 = \frac{\mathbf{R}_{nn}^{-1} \mathbf{S}_0^*}{\mathbf{S}_0^T \mathbf{R}_{nn}^{-1} \mathbf{S}_0^*} \qquad (5.5.72)$$

This equation can be derived directly by minimizing $\mathbf{W}^H \mathbf{R}_{nn} \mathbf{W}$ subject to the constraint of (5.5.71). Because (5.5.72) has the form of (5.5.39), ρ_0 is given by (5.5.40) or by (5.5.51) and (5.5.52) under the appropriate conditions.

5.5.4 Weighted Least-Squares-Error Criterion

The *weighted least-squares-error criterion* makes no assumption of stationarity, but provides an optimal weight that is dependent upon the data values [8]. Assuming that n sets of data samples have been obtained, we define the $n \times N$ matrix $\mathbf{X}_s(n)$ and the n-dimensional vector $\mathbf{D}(n)$ by

$$\mathbf{X}_s(n) = \begin{bmatrix} \mathbf{X}^T(1) \\ \mathbf{X}^T(2) \\ \cdot \\ \cdot \\ \cdot \\ \mathbf{X}^T(n) \end{bmatrix}, \qquad \mathbf{D}(n) = \begin{bmatrix} d(1) \\ d(2) \\ \cdot \\ \cdot \\ \cdot \\ d(n) \end{bmatrix} \tag{5.5.73}$$

where $\mathbf{X}(k)$ and $d(k)$ are the kth samples of the input vector and the reference signal, respectively. If the total number of samples, n, is equal to N, the dimension of $\mathbf{W}$, then an estimate of $\mathbf{W}$ can be obtained from the unique solution of $\mathbf{X}_s(n)\mathbf{W} = \mathbf{D}(n)$. However, if $n > N$, this approach cannot be used without ignoring some of the data. In the spirit of regression analysis, the weighted least-squares-error criterion for real signals and weights requires that the optimal weight vector minimize

$$P(\mathbf{W}) = [\mathbf{X}_s(n)\mathbf{W} - \mathbf{D}(n)]^H \mathbf{A}(n)[\mathbf{X}_s(n)\mathbf{W} - \mathbf{D}(n)]$$

$$= \sum_{k=1}^{n} A_k |\mathbf{X}^T(k)\mathbf{W} - d(k)|^2 \tag{5.5.74}$$

where $\mathbf{A}(n)$ is a diagonal weighting matrix with positive diagonal elements equal to A_k, $k = 1, 2, \ldots, n$.

A plausible form for this matrix, which allows a progressive deemphasis of data as it becomes older, is

$$\mathbf{A}(n) = \begin{bmatrix} \alpha^{n-1} & 0 & \ldots & 0 & 0 \\ 0 & \alpha^{n-2} & \ldots & 0 & 0 \\ \cdot & \cdot & & \cdot & \cdot \\ \cdot & \cdot & & \cdot & \cdot \\ \cdot & \cdot & & \cdot & \cdot \\ 0 & 0 & \ldots & \alpha & 0 \\ 0 & 0 & \ldots & 0 & 1 \end{bmatrix} \tag{5.5.75}$$

where $0 < \alpha \leqslant 1$. For a stationary environment, all the data can be considered equally relevant. Thus, it is reasonable to set $\alpha = 1$, which implies that $\mathbf{A}(n) = \mathbf{I}$. For a nonstationary environment, $\alpha < 1$ is more appropriate.

Setting $\overline{\nabla}_W P(\mathbf{W}) = \mathbf{0}$ gives the optimal weight vector for n sets of data samples:

$$\mathbf{W}(n) = [\mathbf{X}_s^H(n)\mathbf{A}(n)\mathbf{X}_s(n)]^{-1}\mathbf{X}_s^H(n)\mathbf{A}(n)\mathbf{D}(n) \tag{5.5.76}$$

where it is assumed that the inverse exists. By the method used in constructing (5.5.8), it follows that the optimal weight vector is uniquely given by (5.5.76) if $\mathbf{X}_s^H(n)\mathbf{A}(n)\mathbf{X}_s(n)$ is positive definite. Equation (5.5.75) and the use of partitioned matrices yield

$$\mathbf{X}_s^H(n)\mathbf{A}(n)\mathbf{X}_s(n) = \sum_{k=1}^{n} \alpha^{n-k}\mathbf{X}^*(k)\mathbf{X}^T(k) \tag{5.5.77}$$

Therefore, the uniqueness of the optimal weight vector is assured if the positive semidefinite matrix $\mathbf{X}^*(k)\mathbf{X}^T(k)$ is positive definite for some value of k, which is usually a safe assumption. Equation (5.5.76) is entirely in terms of the data and involves no statistical factors.

5.6 WIDROW LMS ALGORITHM

Because the computational difficulty of inverting the correlation matrix is considerable when the number of weights is large, and insofar as time-varying signal statistics may require frequent computations, adaptive algorithms not entailing matrix inversion have been developed. Suppose that a performance measure, $P(\mathbf{W})$, is defined so that it has a minimum value when the weight vector has its optimal value. In the *method of steepest descent,* the weight vector is changed along the direction of the negative gradient of the performance measure. For real valued signals and weights and discrete-time systems with the index k denoting a particular sampling instant or adaptation cycle, the weight vector is determined by

$$\mathbf{W}(k+1) = \mathbf{W}(k) - \mu\nabla_w P(\mathbf{W}(k)) \tag{5.6.1}$$

where the scalar *adaptation constant,* μ, controls the rate of convergence and stability. The adaptation cycle begins with an arbitrary initial weight. For continuous-time systems, the *method of steepest descent* requires that

$$\frac{d}{dt}\mathbf{W}(t) = -\mu\nabla_w P(\mathbf{W}(t)) \tag{5.6.2}$$

If the signals and weights are complex, separate steepest-descent equations can be written for the real and imaginary parts of the weight vector. Combining these equations and using (5.5.12) and (5.5.13), we obtain

$$\mathbf{W}(k+1) = \mathbf{W}(k) - \mu\bar{\nabla}_w P(\mathbf{W}(k)) \tag{5.6.3}$$

for discrete-time signals and

$$\frac{d}{dt}\mathbf{W}(t) = -\mu\bar{\nabla}_w P(\mathbf{W}(t)) \tag{5.6.4}$$

for continuous-time signals.

If the mean square error is the performance criterion for real signals and weights, then $P(\mathbf{W}) = E[\epsilon^2]$. Calculating the gradient of (5.5.2) and using (5.6.1), we obtain the *steepest-descent algorithm:*

$$\mathbf{W}(k+1) = \mathbf{W}(k) - 2\mu[\mathbf{R}_{xx}\mathbf{W}(k) - \mathbf{R}_{xd}] \tag{5.6.5}$$

This ideal algorithm produces a deterministic sequence of weights and does not require a matrix inversion, but it requires the knowledge of $\mathbf{R}_{xx}$ and $\mathbf{R}_{xd}$. However, the possible presence of interference means that $\mathbf{R}_{xx}$ is unknown. In the absence of information about the direction of the desired signal, $\mathbf{R}_{xd}$ is also unknown.

The *Widrow LMS* (least mean square) *algorithm* [3], also known as the *Widrow-Hoff algorithm,* results if $\mathbf{R}_{xx}$ is estimated by $\mathbf{X}(k)\mathbf{X}^T(k)$ and $\mathbf{R}_{xd}$ is estimated by $\mathbf{X}(k)d(k)$ so that

$$\mathbf{W}(k+1) = \mathbf{W}(k) - 2\mu[\mathbf{X}(k)\mathbf{X}^T(k)\mathbf{W}(k) - \mathbf{X}(k)d(k)] \tag{5.6.6}$$

Using (5.5.1), we obtain the discrete-time version of the Widrow LMS algorithm:

$$\mathbf{W}(k+1) = \mathbf{W}(k) + 2\mu\epsilon(k)\mathbf{X}(k) \tag{5.6.7}$$

Equation (5.6.7) can also be derived from (5.5.74) and (5.6.1) with $n = 1$ and $\mathbf{A} = \mathbf{I}$.

According to the Widrow algorithm, the next weight vector is obtained by adding to the present weight vector the input vector scaled by the amount of error. It is shown subsequently that, for an appropriate value of μ, the mean of the weight vector converges to the optimal value given by the Wiener-Hopf equation. Figure 5.11 shows a block-diagram representation of (5.6.7) for one component of the weight vector.

For continuous-time systems, a similar derivation using (5.6.2) yields the Widrow LMS algorithm given by

$$\frac{d}{dt}\mathbf{W}(t) = 2\mu\epsilon(t)\mathbf{X}(t) \tag{5.6.8}$$

Equivalently, if $\mathbf{W}(0) = \mathbf{0}$,

$$\mathbf{W}(t) = 2\mu\int_0^t \epsilon(\tau)\mathbf{X}(\tau)\,d\tau \tag{5.6.9}$$

Figure 5.12 shows an analog realization of one component of this equation.

For complex signals and weights, a suitable performance measure is $P(\mathbf{W}) = E[|\epsilon|^2]$. Using (5.5.15) and (5.6.3), we obtain the complex discrete-time version of the Widrow LMS algorithm:

$$\mathbf{W}(k+1) = \mathbf{W}(k) + 2\mu\epsilon(k)\mathbf{X}^*(k) \tag{5.6.10}$$

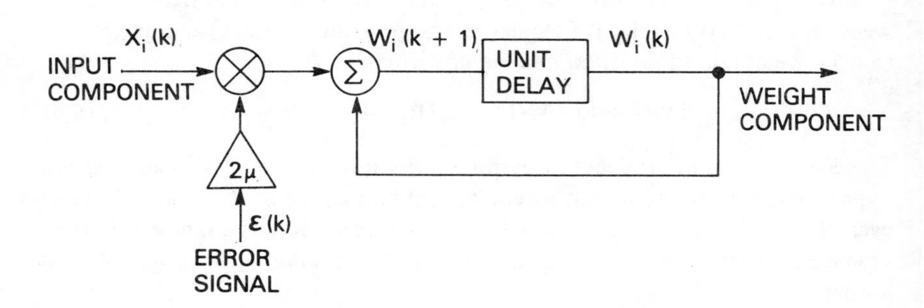

Figure 5.11 Digital implementation of Widrow algorithm.

The complex continuous-time version of the Widrow LMS algorithm is

$$\frac{d}{dt}\mathbf{W}(t) = 2\mu\epsilon(t)\mathbf{X}^*(t) \tag{5.6.11}$$

5.6.1 Convergence of the Mean

No matter how plausible the construction of an adaptive algorithm, it must be checked to verify that the weights converge in some sense to optimal or nearly optimal values. Since the adaptive weights are random if the input signals are random, it is usually appropriate to examine convergence of the mean weights. Consider a matrix $\mathbf{A}(k)$ with elements $a_{ij}(k)$. We define the limit of $\mathbf{A}(k)$ as $k \to \infty$ to be the matrix $\mathbf{B}$ with elements $b_{ij} = \lim a_{ij}(k)$. The limit of a vector is defined similarly. Taking the expected value of both sides of (5.6.6) and rearranging the terms, we obtain

$$E[\mathbf{W}(k+1)] = (\mathbf{I} - 2\mu\mathbf{R}_{xx})E[\mathbf{W}(k)] + 2\mu\mathbf{R}_{xd} - 2\mu E[\mathbf{U}(k)] \tag{5.6.12}$$

where

$$\mathbf{U}(k) = [\mathbf{X}(k)\mathbf{X}^T(k) - \mathbf{R}_{xx}]\mathbf{W}(k) \tag{5.6.13}$$

Taking the limit of both sides of (5.6.12) as $k \to \infty$ and assuming that $E[\mathbf{W}(k)]$ converges, it follows that $E[\mathbf{U}(k)]$ also converges and

$$\lim_{k\to\infty} E[\mathbf{W}(k)] = \mathbf{R}_{xx}^{-1}\left\{\mathbf{R}_{xd} - \lim_{k\to\infty} E[\mathbf{U}(k)]\right\} \tag{5.6.14}$$

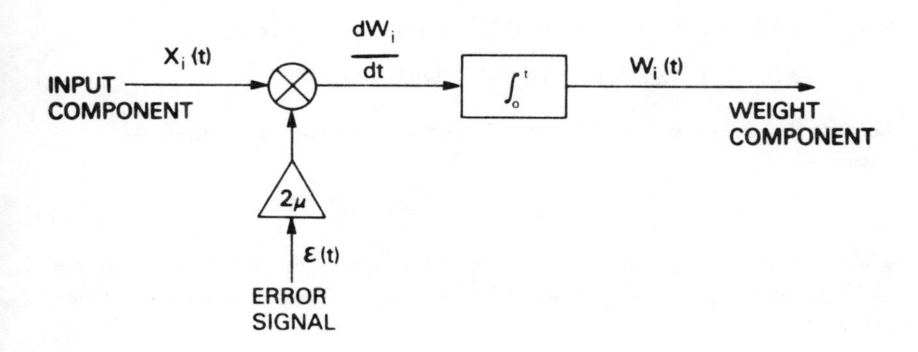

Figure 5.12 Analog implementation of Widrow algorithm.

Therefore, a necessary condition for $E[\mathbf{W}(k)]$ to converge to the optimal weight vector $\mathbf{W}_0$ is

$$\lim_{k\to\infty} E[\mathbf{U}(k)] = 0 \tag{5.6.15}$$

This condition is satisfied if $\mathbf{X}(k)$ and $\mathbf{W}(k)$ are statistically independent. However, it is usually not the only necessary condition, nor is it sufficient for convergence.

We prove convergence for the discrete-time Widrow algorithm under the assumption that samples of the input vector are statistically independent, stationary random vectors. The assumption is valid at least when the input vector is sampled at intervals that are large compared to the correlation time of the input process plus the maximum time delay between input vector components. When the assumption is not valid, the mean weight vector does not necessarily converge to the optimal weight vector. However, it can be shown [9] that under mild conditions $E[\|\mathbf{W}(k) - \mathbf{W}_0\|^2]$ can be made arbitrarily small as $k \to \infty$ by choosing a sufficiently small value of μ.

If $\mathbf{X}(k+1)$ is independent of $\mathbf{X}(n)$ and $d(n)$, $n \leq k$, (5.6.6) implies that $\mathbf{W}(k+1)$ is independent of $\mathbf{X}(k+1)$. Thus, the expected value of the weight vector satisfies

$$E[\mathbf{W}(k+1)] = (\mathbf{I} - 2\mu\mathbf{R}_{xx})E[\mathbf{W}(k)] + 2\mu\mathbf{R}_{xd} \tag{5.6.16}$$

This discrete-time equation is linear and time invariant. Its equilibrium point is easily calculated to be $\mathbf{W}_0$. From (5.5.6) and (5.6.16), it follows that

$$E[\mathbf{W}(k+1)] - \mathbf{W}_0 = (\mathbf{I} - 2\mu\mathbf{R}_{xx})\left\{E[\mathbf{W}(k)] - \mathbf{W}_0\right\} \tag{5.6.17}$$

With an initial weight vector $\mathbf{W}(0)$, this equation implies that

$$E[\mathbf{W}(k+1)] - \mathbf{W}_0 = (\mathbf{I} - 2\mu\mathbf{R}_{xx})^{k+1}[\mathbf{W}(0) - \mathbf{W}_0] \tag{5.6.18}$$

Because $\mathbf{R}_{xx}$ is symmetric and positive definite, it can be represented as (Appendix C)

$$\mathbf{R}_{xx} = \mathbf{Q}\boldsymbol{\Lambda}\mathbf{Q}^{-1} = \mathbf{Q}\boldsymbol{\Lambda}\mathbf{Q}^{T} \tag{5.6.19}$$

where $\mathbf{Q}$ is the orthogonal modal matrix of $\mathbf{R}_{xx}$ with eigenvectors as its columns, and $\boldsymbol{\Lambda}$ is the diagonal matrix of eigenvalues of $\mathbf{R}_{xx}$. Therefore, (5.6.18) can be expressed as

$$\begin{aligned} E[\mathbf{W}(k+1)] - \mathbf{W}_0 &= [\mathbf{I} - 2\mu\mathbf{Q}\boldsymbol{\Lambda}\mathbf{Q}^{-1}]^{k+1}[\mathbf{W}(0) - \mathbf{W}_0] \\ &= \mathbf{Q}[\mathbf{I} - 2\mu\boldsymbol{\Lambda}]^{k+1}\mathbf{Q}^{-1}[\mathbf{W}(0) - \mathbf{W}_0] \end{aligned} \tag{5.6.20}$$

If the diagonal elements of the diagonal matrix $[\mathbf{I} - 2\mu\boldsymbol{\Lambda}]$ have magnitudes less than unity, then

$$\lim_{k\to\infty} [\mathbf{I} - 2\mu\boldsymbol{\Lambda}]^{k} = \mathbf{0} \tag{5.6.21}$$

and consequently

$$\lim_{k\to\infty} E[\mathbf{W}(k)] = \mathbf{W}_0 = \mathbf{R}_{xx}^{-1}\mathbf{R}_{xd} \tag{5.6.22}$$

The result shows that as the number of iterations increases, $E[\mathbf{W}(k)]$ converges to $\mathbf{W}_0$ if and only if the eigenvalues of $\mathbf{I} - 2\mu\Lambda$, which are its diagonal elements, have magnitudes less than unity. Because $\mathbf{R}_{xx}$ is symmetric and positive definite, its eigenvalues, $\lambda_1, \lambda_2, \ldots, \lambda_n$, are positive. Therefore, the eigenvalues of $\mathbf{I} - 2\mu\Lambda$, which are $1 - 2\mu\lambda_1, 1 - 2\mu\lambda_2, \ldots, 1 - 2\mu\lambda_n$, have magnitudes less than unity if

$$|1 - 2\mu\lambda_{max}| < 1 \tag{5.6.23}$$

where λ_{max} is the maximum eigenvalue of $\mathbf{R}_{xx}$. Given the initial assumption of independent input samples, this equation yields the necessary and sufficient convergence condition

$$0 < \mu < \frac{1}{\lambda_{max}} \tag{5.6.24}$$

The sum of the eigenvalues of a square matrix is equal to its trace. Thus,

$$\sum_{i=1}^{N} \lambda_i = \mathrm{tr}(\mathbf{R}_{xx}) = \sum_{i=1}^{N} E[x_i^2] = R_T \tag{5.6.25}$$

where R_T denotes the total input power. Because $\lambda_{max} \leqslant R_T$, (5.6.24) implies that

$$0 < \mu < \frac{1}{R_T} \tag{5.6.26}$$

is sufficient for convergence of $E[\mathbf{W}(k)]$ to $\mathbf{W}_0$. It is easily verified that (5.6.24) and (5.6.26) are also sufficient for the convergence of the deterministic weights of the ideal steepest-descent algorithm given by (5.6.5). To ensure that (5.6.26) is satisfied, an adaptive system may include a device for measuring R_T and adjusting μ accordingly.

For independent input samples, (5.6.24) is a necessary and sufficient condition for the convergence of the complex discrete-time Widrow algorithm. However, if the components of $\mathbf{X}$ are analytic signals, a straightforward calculation similar to (5.6.25), but using (5.5.10) and (5.5.21), indicates that instead of (5.6.26), a sufficient condition for the convergence of the mean weight vector is

$$0 < \mu < \frac{1}{2R_T} \tag{5.6.27}$$

Although stronger convergence results can be proven if the inputs are stationary processes and μ is allowed to decrease with the iteration number, a μ dependent only upon the total input power gives the adaptive system flexibility in processing nonstationary inputs.

For the real and complex continuous-time Widrow algorithms, the assumption that $\mathbf{W}(t)$ varies much more slowly than $\mathbf{X}(t)$ justifies the approximation that $\mathbf{W}(t)$ and $\mathbf{X}(t)$ are independent processes. It then follows from (5.6.11) by the method shown in Section 5.7 that $\mu > 0$ is a necessary and sufficient condition for the convergence of $E[\mathbf{W}(t)]$ to $\mathbf{W}_0$.

According to the theory of linear time-invariant equations, (5.6.16) indicates that during adaptation the weights undergo transients that vary as sums of terms of the form $(1 - 2\mu\lambda_i)^k$. These transients determine the rate of convergence of the mean vector. The remainder in a Taylor-series expansion for the exponential function indicates that $\exp(-x) \geqslant 1 - x$. Therefore, if

$$0 < \mu \leqslant \frac{1}{2\lambda_{max}} \tag{5.6.28}$$

then

$$0 \leqslant (1 - 2\mu\lambda_i)^k \leqslant \exp(-2\mu\lambda_i k) = \exp\left(-\frac{k}{\tau_i}\right), \quad i = 1, 2, \ldots, N \tag{5.6.29}$$

where the convergence rate of each term can be characterized by a time constant

$$\tau_i = \frac{1}{2\mu\lambda_i}, \quad i = 1, 2, \ldots, N \tag{5.6.30}$$

The maximum time constant is

$$\tau_{max} = \frac{1}{2\mu\lambda_{min}} \tag{5.6.31}$$

where λ_{min} is the minimum eigenvalue of $\mathbf{R}_{xx}$ and determines the term of the form $(1 - 2\mu\lambda_i)^k$ that has the slowest rate of convergence. Thus, the convergence rate of the mean weight vector is primarily determined by τ_{max}. Combining (5.6.31) and (5.6.28), we obtain

$$\tau_{max} \geqslant \frac{\lambda_{max}}{\lambda_{min}} \tag{5.6.32}$$

which explicitly demonstrates the dependence of the convergence rate upon the "spread" in eigenvalues.

As an example of the calculation of the lower bound on τ_{max}, consider narrowband complex signals and a single source of interference that is uncorrelated with the desired signal. It follows that

$$\mathbf{R}_{xx} = 2\sigma^2\mathbf{I} + 2R_s\mathbf{S}_0^*\mathbf{S}_0^T + 2R_i\mathbf{J}_0^*\mathbf{J}_0^T \tag{5.6.33}$$

If $\mathbf{S}_0$ and $\mathbf{J}_0$ are spatially orthogonal so that

$$\mathbf{S}_0^T\mathbf{J}_0^* = \mathbf{J}_0^T\mathbf{S}_0^* = 0 \tag{5.6.34}$$

then it is easily verified that $\mathbf{J}_0^*$ is an eigenvector of $\mathbf{R}_{xx}$ with eigenvalue

$$\lambda_1 = 2\sigma^2 + 2R_i \,\|\mathbf{J}_0\|^2 \tag{5.6.35}$$

and $\mathbf{S}_0^*$ is an eigenvector with eigenvalue

$$\lambda_2 = 2\sigma^2 + 2R_s \,\|\mathbf{S}_0\|^2 \tag{5.6.36}$$

Any set of $N - 2$ linearly independent vectors that are orthogonal to both $\mathbf{J}_0$ and $\mathbf{S}_0$ constitute the remaining eigenvectors of $\mathbf{R}_{xx}$ and have the common eigenvalue

$$\lambda_3 = 2\sigma^2 \tag{5.6.37}$$

If (5.5.47) applies, then $\|\mathbf{J}_0\| = \|\mathbf{S}_0\| = N$ and (5.6.32) and (5.6.35) to (5.6.37) yield

$$\tau_{max} \geqslant 1 + N \max(\gamma_i, \gamma_s) \tag{5.6.38}$$

Thus, the convergence rate, as measured by $1/\tau_{max}$, decreases with increases in the number of antenna elements and the larger of γ_i and γ_s. This reasoning can easily be extended to $N - 1$ or fewer mutually orthogonal signals.

5.6.2 Generation of the Reference Signal

Because the desired response, $s(k)$, is not available *a priori,* a reference signal, $d(k)$, that is an estimate of the desired response is used in each iteration of the Widrow algorithm. For effective operation, the reference signal does not have to be a perfect replica of the desired response. If the adaptive system is to be useful, the system output should be a closer facsimile of the desired signal than the reference signal. However, if the mean weight vector is to converge to an approximation of what the optimal weight vector would be if $s(k)$ were available, then $d(k)$ should be such that

$$\mathbf{R}_{xd} \cong \mathbf{R}_{xs} \tag{5.6.39}$$

The reference signal may be transmitted to the adaptive system over a separate channel. A more practical approach is to generate the reference by a feedback loop, which may have one of the forms shown in Figure 5.13. The feedback loop is designed to destroy the correlation between the reference signal and the interference component of the array output. To satisfy (5.6.39), the delay due to the feedback generation of the reference signal should be considerably less than the modulation period.

The limiter in Figure 5.13(b) controls the amplitude of the reference signal. When the limiter input is below the limiter clipping level, the reference amplitude is proportional to the array output amplitude. A weight setting that nulls the desired signal is unstable because the loop can reinforce random weight changes that decrease this nulling. When the limiter input is above the clipping level, stable system operation is achieved because the reference signal is independent of the amplitude of the desired-signal component of the array output. If the reference signal were not, the weights might increase indefinitely or decrease to zero.

The reference-signal generator of an adaptive array for direct-sequence spread-spectrum communications [10] is depicted in Figure 5.14. The output of the synchronization system is a continuous-wave signal modulated by the spreading waveform (Chapter 2). The lowpass filter has a bandwidth wide enough to pass the desired signal, which has only data modulation after synchronization has occurred and the spreading waveform has been removed from the array output by the mixer. The spectrally spread interference is largely eliminated by the filter except for the center portion of its spectrum.

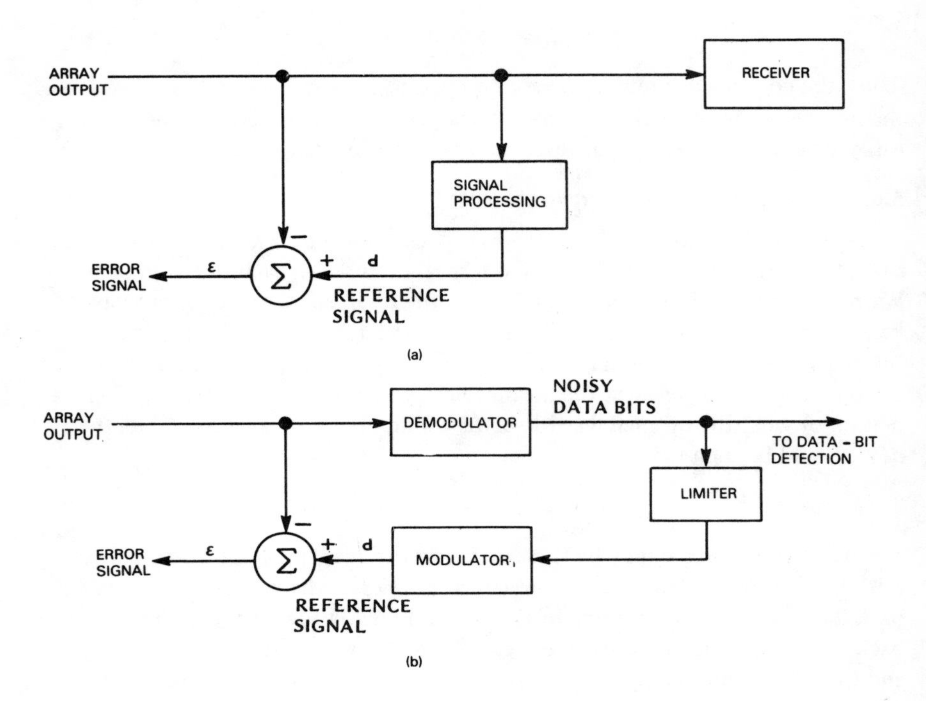

Figure 5.13 Generators of the reference signal.

Before code acquisition, there is no correlation between the reference signal and the desired-signal component of the array output. Therefore, the adaptive signal tends to null both the desired signal and the interference. As the sequence produced by the synchronization system aligns with the received pseudonoise sequence, the reference signal increases its correlation with the desired signal. The resulting changes in the weights cause the desired-signal component of the array output to increase as the local timing approaches its correct value. When acquisition is complete, the interference suppression due to the adaptive array is available in addition to the suppression afforded by the waveform processing.

If the transmission of the desired signal and, hence, the generation of the reference signal are interrupted, then (5.6.22) indicates that the mean weight vector collapses toward zero. Consequently, communications are hindered when the transmission is resumed. A simple solution for this problem is to set one weight at a fixed value. The analysis of section 5.3 indicates that no significant loss of potential performance is likely to result. Other solutions entail extensive hardware [3].

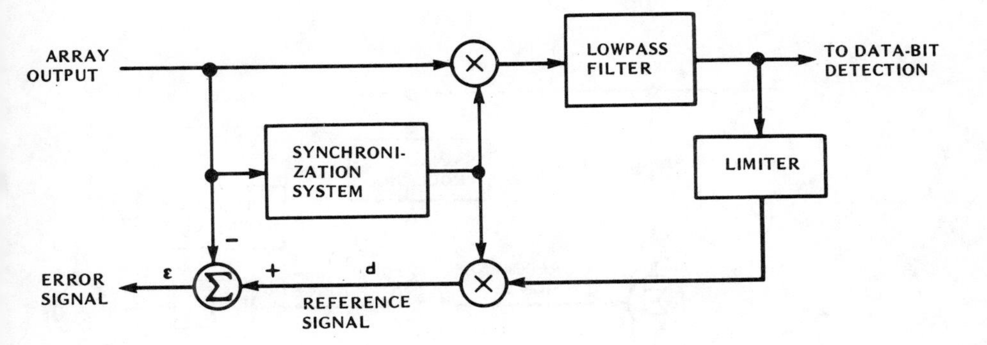

Figure 5.14 Reference-signal generator of an adaptive array for direct-sequence communications.

The delay associated with the reference-signal feedback loop causes an undesirable phase shift in the reference signal with respect to the desired-signal component of the array output. This phase shift causes weight fluctuations, which in turn cause a change in the desired-signal frequency at the array output relative to the array input [11]. An adaptively controlled complex weight in the feedback loop may be necessary to adequately compensate for this effect. Figure 5.15 is a functional block diagram of a possible modification of Figure 5.13(a) to accommodate a complex weight. This weight is generated by an adaptive filter using the Widrow algorithm of (5.6.11). The polarity change preceding the reference-error signal is necessary because the reference signal is the output of an adaptive filter and the array output serves as the reference input to this adaptive filter. Other ways of introducing an adaptive weight into the reference-signal feedback loop are plausible [12].

5.6.3 Misadjustment

According to (5.5.8) and its derivation, the mean square error associated with a deterministic weight vector, $\mathbf{W}(k)$, is

$$E[\epsilon^2(k)] = E_R + \mathbf{V}^T(k)\mathbf{R}_{xx}\mathbf{V}(k) \tag{5.6.40}$$

where

$$\mathbf{V}(k) = \mathbf{W}(k) - \mathbf{W}_0 \tag{5.6.41}$$

When (5.6.24) is satisfied, then as $k \to \infty$ the steepest-descent algorithm of (5.6.5) gives $\mathbf{W}(k) \to \mathbf{W}_0$, and hence $E[\epsilon^2] \to E_R$.

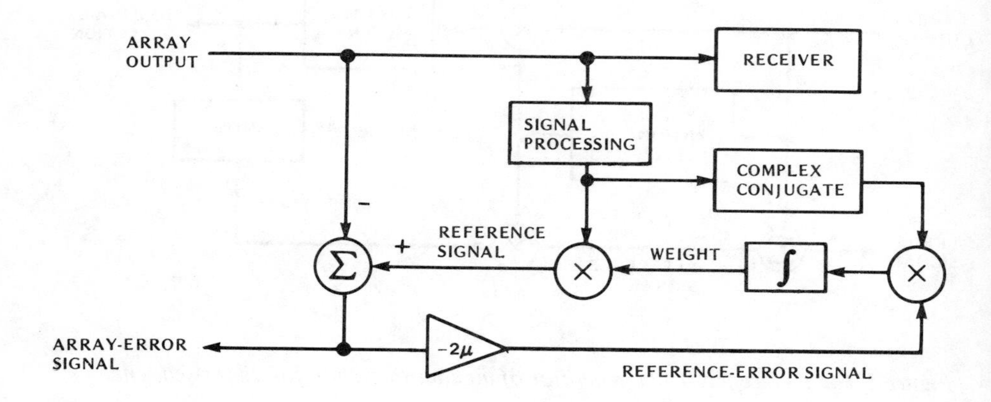

Figure 5.15 Reference-signal generator with adaptive delay compensation.

In the Widrow algorithm, $\mathbf{W}(k)$ is a random vector. If $\mathbf{W}(k)$ and $\mathbf{X}(k)$ are independent, then (5.5.1), (5.5.6), and (5.5.7) can be combined to prove that

$$E[\epsilon^2(k)] = E_R + E[\mathbf{V}^T(k)\mathbf{X}(k)\mathbf{X}^T(k)\mathbf{V}(k)] \tag{5.6.42}$$

Even if $E[\mathbf{W}(k)] \to \mathbf{W}_0$, it does not follow that $E[\epsilon^2] \to E_R$. A measure of the extent to which the Widrow algorithm fails to provide the ideal performance is the excess mean square error, $E[\epsilon^2] - E_R$. A dimensionless measure of the performance loss, called the *misadjustment,* is defined as

$$M = \frac{\lim E[\epsilon^2(k)] - E_R}{E_R} \tag{5.6.43}$$

where the limit is taken as $k \to \infty$.

To derive an expression for the misadjustment, we make the following four assumptions.

1. $\mathbf{X}(k+1)$ is independent of $\mathbf{X}(n)$ and $d(n)$, $n \leqslant k$. It then follows from (5.6.6) that $\mathbf{W}(k)$ is independent of $\mathbf{X}(k)$ and $d(k)$.
2. $0 < \mu < 1/\text{tr}(\mathbf{R}_{xx})$.
3. $E[\|\mathbf{V}(k)\|^2]$ converges as $k \to \infty$.
4. There exists an integer k_0 such that $\epsilon^2(k)$ and $\|\mathbf{X}(k)\|^2$ are statistically independent of each other for $k \geqslant k_0$.

Assumptions 1 and *2* imply convergence of the mean weight vector, as given by (5.6.22). For *Assumption 3* to be true, some restriction on μ in addition to *Assumption 2* is probably necessary. When $\mathbf{W}(k) = \mathbf{W}_0$ and $E[\mathbf{X}(k)] = 0$, then $E[\epsilon(k)\mathbf{X}(k)] = 0$; that is, $\epsilon(k)$ and $\mathbf{X}(k)$ are uncorrelated. Since the independence of $\epsilon(k)$ and $\mathbf{X}(k)$, and hence of $\epsilon^2(k)$ and $\|\mathbf{X}(k)\|^2$, would then follow if they were Gaussian, *Assumption 4* is reasonable. Nevertheless, it is an approximation.

Equations (5.6.7) and (5.6.41) imply that

$$\mathbf{V}(k+1) = \mathbf{V}(k) + 2\mu\epsilon(k)\mathbf{X}(k) \tag{5.6.44}$$

It follows that

$$E[\|\mathbf{V}(k+1)\|^2] = E[\|\mathbf{V}(k)\|^2] + 4\mu E[\epsilon(k)\mathbf{V}^T(k)\mathbf{X}(k)] + 4\mu^2 E[\epsilon^2(k)\|\mathbf{X}(k)\|^2] \tag{5.6.45}$$

Applying *Assumption 4* gives

$$E[\epsilon^2(k)\|\mathbf{X}(k)\|^2] = \mathrm{tr}(\mathbf{R}_{xx})E[\epsilon^2(k)], \quad k \geqslant k_0 \tag{5.6.46}$$

Assumption 1 and (5.5.1), (5.6.22), (5.6.41), and (5.6.42) yield

$$\begin{aligned} E[\epsilon(k)\mathbf{V}^T(k)\mathbf{X}(k)] &= -E[\mathbf{V}^T(k)\mathbf{X}(k)\mathbf{X}^T(k)\mathbf{V}(k)] \\ &= E_R - E[\epsilon^2(k)] \end{aligned} \tag{5.6.47}$$

Combining (5.6.45) to (5.6.47), taking the limit as $k \to \infty$, and using *Assumption 3*, we obtain

$$\lim_{k\to\infty} E[\epsilon^2(k)] = \frac{E_R}{1 - \mu\,\mathrm{tr}(\mathbf{R}_{xx})} \tag{5.6.48}$$

Assumption 2 ensures that the right-hand side of this equation is positive and finite, which could not be guaranteed if (5.6.24) were assumed instead. Substituting (5.6.48) into (5.6.43), we obtain

$$M = \frac{\mu\,\mathrm{tr}(\mathbf{R}_{xx})}{1 - \mu\,\mathrm{tr}(\mathbf{R}_{xx})} = \frac{\mu R_T}{1 - \mu R_T} \tag{5.6.49}$$

This result applies to both the real and complex discrete-time Widrow LMS algorithms. According to (5.6.49) and (5.6.30), increasing μ to improve the convergence rate has the side effect of increasing the misadjustment. For fixed μ, the misadjustment increases with the total input power. Under other assumptions, related convergence results can be derived for the discrete-time Widrow algorithm [13, 14].

5.7 HOWELLS-APPLEBAUM ALGORITHM

The *Howells-Applebaum algorithm* [5, 15] is useful for the adaptive processing of narrowband desired signals with known directions of arrival. The algorithm does not require the generation of a reference signal that is a replica of the desired signal. The algorithm can be derived by first applying the method of steepest descent with the negative of the signal-to-interference-plus-noise ratio as the performance measure.

Assuming that all signals are stationary stochastic processes, using (5.5.14) and (5.5.31), and rearranging factors, we obtain

$$\bar{\nabla}_W \rho_0 = f_1(\mathbf{W})[f_2(\mathbf{W})\mathbf{S}_0^* - R_{nn}\mathbf{W}] \tag{5.7.1}$$

$$f_1(\mathbf{W}) = 4R_s \frac{(\mathbf{W}^H\mathbf{S}_0^*\mathbf{S}_0^T\mathbf{W})}{(\mathbf{W}^H\mathbf{R}_{nn}\mathbf{W})^2} \tag{5.7.2}$$

$$f_2(\mathbf{W}) = \frac{\mathbf{W}^H\mathbf{R}_{nn}\mathbf{W}}{\mathbf{W}^H\mathbf{S}_0^*} \tag{5.7.3}$$

An approximate linearization of (5.7.1) as a function of $\mathbf{W}$ can be used to find simple algorithms. When $\mathbf{W}$ is close to $\mathbf{W}_0$,

$$f_1(\mathbf{W}) \approx f_1(\mathbf{W}_0) = \frac{4R_s}{|\eta|^2} \tag{5.7.4}$$

$$f_2(\mathbf{W}) \approx f_2(\mathbf{W}_0) = \eta \tag{5.7.5}$$

Since η is an arbitrary constant, a reasonable linear approximation to (5.7.1) is obtained by substituting (5.7.4) and (5.7.5) and setting $|\eta|^2 = 2R_s$. The result is

$$\bar{\nabla}_W \rho_0 \approx 2\,[\eta\mathbf{S}_0^* - \mathbf{R}_{nn}\mathbf{W}] \tag{5.7.6}$$

where the factor of two is chosen for later convenience. For complex discrete-time systems, (5.6.3) and (5.7.6) and $P(\mathbf{W}(k)) = -\rho_0(k)$ yield the ideal steepest-descent algorithm:

$$\mathbf{W}(k+1) = \mathbf{W}(k) + 2\mu\,[\eta\mathbf{S}_0^* - \mathbf{R}_{nn}\mathbf{W}(k)] \tag{5.7.7}$$

The *steering vector* is defined as

$$\mathbf{B}^* = \eta\mathbf{S}_0^* \tag{5.7.8}$$

If $\mathbf{R}_{nn}$ is estimated by $\mathbf{X}^*(k)\mathbf{X}^T(k)$ and we set $y(k) = \mathbf{X}^T(k)\mathbf{W}(k)$, we obtain a discrete-time version of the Howells-Applebaum algorithm:

$$\mathbf{W}(k+1) = \mathbf{W}(k) + 2\mu[\mathbf{B}^* - y(k)\mathbf{X}^*(k)] \tag{5.7.9}$$

Figure 5.16 depicts a block-diagram representation of this algorithm for one component of the complex weight vector. The direction of arrival must be known to compute $\mathbf{B}^*$.

A closely related algorithm follows from (5.6.5) if $\mathbf{R}_{xx}$ is replaced by $\mathbf{X}^*(k)\mathbf{X}^T(k)$ and $\mathbf{R}_{xd}$ is directly estimated. The *Griffiths algorithm* is [16]

$$\mathbf{W}(k+1) = \mathbf{W}(k) - 2\mu[y(k)\mathbf{X}^*(k) - \hat{\mathbf{R}}_{xd}] \tag{5.7.10}$$

where $\hat{\mathbf{R}}_{xd}$ is an estimate of $\mathbf{R}_{xd}$. If the desired signal is assumed to be a narrowband stationary process that is uncorrelated with the interference and noise and if $d(t) = \alpha s(t)$, then (5.5.22) holds. Thus, if $\mathbf{S}_0$ is known, we set $\hat{\mathbf{R}}_{xd} = \eta\mathbf{S}_0^* = \mathbf{B}^*$ and (5.7.10) reduces to (5.7.9).

Convergence considerations are greatly facilitated by assuming that $\mathbf{X}(k+1)$ is independent of $\mathbf{X}(n)$, $n \leqslant k$. Equation (5.7.9) then implies that $\mathbf{X}(k+1)$ is independent of $\mathbf{W}(k+1)$. By the method of section 5.6, it then follows from (5.7.9) that

$$\lim_{k\to\infty} E[\mathbf{W}(k)] = \mathbf{R}_{xx}^{-1}\mathbf{B}^* = \eta\mathbf{R}_{xx}^{-1}\mathbf{S}_0^* \tag{5.7.11}$$

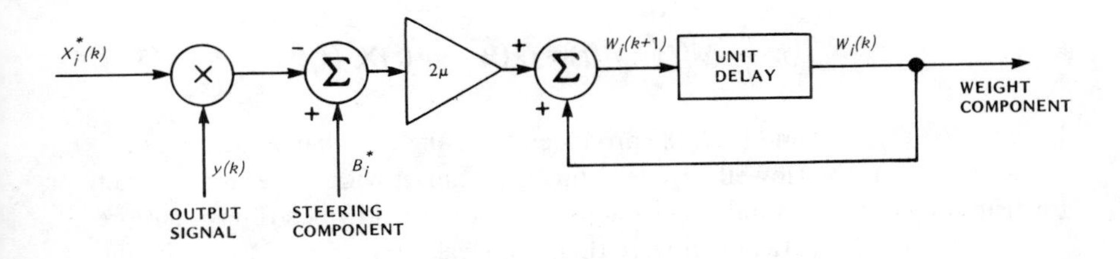

Figure 5.16 Digital implementation of Howells-Applebaum algorithm.

The necessary and sufficient convergence condition is given by (5.6.24). The mean weight vector converges to its optimal value, $\mathbf{W}_0 = \eta\mathbf{R}_{nn}^{-1}\mathbf{S}_0^*$, because $\mathbf{R}_{xx}^{-1}\mathbf{S}_0^*$ is proportional to $\mathbf{R}_{nn}^{-1}\mathbf{S}_0^*$ when the desired signal is narrowband.

The continuous-time or analog algorithm corresponding to (5.7.9) is

$$\frac{d}{dt}\mathbf{W}(t) = 2\mu[\mathbf{B}^* - y(t)\mathbf{X}^*(t)] \tag{5.7.12}$$

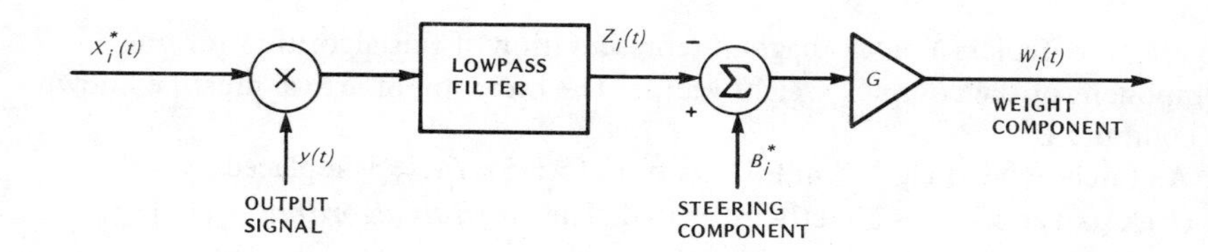

Figure 5.17 Analog implementation of Howells-Applebaum algorithm.

An implementation of one component of this equation would have a form similar to that of Figure 5.16. The usual analog version of the Howells-Applebaum algorithm results from replacing the ideal integrator by a low-pass filter and inserting the steering vector after the filtering rather than before it, as depicted in Figure 5.17. The filter output $Z_i(t)$ is related to the filter input by

$$\tau \frac{d}{dt} Z_i(t) + Z_i(t) = y(t) X_i^*(t) , \quad i = 1, 2, \ldots, N \tag{5.7.13}$$

where τ is the filter time constant. From this equation and the block diagram, it follows that the analog Howells-Applebaum algorithm is

$$\tau \frac{d}{dt} \mathbf{W}(t) + \mathbf{W}(t) = G [\mathbf{B}^* - y(t)\mathbf{X}^*(t)] \tag{5.7.14}$$

If $2\mu = G/\tau$, (5.7.14) and (5.7.12) produce increasingly similar solutions as $\tau \to \infty$. However, the Howells-Applebaum algorithm provides an extra parameter for trimming of the adaptation characteristics, and the lowpass filter is a more realistic model of practical hardware than the ideal integrator. If $\mathbf{B}_i^*$ in Figure 5.17 is replaced by $G\mathbf{B}_i^*$, the final amplifier can be placed before the low-pass filter.

5.7.1 Convergence of the Mean

If τ is sufficiently large that $\mathbf{W}(t)$ varies much more slowly than the input process, then $\mathbf{W}(t)$ is nearly statistically independent of $\mathbf{X}(t)$. Under this assumption, it follows from (5.7.14) that the mean weight vector satisfies

$$\tau \frac{d}{dt} E[\mathbf{W}(t)] + (G\mathbf{R}_{xx} + \mathbf{I}) E[\mathbf{W}(t)] = G\mathbf{B}^* \tag{5.7.15}$$

Because $\mathbf{R}_{xx}$ is Hermitian, it may be represented by

$$\mathbf{R}_{xx} = \mathbf{U}\mathbf{\Lambda}\mathbf{U}^{-1} = \mathbf{U}\mathbf{\Lambda}\mathbf{U}^H \tag{5.7.16}$$

where $\mathbf{U}$ is the unitary modal matrix of $\mathbf{R}_{xx}$ and $\mathbf{\Lambda}$ is the diagonal matrix of eigenvalues of $\mathbf{R}_{xx}$. We define the transformed vectors:

$$\boldsymbol{\omega}(t) = \mathbf{U}^H E\,[\mathbf{W}(t)] \tag{5.7.17}$$

$$\mathbf{L}^* = \mathbf{U}^H \mathbf{B}^* \tag{5.7.18}$$

Equations (5.7.15) to (5.7.18) imply that

$$\tau \frac{d}{dt}\,\boldsymbol{\omega}(t) + (G\mathbf{\Lambda} + \mathbf{I})\,\boldsymbol{\omega}(t) = G\mathbf{L}^* \tag{5.7.19}$$

The components of $\boldsymbol{\omega}(t)$ must satisfy

$$\tau \frac{d}{dt}\,\omega_i(t) + (G\lambda_i + 1)\,\omega_i(t) = GL_i^*\,, \qquad i = 1, 2, \ldots, N \tag{5.7.20}$$

where λ_i is an eigenvalue. This linear differential equation with constant coefficients has the solution

$$\omega_i(t) = \left(\omega_i(0) - \frac{GL_i^*}{G\lambda_i + 1}\right)\exp\left(-\frac{G\lambda_i + 1}{\tau}\,t\right) + \frac{GL_i^*}{G\lambda_i + 1} \tag{5.7.21}$$

where $\omega_i(0)$ is the initial value of $\omega_i(t)$. Because $\mathbf{R}_{xx}$ is Hermitian and positive definite, $\lambda_i > 0$. By definition, $\tau > 0$. Therefore, if $G > 0$, then

$$\lim_{t\to\infty} \omega_i(t) = \frac{GL_i^*}{G\lambda_i + 1} \tag{5.7.22}$$

Consequently, if $G > 0$,

$$\lim_{t\to\infty} \boldsymbol{\omega}(t) = \left(\mathbf{\Lambda} + \frac{1}{G}\,\mathbf{I}\right)^{-1} \mathbf{L}^* \tag{5.7.23}$$

Combining (5.7.23) and (5.7.16) to (5.7.18), we obtain

$$\lim_{t\to\infty} E\,[\mathbf{W}(t)] = \left(\mathbf{R}_{xx} + \frac{1}{G}\,\mathbf{I}\right)^{-1} \mathbf{B}^* \tag{5.7.24}$$

if $G > 0$. This relation indicates that the mean weight vector does not converge to the optimal weight given by (5.5.39), even if the desired signal is absent. However, when the interference is weak, the optimal weight may be unrealistically large for a hardware implementation and a finite value of G in (5.7.24) can ensure that the mean weight converges to a realizable vector. Thus, the Howells-Applebaum algorithm introduces a weight bias in the steady state but allows a greater input dynamic range than the algorithm of (5.7.12), which provides a mean weight vector converging to the optimal value when the desired signal is absent.

According to (5.7.21), the weights undergo transients that vary as sums of exponentials with time constants

$$\tau_i = \left(\frac{G\lambda_i}{\tau} + \frac{1}{\tau}\right)^{-1}, \quad i = 1, 2, \ldots, N \tag{5.7.25}$$

This equation indicates that the convergence speed increases with G/τ. However, the weight variance and, hence, the noise power in the system output also increase with G/τ [17, 18]. Thus, there is a fundamental trade-off between the convergence speed and the amount of residual noise in the output.

The steering vector can be computed if the direction of arrival of the desired signal is known. If the direction is unknown, it can be estimated by using an auxiliary direction-finding system. Alternatively, the steering vector can be estimated by using the Widrow algorithm to adapt to the desired signal during an interval when interference is absent, since then $\mathbf{R}_{nn} = \sigma^2 \mathbf{I}$ and (5.5.26), which holds for narrowband signals, indicates that $\mathbf{W}_0$ is proportional to $\mathbf{S}_0^*$.

The main problem with the Howells-Applebaum algorithm is that the adaptive array may null the desired signal when it arrives from a direction different from that indicated by an erroneous steering vector. The adaptive-array performance degradation due to steering-vector errors increases with the number of antenna elements and the received desired-signal power [19, 20]. For random steering-vector errors, the SINR at the array output is a function of the variance of the steering-vector error. When the SINR must exceed a certain level, the variance that is tolerable decreases with increases in the desired-signal dynamic range that is to be accommodated.

5.8 FROST ALGORITHM

The *Frost Algorithm* [7] is based upon approximating the constrained optimal weight of (5.5.64) by the method of steepest descent. For real signals and weights, the performance measure is given by (5.5.60). The weight vector during adaptation cycle $k + 1$ is related to the weight vector during adaptation cycle k by

$$\begin{aligned} \mathbf{W}(k+1) &= \mathbf{W}(k) - \frac{1}{2}\,\mu\nabla_w H \\ &= \mathbf{W}(k) - \mu\left[\mathbf{R}_{xx}\mathbf{W}(k) + \frac{1}{2}\,\mathbf{C}\Upsilon(k)\right] \end{aligned} \tag{5.8.1}$$

where μ is a constant that regulates the convergence rate, and the factor 1/2 has been inserted for convenience. The *Lagrange-multiplier vector,* $\Upsilon(k)$, is allowed to vary with the adaptation cycle in such a way that the constraint is satisfied for $\mathbf{W}(k+1)$; that is, $\Upsilon(k)$ is chosen so that

$$\mathbf{C}^T\mathbf{W}(k+1) = \mathbf{F} \tag{5.8.2}$$

Substituting (5.8.1) into (5.8.2) and solving for $\Upsilon(k)$ yields

$$\Upsilon(k) = \frac{2}{\mu}(\mathbf{C}^T\mathbf{C})^{-1}[\mathbf{C}^T\mathbf{W}(k) - \mathbf{F}] - 2(\mathbf{C}^T\mathbf{C})^{-1}\mathbf{C}^T\mathbf{R}_{xx}\mathbf{W}(k) \tag{5.8.3}$$

where the indicated inverse is assumed to exist. We define the constant matrices:

$$\mathbf{A} = \mathbf{I} - \mathbf{C}(\mathbf{C}^T\mathbf{C})^{-1}\mathbf{C}^T \tag{5.8.4}$$

$$\mathbf{B} = \mathbf{C}(\mathbf{C}^T\mathbf{C})^{-1}\mathbf{F} \tag{5.8.5}$$

In general, $\mathbf{C}$ is not a square matrix so $(\mathbf{C}^T\mathbf{C})^{-1} \neq \mathbf{C}^{-1}(\mathbf{C}^T)^{-1}$. Substituting (5.8.3) to (5.8.5) into (5.8.1), we obtain

$$\mathbf{W}(k+1) = \mathbf{A}[\mathbf{W}(k) - \mu\mathbf{R}_{xx}\mathbf{W}(k)] + \mathbf{B} \tag{5.8.6}$$

This equation provides a deterministic algorithm that would be used if $\mathbf{R}_{xx}$ were known. Because it is not, we approximate $\mathbf{R}_{xx}$ by $\mathbf{X}(k)\mathbf{X}^T(k)$ and use $y = \mathbf{W}^T\mathbf{X}$. A suitable choice for the initial weight vector is $\mathbf{W}(0) = \mathbf{B}$ because $\mathbf{B}$ must be computed anyway and this choice satisfies the constraint $\mathbf{C}^T\mathbf{W}(0) = \mathbf{F}$. Thus, the Frost algorithm is

$$\mathbf{W}(0) = \mathbf{B} \tag{5.8.7}$$

$$\mathbf{W}(k+1) = \mathbf{A}[\mathbf{W}(k) - \mu y(k)\mathbf{X}(k)] + \mathbf{B} \tag{5.8.8}$$

An important feature of the algorithm is that each iteration automatically corrects for computational errors in the weight vector that prevented exact satisfaction of the constraint during the preceding iteration. These errors often occur during an iteration because of truncation, rounding off, or quantization errors in the computer implementation of the algorithm. If the errors are not corrected, they may have a significant cumulative effect after a few iterations. The error-correcting capability of the algorithm is due to the fact that the factor $(\mathbf{C}^T\mathbf{W}(k) - \mathbf{F})$ was not set to zero in (5.8.3). Thus, apart from other sources of error, $\mathbf{C}^T\mathbf{W}(k+1) = \mathbf{F}$ even if $\mathbf{C}^T\mathbf{W}(k) \neq \mathbf{F}$.

For complex signals and weights, the method of steepest descent yields

$$\mathbf{W}(k+1) = \mathbf{W}(k) - \mu\overline{\nabla}_w H \tag{5.8.9}$$

where H is given by (5.5.68). Applying (5.5.14) and approximating $\mathbf{R}_{xx}$ by $\mathbf{X}^*(k)\mathbf{X}^T(k)$, we obtain the complex Frost algorithm:

$$\mathbf{W}(0) = \mathbf{B} \tag{5.8.10}$$

$$\mathbf{W}(k+1) = \mathbf{A}[\mathbf{W}(k) - \mu y(k)\mathbf{X}^*(k)] + \mathbf{B} \tag{5.8.11}$$

where

$$\mathbf{A} = \mathbf{I} - \mathbf{C}^*(\mathbf{C}^T\mathbf{C}^*)^{-1}\mathbf{C}^T \tag{5.8.12}$$

$$\mathbf{B} = \mathbf{C}^*(\mathbf{C}^T\mathbf{C}^*)^{-1}\mathbf{F} \tag{5.8.13}$$

For the single main-beam constraint specified by (5.5.71), an error in the constraint vector causes a performance degradation in the Frost algorithm similar to that exhibited by the Howells-Applebaum algorithm when there is an error in the steering vector. The addition of a linearized constraint on the first derivative of the main-beam pattern and a nonlinear constraint on the norm of the weight vector can greatly reduce the potential performance degradation due to constraint-vector errors when sidelobe interference is received. However, the ability of the Frost algorithm to cancel main beam interference is greatly reduced and its performance against such interference is inferior to that attained by the Widrow algorithm [20].

5.8.1 Convergence of the Mean

If $\mathbf{X}(k+1)$ is independent of $\mathbf{X}(n)$ and $d(n)$, $n \leqslant k$, then $\mathbf{W}(k)$ and $\mathbf{X}(k)$ are independent and (5.8.11) implies that

$$E[\mathbf{W}(k+1)] = \mathbf{A}[\mathbf{I} - \mu\mathbf{R}_{xx}]E[\mathbf{W}(k)] + \mathbf{B}, \qquad k \geqslant 0 \tag{5.8.14}$$

The equilibrium point is equal to the optimal weight vector given by (5.5.69). We define

$$\mathbf{V}(k) = E[\mathbf{W}(k)] - \mathbf{W}_0 = E[\mathbf{W}(k)] - \mathbf{R}_{xx}^{-1}\mathbf{C}^*(\mathbf{C}^T\mathbf{R}_{xx}^{-1}\mathbf{C}^*)^{-1}\mathbf{F} \tag{5.8.15}$$

Using (5.8.12) to (5.8.15), we obtain

$$\mathbf{V}(k+1) = \mathbf{A}\mathbf{V}(k) - \mu\mathbf{A}\mathbf{R}_{xx}\mathbf{V}(k), \qquad k \geqslant 0 \tag{5.8.16}$$

Direct multiplication verifies that $\mathbf{A}^2 = \mathbf{A}$. It then follows from (5.8.16) that $\mathbf{A}\mathbf{V}(k) = \mathbf{V}(k)$, $k \geqslant 1$. It is easily verified that $\mathbf{A}\mathbf{V}(0) = \mathbf{V}(0)$. Consequently,

$$\begin{aligned}\mathbf{V}(k+1) &= [\mathbf{I} - \mu\mathbf{A}\mathbf{R}_{xx}\mathbf{A}]\mathbf{V}(k) \\ &= [\mathbf{I} - \mu\mathbf{A}\mathbf{R}_{xx}\mathbf{A}]^{k+1}\,\mathbf{V}(0), \qquad k \geqslant 0\end{aligned} \tag{5.8.17}$$

It is easily verified that the matrix $\mathbf{A}\mathbf{R}_{xx}\mathbf{A}$ is Hermitian and thus has a complete set of orthonormal eigenvectors. Direct calculation proves that

$$\mathbf{A}\mathbf{R}_{xx}\mathbf{A}\mathbf{C}^* = \mathbf{0} \tag{5.8.18}$$

If $\mathbf{R}_{xx}$ has dimensions $N \times N$ and $\mathbf{C}$ has dimensions $N \times L$, then (5.8.18) implies that the L columns of $\mathbf{C}^*$ are eigenvectors of $\mathbf{A}\mathbf{R}_{xx}\mathbf{A}$ with eigenvalues equal to zero. Let $\mathbf{e}_i$, $i = 1, 2, \ldots, N - L$, denote that $N - L$ remaining orthonormal eigenvectors. Because the $\mathbf{e}_i$ must be orthogonal to the columns of $\mathbf{C}^*$,

$$\mathbf{C}^T\mathbf{e}_i = \mathbf{0}, \quad i = 1, 2, \ldots, N - L \tag{5.8.19}$$

From this equation and (5.8.12), it follows that

$$\mathbf{A}\mathbf{e}_i = \mathbf{e}_i, \quad i = 1, 2, \ldots, N - L \tag{5.8.20}$$

Let σ_i denote the eigenvalue of $\mathbf{A}\mathbf{R}_{xx}\mathbf{A}$ associated with the unit eigenvector $\mathbf{e}_i$. Using (5.8.20) and the Hermitian character of $\mathbf{A}$, we obtain

$$\sigma_i = \mathbf{e}_i^H\mathbf{A}\mathbf{R}_{xx}\mathbf{A}\mathbf{e}_i = \mathbf{e}_i^H\mathbf{R}_{xx}\mathbf{e}_i, \; i = 1, 2, \ldots, N - L \tag{5.8.21}$$

It follows from (C.2.7) of Appendix C that if $\mathbf{e}_i$ is a unit vector, then

$$\lambda_{min} \leqslant \mathbf{e}_i^H\mathbf{R}_{xx}\mathbf{e}_i \leqslant \lambda_{max} \tag{5.8.22}$$

where λ_{min} and λ_{max} are the minimum and maximum eigenvalues of $\mathbf{R}_{xx}$. Combining (5.8.21) and (5.8.22) yields

$$\lambda_{min} \leqslant \sigma_i \leqslant \lambda_{max}, \; i = 1, 2, \ldots, N - L \tag{5.8.23}$$

If we assume that $\mathbf{R}_{xx}$ is positive definite, $\lambda_{min} > 0$ and hence $\sigma_i > 0$, $i = 1, 2, \ldots, N - L$. We conclude that the $\mathbf{e}_i$ correspond to non-zero eigenvalues.

Equations (5.8.10), (5.8.13), and (5.8.15) indicate that $\mathbf{C}^T\mathbf{V}(0) = \mathbf{C}^T(\mathbf{B} - \mathbf{W}_0) = \mathbf{0}$. Because $\mathbf{C}^T = \mathbf{C}^{*H}$, the initial vector $\mathbf{V}(0)$ must be orthogonal to the columns of $\mathbf{C}^*$. Therefore, $\mathbf{V}(0)$ is equal to a linear combination of the $\mathbf{e}_i$, $i = 1, 2, \ldots, N - L$, which are the eigenvectors of $\mathbf{A}\mathbf{R}_{xx}\mathbf{A}$ corresponding to the non-zero eigenvalues. If $\mathbf{V}(0)$ is equal to the eigenvector $\mathbf{e}_j$ with eigenvalue σ_j, then (5.8.17) gives

$$\mathbf{V}(k + 1) = (1 - \mu\sigma_j)^{k+1}\mathbf{e}_j, \; k \geqslant 0 \tag{5.8.24}$$

Therefore, a necessary and sufficient condition for the convergence of the mean weight vector is that $|1 - \mu\sigma_i| < 1$ for $i = 1, 2, \ldots, N - L$. Because $\sigma_i > 0$, the necessary and sufficient condition for convergence is

$$0 < \mu < \frac{2}{\sigma_{max}} \tag{5.8.25}$$

From (5.8.23), it follows that a sufficient condition for the convergence of the Frost algorithm is

$$0 < \mu < \frac{2}{\lambda_{max}} \tag{5.8.26}$$

Analogously to (5.6.30), the convergence of the mean weight vector of the Frost algorithm along any eigenvector $\mathbf{e}_i$ can be characterized by the time constant

$$\tau_i = \frac{1}{\mu \sigma_i}, \quad i = 1, 2, \ldots, N - L \tag{5.8.27}$$

when

$$0 < \mu \leqslant \frac{1}{\sigma_{max}} \tag{5.8.28}$$

It follows that

$$\tau_{max} \geqslant \frac{\sigma_{max}}{\sigma_{min}} \tag{5.8.29}$$

Relations (5.8.25) through (5.8.29), derived assuming complex signals and weights, are obviously valid for real signals and weights.

5.9 ADAPTIVE NOISE CANCELLER

An *adaptive noise canceller* is a system with the form of Figure 5.18 and an adaptation that minimizes the mean-square-error signal or the output power [21]. The weight in the primary path is a constant and is set equal to unity for convenience. If the primary input is derived from either an antenna or the weighted sum of a group of elements in a phased array, and if the auxiliary inputs are derived from either separate antennas or phased-array elements, then the adaptive noise canceller is called a *sidelobe canceller.* When the number of auxiliary inputs exceeds one, the canceller is often called a multiple sidelobe canceller. Figure 5.18 has the form of Figure 5.5(a) with a fixed value for one of the weights.

The system output for the adaptive noise canceller is identical to the error signal. We show that minimizing the mean square output is approximately equivalent to causing the output to be a minimum mean-square-error estimate of the desired signal, s, if the adaptive-filter output, y, is such that it has negligible correlation with s and if the primary input is the sum of the desired signal and uncorrelated noise, n_0. Referring to Figure 5.18, the canceller output is

$$\epsilon = s + n_0 - y \tag{5.9.1}$$

We assume that all signals are derived from stationary stochastic processes. Taking the expected value of the squares of both sides of (5.9.1) gives

$$E[\epsilon^2] = E[s^2] + E[(n_0 - y)^2] + 2E[sn_0] - 2E[sy] \tag{5.9.2}$$

If s is uncorrelated with n_0 and has negligible correlation with y, and if $E[n_0] = 0$ and $E[y] \cong 0$, then

$$\begin{aligned} E[\epsilon^2] &\cong E[s^2] + E[(n_0 - y)^2] \\ &= E[s^2] + E[(\epsilon - s)^2] \end{aligned} \tag{5.9.3}$$

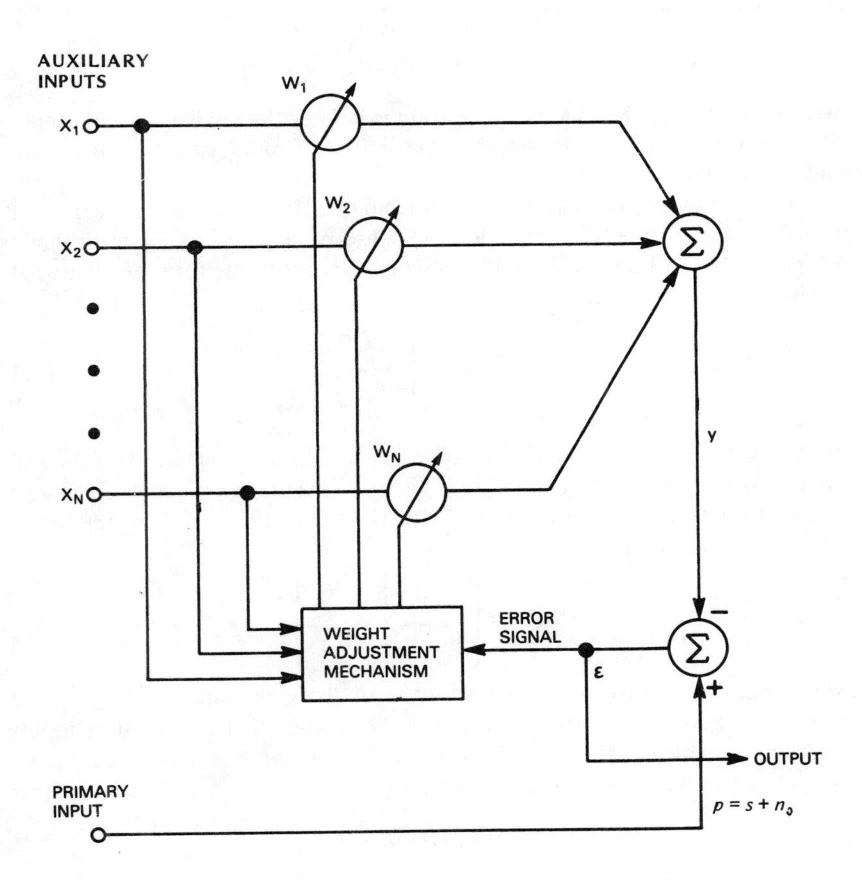

Figure 5.18 Adaptive noise canceller.

The signal power, $E[s^2]$, is unaffected as the adaptive filter minimizes $E[\epsilon^2]$. Consequently, $E[(\epsilon - s)^2]$ is minimized when $E[\epsilon^2]$ is minimized. We conclude that the adaptive-noise-canceller output is an approximate minimum mean-square-error estimate of the desired signal. The adaptive-filter output, y, is an approximate minimum mean-square-error estimate of n_0, the noise in the primary input.

Let $\mathbf{W}_0$ denote the optimal weight vector excluding the unit primary weight. If the primary input is regarded as a reference signal that approximates the desired response, then the $\mathbf{W}_0$ for minimizing the mean-square error signal is given by the Wiener-Hopf equation,

$$\mathbf{W}_0 = \mathbf{R}_{xx}^{-1}\mathbf{R}_{xp} \tag{5.9.4}$$

where we interpret $\mathbf{R}_{xx}$ as the correlation matrix of the auxiliary inputs and $\mathbf{R}_{xp}$ is the cross-correlation vector of the auxiliary inputs with the primary input, $p = s + n_0$.

Equation (5.9.4) can also be derived from (5.5.39), which gives the optimal weight vector for the *SINR* criterion [15]. Let $\mathbf{W}_0'$, $\mathbf{R}_{nn}'$, and $\mathbf{S}_0$ refer to the adaptive filter of Figure 5.5(a). If this figure is put into the form of Figure 5.18, we have

$$\mathbf{W}_0' = \left[\begin{array}{c} W_p \\ \hline -\mathbf{W}_0 \end{array}\right] \tag{5.9.5}$$

where $\mathbf{W}_0$ is the optimal weight vector of the auxiliary branches, W_p is the scalar weight of the primary branch ($W_p = 1$ in Figure 5.18), and the minus sign accounts for the subtraction in Figure 5.18. The correlation matrix of the interference plus noise can be partitioned as

$$\mathbf{R}_{nn}' = \left[\begin{array}{c|c} E[n_0^2] & \mathbf{R}_{nn_0}^H \\ \hline \mathbf{R}_{nn_0} & \mathbf{R}_{nn} \end{array}\right] \tag{5.9.6}$$

where $\mathbf{R}_{nn}$ is the correlation matrix for the auxiliary branches and $\mathbf{R}_{nn_0}$ is the cross-correlation vector between $\mathbf{n}$ and n_0. We assume that the auxiliary inputs make a negligible contribution to the desired-signal component of the system output. Thus, $\mathbf{S}_0$ can be approximated by

$$\mathbf{S}_0 = [1 \quad 0 \quad 0 \quad \dots \quad 0]^T \tag{5.9.7}$$

Equation (5.5.39) implies that

$$\mathbf{R}_{nn}'\mathbf{W}_0' = \eta\mathbf{S}_0^* \tag{5.9.8}$$

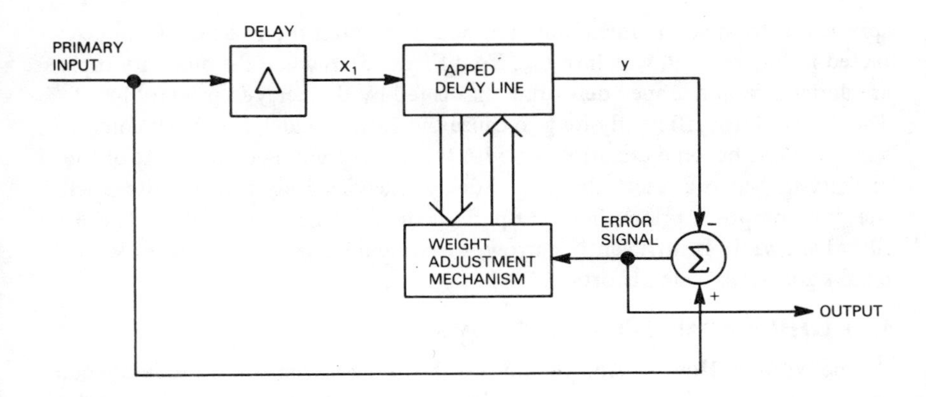

Figure 5.19 Adaptive notch filter.

The substitution of (5.9.5) to (5.9.7) into (5.9.8) yields the scalar equation

$$E\,[n_0^2]\,W_p - \mathbf{R}_{nn_0}^H\,\mathbf{W}_0 = \eta \tag{5.9.9}$$

and the matrix equation

$$W_p\,\mathbf{R}_{nn_0} - \mathbf{R}_{nn}\,\mathbf{W}_0 = \mathbf{0} \tag{5.9.10}$$

Solving (5.9.10) for $\mathbf{W}_0$ and substituting the result into (5.9.9) gives

$$W_p(E\,[n_0^2] - \mathbf{R}_{nn_0}^H\,\mathbf{R}_{nn}^{-1}\,\mathbf{R}_{nn_0}) = \eta \tag{5.9.11}$$

We assume that the quantity in parentheses is non-zero. Because η can take any non-zero value, its value can be chosen so that $W_p = 1$. Equation (5.9.10) then implies that

$$\mathbf{W}_0 = \mathbf{R}_{nn}^{-1}\,\mathbf{R}_{nn_0} \tag{5.9.12}$$

If the desired-signal components of the auxiliary inputs are sufficiently small, then $\mathbf{R}_{nn_0} \cong \mathbf{R}_{xp}$, $\mathbf{R}_{nn} \cong \mathbf{R}_{xx}$, and (5.9.12) becomes (5.9.4).

The Widrow algorithm may be used to adapt the weights in the auxiliary branches. For an analog system with a two-dimensional auxiliary vector $\mathbf{X}^T = [X_1, X_2]$, where $X_2(t)$ is the result of passing $X_1(t)$ through a quarter-wavelength delay, the Widrow algorithm leads to the implementation of Figure 5.1.

The adaptive noise canceller can be used as an *adaptive notch filter,* which is useful in applications such as removing periodic interference from direct-sequence spread-spectrum communications. The adaptive notch filter has the form depicted in Figure 5.19, which results from Figure 5.18 when the auxiliary inputs are derived from a tapped delay line that is fed by the delayed primary input. The delay, Δ, is sufficiently long to cause the desired wideband signal component of X_1 to become uncorrelated with the corresponding component of the primary input. As a result, the tapped delay line adaptively forms a filter such that y is composed primarily of the periodic interference components. If the filter bandwidth is sufficiently narrow, the system output is the desired wideband signal with little distortion.

5.10 OTHER ADAPTIVE ALGORITHMS

The Widrow, Howells-Applebaum, and Frost algorithms offer moderate convergence rates and have moderate implementation requirements. Although they are the most important adaptive algorithms, many others with special features have been proposed [22]. In this section, the main features of direct matrix inversion, the Gram-Schmidt preprocessor, recursive algorithms, and perturbation algorithms are briefly summarized. Direct matrix inversion and recursive algorithms offer fast convergence, but require a large amount of computation. At the other extreme, perturbation algorithms offer implementation simplicity, but are slow to converge.

5.10.1 Direct Matrix Inversion

Direct matrix inversion or *sample matrix inversion* entails the direct implementation of (5.5.16), (5.5.39), or (5.5.69) after suitable approximations of the matrix elements. The associated adaptive system has the form of Figure 5.5(a) without the feedback from the output. Thus, the adaptation is based upon the input vector alone.

For n samples of the input vector, the *sample correlation matrix* is defined as

$$\hat{\mathbf{R}}_{xx} = \frac{1}{n} \sum_{i=1}^{n} \mathbf{X}^*(i)\mathbf{X}^T(i) \tag{5.10.1}$$

If a reference signal is available, n samples of it are used in calculating the *sample cross-correlation vector* defined as

$$\hat{\mathbf{R}}_{xd} = \frac{1}{n} \sum_{i=1}^{n} \mathbf{X}^*(i)\, d(i) \tag{5.10.2}$$

Equation (5.5.76) with $\mathbf{A}(n) = \mathbf{I}$ indicates that the optimal weight for the least-squares-error criterion of (5.5.74) is given by

$$\mathbf{W}_0(n) = \hat{\mathbf{R}}_{xx}^{-1} \hat{\mathbf{R}}_{xd} \tag{5.10.3}$$

This equation represents an approximate implementation of (5.5.16). If an accurate estimate of $\mathbf{R}_{xd}$ is known, it can be used in place of (5.10.2). For real signals, $\mathbf{X}^*(i)$ is replaced by $\mathbf{X}(i)$ in (5.10.1) and (5.10.2).

To approximately implement (5.5.69), (5.10.1) is substituted in place of $\mathbf{R}_{xx}$. For a narrowband desired signal, (5.5.23) and (5.5.25) indicate that $\mathbf{R}_{xx}^{-1}\mathbf{S}_0^*$ is proportional to $\mathbf{R}_{nn}^{-1}\mathbf{S}_0^*$. Thus, an approximate implementation of (5.5.39) is given by

$$\mathbf{W}_0(n) = \eta \hat{\mathbf{R}}_{xx}^{-1}\mathbf{S}_0^* \tag{5.10.4}$$

where $\hat{\mathbf{R}}_{xx}$ is determined from (5.10.1).

If the components of the input vector have stationary statistics, then the performance of an adaptive system using direct matrix inversion converges toward the optimal performance as the number of data samples increases. The rapid convergence is due to the absence of iterative operations. Satisfactory performance usually requires that $n > 2N$, where N denotes the dimension of the input vector $\mathbf{X}(i)$. Although in principle the convergence time does not depend upon the noise environment or the eigenvalue spread of $\mathbf{R}_{xx}$, computational limitations when the matrix to be inverted has a small determinant, may cause the performance to depend upon these factors. To avoid a small determinant, the diagonal elements of a matrix may be increased by a constant amount, which is equivalent to increasing the uncorrelated noise. The total number of multiplications required to obtain a weight vector by direct matrix inversion is on the order of N^3 except for special cases, and the computation must be repeated whenever there is a significant change in the input statistics [23].

In contrast to direct matrix inversion, the Widrow algorithm has a convergence rate that depends directly upon the eigenvalue spread and the initial weight vector, $\mathbf{W}(0)$. Except for fortuitous choices of $\mathbf{W}(0)$, the iterative nature of the Widrow algorithm causes it to converge much more slowly than direct matrix inversion. On the order of N multiplications are required to compute the weight vector. Thus, direct matrix inversion offers fast convergence, but requires many processing steps and a high computational accuracy compared with the Widrow algorithm.

5.10.2 Gram-Schmidt Preprocessor

The convergence rates of the Widrow, Howells-Applebaum, and Frost algorithms are sensitive to the *eigenvalue spread* of the input correlation matrix, as indicated by (5.6.32) and (5.8.29). This spread can be greatly reduced through the use of a cascade preprocessor. The *Gram-Schmidt preprocessor,* which appears to be the most effective of the cascade preprocessors, is based upon the Gram-Schmidt orthogonalization of vectors. It requires a relatively small number of iterations for the preprocessor to adapt its parameters to their desired values. After this adaptation, the main adaptive algorithm converges at an increased rate so that the overall convergence rate is often greatly increased. However, the overall convergence rate of an adaptive algorithm and a Gram-Schmidt preprocessor may not match that provided by direct matrix inversion [22].

The Gram-Schmidt preprocessor, which is depicted in Figure 5.20, reduces the eigenvalue spread by producing outputs that are nearly orthogonal to each other and have nearly the same mean power. Thus, $E[x_i^* x_j] \cong 0, i \neq j$, for the inputs entering the main processor, and $\mathbf{R}_{xx}$ is nearly the identity matrix. Each of the blocks labeled ANC is an adaptive noise canceller consisting of a single correlation loop with a primary input from the vertical direction and a single auxiliary input from the horizontal direction. Let p, x, and ϵ denote the primary input, the auxiliary input, and the output, respectively, of one of the cancellers. Because the optimal weight is the scalar $W_0 = R_{xp}/R_{xx}$ and $\epsilon = p - W_0 x$ after convergence, we have

$$E[x^* \epsilon] = 0 \tag{5.10.5}$$

which indicates that the output of an adaptive noise canceller is orthogonal to its auxiliary input after convergence of the weight. Further straightforward calculations indicate that the outputs of the final cancellers in the vertical columns become orthogonal to each other following ideal convergence. The automatic-gain-control devices, labeled AGC in the diagram, then equalize the mean powers before the outputs pass to the main processor. For N inputs to the preprocessor, $N(N-1)/2$ cancellers are required to perform the sequential orthogonalization.

The computational requirements of direct matrix inversion are greatly reduced by using the output of the Gram-Schmidt preprocessor since $\mathbf{R}_{xx}^{-1} \cong \mathbf{I}$ and hence only the sample cross-correlation vector needs to be computed. If the automatic-gain-control devices are eliminated, the computational requirements only slightly increase because the inversion of the diagonal sample correlation matrix is trivial.

5.10.3 Recursive Algorithms

Recursive algorithms perform recursive, rather than direct, approximate computations of the inverse of the covariance matrix. The associated adaptive system has the form of Figure 5.5(a) without the feedback from the output. Recursive algorithms allow the continual updating of the weight vector, whereas direct matrix inversion requires the accumulation of n samples of the input vector. Let $\hat{\mathbf{R}}_{xx}(k)$ denote an estimate of $\mathbf{R}_{xx}$ based upon k samples of $\mathbf{X}$. A plausible form for a complex recursion equation is

$$\hat{\mathbf{R}}_{xx}(k+1) = \alpha \hat{\mathbf{R}}_{xx}(k) + \beta \mathbf{X}^*(k+1)\mathbf{X}^T(k+1), \qquad k \geqslant 0 \tag{5.10.6}$$

where α and β are constants. It is convenient to set $\hat{\mathbf{R}}_{xx}(0)$, the initial correlation matrix, equal to a constant times the identity matrix. If we take $\alpha = k/(k+1)$

and $\beta = 1/(k+1)$, then (5.10.6) is equivalent to (5.10.1) when $k = n$. This choice is appropriate if the inputs have stationary statistics. For nonstationary inputs, if we take $\alpha \leqslant \beta$, then the past data are never more important than the current data. To prevent $\hat{\mathbf{R}}_{xx}(k)$ from increasing monotonically with k, it is reasonable to require that $\alpha + \beta = 1$.

Applying the matrix inversion identity of (5.5.25) to (5.10.6) yields the recursive equation

$$\hat{\mathbf{R}}_{xx}^{-1}(k+1) = \frac{1}{\alpha}\left[\hat{\mathbf{R}}_{xx}^{-1}(k) - \frac{\beta\hat{\mathbf{R}}_{xx}^{-1}(k)\mathbf{X}^*(k+1)\mathbf{X}^T(k+1)\hat{\mathbf{R}}_{xx}^{-1}(k)}{\alpha + \beta\mathbf{X}^T(k+1)\hat{\mathbf{R}}_{xx}^{-1}(k)\mathbf{X}^*(k+1)}\right], k \geqslant 0 \quad (5.10.7)$$

where $\hat{\mathbf{R}}_{xx}^{-1}(0)$ is proportional to the identity matrix. The optimal weight vector of (5.5.16) is estimated by

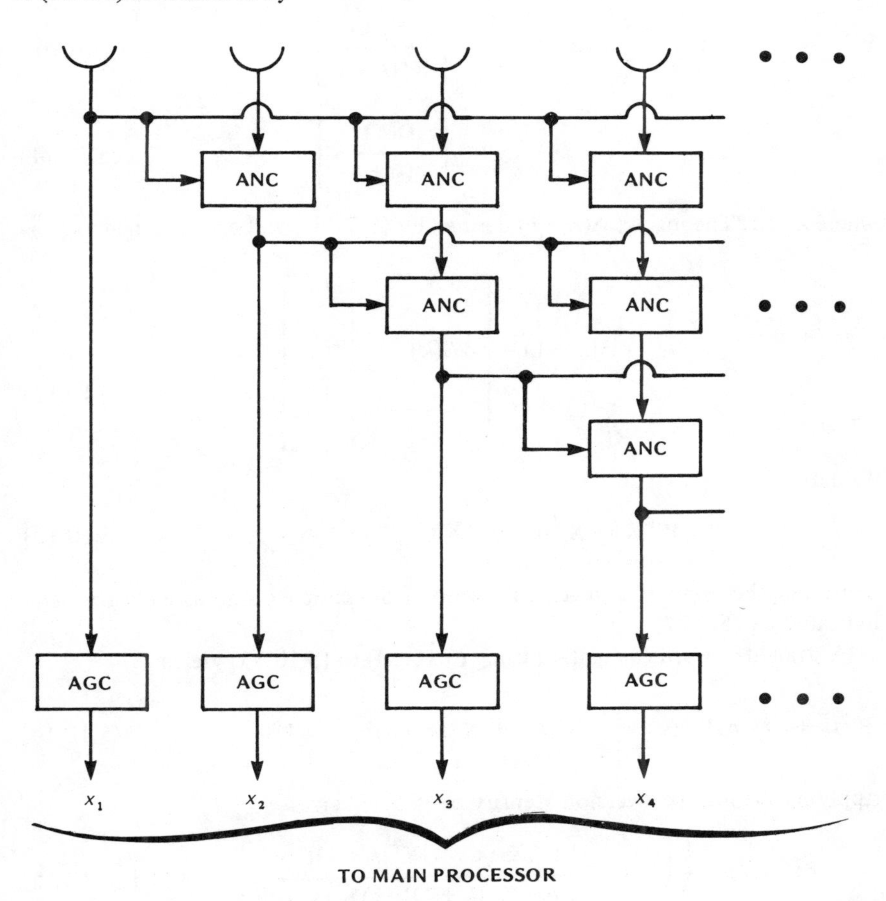

Figure 5.20 Gram-Schmidt preprocessor.

$$\mathbf{W}_0(k) = \hat{\mathbf{R}}_{xx}^{-1}(k)\hat{\mathbf{R}}_{xd}(k) \tag{5.10.8}$$

where $\hat{\mathbf{R}}_{xd}(k)$ is the sample cross-correlation vector of (5.10.3), which can be computed recursively, or is a known constant vector. A recursive algorithm that approximately implements (5.5.69) is given by (5.10.7) and

$$\mathbf{W}_0(k) = \hat{\mathbf{R}}_{xx}^{-1}(k)\mathbf{C}^*[\mathbf{C}^T\hat{\mathbf{R}}_{xx}^{-1}(k)\mathbf{C}^*]^{-1}\mathbf{F} \tag{5.10.9}$$

Other recursive algorithms based upon the constrained minimum-power criterion have been developed [24].

The *Baird recursive least-squares algorithm* results from implementing (5.5.76) recursively [8]. The matrix $\mathbf{X}_s(k+1)$ and the vector $\mathbf{D}(k+1)$ defined by (5.5.73) can be partitioned as

$$\mathbf{X}_s(k+1) = \begin{bmatrix} \mathbf{X}_s(k) \\ \hline \mathbf{X}^T(k+1) \end{bmatrix} \tag{5.10.10}$$

$$\mathbf{D}(k+1) = \begin{bmatrix} \mathbf{D}(k) \\ \hline d(k+1) \end{bmatrix} \tag{5.10.11}$$

where $k \geqslant 1$. The matrix $\mathbf{A}(k+1)$ defined by (5.5.75) can be partitioned as

$$\mathbf{A}(k+1) = \left[\begin{array}{c|c} & 0 \\ & \vdots \\ \alpha\mathbf{A}(k) & \vdots \\ & 0 \\ \hline 0 \; . \; . \; . \; 0 & 1 \end{array}\right] \tag{5.10.12}$$

We define

$$\mathbf{P}^{-1}(k) = \mathbf{X}_s^H(k)\mathbf{A}(k)\mathbf{X}_s(k), \qquad k \geqslant 1 \tag{5.10.13}$$

which may be regarded as a generalization of the sample covariance matrix, as indicated by (5.5.77).

A straightforward calculation using (5.10.10) to (5.10.13) yields

$$\mathbf{P}^{-1}(k+1) = \alpha[\mathbf{P}^{-1}(k) + \frac{1}{\alpha}\mathbf{X}^*(k+1)\mathbf{X}^T(k+1)], \qquad k \geqslant 1 \tag{5.10.14}$$

Applying the matrix inversion identity of (5.5.25) gives

$$\mathbf{P}(k+1) = \frac{1}{\alpha}\left[\mathbf{P}(k) - \frac{\mathbf{P}(k)\mathbf{X}^*(k+1)\mathbf{X}^T(k+1)\mathbf{P}(k)}{\alpha + \mathbf{X}^T(k+1)\mathbf{P}(k)\mathbf{X}^*(k+1)}\right], \qquad k \geqslant 1 \tag{5.10.15}$$

Using (5.10.15) and (5.10.10) to (5.10.12), (5.5.76) implies that

$$\mathbf{W}(k+1) = \mathbf{W}(k) + \frac{\mathbf{P}(k)\mathbf{X}^*(k+1)[d(k+1) - \mathbf{W}^T(k)\mathbf{X}(k+1)]}{\alpha + \mathbf{X}^T(k+1)\mathbf{P}(k)\mathbf{X}^*(k+1)}, \quad k \geqslant 1 \quad (5.10.16)$$

Equations (5.10.15) and (5.10.16) define the Baird recursive least-squares algorithm. Equations (5.10.13) and (5.5.76) indicate that suitable initial values for the algorithm are given by

$$\mathbf{P}(1) = [\mathbf{X}^*(1)\mathbf{X}^T(1)]^{-1}, \quad \mathbf{W}(1) = \mathbf{P}(1)\mathbf{X}^*(1)d(1) \qquad (5.10.17)$$

provided that the indicated inverse exists. Another way to begin the algorithm, which avoids the computation of an inverse, is to specify $\mathbf{P}(0)$ and $\mathbf{W}(0)$ and then compute $\mathbf{P}(1)$ and $\mathbf{W}(1)$ by (5.10.15) and (5.10.16). The initial weight vector $\mathbf{W}(0)$ is arbitrary. To ensure that $\mathbf{P}(k)$ is Hermitian, as required by (5.10.13), the initial matrix $\mathbf{P}(0)$ must be Hermitian. Thus, suitable choices are

$$\mathbf{W}(0) = [1 \;\; 0 \;\; 0 \;\; \ldots \;\; 0]^T, \qquad \mathbf{P}(0) = \mathbf{I} \qquad (5.10.18)$$

If more information about the environment and the desired signal is available, recursive algorithms related to *Kalman filtering* become attractive. The Baird algorithm can be regarded as a special case of a more general algorithm based on Kalman filtering [8, 22].

For both direct matrix inversion and recursive algorithms, the weights are not functions of the adaptive array output. The absence of feedback precludes adaptive compensation for systematic errors in the weights and the circuit elements.

5.10.4 Perturbation Algorithms

Perturbation algorithms optimize the weights by trial and error. They offer simplicity in implementation but slow convergence speeds. Each iteration of these algorithms usually consists of two phases. During the first phase, the gradient of the performance measure is estimated by temporarily perturbing the weight vector and measuring the resulting performance. During the second phase, the weight vector is updated by using the gradient estimate, and the output of the adaptive system is passed to the next receiver stage. If it is not feasible to interrupt the data processing during the perturbation phase, then either a degraded output must be tolerated, or two parallel sets of weights must be used. The perturbations may be random or a sequence of deterministic vectors. If they are random, the perturbation algorithm is called a *random search.*

An adaptive system that uses a perturbation algorithm has one of the forms of Figure 5.5, but the inputs are not applied to the weight-adjustment mechanism.

Because the inputs are not needed in the iterative computation of the weights, a major simplification in the system implementation is usually possible.

The *linear random-search algorithm* [25] entails temporarily adding a small random change to the weight vector at the beginning of each iteration. The corresponding change in an estimate of the performance measure is observed. Then the weight vector is permanently changed according to

$$\mathbf{W}(k+1) = \mathbf{W}(k) + \beta\left\{\hat{P}[\mathbf{W}(k)] - \hat{P}[\mathbf{W}(k) + \mathbf{U}(k)]\right\} \mathbf{U}(k) \tag{5.10.19}$$

where $\mathbf{U}(k)$ is a random vector designed to have covariance $\sigma^2\mathbf{I}$, β and σ^2 are constants affecting the stability and convergence speed, and $\hat{P}[\mathbf{W}(k)]$ and $\hat{P}[\mathbf{W}(k) + \mathbf{U}(k)]$ are estimates of the performance measured based on one or more samples of the error with $\mathbf{W} = \mathbf{W}(k)$ and $\mathbf{W} = \mathbf{W}(k) + \mathbf{U}(k)$, respectively. The algorithm is called "linear" because the weight change is proportional to the change in the estimated performance measure.

If the mean output power is the performance measure and constraints are imposed, then (5.8.6) provides a deterministic algorithm. The factor $\mathbf{R}_{xx}\mathbf{W}(k)$ is the gradient of the mean output power. Replacing this factor, we obtain

$$\mathbf{W}(k+1) = \mathbf{A}[\mathbf{W}(k) - \mu\mathbf{G}(k)] + \mathbf{B} \tag{5.10.20}$$

where $\mathbf{G}(k)$ is an N-dimensional estimate of the gradient of the mean output power and $\mathbf{A}$ and $\mathbf{B}$ are given by (5.8.4) and (5.8.5). Let $P_m(\mathbf{W})$ denote the measured output power when $\mathbf{W}$ is the weight vector. Let $\boldsymbol{\delta}_i$ denote a perturbation vector of the form

$$\boldsymbol{\delta}_i = [0 \;\; 0 \;\; \ldots \;\; 0 \;\; \epsilon \;\; 0 \;\; \ldots \;\; 0]^T, \qquad i = 1, 2, \ldots, N \tag{5.10.21}$$

where the ϵ occupies position i. Equation (5.10.20) provides a perturbation algorithm if component i of $\mathbf{G}(k)$ is [26]

$$G_i(k) = \frac{P_m(\mathbf{W}(k) + \boldsymbol{\delta}_i) - P_m(\mathbf{W}(k) - \boldsymbol{\delta}_i)}{2\epsilon}, \quad i = 1, 2, \ldots, N \tag{5.10.22}$$

The symmetrical perturbations are used to prevent bias in the results. A total of $2N$ samples of the output power are required to determine $\mathbf{G}(k)$. After the gradient is estimated, a single iteration of (5.10.20) is performed, and then gradient estimation begins anew. Other sets of orthogonal perturbation vectors that contain more than the single nonzero component of (5.10.21) are sometimes useful [27].

Use of a perturbation algorithm may be attractive when the performance measure is an unknown function of the weights, is a discontinuous function, or has both local and global optima. In contrast, gradient-based algorithms are suitable for unimodal performance measures, but may yield poor results for multimodal ones.

5.11 ADAPTIVE POLARIZATION DISCRIMINATION

Adaptive antenna systems are potentially useful when the desired signal and interference are distinguishable *a priori.* The discriminant may be the direction of arrival, a waveform characteristic, or signal polarization. *Polarization discrimination* is possible even if the two signals arrive from the same direction provided that the vector of antenna-element voltages produced by an interference signal is not a scalar multiple of the voltage vector produced by the desired signal.

As a simple example, we consider an adaptive system that uses a pair of crossed dipoles, which are depicted in Figure 5.21 [28]. Each dipole provides an output that is proportional to the electric field component along the dipole. This two-element array adapts to polarization alone, but could be part of an array of spatially dispersed crossed-dipole pairs that adapt to both the polarization and the direction of arrival [29]. In this example, we assume that one of the dipoles serves as a primary antenna and the other as an auxiliary antenna. The outputs of the primary and auxiliary antennas provide the primary and auxiliary inputs to an adaptive noise canceller, as shown in Figure 5.22. We assume that the desired signal has a linearly polarized electric field that lies in the plane of the crossed dipoles. The primary antenna is aligned as accurately as possible with the polarization of the desired signal. If the alignment is perfect and the auxiliary antenna is perpendicular to the primary antenna, the desired signal produces no response in the auxiliary antenna. Thus, (5.2.36) implies that, if the thermal noise is negligible, a single linearly polarized interference signal can be completely rejected by the adaptive system. If the antennas are not perpendicular or if the primary antenna and the desired-signal polarization are offset by a known positive angle, then an auxiliary input with no signal component can be produced by subtracting the weighted primary signal from the auxiliary signal, as illustrated in Figure 5.22. In the following analysis, it is assumed that the dipoles are perpendicular and that the optional circuitry is absent.

Interference and thermal noise cause an error in the alignment of the primary antenna with the polarization of the desired signal. Let α denote the unknown polarization angle of the desired signal relative to the primary antenna in the plane of the dipoles, as indicated in Figure 5.21. Suppose that the desired signal $s(t)$, is received along with various interference signals, $I_k(t)$, $k = 1, 2, \ldots$ that are linearly polarized in the plane of the dipoles and have polarization angles β_k relative to the primary antenna. If we neglect the thermal noise, the primary input to the adaptive noise canceller is

$$X_0(t) = s(t) \cos \alpha + \sum_k I_k(t) \cos \beta_k \tag{5.11.1}$$

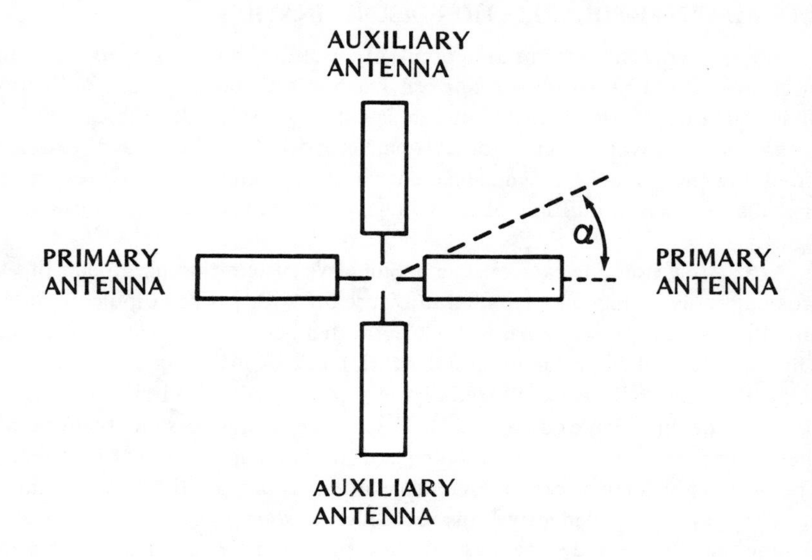

Figure 5.21 Pair of crossed dipoles.

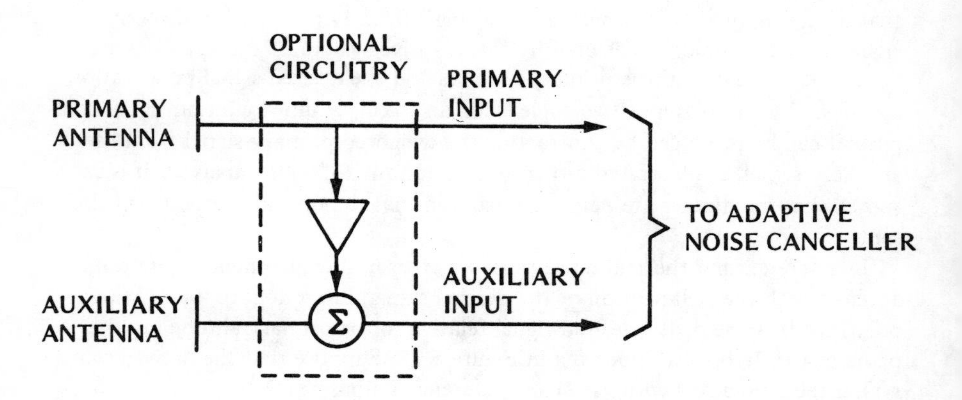

Figure 5.22 Adaptive system for polarization discrimination.

The auxiliary input is

$$X_1(t) = s(t) \sin \alpha + \sum_k I_k(t) \sin \beta_k \tag{5.11.2}$$

The optimal weight vector is a scalar given by (5.9.4). We model $s(t)$ and the $I_k(t)$ as uncorrelated, stationary, zero-mean, stochastic processes. It follows that

$$W_0 = \frac{R_s \sin \alpha \cos \alpha + \sum_k R_{ik} \sin \beta_k \cos \beta_k}{R_s \sin^2\alpha + \sum_k R_{ik} \sin^2\beta_k} \tag{5.11.3}$$

where $R_s = E[s^2(t)]$ and $R_{ik} = E[I_k^2(t)]$. If ideal adaptation occurs, the output of the adaptive noise canceller (Figure 5.18) is

$$\begin{aligned} \epsilon &= X_0 - W_0 X_1 \\ &= s(t)(\cos \alpha - W_0 \sin \alpha) + \sum_k I_k(t)(\cos \beta_k - W_0 \sin \beta_k) \end{aligned} \tag{5.11.4}$$

Thus, the signal-to-interference ratio at the output is

$$\rho_0 = \frac{R_s (\cos \alpha - W_0 \sin \alpha)^2}{\sum_k R_{ik} (\cos \beta_k - W_0 \sin \beta_k)^2} \tag{5.11.5}$$

Substituting (5.11.3) into (5.11.5) and using algebraic and trigonometric simplification, we obtain

$$\rho_0 = \frac{R_s [\sum_k R_{ik} \sin \beta_k \sin(\beta_k - \alpha)]^2}{\sum_n R_{in} [R_s \sin \alpha \sin(\alpha - \beta_n) + \sum_k R_{ik} \sin \beta_k \sin(\beta_k - \beta_n)]^2} \tag{5.11.6}$$

If only one interference signal is received, this equation reduces to

$$\rho_0 = \frac{R_{i1} \sin^2\beta_1}{R_s \sin^2\alpha} \tag{5.11.7}$$

This result, which is similar to what would have been obtained by applying (5.11.2) to (5.2.36), shows explicitly that imperfect knowledge of the desired signal's polarization degrades the interference rejection capability of the adaptive system. If $\alpha < \beta_1$, the interference cancellation is beneficial, but if $\alpha > \beta_1$, the cancellation of the desired signal exceeds that of the interference.

Even if the antenna alignment is perfect, the interference cancellation is usually greatly diminished when two or more interference signals with different polarizations are received because the dipole pair does not have enough degrees of freedom. Setting $\alpha = 0$ in (5.11.6) yields

$$\rho_0 = \frac{R_s\left[\sum_k R_{ik} \sin^2\beta_k\right]^2}{\sum_n R_{in}\left[\sum_k R_{ik} \sin\beta_k \sin(\beta_k - \beta_n)\right]^2} \tag{5.11.8}$$

This equation indicates that complete interference rejection is impossible unless the interference signals have identical polarizations in the plane of the dipoles.

Consider two uncorrelated interference signals with $\beta_1 = \beta_2 + (\pi/2)$ radians and $R_{i1} = R_{i2} = R_{it}/2$, where R_{it} is the total interference power. Equation (5.11.8) becomes

$$\rho_0 = \frac{2R_s}{R_{it}} \tag{5.11.9}$$

Thus, the interference cancellation of the adaptive system is small in this case. Equation (5.11.3) indicates that $W_0 = 0$. Because the array cannot cancel the two uncorrelated interference signals, it turns off the auxiliary dipole, thereby eliminating half of the interference power.

A performance analysis of an adaptive system with three mutually perpendicular dipoles indicates that the system protects an elliptically polarized desired signal from almost any single interference signal unless it both arrives from the same direction and has the same polarization as the desired signal [30]. However, if the desired signal is linearly polarized, the adaptive system provides little protection against an interference signal with parallel polarization, even if it arrives from a different direction. The adaptive system is least susceptible to *cross-polarized jamming,* which consists of two statistically independent signals transmitted on orthogonal polarizations from the same site, if the desired signal is circularly polarized. The array is then effective except when the jamming arrives within a small solid angle around either the desired-signal direction or the direction opposite to the desired signal.

5.12 SPECIAL TYPES OF INTERFERENCE

The signal cancellation phenomenon appearing in (5.2.38) is not restricted to adaptive noise cancellers, but has been observed [31] in applications of the Widrow, Howells-Applebaum, and Frost algorithms. Although tone interference may be largely eliminated by the adaptation, enough power remains in the residual tone to nearly cancel any signal component at the same frequency. Thus, tone interference causes the transfer function for the desired signal to have a notch at the tone frequency. The bandwidth of the notch is proportional to the adaptation constant μ and to the interference power.

Pulsed interference causes an array weight vector to alternately converge toward each of two values that correspond to whether the interference pulse is present or not. If the pulse duration is less than or comparable to the weight time constants, the mean weight vector may never attain its optimal value. An analysis [32] indicates that periodic pulsed interference causes amplitude modulation of the desired signal at the array output and a time-varying *SINR*. If the array is used in a digital communication system, the *SINR* variation increases the bit error probability, but the increase due to periodic pulsed interference relative to continuous interference does not appear to be severe. Furthermore, pulsed interference is significantly more damaging than tone interference only if the pulse rate, width, and power are properly chosen.

A *multipath signal* arrives at an adaptive antenna system from a different direction and at a later time than the associated direct signal that arrives after traversing a direct path. If the delay exceeds the inverse of the signal bandwidth, the multipath signal is largely decorrelated from the direct signal. Consequently, a multipath signal of sufficient power, whether associated with the desired signal or with interference, may act as an independent source of interference. A multipath interference signal entering the main beam of the primary antenna of a sidelobe canceller can be more damaging than the associated direct interference signal that enters a sidelobe. An adaptive tapped-delay-line filter can compensate for some of the degradation due to a multipath signal with a delay less than the inverse of the signal bandwidth, especially if the intertap delay is less than or approximately equal to the multipath delay [22].

5.13 ADAPTIVE ANTENNA SYSTEMS AND FREQUENCY HOPPING

Adaptive antenna processing and frequency hopping are two of the most powerful methods for interference rejection. Because the total hopping bandwidth is usually large, frequency compensation is required if the two methods are to be efficiently combined. Three potentially effective frequency-compensation methods are parameter-dependent processing, spectral processing, and anticipative processing.

Parameter-dependent processing uses an adaptive filter behind each antenna element. Each adaptive filter has enough adjustable parameters to allow the formation of nulls in the direction of the sources of interference for all frequencies with significant interference power. Ideal parameter-dependent processing for a two-element array is analyzed in section 5.3.

If each adaptive filter is a tapped delay line with uniformly spaced taps, and the total hopping bandwidth to be accommodated is W, then the time delay between taps must be approximately $1/W$ or less. The total delay of the line must exceed the inverse of the frequency resolution desired in the filter transfer function after adaptation [22]. Therefore, many taps or other adaptive-filter parameters may be necessary for frequency-hopping systems, especially if the interference may occupy only a small and possibly non-contiguous part of the total hopping bandwidth.

For parameter-dependent processing to form an appropriate spectral response over the total hopping bandwidth at the maximum convergence rate, it appears that the entire bandwidth should be monitored continuously. Therefore, it is probably desirable that the dehopping follow, rather than precede, the adaptive filtering and the extraction of any feedback signal to control the adaptive weight adjustment.

Figure 5.23 shows the form of a parameter-dependent adaptive system using the Widrow LMS algorithm. Following frequency-hopping code synchronization, the reference signal is produced as an approximate replica of the desired frequency-hopping signal. The error signal controls the adaptive weight adjustment.

Figure 5.24 depicts a hypothetical system in the configuration of a sidelobe canceller. The primary input, which provides the reference signal, may be derived from a single antenna, a group of phased-array antenna elements, or an entire antenna array. The system has the advantage that a reference signal including the data modulation is available and does not have to be generated internally. However, the primary input may require monitoring to verify that the desired signal is present.

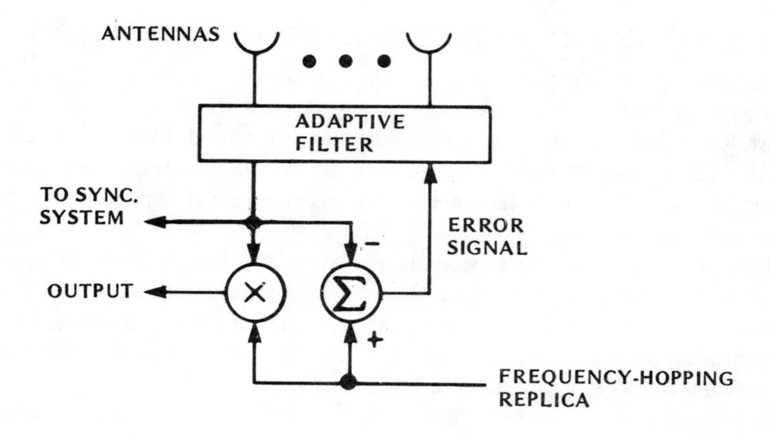

Figure 5.23 Adaptive system using Widrow LMS algorithm.

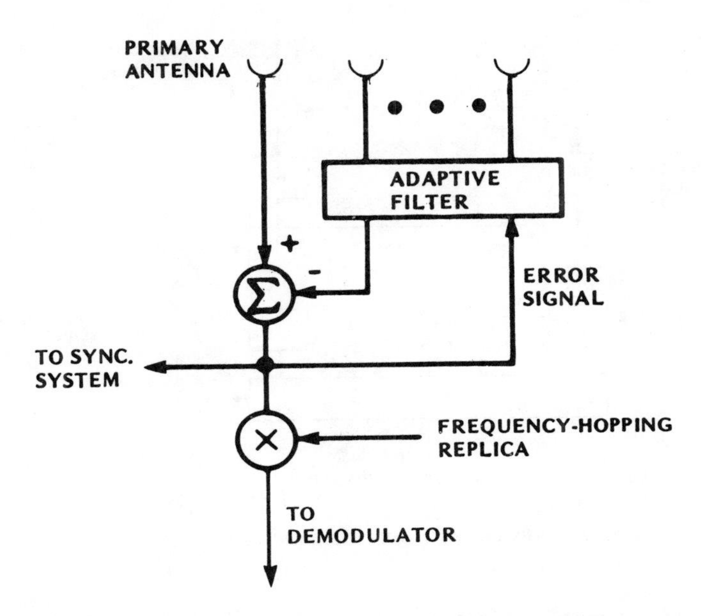

Figure 5.24 Sidelobe canceller for frequency-hopping communications.

Spectral processing is based upon dividing the total hopping bandwidth into a number of spectral regions and adapting independently when the hopping frequency is in one of the regions. Figure 5.25 shows a hypothetical system with a separate adaptive filter for each of the ν spectral regions. Following code synchronization, the receiver-generated frequency-hopping code controls switches that route the dehopped signals through the appropriate adaptive filter associated with the hopping frequency being used. The dehopping converts the frequency-hopping signals to a single intermediate frequency, but information essential to the adaptation is preserved in the phase angles of the dehopped signals. Each of the adaptive filters forms nulls against interference within a fixed region of the total hopping band. An advantage of spectral processing relative to parameter-dependent processing is that the dehopping and bandpass filtering eliminate most of the noise and potential interference before they can enter the adaptive processor.

A practical implementation of spectral processing is depicted in Figure 5.26. The weights associated with each spectral region are stored in a memory. At the end of the signal time interval at a specific frequency, the weights of the adaptive filter are transferred to the memory. Then the weights associated with the new hopping frequency are transferred from the memory to the filter. These weights

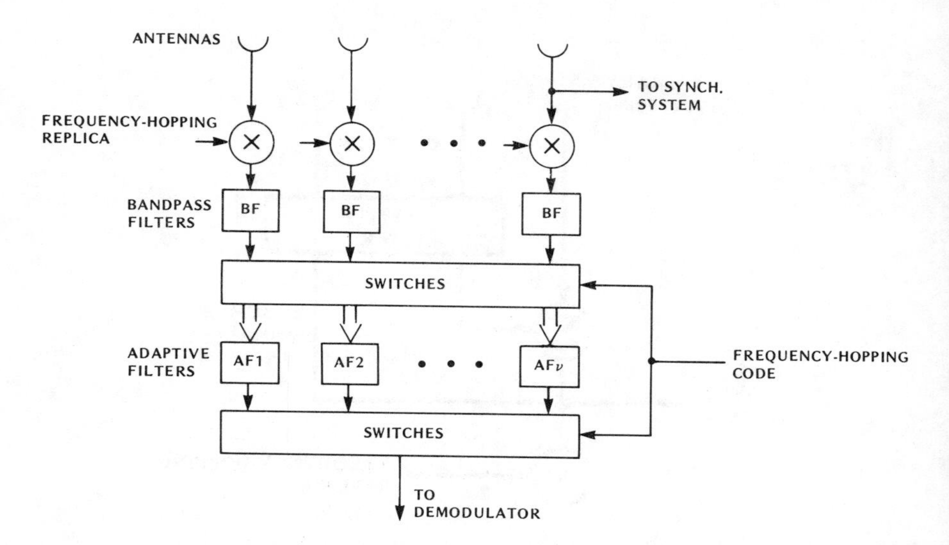

Figure 5.25 Adaptive system with spectral processing.

are updated during the time interval at the new frequency. Transfers between the memory and the filter are controlled by the frequency-hopping code.

An *anticipative adaptive system* begins adaptation toward the optimal weights for a hopping frequency before that frequency is transmitted. Parallel anticipative processing is illustrated by Figure 5.27. A time-advanced frequency-hopping replica hops approximately one hopping period ahead of the replica used for dehopping the desired signal. While the main adaptive filter produces the output, the auxiliary filter adapts the weights corresponding to the next hopping frequency. After each hop, the weights associated with the new hopping frequency are transferred from the auxiliary filter to the main filter. Transfers are controlled by the frequency-hopping code.

If the frequency synthesizer in the receiver produces signals with less switching time than the synthesizer in the transmitter, the differential switching time can be used for serial anticipative processing. For effective interference nulling, the convergence time of the adaptive weights must be less than the differential switching time. When a frequency-hopping pulse appears, the convergence speed may be decreased to reduce the residual output noise.

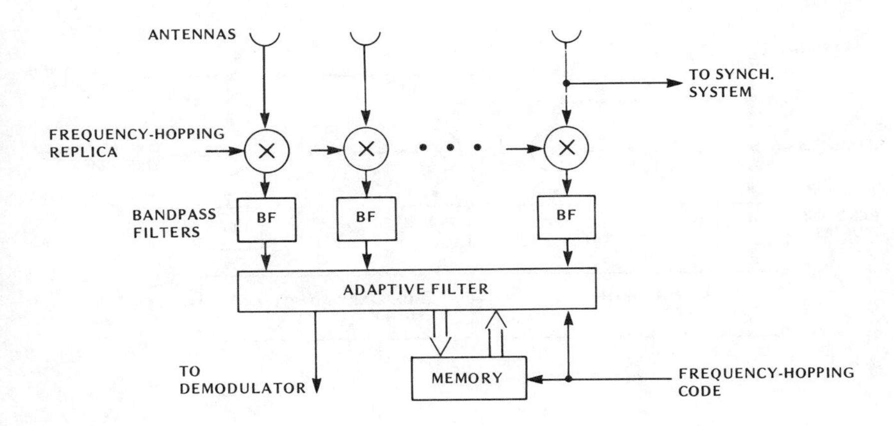

Figure 5.26 Adaptive system with memory.

Combined anticipative and spectral processing can be implemented by adding memory to the anticipative system of Figure 5.27, by exploiting the switching time while using the spectral processor of Figure 5.25 or Figure 5.26, or by combining the systems of Figures 5.25 and 5.27.

The Howells-Applebaum algorithm and other algorithms entail the use of a steering vector. Because the steering vector is frequency dependent, a set of steering vectors have to be stored and controlled by the frequency-hopping code. Because the Howells-Applebaum algorithm is highly sensitive to errors in the stored vectors, it may require as many stored vectors as the number of frequency channels for hopping. However, if the algorithm is preceded by steering delays, which are depicted in Figures 5.6 and 5.7, then the components of the steering vector are all equal to unity regardless of the frequency. The Frost algorithm in its usual configuration uses steering delays.

The Widrow algorithm and other algorithms require a reference signal, which may sometimes be generated from the array output. Because of the complexity of the data-modulated frequency-hopping waveform, the sidelobe canceller, which uses the primary antenna output as the reference signal, may well provide the most practical configuration for using these algorithms.

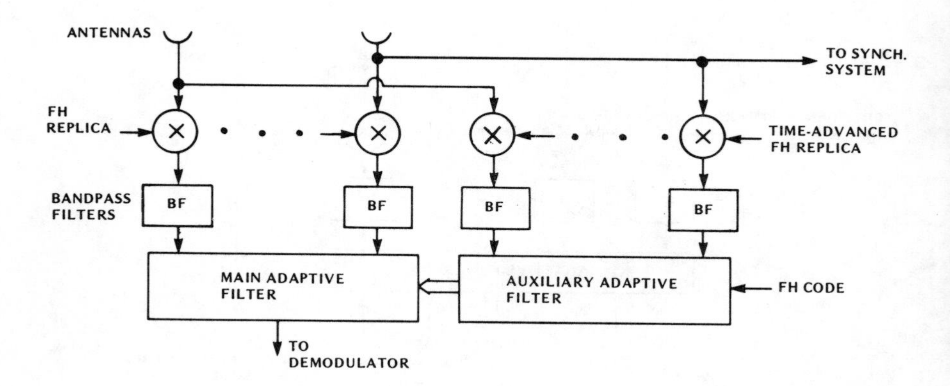

Figure 5.27 Anticipative adaptive system.

Anticipative processing allows the adaptive antenna system to observe the interference and noise without the simultaneous presence of the desired signal. This feature is beneficial for the Howells-Applebaum algorithm, but may cause problems for the Widrow algorithm.

In anticipative processing, convergence of the weights begins again following every signal pulse. Consequently, the convergence rate must be faster than the hopping rate. The rapid adaptation required by anticipative processing may make direct matrix inversion or a recursive algorithm preferable to the Widrow algorithm. The requirements of the convergence rate can be greatly alleviated by using combined anticipative and spectral processing.

An alternative to using one of the standard adaptive algorithms is to use one that is specifically designed to exploit the characteristics of the frequency-hopping waveform. The *Maximin algorithm* [33] discriminates between a frequency-hopping signal and any interference on the basis of the spectral characteristics. The algorithm requires neither a steering vector nor a reference signal.

REFERENCES

1. A. Ishide and R.T. Compton, "On Grating Nulls in Adaptive Arrays," *IEEE Trans. Antennas Propag.* AP-28, 467, July 1980.
2. W.E. Rodgers and R.T. Compton, "Adaptive Array Bandwidth with Tapped Delay-Line Processing," *IEEE Trans. Aerosp. Electr. Syst.* AES-15, 21, January 1979.
3. B. Widrow, *et al.*, "Adaptive Antenna Systems," *Proc. IEEE* 55, 2143, December 1967.
4. D.H. Brandwood, "A Complex Gradient Operator and its Application in Adaptive Array Theory," *IEE Proc.* 130, Pts. F and H, 11, February 1983.
5. L.E. Brennan and I.S. Reed, "Theory of Adaptive Radar," *IEEE Trans. Aerosp. Electr. Syst.* AES-9, 237, March 1973.
6. I.J. Gupta and A.A. Ksienski, "Effect of Mutual Coupling on the Performance of Adaptive Arrays," *IEEE Trans. Antennas Propag.* AP-31, 785, September 1983.
7. O.L. Frost, "An Algorithm for Linearly Constrained Adaptive Array Processing," *Proc. IEEE* 60, 926, August 1972.
8. C.A. Baird, "Recursive Algorithms for Adaptive Array Antennas," Rome Air Development Center RADC-TR-74-46, Nat. Tech. Inf. Serv. AD-778-947, 1974.
9. O. Macchi and E. Eweda, "Second-Order Convergence Analysis of Stochastic Adaptive Linear Filtering," *IEEE Trans. Automat. Contr.* AC-28, 76, January 1983.
10. R.T. Compton, "An Adaptive Array in a Spread-Spectrum Communication System," *Proc. IEEE* 66, 289, March 1978.
11. D.M. DiCarlo, "Reference Loop Phase Shift in an N-Element Adaptive Array," *IEEE Trans. Aerosp. Electr. Syst.* AES-15, 576, July 1979.
12. Y. Bar-Ness, "Eliminating Reference Loop Phase Shift in Adaptive Arrays," *IEEE Trans. Aerosp. Electr. Syst.* AES-18, 115, January 1982.
13. B. Widrow, *et al.*, "Stationary and Nonstationary Learning Characteristics of the LMS Adaptive Filter," *Proc. IEEE* 64, 1151, August 1976.
14. L.L. Horowitz and K.D. Senne, "Performance Advantage of Complex LMS for Controlling Narrow-Band Adaptive Arrays," *IEEE Trans. Acoust., Speech, Signal Proc.* ASSP-29, 722, June 1981.
15. S.P. Applebaum, "Adaptive Arrays," *IEEE Trans. Antennas Propag.* AP-24, 585, September 1976.
16. L.J. Griffiths, "A Simple Adaptive Algorithm for Real-Time Processing in Antenna Arrays," *Proc. IEEE* 57, 1696, October 1969.
17. E. Brennan, E.L. Pugh, and I.S. Reed, *et al.*, "Control Loop Noise in Adaptive Array Antennas," *IEEE Trans. Aerosp. Electr. Syst.* AES-7, 254, March 1971.

18. A.J. Berni, "Weight Jitter Phenomena in Adaptive Array Control Loops," *IEEE Trans. Aerosp. Electr. Syst.* AES-13, 355, July 1977.
19. R.T. Compton, "The Effect of Random Steering Vector Errors in the Applebaum Adaptive Array," *IEEE Aerosp. Electr. Syst.* AES-18, 392, September 1982.
20. J.E. Hudson, *Adaptive Array Principles.* London: Peter Peregrinus Ltd., 1981.
21. B. Widrow, *et al.,* "Adaptive Noise Cancelling: Principles and Applications," *Proc. IEEE* 63, 1692, December 1975.
22. R.A. Monzingo and T.W. Miller, *Introduction to Adaptive Arrays.* New York: John Wiley and Sons, 1980.
23. I.S. Reed, J.D. Mallett, and L.E. Brennan, "Rapid Convergence Rates in Adaptive Arrays," *IEEE Trans. Aerosp. Electr. Syst.* AES-10, 853, November 1974.
24. A. Cantoni and L.C. Godara, "Fast Algorithms for Time Domain Broadband Adaptive Array Processing," *IEEE Trans. Aerosp. Electr. Syst.* AES-18, 682, September 1982.
25. B. Widrow and J.M. McCool, "A Comparison of Adaptive Algorithms Based on the Methods of Steepest Descent and Random Search," *IEEE Trans. Antennas Propag.* AP-24, 615, September 1976.
26. J.E. Hudson, A.S. Ratner, and A.C. Cantoni, "Coefficient Perturbation Adaptive HF Array," *IEE Proc.* 130, Pts. F and H, 91, February 1983.
27. L.C. Godara and A. Cantoni, "Analysis of the Performance of Adaptive Beam Forming Using Perturbation Sequences," *IEEE Trans. Antennas Propag.* AP-31, 268, March 1983.
28. B.D. Steinberg, *Principles of Aperture and Array System Design.* New York: John Wiley and Sons, 1976.
29. R.T. Compton, "On the Performance of a Polarization Sensitive Adaptive Array," *IEEE Trans. Antennas Propag.* AP-29, 718, September 1981.
30. R.T. Compton, "The Performance of a Tripole Adaptive Array Against Cross-Polarized Jamming," *IEEE Trans. Antennas Propag.* AP-31, 682, July 1983.
31. B. Widrow, *et al.,* "Signal Cancellation Phenomena in Adaptive Antennas: Causes and Cures," *IEEE Trans. Antennas Propag.* AP-30, 469, May 1982.
32. R.T. Compton, "The Effect of a Pulsed Interference Signal on an Adaptive Array," *IEEE Trans. Aerosp. Electr. Syst.* AES-18, 297, May 1982.
33. K. Bakhru and D.J. Torrieri, "The Maximin Algorithm for Adaptive Arrays and Frequency-Hopping Communications," *IEEE Trans. Antennas Propag.* AP-32, September 1984.

Chapter 6
Cryptographic Communications

6.1 DIGITAL CIPHERS AND CRYPTANALYSIS

Cryptography is the set of techniques used to ensure the secrecy of messages when hostile personnel have the technical capability to intercept and correctly interpret an unprotected message, which is called the *plaintext.* After transformation to secret form, a message is called the *ciphertext* or a *cryptogram.* The process of transforming the plaintext into ciphertext is called *encryption. Cryptanalysis* is the unauthorized extraction of information from the ciphertext.

A practical cryptographic communication system must produce ciphertext that not only resists cryptanalysis, but also can be efficiently deciphered by authorized personnel. These requirements have been more fully achieved by digital encryption and the transmission of a digital waveform than by analog encryption and transmission. Thus, if secure voice or other analog information is to be communicated at a high level of security, analog-to-digital conversion and subsequent digital operations are desirable.

In a cryptographic system, a set of parameters that determines a specific cryptographic transformation or its inverse is called a *key*. In most applications, a single key determines both a cryptographic transformation and its inverse. Cryptographic systems are designed so that their security depends upon the inability of the cryptanalyst to determine the key even if he knows the structure of the cryptographic system.

There are two distinct types of digital encryption: encoding and enciphering. *Encoding* consists of the substitution of groups of bits of variable length for plaintext groups of variable length. *Enciphering* consists of the substitution of fixed-length groups of bits for fixed-length plaintext groups. Encoding is difficult to automate and the frequent code changes necessary for secrecy are often difficult to execute. Because enciphering systems are easily automated and modified, they are nearly always used.

An enciphering algorithm or transformation is called a *cipher.* Messages can be enciphered by a block cipher, a stream cipher, or a combination of these ciphers. A *block cipher* divides the plaintext into separate blocks of m bits and associates with each block $(x_i, x_{i+1}, \ldots, x_{i+m-1})$ of plaintext a block $(y_i, y_{i+1}, \ldots, y_{i+n-1})$ of n bits of ciphertext such that

$$y_k = f_k(x_i, x_{i+1}, \ldots, x_{i+m-1}), \quad i \leqslant k \leqslant i+n-1 \tag{6.1.1}$$

where the f_k are functions. Messages are deciphered with inverse functions. For unambiguous deciphering, $n \geqslant m$ is necessary; for ease of automation, $n = m$ in nearly all practical block ciphers.

A *stream cipher* performs bit-by-bit modulo-two addition of the plaintext and a set of bits called the *keystream.* The keystream may be a function of the plaintext itself. Thus, a stream cipher associates with each stream $(x_1, x_2, \ldots, x_i, x_{i+1}, \ldots)$ of plaintext a stream $(y_1, y_2, \ldots, y_i, y_{i+1}, \ldots)$ of ciphertext such that

$$y_k = \begin{cases} x_k + f_k(x_{k-1}, x_{k-2}, \ldots, x_{k-n}), & k \geqslant n \\ x_k + f_k(x_{k-1}, x_{k-2}, \ldots, x_1), & k < n \end{cases} \tag{6.1.2}$$

In addition, y_k may be a function of the initial conditions of the cryptographic system. Modulo-two addition is used because $a + b = c$ implies $a = b + c$. Thus, ciphertext is deciphered by adding the keystream bits to the corresponding enciphered bits. The inverse of f_k is not needed and may not exist despite the one-to-one relation between the plaintext and the ciphertext.

A *known-plaintext attack* is an attempted cryptanalysis based upon some knowledge of corresponding ciphertext and plaintext. An *unconditionally secure cipher* is one that is invulnerable to cryptanalysis no matter how much ciphertext is available. A *computationally secure cipher* is one that is vulnerable to cryptanalysis only if a prohibitively large amount of computation is performed. Most practical ciphers are at best computationally secure.

6.1.1 Block Ciphers

Block ciphers are basically digital substitution ciphers. Hence, they are potentially susceptible to classical frequency analysis of the blocks. Security requires the use of large blocks that include at least four data words. The generic form of a system for enciphering or deciphering blocks is shown in Figure 6.1.

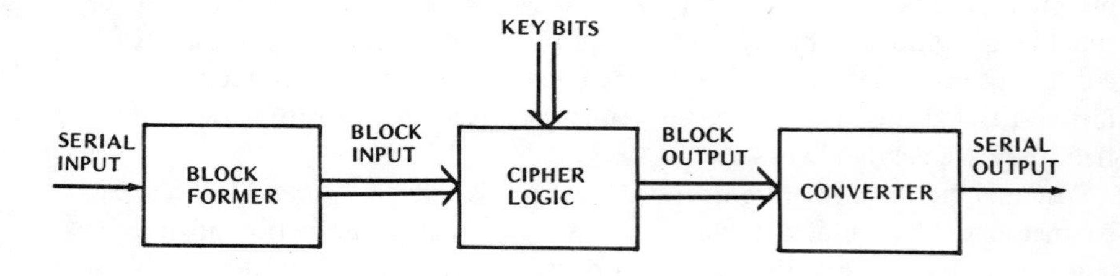

Figure 6.1 Generic form of system for enciphering or deciphering blocks.

To ensure an integral number of blocks in a message, *padding bits* may have to be appended to the final message bits to form a final block of the standard size. The transmitted bits may include bits that indicate the number of padding bits. When the presence of padding bits makes the final block more vulnerable to cryptanalysis, it may sometimes be desirable to encipher these bits with a stream cipher.

A practical difficulty of secure block ciphers is the large number of key bits required to specify the block transformations. If the plaintext and ciphertext blocks are n bits long, there are 2^n distinct blocks and $2^n!$ possible block ciphers for which inverses exist. The number of key bits needed to specify a particular block cipher is $[\log_2(2^n!)] + 1$, where $[x]$ is the largest integer less than or equal to x, and is approximately equal to $n2^n$. To reduce the number of required key bits, one must use a fixed form of cipher logic that restricts the block transformations that are possible. As an example, Figure 6.2 illustrates a specific block enciphering system [1]. The plaintext block is transformed by successive stages

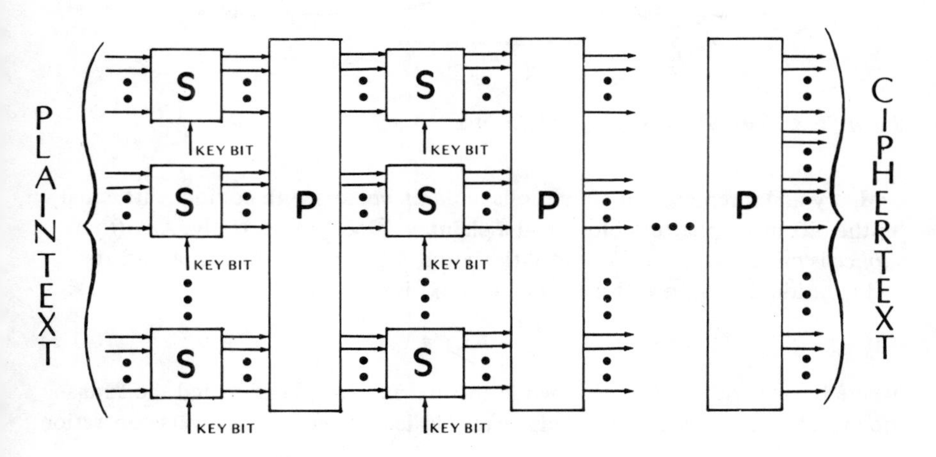

Figure 6.2 Block enciphering system with alternating stages.

that alternate between a set of substitution (S) boxes and a single permutation (P) box. A *substitution box* consists of digital logic and produces output bits that are Boolean functions of the input bits. A key bit determines which one of two possible transformations each substitution box performs. A *permutation box* rearranges or transposes its input bits as illustrated in Figure 6.3. If the permutation boxes are not defined by the key, the number of required bits for this cipher is equal to the total number of substitution boxes. Deciphering is accomplished by running the data in reverse order through the inverses of the boxes. The permutation boxes ensure the cryptographic complexity of the cipher by preventing the overall transformation from being equivalent to a single stage of substitution boxes. A practical constraint on the number of stages in a block-cipher system is the time available for enciphering or deciphering each block.

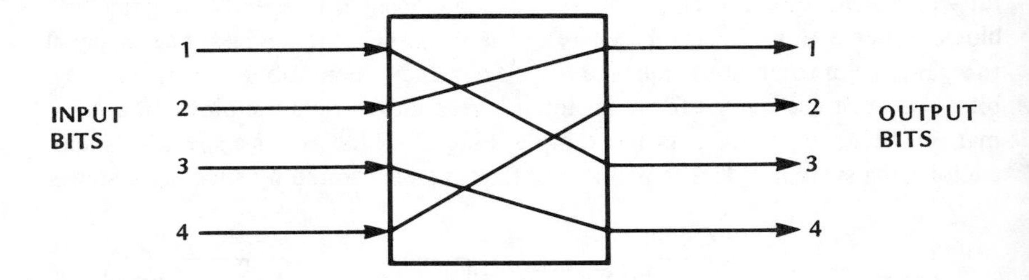

Figure 6.3 Permutation box for four bits.

Let **y** and **x** represent n-dimensional column vectors with components equal to the bits in a ciphertext block and a plaintext block, respectively. An *affine cipher* is one in which the key specifies a nonsingular $n \times n$ matrix **A** and an n-dimensional column vector **t**. The plaintext is enciphered using

$$\mathbf{y} = \mathbf{A}\mathbf{x} + \mathbf{t} \tag{6.1.3}$$

where binary arithmetic is assumed in the matrix multiplication and the addition (Section 2.2). The received ciphertext is deciphered by the inverse operation:

$$\mathbf{x} = \mathbf{A}^{-1}(\mathbf{y} + \mathbf{t}) \tag{6.1.4}$$

where modulo-two arithmetic is used in computing the inverse matrix. If $\mathbf{A} = \mathbf{I}$, the block cipher is a translation cipher; if $\mathbf{t} = 0$ the block cipher is a linear cipher. It is easily verified that the product of successive affine transformations is affine.

Affine block ciphers would be extremely convenient, but they are highly susceptible to cryptanalysis. Only $n + 1$ pairs of ciphertext blocks and corresponding plaintext blocks are needed to enable the cryptanalysis of all messages. Suppose that $\mathbf{x}_i$ and $\mathbf{y}_i$, $i = 0, 1, \ldots, n$, are the corresponding pairs known to the cryptanalyst. Equation (6.1.3) gives

$$\mathbf{y}_i = \mathbf{A}\,\mathbf{x}_i + \mathbf{t}, \quad i = 0, 1, \ldots, n \tag{6.1.5}$$

Therefore,

$$\mathbf{y}_i + \mathbf{y}_0 = \mathbf{A}(\mathbf{x}_i + \mathbf{x}_0)\,, \quad i = 1, 2, \ldots, n \tag{6.1.6}$$

We define the $n \times n$ matrices

$$\mathbf{Y} = [\mathbf{y}_1 + \mathbf{y}_0 \quad \mathbf{y}_2 + \mathbf{y}_0 \quad \cdots \quad \mathbf{y}_n + \mathbf{y}_0] \tag{6.1.7}$$

$$\mathbf{X} = [\mathbf{x}_1 + \mathbf{x}_0 \quad \mathbf{x}_2 + \mathbf{x}_0 \quad \cdots \quad \mathbf{x}_n + \mathbf{x}_0] \tag{6.1.8}$$

Equations (6.1.6) to (6.1.8) imply that

$$\mathbf{Y} = \mathbf{A}\,\mathbf{X} \tag{6.1.9}$$

Consequently, if **X** is invertible, **A** can be determined from

$$\mathbf{A} = \mathbf{Y}\,\mathbf{X}^{-1} \tag{6.1.10}$$

Subsequently, **t** can be determined from

$$\mathbf{t} = \mathbf{y}_0 + \mathbf{A}\,\mathbf{x}_0 \tag{6.1.11}$$

Once **A** and **t** are known, messages can be cryptanalyzed even though the key is still unknown. If **X** is not invertible, more than $n + 1$ corresponding pairs of ciphertext and plaintext blocks are needed.

6.1.2 Synchronous Ciphers

There are two types of stream ciphers: those for which the keystream is independent of the plaintext and those for which the keystream is not. Independently keyed ciphers are often called *synchronous ciphers* because they require synchronization between the keystream and the ciphertext for successful deciphering. Stream ciphers for which the keystream is a function of the plaintext are often called *auto-key ciphers.* They are also called *self-synchronizing ciphers* because, as explained subsequently, an erroneously added or lost bit causes only a fixed number of errors in the deciphered plaintext, after which correct plaintext is again produced.

The only unconditionally secure cipher is a synchronous cipher that generates a completely random keystream as long as the message. Because the key, which is the same as the keystream in this case, cannot be reused, this cipher is called the *one-time tape* or *one-time pad.* The necessity of long, frequently changed keys renders the one-time tape impractical for most applications. Thus, most synchronous ciphers use a moderate number of key bits and feedback shift registers to generate pseudorandom keystreams with long periods (see Section 2.2). The generic form of synchronous enciphering or deciphering systems is shown in Figure 6.4, where the indicated addition is modulo-two. The output bits can be formed sequentially or in parallel. The number of parallel bits is determined by the arrival rate of the input bits and by the number arriving simultaneously.

Linear feedback shift registers can generate long keystream sequences with desirable randomness properties such as a nearly even balance of the symbols 0 and 1. However, the structure of these linear generators makes them susceptible to cryptanalysis. Consider a keystream produced by a linear feedback shift register. The feedback coefficients are derived from the key. Let

$$\mathbf{c} = [c_1 \ c_2 \ \dots \ c_n]^T \tag{6.1.12}$$

denote the column vector of feedback coefficients. The column vector of successive keystream bits $a_i, a_{i+1}, \dots, a_{i+n-1}$ is denoted by

$$\mathbf{a}_i = [a_i \ a_{i+1} \ \dots \ a_{i+n-1}]^T, \quad i \geqslant 0 \tag{6.1.13}$$

Let $\mathbf{A}(i)$ denote the $n \times n$ matrix with columns consisting of the $\mathbf{a}_j, j = i, i+1, \dots, i+n-1$:

$$\mathbf{A}(i) = \begin{bmatrix} a_{i+n-1} & a_{i+n-2} & \dots & a_i \\ a_{i+n} & a_{i+n-1} & \dots & a_{i+1} \\ \cdot & \cdot & & \cdot \\ \cdot & \cdot & & \cdot \\ \cdot & \cdot & & \cdot \\ a_{i+2n-2} & a_{i+2n-3} & \dots & a_{i+n-1} \end{bmatrix} \tag{6.1.14}$$

From (2.2.6) of Chapter 2, it follows that the keystream bits and feedback coefficients are related by

$$\mathbf{a}_{i+n} = \mathbf{A}(i)\,\mathbf{c}, \quad i \geqslant 0 \tag{6.1.15}$$

Suppose that $2n$ consecutive bits of ciphertext and the equivalent plaintext allow calculation of $2n$ consecutive keystream bits, starting at bit i. Then all elements of $\mathbf{a}_{i+n}$ and $\mathbf{A}(i)$ in (6.1.15) are known. If $\mathbf{A}(i)$ is invertible, then the feedback coefficients are determined from

$$\mathbf{c} = \mathbf{A}^{-1}(i)\,\mathbf{a}_{i+n}, \quad i \geqslant 0 \tag{6.1.16}$$

Any n successive keystream bits determine a state vector of the linear feedback shift register. The keystream or output sequence of the shift register is completely determined by the feedback coefficients and any state vector. Thus, the $2n$ pairs of ciphertext and plain text bits provide enough information for a complete cryptanalysis unless $\mathbf{A}(i)$ is not invertible. In that case, one or more additional pairs are necessary. If it is known that the shift register uses exactly n stages, then $c_n = 1$ is known. However, it is more likely that only an upper bound on the number of stages is known to the cryptanalyst, who must solve for c_n along with the other coefficients.

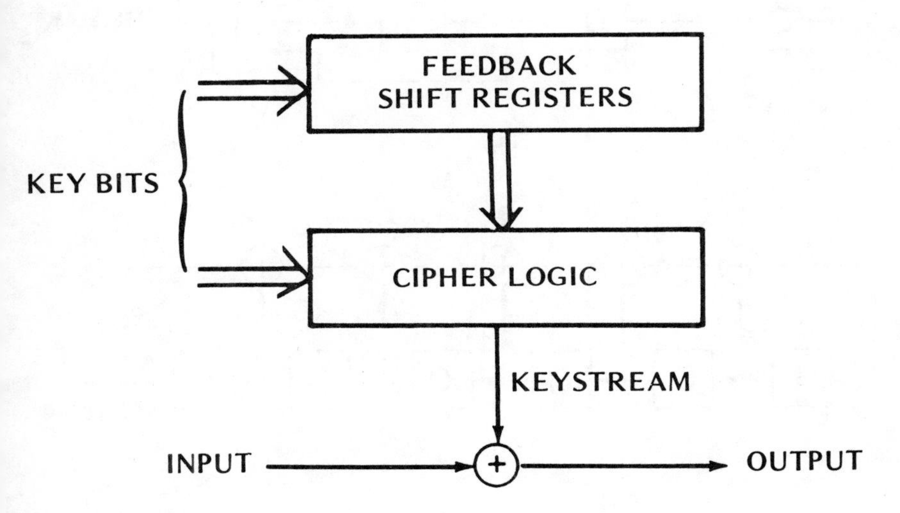

Figure 6.4 Generic form of synchronous enciphering or deciphering system.

Because there are practical limits to the number of shift-register stages, one or more shift registers with nonlinear operations on the outputs or nonlinear feedback are necessary to produce secure keystreams. If a binary sequence has period p, it can always be generated by a p-stage linear shift register by connecting the output of the last stage to the input of the first stage. The *linear equivalent* of the generator of a sequence is the linear shift register with the fewest stages that produces the sequence. For example, it follows from the properties of generating functions and characteristic polynomials (Chapter 2) that the linear equivalent of a Gold code generator using two n-stage registers has $2n$ stages. Algorithms are known for constructing the linear equivalent of the generator of any periodic binary sequence. However, nonlinear generators with relatively few shift-register stages can produce sequences that require an enormous number of stages in the linear equivalent.

Figures 6.5 and 6.6 depict nonlinear keystream generators and their linear equivalents [2]. The initial contents of the shift-register stages are indicated by binary numbers. In Figure 6.5(a), a linear feedback shift register is used, but the output of two stages are nonlinearly combined to produce the keystream. In

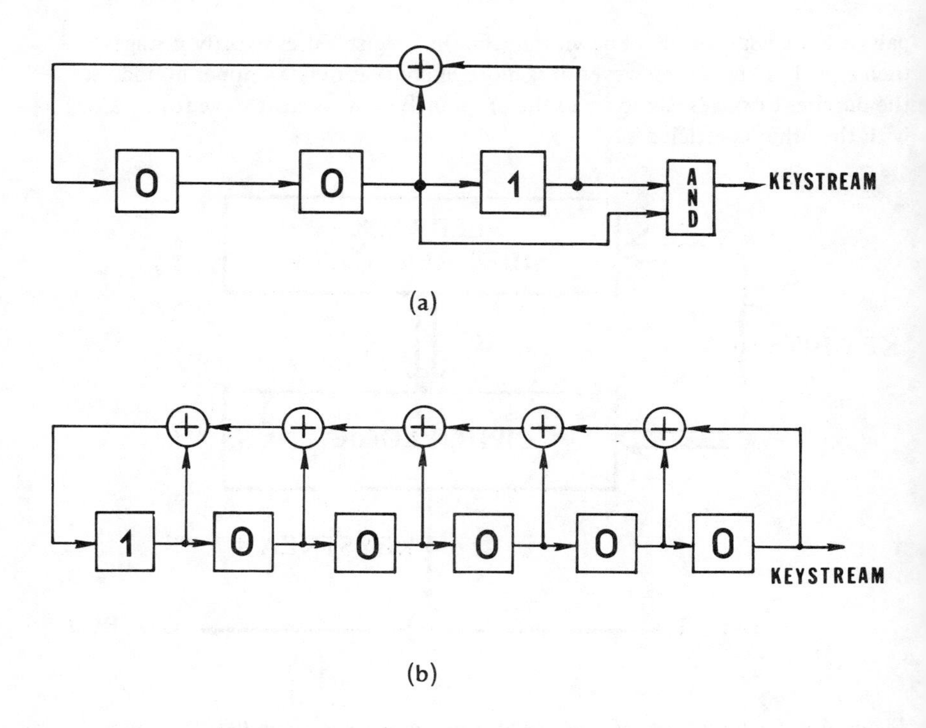

Figure 6.5 (a) Nonlinear generator using single shift register and (b) its linear equivalent.

Figure 6.6(a), two linear feedback shift registers have their outputs nonlinearly combined to produce the keystream.

The susceptibility of a synchronous cipher to cryptanalysis is greatly increased if the same part of the keystream is used to produce two or more different ciphertexts. Consequently, a stream cipher should produce different keystreams at each iteration of the algorithm. When the algorithm and the key are fixed, a variable keystream is produced by varying the initial state of the enciphering and deciphering systems at the start of each message. An n-stage shift register permits 2^n different initial stages. Thus, if n is large, it is unlikely that the same initial state will be often repeated if the initial states are chosen randomly or pseudorandomly. The initial state for each message may be separately generated in the enciphering and the deciphering systems provided that synchronization between the generators is maintained. Alternatively, the initial state may be transmitted in plaintext form to the deciphering system, but an error-correcting code may be required to ensure correct reception.

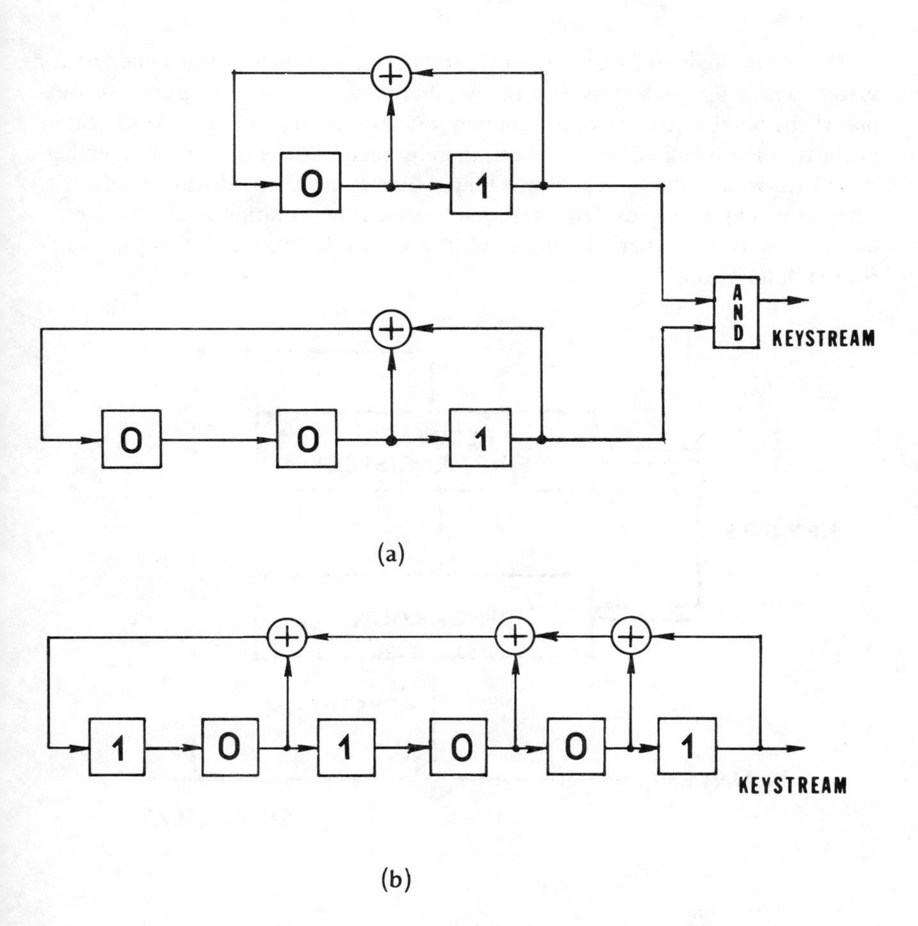

Figure 6.6 (a) Nonlinear generator using two shift registers and (b) its linear equivalent.

6.1.3 Auto-Key Ciphers

The auto-key cipher provides cryptographic strength by diffusing the statistical properties of the plaintext over the ciphertext. The auto-key enciphering system usually employs ciphertext feedback and has the form of Figure 6.7. The corresponding auto-key deciphering system has the form of Figure 6.8. Because of the ciphertext feedback in the enciphering system, the keystream depends upon the ciphertext and, thus, upon the plaintext.

The cipher logic of Figures 6.4, 6.7, and 6.8 may include a block-enciphering system with a block size equal to the number of shift-register outputs. The output of the block-enciphering system may provide several successive keystream bits after each block-cipher iteration, thereby reducing the rate of block-cipher iterations, which usually take much longer than those of the shift registers. If ν keystream bits are provided after each block-cipher iteration in an auto-key enciphering system, then ν ciphertext bits are provided to the shift registers before each iteration.

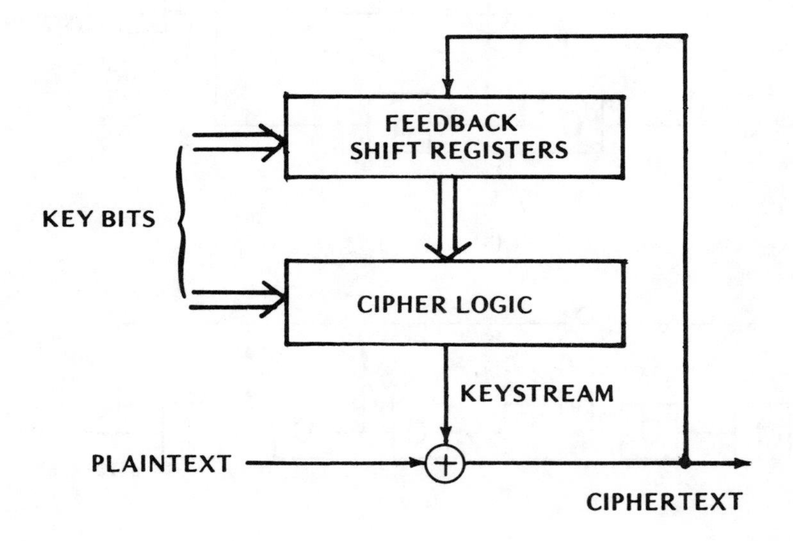

Figure 6.7 Generic form of auto-key enciphering system with ciphertext feedback.

Consider linear auto-key enciphering and deciphering systems with n shift-register stages. Let Z_i denote the initial content of stage i for $i = 1, 2, \ldots, n$. Let $x_0, x_1, x_2, \ldots$ denote the plaintext bits and $y_0, y_1, y_2, \ldots$ denote the corresponding ciphertext bits. The feedback coefficients c_i, $i = 1, 2, \ldots, n$, may take the values 0 or 1, depending on whether the corresponding switches in the enciphering and deciphering systems are open or closed, respectively. After each of the first n outputs of the enciphering system, one of the Z_i is replaced by a ciphertext bit. Thus, for an enciphering system having the form of Figure 6.9,

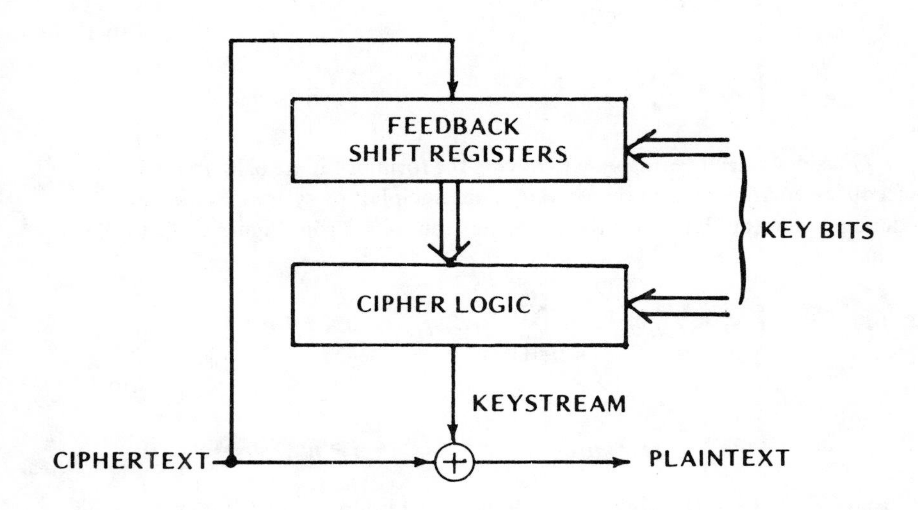

Figure 6.8 Generic form of auto-key deciphering system corresponding to Figure 6.7.

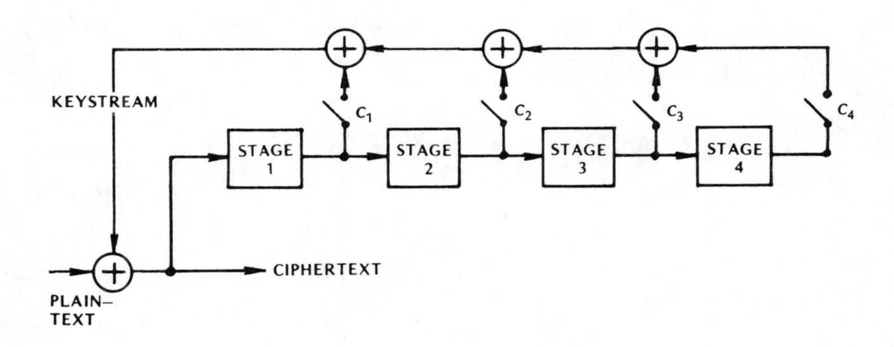

Figure 6.9 Linear auto-key enciphering system.

$$y_i = \begin{cases} x_i + \sum_{j=1}^{i} c_j y_{i-j} + \sum_{j=i+1}^{n} c_j Z_{j-i}\,, & 0 \leqslant i \leqslant n-1 \\ x_i + \sum_{j=1}^{n} c_j y_{i-j}\,, & i \geqslant n \end{cases} \tag{6.1.17}$$

Consider a deciphering system having the form of Figure 6.10 and the same feedback coefficients and initial state as the enciphering system. Let $v_1, v_2, \ldots$ denote the output bits when the y_i are the input bits. From Figure 6.10, it follows that

$$v_i = \begin{cases} y_i + \sum_{j=1}^{i} c_j y_{i-j} + \sum_{j=i+1}^{n} c_j Z_{j-i}, & 0 \leqslant i \leqslant n-1 \\ y_i + \sum_{j=1}^{n} c_j y_{i-j}\,, & i \geqslant n \end{cases} \tag{6.1.18}$$

From (6.1.17) and (6.1.18) and modulo-two addition, we obtain $v_i = x_i$ for all i. Thus, the plaintext is reproduced by the deciphering system.

Let $\mathbf{x}_i$ and $\mathbf{y}_i$ denote the n-dimensional column vectors of plaintext and ciphertext bits:

$$\mathbf{x}_i = [x_i \;\; x_{i+1} \;\ldots\; x_{i+n-1}]^T\,, \qquad i \geqslant 0 \tag{6.1.19}$$

$$\mathbf{y}_i = [y_i \;\; y_{i+1} \;\ldots\; y_{i+n-1}]^T\,, \qquad i \geqslant 0 \tag{6.1.20}$$

Let $\mathbf{Y}(i)$ denote the $n \times n$ matrix

$$\mathbf{Y}(i) = \begin{bmatrix} y_{i+n-1} & y_{i+n-2} & \cdots & y_i \\ y_{i+n} & y_{i+n-1} & \cdots & y_{i+1} \\ \cdot & \cdot & & \cdot \\ \cdot & \cdot & & \cdot \\ \cdot & \cdot & & \cdot \\ y_{i+2n-2} & y_{i+2n-3} & \cdots & y_{i+n-1} \end{bmatrix} \tag{6.1.21}$$

If $\mathbf{c}$ is defined by (6,1.12), then (6.1.17) implies that

$$\mathbf{y}_{i+n} = \mathbf{x}_{i+n} + \mathbf{Y}(i)\mathbf{c}\,, \qquad i \geqslant 0 \tag{6.1.22}$$

If $2n$ pairs of corresponding plaintext and ciphertext bits are known and $\mathbf{Y}(i)$ is invertible, then the feedback coefficients are determined from

$$\mathbf{c} = \mathbf{Y}^{-1}(i)\,(\mathbf{y}_{i+n} + \mathbf{x}_{i+n}), \qquad i \geqslant 0 \tag{6.1.23}$$

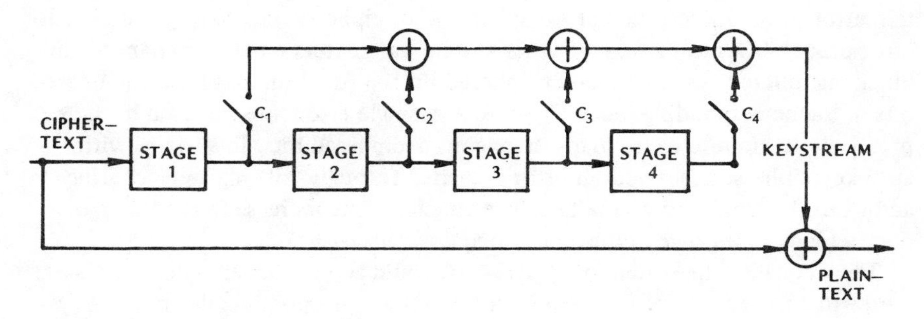

Figure 6.10 Linear auto-key deciphering system.

Additional pairs are necessary if $\mathbf{Y}(i)$ is not invertible. A comparison of (6.1.23) and (6.1.16) indicates that linear auto-key ciphers are no more secure than linear stream ciphers.

As an example, consider the case where $n = 4$ and we acquire the following corresponding sequences of $2n = 8$ plaintext and ciphertext bits:

$$\begin{aligned} &\text{plaintext:}\quad 1\ 0\ 1\ 0\ 1\ 0\ 1\ 0 \\ &\text{ciphertext:}\ 1\ 0\ 1\ 1\ 0\ 0\ 0\ 1 \end{aligned} \tag{6.1.24}$$

Equation (6.1.23) and binary arithmetic yield

$$\begin{bmatrix} c_1 \\ c_2 \\ c_3 \\ c_4 \end{bmatrix} = \begin{bmatrix} 1 & 1 & 1 & 0 \\ 0 & 1 & 1 & 1 \\ 0 & 0 & 1 & 1 \\ 0 & 0 & 0 & 1 \end{bmatrix} \begin{bmatrix} 1 \\ 0 \\ 1 \\ 1 \end{bmatrix} = \begin{bmatrix} 0 \\ 0 \\ 0 \\ 1 \end{bmatrix} \tag{6.1.25}$$

Thus, the key has been completely determined. If it is known that the sequences begin at $i = 0$, then (6.1.17) can be used to determine the initial state. We obtain

$$Z_1 = 1, \quad Z_2 = 0, \quad Z_3 = 0, \quad Z_4 = 0 \tag{6.1.26}$$

The initial contents of the shift register in auto-key systems are irrelevant after the first n clock pulses shift the initial contents out. Consequently, rather than storing or generating the initial contents, it may be preferable for a communicator to transmit n random or pseudorandom bits before the ciphertext corresponding to the message. The receiver ignores the first n deciphered bits, but interprets the subsequent deciphered bits as the message in plaintext.

In any digital communication system, the received bits and words have certain error probabilities. Except for synchronous ciphers, enciphering causes these error probabilities to increase if other system parameters remain unchanged. In block-enciphered systems, each deciphered bit is a function of all the enciphered bits in the corresponding block. Therefore, a single erroneous received bit is practically certain to cause many erroneous deciphered bits. In systems with auto-key ciphers, a received bit error is carried through shift registers, causing additional bit errors down the line. The characteristic increase in the bit errors due to block or auto-key ciphers is called *error propagation.*

Theoretically, the system of Figure 6.10 could serve as an enciphering system, while the system of Figure 6.9 serves as the corresponding deciphering system. However, this choice is not a good one in practice because a single bit error at the deciphering system input would cause an indefinite number of errors at the output. In the original configuration, at most four output bits are affected by a single bit error in the deciphering systems input.

6.1.4 Cipher-Block Chaining

A cryptographic system must protect itself against false ciphertext generated by an opponent to confuse the victim receiver. One type of false ciphertext results from *playback*, which is the retransmission of intercepted cryptographic communications at a later time. Because synchronous ciphers use continually changing keystreams that are independent of the plaintext, false ciphertext is often rendered meaningless, and hence easily identifiable, after decipherment. However, the transmission of numerical data or the use of block or auto-key ciphers requires special protective measures. If the key of an auto-key cipher has not been changed since the interception, the first n receiver output bits due to playback are affected by error propagation, but subsequent output bits constitute an erroneous, but comprehensible, message. An effective countermeasure for stream ciphers is to intersperse the plaintext with the date and time of day or other time-varying authentication information.

This approach may be effective for block ciphers unless an opponent can isolate the specific blocks that contain the authentication information and appropriately alter these blocks before retransmission. To preclude this possibility and similar threats, interdependence between adjacent blocks may be introduced. In *block chaining,* each successive plaintext block contains a number of bits from the previous ciphertext block. A deciphered block is checked to ensure that these authentication bits match the corresponding bits of the previous ciphertext block. The problem with block chaining is that an adequate number of authentication bits may significantly reduce the transmission rate of information bits. To eliminate this problem, *cipher-block chaining* may be used. In this method, a ciphertext block is a function of both the plaintext block and the previous ciphertext block. Let $\mathbf{x}_i$ and $\mathbf{y}_i$ denote the vectors of the ith plaintext and ciphertext blocks, respectively. Enciphering is described by

$$\mathbf{y}_i = \mathbf{f}(\mathbf{x}_i + \mathbf{y}_{i-1}), \qquad i \geq 1 \tag{6.1.27}$$

where $\mathbf{f}(\ )$ is a block cipher that maps one vector into another vector. The first block of the chain, $\mathbf{y}_0$, may consist partly or entirely of random bits. Equation (6.1.27) indicates that cipher-block chaining can be interpreted as a hybrid of a stream cipher and a block cipher. Deciphering is described by the inverse of (6.1.27):

$$\mathbf{x}_i = \mathbf{f}^{-1}(\mathbf{y}_i) + \mathbf{y}_{i-1}, \qquad i \geq 1 \tag{6.1.28}$$

The deciphering system must generate or receive $\mathbf{y}_0$. The forms of the enciphering and deciphering systems are depicted in Figures 6.11 and 6.12. Cipher-block chaining eliminates repetitive patterns in the ciphertext because identical plaintext blocks become different ciphertext blocks.

In addition to error propagation within blocks, there is a limited error propagation between adjacent blocks. For example, suppose there are ν bit errors in a received ciphertext block and no errors in the next one. After deciphering, the first corresponding plaintext block will usually contain many bit errors, but the second plaintext block will contain exactly ν errors.

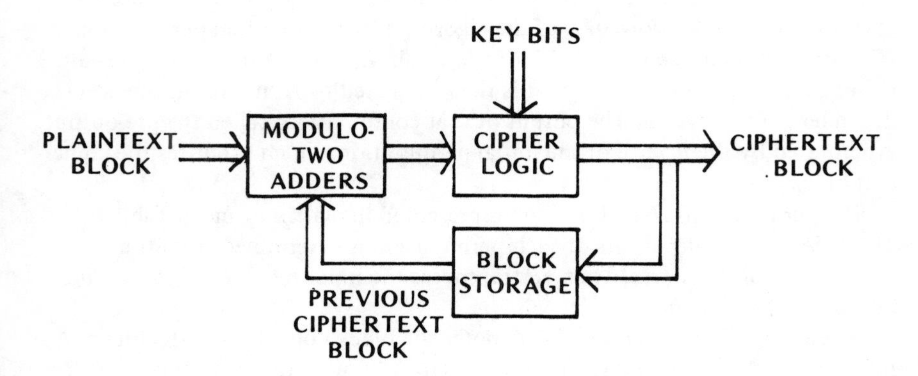

Figure 6.11 Enciphering system for cipher-block chaining.

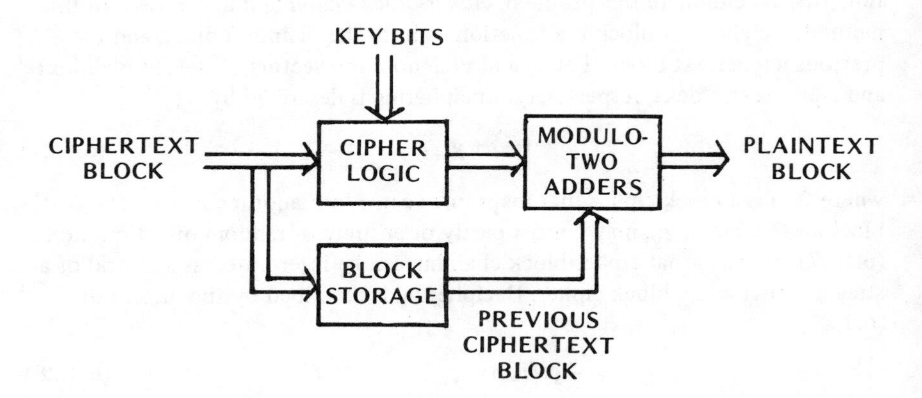

Figure 6.12 Deciphering system for cipher-block chaining.

6.2 DATA ENCRYPTION STANDARD

The *Data Encryption Standard* (DES) is a cryptographic algorithm used in un classified US Government and commercial applications [3, 4, 5]. It provides an excellent example of a strong block cipher with realistic implementation requirements.

The DES enciphers 64-bit blocks of plaintext into 64-bit blocks of ciphertext. Figure 6.13 is a flowchart of the algorithm. After an initial permutation, *IP*, the plaintext block is divided into a left half, L_0, of 32 bits and a right half, R_0, of 32 bits. The permuted input is then subjected to 16 iterations of a key-dependent computation. The output of that computation, called the pre-output, is permuted by the inverse of the initial permutation, which produces the ciphertext block.

The permutations *IP* and IP^{-1} are represented in matrix forms in Table 6.1. The order of the output bits of each permutation is determined from the successive rows of its matrix. Thus, *IP* transposes the input block $B = b_1 b_2 \ldots b_{64}$ into $B_0 = b_{58} b_{50} \ldots b_7$.

Given two blocks, L and R, LR denotes the block consisting of the bits of L followed by the bits of R. The output $L_{i+1}R_{i+1}$ of an iteration with input L_iR_i is defined by

$$L_{i+1} = R_i \qquad i = 0, 1, \ldots, 15 \tag{6.2.1}$$

$$R_{i+1} = L_i + f(R_i, K_{i+1}), \quad i = 0, 1, \ldots, 15 \tag{6.2.2}$$

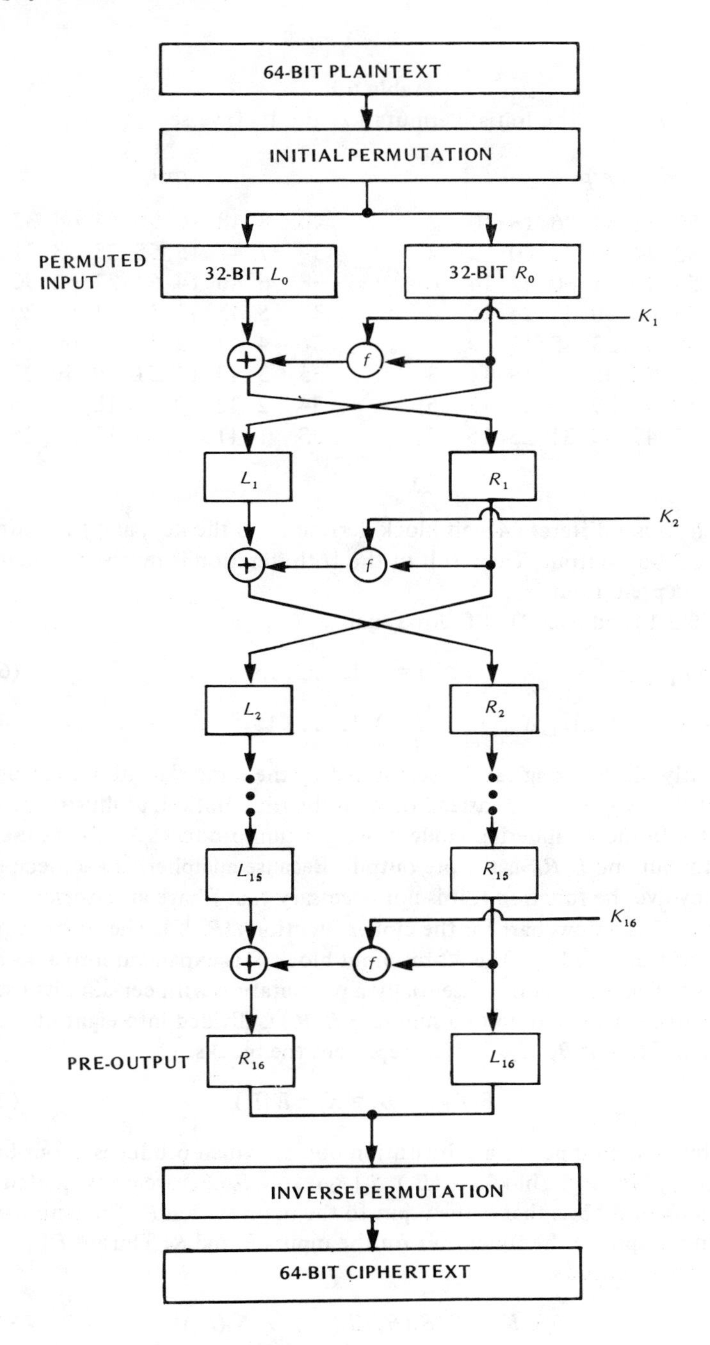

Figure 6.13 Flowchart for DES enciphering.

Table 6.1
The Initial Permutation and Its Inverse

IP

58	50	42	34	26	18	10	2
60	52	44	36	28	20	12	4
62	54	46	38	30	22	14	6
64	56	48	40	32	24	16	8
57	49	41	33	25	17	9	1
59	51	43	35	27	19	11	3
61	53	45	37	29	21	13	5
63	55	47	39	31	23	15	7

IP^{-1}

40	8	48	16	56	24	64	32
39	7	47	15	55	23	63	31
38	6	46	14	54	22	62	30
37	5	45	13	53	21	61	29
36	4	44	12	52	20	60	28
35	3	43	11	51	19	59	27
34	2	42	10	50	18	58	26
33	1	41	9	49	17	57	25

where each K_i is a different 48-bit block derived from the key, and f is a function with a 32-bit output. The result of the 16th iteration is reversed so that $R_{16}L_{16}$ is the pre-output.

From (6.2.1) and (6.2.2), it follows that

$$R_i = L_{i+1}\,, \qquad i = 0, 1, \ldots, 15 \tag{6.2.3}$$

$$L_i = R_{i+1} + f(L_{i+1}, K_{i+1})\,, \qquad i = 0, 1, \ldots, 15 \tag{6.2.4}$$

Consequently, deciphering can be performed by the same algorithm as enciphering except that K_{17-i} is used instead of K_i in the ith iteration, as illustrated in Figure 6.14. In the deciphering mode, the algorithm produces $R_{16}L_{16}$ as the permuted input and L_0R_0 as the pre-output. Because enciphering and deciphering both involve the function f, it is not necessary that f have an inverse.

Figure 6.15 is a flowchart for the cipher function $f(R, K)$. The 48-bit input K_i is derived from the key. The 32-bit input block R is expanded into a 48-bit block $E(R)$. The expansion is essentially a permutation with certain bits used more than once. The modulo-two sum $K_i + E(R)$ is divided into eight blocks of six bits each. Thus, if $B_1, B_2, \ldots, B_8$ represent the blocks,

$$B_1B_2 \ldots B_8 = K_i + E(R) \tag{6.2.5}$$

Each six-bit B_i is an input to a substitution box S_i, which produces a four-bit output $S_i(B_i)$. The eight blocks $S_1(B_1), S_2(B_2), \ldots, S_8(B_8)$ are consolidated into a single block of 32 bits that is the input to the permutation P. The permuted block is the output of the function f for the inputs R and K. Thus, if $P[\;\;]$ represents the permutation,

$$f(R,K) = P[S_1(B_1)S_2(B_2) \ldots S_8(B_8)] \tag{6.2.6}$$

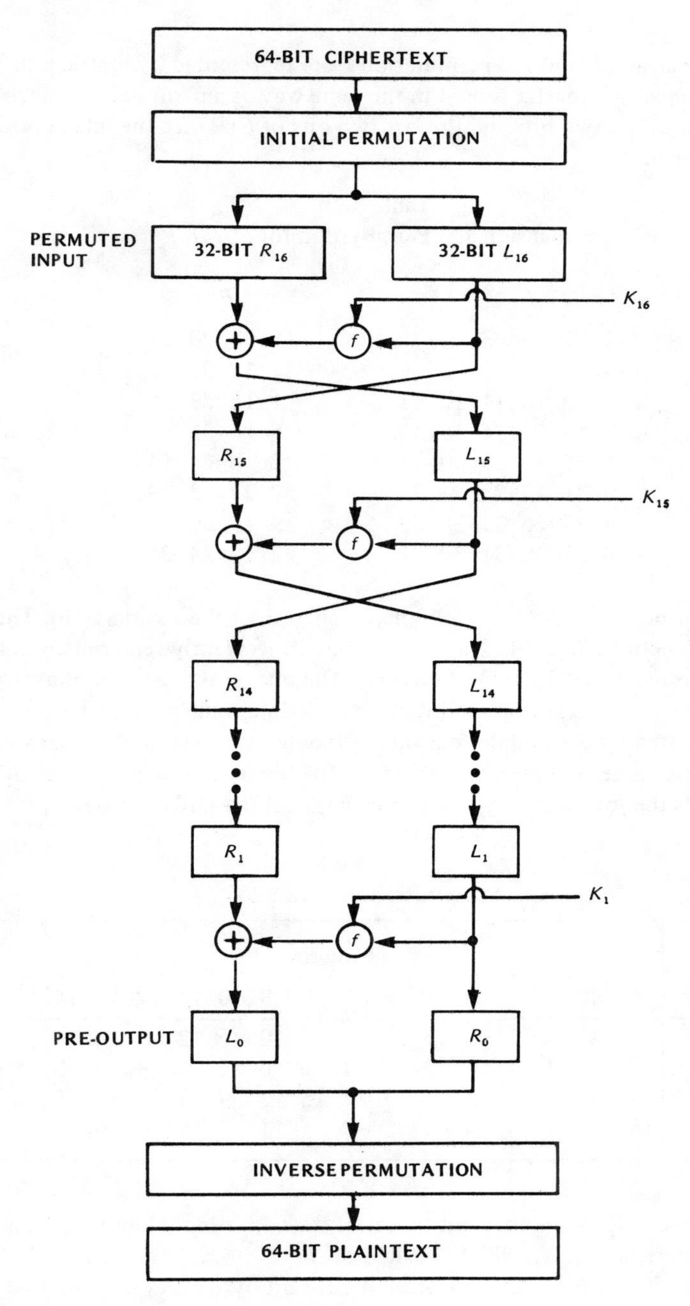

Figure 6.14 Flowchart for DES deciphering.

The expansion E and the permutation P are represented by matrices in Table 6.2. The expansion matrix is used in the same way as a permutation matrix. For example, the first two bits and the last two bits of $E(R)$ are the bits in positions 32 and 1 of R.

Table 6.2
Expansion and Permutation for $f(R,K)$

E

32	1	2	3	4	5
4	5	6	7	8	9
8	9	10	11	12	13
12	13	14	15	16	17
16	17	18	19	20	21
20	21	22	23	24	25
24	25	26	27	28	29
28	29	30	31	32	1

P

16	7	20	21
29	12	28	17
1	15	23	26
5	18	31	10
2	8	24	14
32	27	3	9
19	13	30	6
22	11	4	25

Each output bit of $S_i(B_i)$ is a Boolean function of the six bits in B_i. The substitution functions S_i, $i = 1, 2, \ldots, 8$, can be conveniently represented in tabular form, as illustrated in Table 6.3 for S_1. The integer represented in binary form by the first and last bits of input block B determine a row, while the integer represented by the middle four bits of B determine a column. The row and column determine an integer in Table 6.3. The binary representation of this integer yields the four bits of $S_1(B)$. For example, if $B = 010101$, then $S_1 = 1100$.

Table 6.3
Substitution Function S_1

	Column															
Row	**0**	**1**	**2**	**3**	**4**	**5**	**6**	**7**	**8**	**9**	**10**	**11**	**12**	**13**	**14**	**15**
0	14	4	13	1	2	15	11	8	3	10	6	12	5	9	0	7
1	0	15	7	4	14	2	13	1	10	6	12	11	9	5	3	8
2	4	1	14	8	13	6	2	11	15	12	9	7	3	10	5	0
3	15	12	8	2	4	9	1	7	5	11	3	14	10	0	6	13

Each 48-bit K_i uses a different subset of the 64-bit input key K_0. The algorithm for deriving the K_i, which is called the *key schedule calculation*, is diagrammed in Figure 6.16. One bit of each eight-bit byte of K_0 may be used for error detection in the key generation, distribution, and storage. The permutation $P1$ disregards these parity bits, permutes the remaining 56 bits, and produces two 28-bit blocks C_0 and D_0. These blocks are then subjected to 16 iterations of computation. Each iteration involves one or two left shifts and then a permutation $P2$ to produce one of the K_i. A single left shift rotates the bits one place to the left so that the bits in the 28 positions are the bits that were previously in positions 2, 3, . . . , 28, 1. Let LS_i denote the left-shift function of iteration i. The functions LS_1, LS_2, LS_9, and LS_{16} produce single left shifts, while the other LS_i produce two left shifts. Let C_i and D_i denote the blocks that are permuted to produce K_i. As shown in Figure 6.16,

$$C_0D_0 = P1(K_0) \tag{6.2.7}$$

$$C_{i+1} = LS_{i+1}(C_i)\,, \quad i = 0, 1, \ldots, 15 \tag{6.2.8}$$

$$D_{i+1} = LS_{i+1}(D_i)\,, \quad i = 0, 1, \ldots, 15 \tag{6.2.9}$$

$$K_i = P2(C_iD_i)\,, \quad i = 0, 1, \ldots, 15 \tag{6.2.10}$$

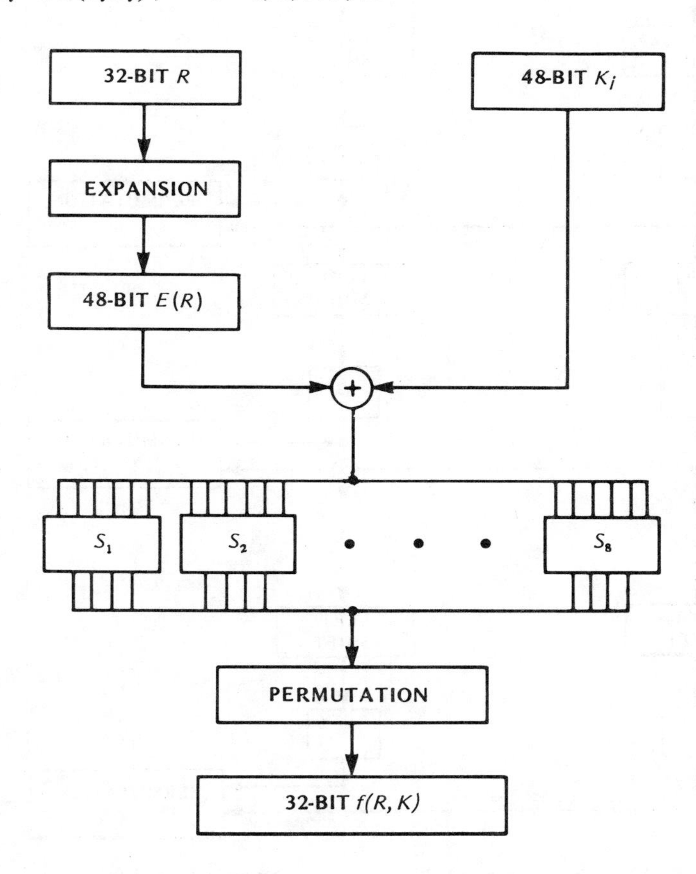

Figure 6.15 Flowchart for $f(R, K_i)$.

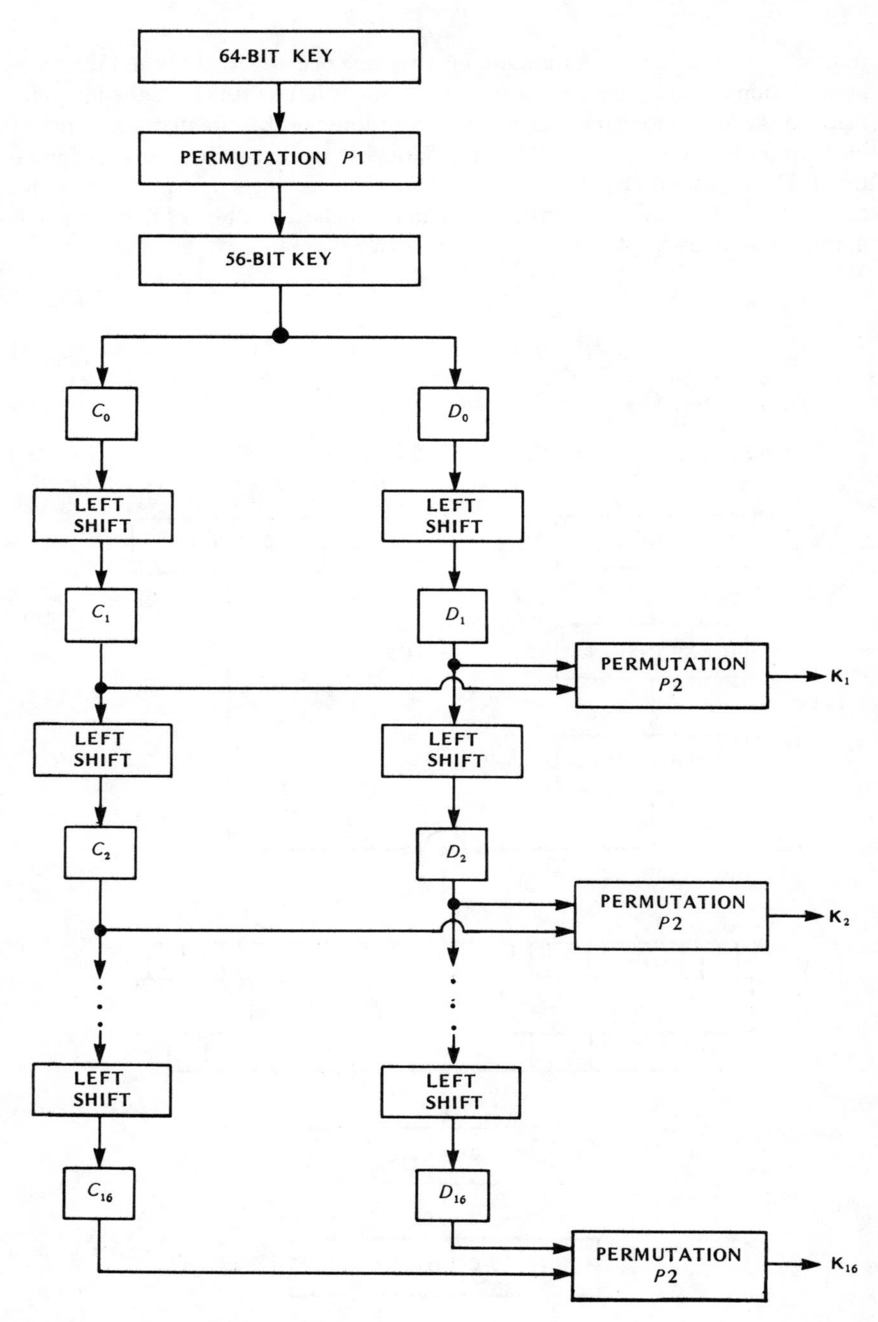

Figure 6.16 Flowchart for key schedule calculation.

Table 6.4
Permutations for Key Schedule Calculation

P_1

57	49	41	33	25	17	9
1	58	50	42	34	26	18
10	2	59	51	43	35	27
19	11	3	60	52	44	36
63	55	47	39	31	23	15
7	62	54	46	38	30	22
14	6	61	53	45	37	29
21	13	5	28	20	12	4

P_2

14	17	11	24	1	5
3	28	15	6	21	10
23	19	12	4	26	8
16	7	27	20	13	2
41	52	31	37	47	55
30	40	51	45	33	48
44	49	39	56	34	53
46	42	50	36	29	32

Permutation $P2$ disregards eight bits of its 56-bit input and permutes the remaining 48 bits. The permutations $P1$ and $P2$ are represented by matrices in Table 6.4. Thus, the bits of C_0 are bits 57, 49, . . . , 36 of K_0, while the bits of D_0 are bits 63, 55, . . . , 4 of K_0. The bits of K_i are bits 14, 17, . . . , 32 of C_iD_i.

After removal of the parity bits, the 56-bit key allows the formation of 2^{56} different block ciphers, a large number but a minute fraction of the $2^{64}!$ that are possible for 64-bit data blocks.

6.3 ERROR PROBABILITY BOUNDS AND ENSEMBLE AVERAGES FOR STREAM CIPHERS

A *word* is defined as a group of bits that represent a unit of information such as a letter or a number. In this and the following two sections, it is assumed that no error-correcting code is used and that the bit errors at a receiver output occur independently of each other. Consequently, the word error probability at a receiver output is

$$P_w = 1 - (1 - P_b)^k \tag{6.3.1}$$

where k denotes the number of bits per word and P_b denotes the bit error probability. If the receiver output is ciphertext, it is applied to a deciphering system. The k bits of a ciphertext word entering the deciphering system are referred to as the *input word.* The corresponding k plaintext bits emerging from the deciphering system are designated the *output word.* The probability of one or more erroneous bits in the output word is denoted by P_{cw}. For synchronous ciphers, $P_{cw} = P_w$. For auto-key ciphers, error propagation causes an increase in P_{cw} relative to P_w.

Suppose a ciphertext bit is erroneously received as a result of random noise or other interference. As the erroneous bit proceeds through the deciphering system, each of n consecutive output bits may be affected. This set of n consecutive bits emerging from the deciphering system is called a *train.* The parameter n defines a subset of stream ciphers. For a synchronous cipher, $n = 1$; for an auto-key cipher, $n > 1$.

We say that a train is of *external origin* with respect to an output word if the first bit of the train occurs before the first bit of the word. The joint probability of a word error and a train of external origin extending into the word is denoted by $P(w,t)$. If no train of external origin extends into the word, the conditional probability of an output word error is equal to the probability of an input word error, P_w, because an error in one of the bits of the input word causes an error in the corresponding bit of the output word. The probability that a train of external origin does not extend into a word is denoted by $P(\bar{t})$. From the theorem of total probability,

$$P_{cw} = P(w,t) + P_w P(\bar{t}) \tag{6.3.2}$$

A train extends into an output word if and only if one of the $n - 1$ input bits immediately preceding the corresponding input word is in error. Thus, assuming that input bit errors are independent,

$$P(\bar{t}) = (1 - P_b)^{n-1} \tag{6.3.3}$$

The event that i bits of a train of external origin extend into a k-bit word is denoted by the symbols $tb = i$. For example, $P(tb = i)$ denotes the probability that a word contains i externally generated train bits. Because $P(w, t \mid tb = i) = P(w \mid tb = i)$.

$$P(w, t) = P(w \mid tb = k) P(tb = k) + \sum_{i=1}^{k-1} P(w \mid tb = i) P(tb = i) \tag{6.3.4}$$

If at least one of the $n - k$ bits preceding an input word is in error and $n > k$, then $tb = k$. Thus,

$$P(tb = k) = \begin{cases} 1 - (1 - P_b)^{n-k}, & n > k \\ 0, & n \leqslant k \end{cases} \tag{6.3.5}$$

For $tb = i$, where $1 \leqslant i < k$, it is necessary that there be an input error precisely $n - i$ bits prior to the word but not erroneous input bits among the next $n - i - 1$ bits. Therefore, for $1 \leqslant i < k$,

$$P(tb = i) = \begin{cases} P_b(1 - P_b)^{n-i-1}, & n > i \\ 0, & n \leqslant i \end{cases} \tag{6.3.6}$$

Substitution of (6.3.1) to (6.3.6) into (6.3.2) yields the decomposition

$$P_{cw} = P(w|tb = k)\left[1 - (1 - P_b)^{n-k}\right]u(n - k)$$
$$+ \sum_{i=1}^{\min(k-1,n-1)} P(w|tb = i)P_b(1 - P_b)^{n-i-1}$$
$$+ \left[1 - (1 - P_b)^k\right](1 - P_b)^{n-1} \tag{6.3.7}$$

where $u(n - k)$ is a step function; that is, $u(n - k) = 0$ for $n < k$ and is 1 for $n \geq k$. In the summation term, i extends to the smaller of the two integers $k - 1$ and $n - 1$.

To evaluate the decomposition, the exact configuration of the cryptographic system must be specified. However, an upper bound can be obtained by simply observing that $P(w|tb = k)$ and $P(w|tb = i)$ must be less than or equal to unity. After some algebraic simplification, (6.3.7) yields the bound (which can be derived directly)

$$P_{cw} \leq 1 - (1 - P_b)^{n+k-1} \tag{6.3.8}$$

We show that there is a simpler bound:

$$P_{cw} \leq (n + k - 1)P_b \tag{6.3.9}$$

Consider the function of x defined by

$$y = mx + (1 - x)^m - 1, \quad 0 \leq x \leq 1, \quad m \geq 1 \tag{6.3.10}$$

Clearly, $y = 0$ at $x = 0$. Because $m \geq 1$, y has a nonnegative derivative for all x such that $0 \leq x \leq 1$. Thus, $y \geq 0$. We conclude that

$$mx \geq 1 - (1 - x)^m, \quad 0 \leq x \leq 1, \quad m \geq 1 \tag{6.3.11}$$

Applying this inequality to (6.3.8) yields (6.3.9).

A second measure of performance for auto-key ciphers is obtained by averaging P_{cw} over an *ensemble of deciphering systems for which at most n output bits are affected by an input bit error.* Suppose that an erroneous bit starts a train. Let X denote the ensemble-average probability that a subsequent train bit, called an *interior train bit,* is in the correct state. Let Y denote the ensemble-average probability that a subsequent keystream bit corresponding to an interior train bit is in the correct state. Because an input bit is added to a keystream bit to produce a train bit,

$$X = (1 - P_b)Y + P_b(1 - Y) \tag{6.3.12}$$

To determine the value of Y, we first consider the ensemble of deciphering systems of the form of Figure 6.10, where $n = 5$. It is equally likely that c_1 is open or closed. Suppose that after three or more correct input bits, an erroneous bit is received. If c_1 is closed, the next keystream bit will be in error. Thus, the ensemble average probability that the next keystream bit will be correct is one-half. Regardless of the presence or absence of input errors, each of the four successive keystream bits corresponding to the interior train bits has an ensemble-average probability of being correct equal to one-half. Thus, it is plausible that $Y = 1/2$ for auto-key ciphers. For a restricted ensemble of nonlinear ciphers, Y may be different from $1/2$. However, the most important cipher ensembles are those for which cryptanalysis is very difficult. Setting Y equal to $1/2$ for these ensembles appears to be an excellent approximation. If $Y = 1/2$, (6.3.12) indicates that $X = 1/2$.

We indicate an ensemble-average probability by a bar over a P. When $k = 1$, $\bar{P}(w|tb = k) = 1 - X$. Thus, (6.3.7) yields the ensemble-average cryptographic bit error probability:

$$\bar{P}_{cb} = (1 - X)\left[1 - (1 - P_b)^{n-1}\right] + P_b(1 - P_b)^{n-1} \tag{6.3.13}$$

Substituting (6.3.12) into (6.3.13) gives

$$\bar{P}_{cb} = (1 - Y)\left[1 - (1 - P_b)^{n-1} + 2P_b(1 - P_b)^{n-1}\right] - P_b(1 - 2Y) \tag{6.3.14}$$

In this equation, the unspecified parameter Y is retained because it does not complicate the expression significantly. However, for the reasons mentioned and for an easier derivation, we always assume that $Y = 1/2$ and hence $X = 1/2$ in determining the ensemble-average word error probability.

Given $tb = i$, we denote the event that one or more of the first i bits of an input word is in error by the symbol α and the opposite event by $\bar{\alpha}$. Using the theorem of total probability, we find that

$$\bar{P}(w \mid tb = i) = \bar{P}(w,\alpha \mid tb = i) + \bar{P}(w \mid tb = i,\bar{\alpha})P(\bar{\alpha} \mid tb = i) \tag{6.3.15}$$

If $tb = i$ and α is false, the ensemble-average probability of no error in the first i bits of the output word is $(1/2)^i$. The last $k - i$ output bits are not part of a train generated by the first i input bits. Consequently, the first error in the last $k - i$ input bits causes an output bit error. Therefore, the probability of no error in the last $k - i$ output bits is equal to the probability of no error in the corresponding input bits. From the independence of input bit errors, we conclude that

$$\bar{P}(w|tb = i, \bar{\alpha}) = 1 - 2^{-i}(1 - P_b)^{k-i} \tag{6.3.16}$$

and

$$P(\bar{\alpha}|tb = i) = P(\bar{\alpha}) = (1 - P_b)^i \tag{6.3.17}$$

Given $tb = i$, let the symbol $\beta = l$ denote the event that the last bit error among the first i input bits occurs at input bit l, where $1 \leqslant l \leqslant i$. If $\beta = l$, then

α is true; thus, we make the decomposition

$$\bar{P}(w,\alpha|tb=i)=\sum_{l=1}^{i}\bar{P}(w|tb=i,\beta=l)P(\beta=l|tb=i) \tag{6.3.18}$$

The probability that $\beta = l$ is equal to the probability that input bit l is erroneous and input bits $l+1$ through i are correct. We conclude that

$$P(\beta=l|tb=i)=P(\beta=l)=P_b(1-P_b)^{i-l}, \quad 1\leqslant l\leqslant i \tag{6.3.19}$$

Given $tb = i$, the probability that the first i output bits are correct has an ensemble average equal to $(1/2)^i$. The probability that the last $k - i$ output bits are correct depends on the event $\beta = l$, which implies that $n + l - i - 1$ train bits extend into the final $k - i$ bits. Let w_{k-i} denote the event that an error occurs in a word consisting of $k - i$ output bits. From the previous discussion, it follows that

$$\bar{P}(w|tb=i,\beta=l)=1-2^{-i}[1-\bar{P}(w_{k-i}|tb=n+l-i-1)] \tag{6.3.20}$$

Combining (6.3.15) to (6.3.20), we obtain

$$\bar{P}(w|tb=i)=1-2^{-i}(1-P_b)^k-2^{-i}[1-(1-P_b)^i]$$
$$+2^{-i}P_b\sum_{l=1}^{i}(1-P_b)^{i-l}\bar{P}(w_{k-i}|tb=n+l-i-1) \tag{6.3.21}$$

The factor $\bar{P}(w_{k-i}|\,tb=n+l-i-1)$ can be evaluated by the same procedures as that leading to (6.3.21) itself. If $n < k$, a finite hierarchy of equation is obtained. The *ensemble-average cryptographic word error probability*, $\bar{P}_{cw}$, follows on substitution of (6.3.21) into the ensemble average of (6.3.7).

In nearly all practical communication systems with auto-key ciphers, $n \geqslant k$. In this case, all of the final $k - i$ bits are part of a train when $tb = n + l - i - 1$. Thus,

$$\bar{P}(w_{k-i}\,|\,tb=n+l-i-1)=1-2^{-(k-i)}, \quad 1\leqslant l\leqslant i\leqslant k\leqslant n \tag{6.3.22}$$

Substituting (6.3.21) and (6.3.22) into the ensemble average of (6.3.7), performing summations, and regrouping terms, we obtain for $n \geqslant k$,

$$\bar{P}_{cw}=\begin{cases}1-2^{-k}+k2^{-k}P_b(1-P_b)^{n-1}-\dfrac{(1-P_b)^n[(1-P_b)^k-2^{-k}]}{1-2P_b}, & P_b\neq 1/2\\[2ex] 1-2^{-k}, & P_b=1/2\end{cases} \tag{6.3.23}$$

A simple approximation to this equation, which is accurate over the usual range of interest, can be obtained by deriving a Taylor-series expansion about the point $P_b = 0$ and retaining only the first-order term. The result is

$$\overline{P}_{cw} \approx [n + k - 2 - 2^{-k}(n - k - 2)]P_b\,, \quad n \geqslant k \tag{6.3.24}$$

From the remainder terms in the Taylor-series expansions of the factors in (6.3.23), it is found that the simple condition

$$P_b \leqslant (n + k - 2)^{-1} \tag{6.3.25}$$

is sufficient for the validity of (6.3.24). Using the same method on (6.3.14), we obtain

$$\overline{P}_{cb} \approx [n(1 - Y) + Y]P_b \tag{6.3.26}$$

where (6.3.25) with $k = 1$ provides a sufficient condition for validity.

To obtain an approximate equation for $\overline{P}_{cw}$ when $n < k$, we note that the last term in (6.3.21) cannot contribute to the first-order term in a Taylor-series expansion of $\overline{P}_{cw}$. Thus, (6.3.21) and (6.3.7) give

$$\overline{P}_{cw} \approx [n + k - 2(1 - 2^{-n})]P_b\,, \quad n < k \tag{6.3.27}$$

where (6.3.25) provides a sufficient condition for validity. For a synchronous cipher, $\overline{P}_{cw} = P_{cw} = P_w$. If P_b is small, (6.3.1) yields $P_{cw} \approx kP_b$, which also results when $n = 1$ in (6.3.24) and (6.3.27) because $k \geqslant 1$.

6.4 ERROR PROBABILITY BOUNDS AND ENSEMBLE AVERAGES FOR BLOCK CIPHERS

In the conventional block cipher, a plaintext block of n total bits, usually comprising an integral number of words of k bits each, is enciphered as a block of n total bits. After transmission and reception, the plaintext block is restored as the output of the deciphering system. No output words are in error unless the received ciphertext block contains an error in at least one of its n bits. Assuming the independence of input bit errors,

$$P_{cw} = P(w\,|\,be)[1 - (1 - P_b)^n] \tag{6.4.1}$$

where $P(w\,|\,be)$ is the probability of an error in an output word, given that there is a block error at the input of the deciphering system. Setting $P(w\,|\,be) = 1$ and using (6.3.11), we obtain

$$P_{cw} \leqslant 1 - (1 - P_b)^n \leqslant nP_b \tag{6.4.2}$$

Due to the one-to-one correspondence between the ciphertext blocks and plaintext blocks, an error in a received ciphertext block is certain to cause at least one erroneous bit in the output block. Consequently, over the ensemble of block ciphers of size n, there are $2^n - 1$ equally likely output blocks corresponding to an erroneous ciphertext block. Consider any fixed bit in these output blocks. In $2^{n-1} - 1$ of the possible output blocks, this bit is correct, that is, in the same state that it would have been in if no error had occurred in the enciphered block. We conclude that given a block error, there is an ensemble-average probability that a bit is correct equal to $(2^{n-1} - 1)/(2^n - 1)$. Consider a second fixed output bit. Given that there is a block error and that the first fixed output bit is correct, it follows from an extension of the previous reasoning that there is an ensemble-average probability that the second fixed bit is correct equal to $(2^{n-2} - 1)/(2^{n-1} - 1)$. If $x_1, x_2, \ldots, x_k$ are events, the probability of all these events is equal to the product of conditional probabilities:

$$P(x_1, x_2, \ldots, x_k) = P(x_k \mid x_{k-1}, \ldots, x_1) \ldots P(x_2 \mid x_1) P(x_1) \tag{6.4.3}$$

Using this equation and repeating the analysis for successive output bits, we conclude that for a k-bit word contained within a single block,

$$\bar{P}(w \mid be) = 1 - \prod_{i=1}^{k} \frac{2^{n-i} - 1}{2^{n+1-i} - 1}$$

$$= \frac{1 - 2^{-k}}{1 - 2^{-n}} \tag{6.4.4}$$

Combining this relation with (6.4.1), we obtain the ensemble-average cryptographic word error probability for block ciphers,

$$\bar{P}_{cw} = (1 - 2^{-n})^{-1}(1 - 2^{-k})[1 - (1 - P_b)^n] \tag{6.4.5}$$

A Taylor-series expansion yields

$$\bar{P}_{cw} \approx (1 - 2^{-n})^{-1}(1 - 2^{-k}) n P_b \tag{6.4.6}$$

which is accurate if

$$P_b \ll 2(n - 1)^{-1} \tag{6.4.7}$$

The ensemble-average cryptographic bit error probability for block ciphers is obtained by setting $k = 1$ in (6.4.5) or (6.4.6). Although these equations hold for all values of n and k, $n \geq 4k$ is usually required to safeguard against the frequency analysis of block patterns.

6.5 DEGRADATION DUE TO CRYPTOGRAPHY

The bit error probability for uncoded binary communications is a function of the average energy per bit, E_b; that is,

$$P_b = f(E_b) \tag{6.5.1}$$

where f is a function. If this equation is substituted into the equations for the cryptographic error probabilities, there result equations in terms of E_b. By comparing these equations with (6.3.1) and (6.5.1), we can determine the increase in energy required to obtain the same error probability from a cryptographic system as provided by the corresponding unenciphered system. This increase provides a quantitative measure of *cryptographic degradation.* The degradation in decibels is defined to be

$$D = 10 \log_{10} E_{b1} - 10 \log_{10} E_b = 10 \log_{10} \left(\frac{E_{b1}}{E_b} \right) \tag{6.5.2}$$

where E_{b1} is the energy required to produce a value of $\bar{P}_{cw}$ that equals the value of P_w when the energy is E_b.

It is often convenient to have a simple approximate formula for the degradation. To derive such a formula, note that for small values of P_b, (6.3.1) becomes

$$P_w \cong kP_b \, , \quad P_b \ll 2(k-1)^{-1} \tag{6.5.3}$$

With this approximation, the ensemble-average word error probabilities for small values of P_b have the form

$$\bar{P}_{cw} \cong g(n,k) P_w \, , \quad P_b \ll (n+k-2)^{-1} \tag{6.5.4}$$

The *cipher memory, n,* is the number of output bits that may be affected by an error in a received ciphertext bit. For stream ciphers,

$$g(n,1) = n(1-Y) + Y \tag{6.5.5}$$

for bits and

$$g(n,k) = \begin{cases} \dfrac{n+k-2-2^{-k}(n-k-2)}{k} \, , & n \geqslant k \\[2ex] \dfrac{n+k-2(1-2^{-n})}{k} \, , & n < k \end{cases} \tag{6.5.6}$$

for words with $Y = 1/2$. For block ciphers, the cipher memory is equal to the block length, and

$$g(n,k) = \frac{(1 - 2^{-k})n}{(1 - 2^{-n})k} \tag{6.5.7}$$

According to the definition of E_{b1}, it is implicitly related to E_b by

$$\overline{P}_{cw}(E_{b1}) = P_w(E_b) \tag{6.5.8}$$

Combining (6.5.1), (6.5.3), (6.5.4), and (6.5.8), we find that the degradation can be determined analytically by solving

$$g(n,k)f(E_{b1}) = f(E_b) \tag{6.5.9}$$

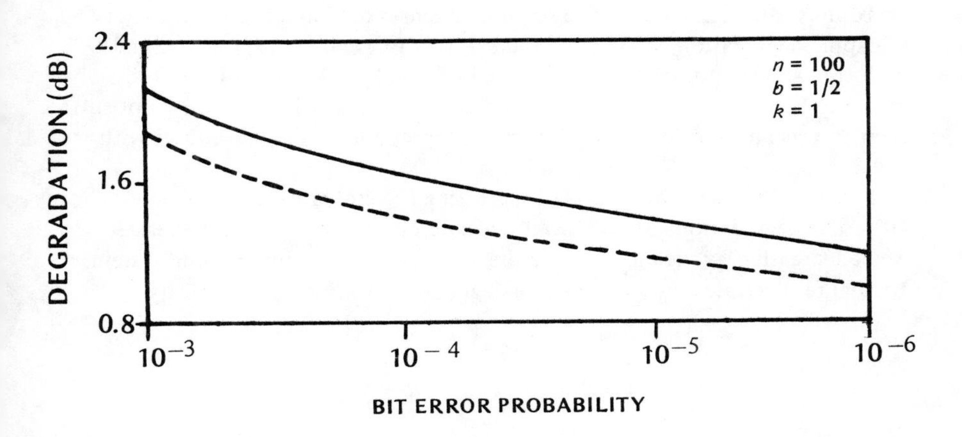

Figure 6.17 Cryptographic degradation as function of bit error probability. Solid curve: block cipher or stream cipher with Y = 1/2. Dashed curve: stream cipher with Y = 3/4.

Many noncoherent and differentially coherent uncoded systems operating in white Gaussian noise or interference have bit error probabilities of the form

$$P_b = f(E_b) = b \exp\left(-\frac{\xi E_b}{N_0}\right) \tag{6.5.10}$$

where b and ξ are independent of E_b, but depend on the modulation type (Section 1.4). Substituting (6.5.10) into (6.5.9), solving for E_{b1}/E_b, and using the result in (6.5.2), we obtain

$$D = 10 \log_{10} \left[1 + \frac{\ln g(n,k)}{\xi E_b/N_0} \right] \tag{6.5.11}$$

Substituting (6.5.3) and (6.5.10) into (6.5.11) yields

$$D = 10 \log_{10} \left[1 - \frac{\ln g(n,k)}{\ln \left(\frac{P_w}{bk} \right)} \right] \tag{6.5.12}$$

In this form, D can be calculated without specifying ξ. If (6.5.10) is valid for small values of E_b/N_0, then usually $b = 1/2$.

To illustrate various aspects of ciphers, Figures 6.17 and 6.18 plot the cryptographic degradation when $k = 1$ and $b = 1/2$. In Figure 6.17, $n = 100$ and D is plotted as a function of P_b. The impact of changing $Y = 1/2$ to $Y = 3/4$ is observed to be relatively minor. Thus, the reduced degradation is usually not sufficient to warrant limiting the cryptographic ensemble to a cipher subset with $Y = 3/4$.

Figure 6.18 shows the effects of increasing the cipher memory when $P_b = 10^{-3}$. The degradations of block and stream ciphers converge as n increases. Since increasing n tends to enhance the security of the cryptographic system, the figure illustrates the price paid in degradation for increased security.

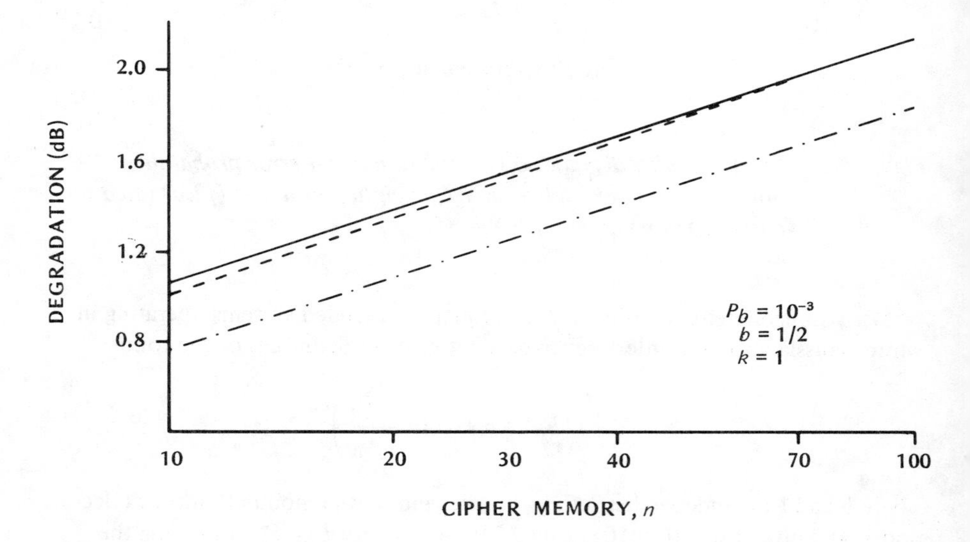

Figure 6.18 Cryptographic degradation as function of cipher memory. Solid curve: stream cipher with $Y = 1/2$. Dashed curve: block cipher. Dot-dashed curve: stream cipher with $Y = 3/4$.

Figure 6.19 illustrates the cryptographic degradation as a function of the number of bits in a word. The curves, which are represented as continuous lines for convenience, are determined from (6.5.11) with $n = 100$ and $\xi E_b/N_0 = 10$. For the stream cipher, $Y = 1/2$. If an integral number of words are contained in a block, then n/k must be an integer and only a few points of the block-cipher curve are relevant.

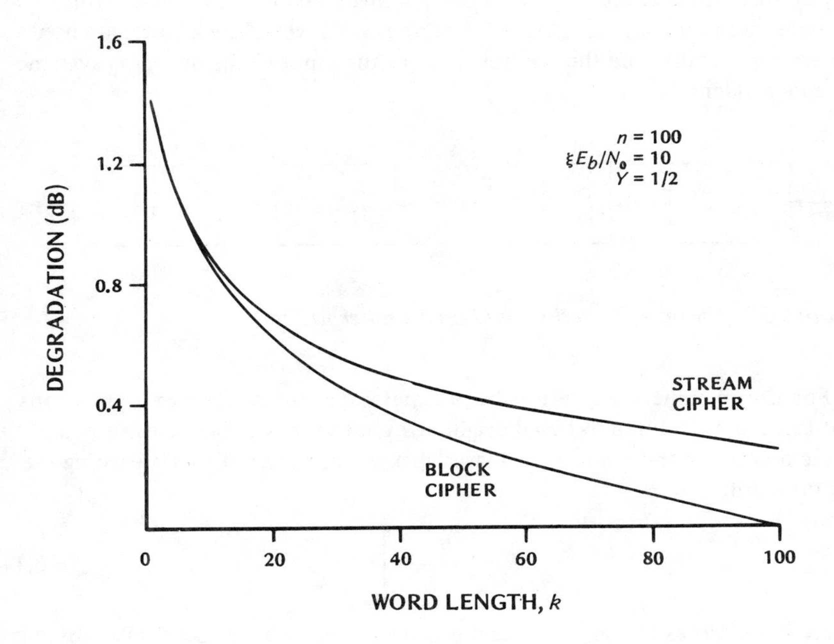

Figure 6.19 Cryptographic degradation as function of word length.

6.6 ERROR CORRECTION

To improve the performance of a cryptographic communication system, an error-correcting code may be superimposed on the enciphering. Figure 6.20 illustrates the system configuration that is usually preferable. The data bits are first enciphered and then encoded; after transmission, the received symbols are first decoded and then deciphered. By encoding after the enciphering, no cryptanalytically useful redundancy is added to the plaintext. If the inner and outer blocks of Figure 6.20 are interchanged so that the encoding precedes the enciphering and the deciphering precedes the decoding, then the error propagation, if present, might overwhelm the error-correcting capability of the decoding system. However, the error propagation may sometimes be useful for detecting the false ciphertext transmitted by an opponent.

The preceding derivations of cryptographic error probabilities depend upon the assumption of independent bit errors at the input to the deciphering system. When this input is the output from a decoding system that corrects word errors, the input bit errors are not independent, but occur in clusters. Thus, the preceding equations for the cryptographic error probabilities do not apply. However, assuming the independence of the input word errors, we can relate the word error probabilities at the outputs of deciphering systems to the word error probabilities at the inputs. This assumption is valid when block codes are used for error correction and the symbol errors at the input to the decoding system are independent.

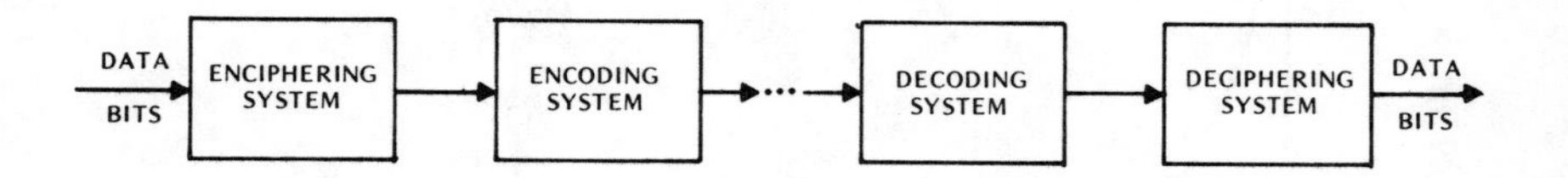

Figure 6.20 Encoding superimposed on enciphering.

For stream ciphers, we define the parameter h as the number of input words that can cause a train of external origin in an output word. Because h is equal to the number of input words that overlap some of the $n - 1$ bits preceding the output word,

$$h = \left[\frac{n-2}{k}\right] + 1 \tag{6.6.1}$$

where $[x]$ denotes the largest integer less than or equal to x. Let $P(t)$ denote the probability that a train of external origin extends into an output word. From (6.3.2), we obtain $P_{cw} \leqslant P(t) + P_w$. Therefore,

$$P_{cw} \leqslant 1 - (1 - P_w)^h + P_w \leqslant (h + 1)P_w \tag{6.6.2}$$

where P_w and P_{cw} are the error probabilities of the input and output words, respectively. These upper bounds can be used in approximate calculations of the cryptographic degradation. An expression for $\overline{P}_{cw}$ is very complicated.

For block ciphers, the methods of section 6.4 yield

$$P_{cw} \leqslant 1 - (1 - P_w)^{n/k} \leqslant \frac{n}{k} P_w \tag{6.6.3}$$

and

$$\overline{P}_{cw} = (1 - 2^{-n})^{-1}(1 - 2^{-k})[1 - (1 - P_w)^{n/k}] \tag{6.6.4}$$

where the integer n/k is the number of words in a block. A Taylor-series expansion yields the approximation

$$\bar{P}_{cw} \approx (1 - 2^{-n})^{-1}(1 - 2^{-k}) \frac{n}{k} P_w \tag{6.6.5}$$

which is accurate if

$$P_w \ll 2k(n - k)^{-1} \tag{6.6.6}$$

6.7 SYNCHRONIZATION AND INTERFERENCE

The operation of communication systems using synchronous ciphers depends upon the perfect alignment of the keystream and received bits in the deciphering system. Once misalignment occurs, special measures must be employed to restore synchronization. In contrast, communication systems with auto-key ciphers restore alignment of the bits automatically since the keystream is continually produced by the received bits. Synchronization is lost in an auto-key system whenever the receiver incorrectly identifies the word boundaries. Synchronization is lost in a block-cipher system whenever the receiver incorrectly identifies the block boundaries. Both auto-key and block-cipher systems often can resynchronize automatically as soon as the next frame identification bits are received.

Loss of synchronization in a receiver for a synchronous cipher can occur when interference causes the clock output of a bit synchronizer to skip a pulse or generate an extra pulse. Alternatively, synchronization can be lost when interference causes a sufficient number of frame synchronization bits to be received erroneously. When this event is recognized, the receiver assumes that a misalignment has occurred and initiates the resynchronization procedure.

Suppose that pulsed interference of duration T_D occurs every T_B seconds during the reception of the uncoded ciphertext produced by a synchronous cipher. If a pulse causes a loss of synchronization in the receiver, time is lost while the communication system recognizes the loss of synchronization, initiates the resynchronization procedure, and reestablishes synchronization between the received ciphertext and the keystream. We call this lost time the ***reacquisition time*** and denote its average duration by T_R. Because reacquisition cannot be completed until the pulse has ceased, $T_R > T_D$, as illustrated in Figure 6.21. In general, T_R is a function of T_D. When T_D is sufficiently large, it is reasonable to expect that $T_R \cong T_D + C$, where C is a constant.

During the reacquisition time, the probability of error of the deciphered bits in the receiver is one-half because the transmitted information has been entirely destroyed. (If the receiver output is disabled upon recognition of a synchronization loss, the missing bits must be guessed, so the equivalent bit error probability is one-half.) After reacquisition, assuming there is no further loss of synchronization before the occurrence of the next interference pulse, the bit error probabili-

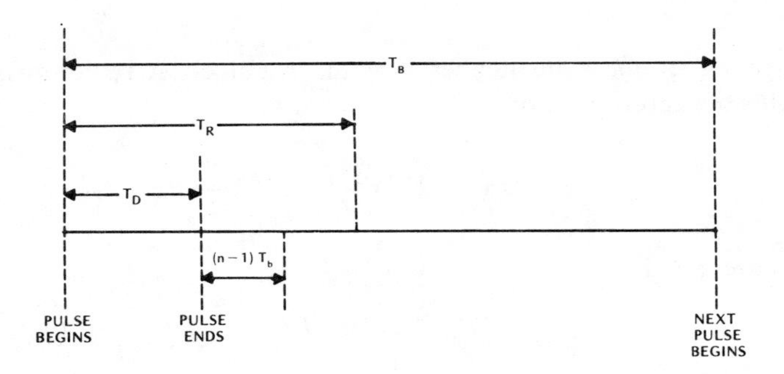

Figure 6.21 Timing diagram for pulsed interference.

ty at the output of the deciphering system becomes P_b, the usual channel-bit error probability. To ignore the relative time alignment of the bit edges and the pulses, we assume that

$$T_R \gg T_b \tag{6.7.1}$$

$$T_B - T_R \gg T_b \tag{6.7.2}$$

where T_b is the bit duration. It follows that the expected number of bit errors over a pulse period for a synchronous-cipher system is

$$N_S \cong \left(\frac{T_R}{T_b}\right)\frac{1}{2} + \left(\frac{T_B - T_R}{T_b}\right) P_b \tag{6.7.3}$$

Suppose that the same pulsed interference temporarily disrupts the reception of the uncoded ciphertext produced by an auto-key cipher with $Y = 1/2$, but that word synchronization is maintained. For the pulse duration and $n - 1$ bits following the cessation of the pulse, the bit error probability is one-half. The remaining bits before the next pulse have an ensemble-average error probability equal to $\bar{P}_{cb}$. As seen in Figure 6.21, the relative time alignments of the bit edges and the pulses may be ignored by assuming that

$$T_D + (n-1)T_b \gg T_b \tag{6.7.4}$$

$$T_B - T_D - (n-1)T_b \gg T_b \tag{6.7.5}$$

It follows that the expected number of bit errors over a pulse period for an auto-key system is

$$N_A \cong \left[\frac{T_D + (n-1)T_b}{T_b}\right]\frac{1}{2} + \left[\frac{T_B - T_D - (n-1)T_b}{T_b}\right]\bar{P}_{cb} \tag{6.7.6}$$

Assuming that word synchronization is maintained, an auto-key system produces fewer bit errors than a synchronous system if $N_A < N_S$. From (6.7.3) and

(6.7.6), we see that this situation exists if

$$T_B < \frac{T_R\left(\frac{1}{2} - P_b\right) - \left[T_D + (n-1)T_b\right]\left(\frac{1}{2} - \overline{P}_{cb}\right)}{\overline{P}_{cb} - P_b} \tag{6.7.7}$$

For most practical communication systems using an auto-key cipher, this inequality can be approximated by

$$T_B < \frac{T_R - T_D - (n-1)T_b}{2\overline{P}_{cb}}, \quad P_b \ll \overline{P}_{cb} \ll \frac{1}{2} \tag{6.7.8}$$

Equations (6.7.1), (6.7.2), (6.7.4), (6.7.5), and (6.7.7) or (6.7.8) are sufficient conditions for an auto-key system to outperform a synchronous system when pulsed interference is present and word synchronization is maintained. If word synchronization is not maintained in an auto-key system, the sufficient conditions are obtained by substituting T_F in place of $(n-1)T_b$, where T_F is the average time until frame identification bits allow resynchronization.

As an example, suppose $n = 101$, $T_D = 900\ T_b$, $T_R = 10^4\ T_b$, and $P_b = 10^{-4}$. Since $k = 1$ and (6.3.25) is satisfied, (6.3.26) with $Y = 1/2$ gives $\overline{P}_{cb} = 5.1 \times 10^{-3}$ Inequality (6.7.8) applies and yields $T_B < 0.9 \times 10^6\ T_b$. Edge effects are negligible if $T_B > 1.1 \times 10^4\ T_b$.

6.8 SECURITY

It is a fundamental tenet of cryptography that the particular cryptographic algorithm cannot be kept secret, and therefore security depends upon the integrity of the key. A key must be kept secret for the duration of its use and must resist cryptanalysis until the protected data loses its importance. For example, tactical military messages normally require a few hours or at most a few days of cryptographic protection. Keys should be randomly generated or appear to be random and have enough bits to thwart cryptanalysis by exhaustive search of the possible keys. One way to generate a key is to apply a block cipher to an arbitrary or pseudorandom input. The frequency with which a key is changed depends on the amount of data it encrypts and the cost of implementing a change.

A key may be transmitted to potential communicators while being protected by encipherment under a different key called a *key-encrypting key.* A key used to encipher and decipher data but not other keys is called a *data-encrypting key.* A key hierarchy may be established in which key-encrypting keys are protected by additional levels of key-encrypting keys. Ultimately, the key at the top of the hierarchy is distributed to communicators by means that are the most secure possible, such as couriers, but that are too slow for the transmission of all messages. If a key-encrypting key is used only for protecting randomly generated keys, then a cryptanalyst must learn a number of the protected keys to have any chance of determining the key-encrypting key. It is not safe to transmit each new data-encrypting key after enciphering it using the preceding data-encrypting key because then all security depends upon the first key in the sequence.

There are several ways to allocate keys in a network. To minimize the resulting damage if a key should be compromised, all potential pairs of communicators should use different keys. In a network of n mutually communicating elements, $\binom{n}{2} = n(n-1)/2$ different keys are needed and each element must store $n-1$ different keys. The keys may be key-encrypting keys that are used for the exchange of data-encrypting keys prior to communicating.

If this procedure entails too many different keys, an alternative is to distribute a single *master key* to each network element. Each master key enables an element to communicate with a *key distribution center.* When two elements wish to communicate, they request a *session key* from the key distribution center, which transmits the session key to the elements in their master keys. In this configuration, the overall security of the network depends upon the security of the key distribution center.

It is sometimes necessary for network communications to pass through intermediate nodes. If the message is enciphered at the source and deciphered only at its destination, with only the address or identification of the destination transmitted as plaintext, the procedure is called *end-to-end encryption.*

Another way to allocate keys and communicate within a network, called *link encryption,* is to provide each element with a single key for communicating with a central node or a local node. A message is successively deciphered and then enciphered by use of the node key as the message passes through each successive node until it reaches its final destination. The security of a message depends upon the security of the intermediate nodes.

With end-to-end encryption, only those nodes where elements originate or receive enciphered messages require cryptographic devices. End-to-end encryption offers greater security than link encryption because messages are not deciphered until they reach their destinations. Thus, end-to-end encryption appears to be usually preferable to link encryption when a communications network has many links that must be protected. A disadvantage of end-to-end encryption is that a separate key is needed for each pair of communicating elements.

The pattern of messages in a communication network can sometimes yield useful information to an opponent. The extraction of such information from statistics on the number of messages transmitted to and from certain nodes is called *traffic analysis.* It can be prevented by creating enough artificial messages to obscure any occurrence of an unusual number of meaningful messages. However, increasing the amount of network traffic often increases the operating costs.

Conventional ciphers use the same key for both enciphering and deciphering. In *public-key cryptography,* two distinct keys are used for these functions. Each user has both a public and a private key. The *public key* is used for enciphering and is not kept secret. It may be transmitted as plaintext when requested. The *private key*, which is kept secret, is used for deciphering and is presumed to be computationally infeasible to determine from the public key [3, 4, 5, 6]. Consequently, two users can communicate when they know only each other's public keys. Because of their large computational requirements, public-key systems appear to be primarily useful as key-encrypting systems.

6.9 SCRAMBLERS

A *scrambler* is a device that alters an analog signal before transmission so that only the intended receivers can extract useful information from it. A *descrambler* is a device used in the receiver to restore the original analog signal. Scramblers, which usually protect speech communications, provide privacy against inadvertent or casual eavesdropping, but do not appear to be able to offer the level of security possible with digital ciphers while still providing acceptable speech quality. However, scramblers have applications when the security requirements are not too stringent and the bandwidth available for transmission cannot support the bit rate required by a speech digitizer that provides digital speech communications of adequate quality. Although the final output of a scrambler is an analog signal, digital processing is often used inside the scrambler.

The security provided by a scrambler is undermined by any *residual intelligibility,* which is the fraction of the original speech that can be understood by producing the audio equivalent of the scrambled signal without first descrambling it. Residual intelligibility is not a problem for digital ciphers because they operate on much smaller information elements than scramblers.

One of the earliest scramblers was the *frequency inverter.* This scrambler inverts the spectrum of the speech signal, interchanging high and low frequencies as illustrated in Figure 6.22. The inversion may be implemented by first mixing the speech signal with a tone to produce a double-sideband signal, extracting the lower sideband by filtering, and then translating the lower sideband in frequency. The problem with the frequency inverters is that the scrambled output has a high residual intelligibility and is easily descrambled by using the same device. Consequently, the primary role of a frequency inverter is that of a building block in more effective scramblers.

Some reduction in residual intelligibility is provided by the *band-shift inverter.* This device divides the inverted spectrum into two unequal bands that are then interchanged, as illustrated in Figure 6.22. The dividing point indicated by f_3 in the figure can be changed periodically under the control of a key. However, the original signal can be recovered by an opponent after a search over the possible dividing points. The key need not be determined.

A *band scrambler* or *band splitter* divides the speech spectrum into n equal frequency bands that are permuted and possibly inverted, as illustrated in Figure 6.23. Because there are $n!$ possible permutations and 2^n ways to decide which bands are inverted, the scrambler can produce $2^n n!$ different rearrangements, each of which is determined by a key. However, the number of useful rearrangements or keys in a practical band scrambler is modest for two reasons. First, implementation losses that degrade the quality of the speech communications increase rapidly with n. Second, for a given value of n, the number of keys that produce a low residual intelligibility is much less than $2^n n!$. For example, keys that leave one or more of the bands in their original positions tend to produce a high residual intelligibility. The limited number of useful keys makes feasible an exhaustive search with a human listener deciding when the correct key is being tried.

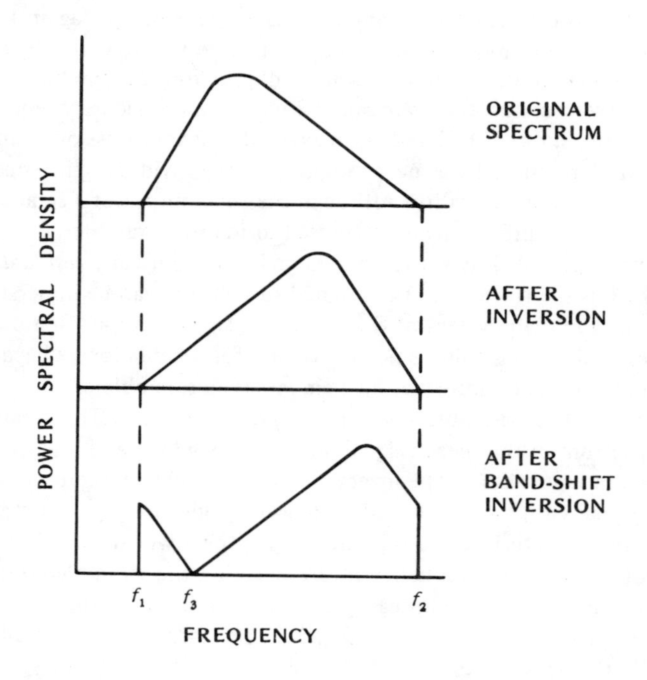

Figure 6.22 Frequency inversion and band-shift inversion.

If the permutations and inversions are changed periodically, the band scrambler is often called a *rolling-code scrambler.* The useful rearrangements can be

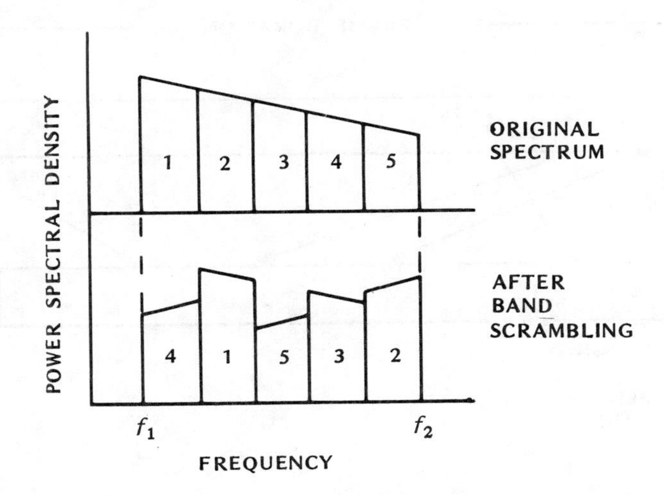

Figure 6.23 Band scrambling.

stored as matrices in a ROM (read-only-memory). Parallel output bits from a stream-cipher system or groups of bits from its output sequence can be used to determine the sequence of rearrangements used by the scrambler. The stream cipher is determined by the key, which can be made long enough to thwart an exhaustive search of the possible keys. Although the rolling-code scrambler provides greatly increased security, partial descrambling is often adequate to extract significant information from communications.

Time-division scramblers usually divide a speech signal into equal time periods called *frames,* divide each frame into equal time periods called *segments,* and permute the segments within each frame, as diagrammed in Figure 6.24. To implement this scrambler, the speech signal can be sampled and each sample represented digitally. The permuted segments can be used to generate the scrambled analog signal that is transmitted.

Because the samples within a segment are not altered, the residual intelligibility decreases as the segment duration decreases. However, if a segment contains only one sample, there is usually a *bandwidth expansion,* which is the increase in bandwidth required for the transmission of the scrambled signal instead of the original signal. A bandwidth expansion diminishes one of the principal advantages in using analog rather than digital cryptographic communications. If a segment contains many samples, the bandwidth expansion is often negligible.

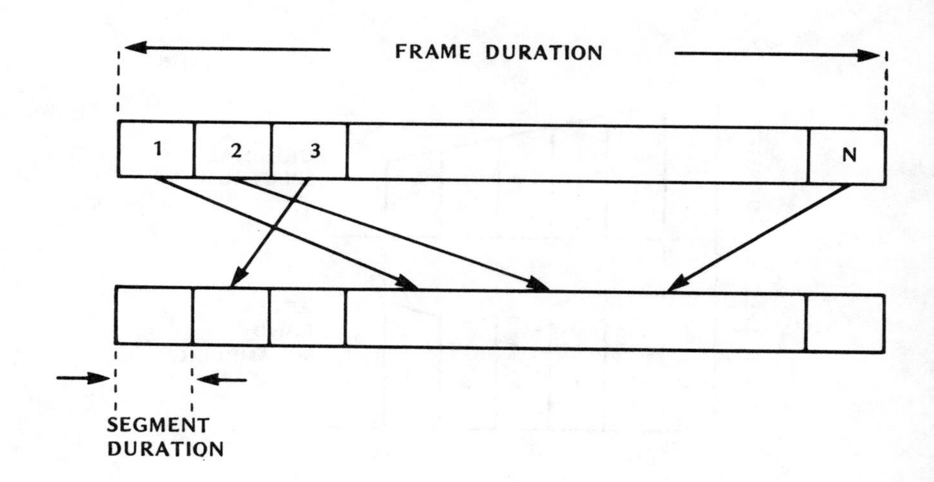

Figure 6.24 Time-division scrambling.

Since the speech signal is a bandpass signal, its highest significant frequency, W, exceeds its bandwidth, B. Because the Nyquist sampling rate is $2W$, the bandwidth of the scrambled bandpass signal is W. To avoid a bandwidth expansion, the speech signal may be translated to baseband so that $W = B$. In principle, a sample permutation at baseband does not cause a bandwidth expansion. However, because the high frequency components may be accentuated, an increased transmission bandwidth may be required to accommodate the scrambled signal without causing excessive distortion. Thus, a bandwidth expansion usually results from a sample permutation.

Security is enhanced by using a long frame with many segments so that many different sounds are mixed together by the scrambling. However, there are practical limitations to the frame duration, T_f. If it is possible to permute the segments so that the final segment of a frame becomes the first segment of the scrambled frame, then the entire frame must enter the scrambler before an output can be produced. Thus, a delay of T_f seconds occurs in the scrambler. Similarly, a delay of T_f seconds occurs in the descrambler. Therefore, $2T_f$ seconds of delay occur in the complete communication system relative to the same system without the scrambling. This system delay can be reduced by restricting the possible permutations to those that do not delay any segment more than some fixed amount relative to its initial position.

Not all permutations are acceptable in a time-division scrambler if a low residual intelligibility is to result. The average displacement of a segment after a permutation must be high. Consecutive segments should not appear in the same order after a permutation, even if one or two other segments are inserted between them.

A key can be used to select a fixed permutation that is applied to every frame. However, security is greatly enhanced by the selection of different permutations for successive frames. The set of permutations, which is some subset of the acceptable ones, can be stored in a ROM that is addressed by the output bits of a stream-cipher system. However, practical limitations on the memory size may not allow a large enough number of different permutations for adequate security. An alternative is to calculate each permutation from the cipher-system output, but this method causes a delay in the scrambled speech transmission.

Time-division scramblers require synchronization between the receiver and the transmitter so that the receiver can recognize the frame boundaries. When the permutations are time varying, synchronization is required to ensure that the sequence of permutations used by the receiver coincides with the sequence used by the transmitter. Synchronization information may be transmitted in a different frequency band or time interval than the speech signal. Alternatively, it may be mixed with the speech signal by removing the portion of the speech signal in a small frequency band and inserting the synchronization information into this band.

A *two-dimensional scrambler* combines two scramblers in series. A carefully designed two-dimensional scrambler produces cryptograms with much less residual intelligibility than those produced by the constituent scramblers. An example is the combination of a time-division scrambler and a frequency inverter. This two-dimensional scrambler requires no more key bits and no more bandwidth than the time-division scrambler alone.

Scrambling can be implemented with little residual intelligibility by combining the speech signal with a *masking signal.* However, *multiplicative masking* leads to bandwidth expansion. *Additive masking* may not produce bandwidth expansion, but it usually diverts some of the available power to the masking signal and requires the production of a synchronized replica of the masking signal at the receiver.

The diversion of power is avoided by the *sample-masking algorithm,* in which the masking signal is a pseudorandom sequence of numbers that are distributed over an interval $(-\alpha, \alpha)$ and are added modulo-2α to the sequence of speech samples. The masking sequence, which has elements denoted by $M_1, M_2, \ldots$, can be produced from the output bits of a stream-cipher system. The speech sample values are denoted by $V_1, V_2, \ldots$. The scrambled sequence, $S_1, S_2, \ldots$, is given by

$$S_i = \begin{cases} V_i + M_i \,, & |V_i + M_i| \leqslant \alpha \,, \\ V_i + M_i + 2\alpha \,, & V_i + M_i < -\alpha \,, \\ V_i + M_i - 2\alpha \,, & V_i + M_i > \alpha \,. \end{cases} \tag{6.9.1}$$

The scrambled sequence is converted into an analog waveform and transmitted. In the receiver, the scrambled sequence is restored. If the V_i are restricted to the interval $(-\alpha,\alpha)$, then the inverse of (6.9.1) is

$$V_i = \begin{cases} S_i - M_i \,, & |S_i - M_i| \leqslant \alpha \\ S_i - M_i + 2\alpha \,, & S_i - M_i < -\alpha \\ S_i - M_i - 2\alpha \,, & S_i - M_i > \alpha \end{cases} \tag{6.9.2}$$

Thus, the speech signal can be recovered when noise is absent if the receiver computes the right-hand side of this equation. Observe that $|S_i - M_i| \neq \alpha$ when noise is absent because $|V_i| < \alpha$.

From elementary probability, it follows that if M_i has a density function, V_i has a continuous density function, and V_i is statistically independent of M_i, then the density function of S_i is

$$f(y) = \begin{cases} \int_{-\alpha}^{\alpha} f_1(x)[f_2(y - x) + f_2(y + 2\alpha - x) + f_2(y - 2\alpha - x)]dx, & -\alpha \leqslant y \leqslant \alpha \\ 0 \,, & \text{otherwise} \end{cases} \tag{6.9.3}$$

where $f_1(\;)$ and $f_2(\;)$ are the density functions of M_i and V_i, respectively. If $f_1(x)$ is uniformly distributed over $(-\alpha,\alpha)$ and V_i is restricted to this interval, then a straightforward calculation using (6.9.3) yields

$$f(y) = \begin{cases} \dfrac{1}{2\alpha} \,, & -\alpha \leqslant y \leqslant \alpha \\ 0 \,, & \text{otherwise} \end{cases} \tag{6.9.4}$$

This result indicates that S_i is uniformly distributed over $(-\alpha,\alpha)$. Thus, the scrambled signal produces a sound without residual intelligibility. However, a bandwidth expansion may result unless some type of filtering is used [7]. This type of additive masking is similar to digital encryption, which essentially masks bits by modulo-two addition. Thus, the security level approaches that of digital encryption at the cost of a considerable implementation complexity, especially if a bandwidth expansion is to be avoided or mitigated.

Scramblers that manipulate the transform of the speech signal can provide very low residual intelligibility. Successive blocks of speech samples are mapped into successive blocks of transform coefficients. Each block of transform coefficients is subjected to a permutation or masking, and then an inverse transform operation produces the scrambled signal for transmission. Because of the efficiency of the fast Fourier transform (FFT) algorithm, the discrete Fourier transform is an attractive choice for the *transform-domain scrambler.* The frequency range and, hence, the bandwidth are unchanged by the transform-domain manipulations, but the bandwidth required for transmission may be increased because high frequency components are strengthened. The essential elements of a scrambler and a descrambler using the FFT are shown in Figure 6.25. The equalizer may be necessary to compensate for the group delay distortion [8].

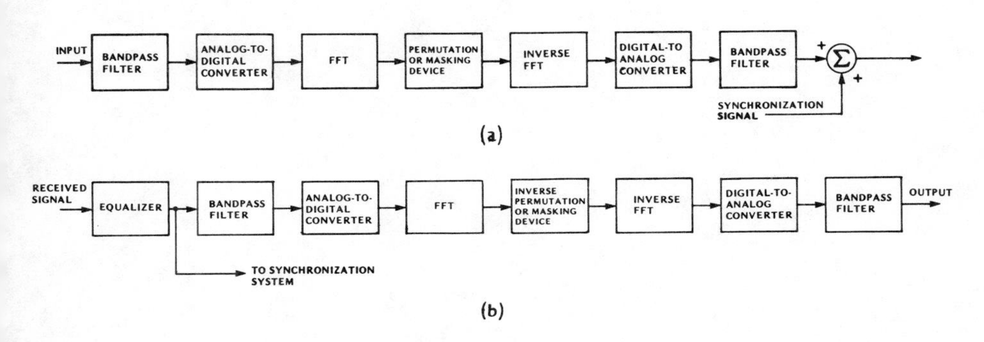

Figure 6.25 Transform-domain system: (a) scrambler and (b) descrambler.

REFERENCES

1. H. Feistel, W.A. Notz, and J.L. Smith, "Some Cryptographic Techniques for Machine-to-Machine Data Communications," *Proc. IEEE* 63, 1545, November 1975.
2. E.L. Key, "An Analysis of the Structure and Complexity of Nonlinear Binary Sequence Generators," *IEEE Trans. Inform. Theory* IT-22, 732, November 1976.
3. H. Beker and F. Piper, *Cipher Systems, The Protection of Communications.* New York: John Wiley and Sons, 1982.
4. D.E. Denning, *Cryptography and Data Security.* Reading, MA: Addison-Wesley, 1982.
5. C.H. Meyer and S.M. Matyas, *Cryptography: A New Dimension in Computer Data Security.* New York: John Wiley and Sons, 1982.
6. W. Diffie and M.E. Hellman, "Privacy and Authentication: An Introduction to Cryptography," *Proc. IEEE* 67, 397, March 1979.

7. R.J. Cosentino and S.J. Meehan, "An Efficient Technique for Sample-Masked Voice Transmission," *IEEE J. Selected Areas Commun.* SAC-2, 426, May 1984.
8. K. Sakurai, K. Koga, and T. Muratani, "A Speech Scrambler Using the Fast Fourier Transform Technique " *IEEE J. Selected Areas Commun.* SAC-2, 434, May 1984.
9. S.C. Kak, "Overview of Analogue Signal Encryption," *IEE Proc.* 130, Pt. F, 399, August 1983.
10. N.S. Jayant, "Analog Scramblers for Speech Privacy," in *Computers and Security 1.* Amsterdam, The Netherlands: North-Holland, 1982.
11. N.S. Jayant, *et al.,* "A Comparison of Four Methods for Analog Speech Privacy," *IEEE Trans. Commun.* COM-29, 18, January 1981.

Appendix A
Signal Representations

A.1 HILBERT TRANSFORM

Consider a real-valued function $g(t)$ defined in the time interval $-\infty < t < \infty$. The *Hilbert transform* of $g(t)$, which is denoted by $\breve{g}(t)$, is defined by

$$\breve{g}(t) = \frac{1}{\pi} \int_{-\infty}^{\infty} \frac{g(u)}{t - u} \, du \tag{A.1.1}$$

Because this integrand has a singularity, we define the integral as its Cauchy principal value. Thus,

$$\int_{-\infty}^{\infty} \frac{g(u)}{t - u} \, du = \lim_{\epsilon \to 0} \left[\int_{-\infty}^{t-\epsilon} \frac{g(u)}{t - u} \, du + \int_{t+\epsilon}^{\infty} \frac{g(u)}{t - u} \, du \right] \tag{A.1.2}$$

provided that the limit exists. Subsequently, integrals are to be interpreted as Cauchy principal values if they contain singularities.

The definition of the Hilbert transform indicates that $\breve{g}(t)$ may be interpreted as the convolution of $g(t)$ with $1/\pi t$. Therefore, $\breve{g}(t)$ results from passing $g(t)$ through a linear filter with an impulse response equal to $1/\pi t$. The transfer function of the filter is given by the Fourier transform

$$F\left\{\frac{1}{\pi t}\right\} = \int_{-\infty}^{\infty} \frac{\exp(-j2\pi f t)}{\pi t} \, dt \tag{A.1.3}$$

where $j = \sqrt{-1}$. This integral can be rigorously evaluated by using contour integration and Jordan's lemma. Alternatively, we observe that since $1/t$ is an odd function,

$$F\left\{\frac{1}{\pi t}\right\} = -j \int_{-\infty}^{\infty} \frac{\sin 2\pi f t}{\pi t} \, dt \tag{A.1.4}$$

which is a standard integral. Thus,

$$F\left\{\frac{1}{\pi t}\right\} = -j \operatorname{sgn}(f) \tag{A.1.5}$$

where sgn(f) is the signum function defined by

$$\text{sgn}(f) = \begin{cases} 1 \ , & f > 0 \\ 0 \ , & f = 0 \\ -1 \ , & f < 0 \end{cases} \tag{A.1.6}$$

Let $G(f)$ denote the Fourier transform of $g(t)$ and let $\breve{G}(f)$ denote the Fourier transform of $\breve{g}(t)$. Equations (A.1.1) and (A.1.5) and the convolution theorem imply that

$$\breve{G}(f) = -j \ \text{sgn}(f) \ G(f) \tag{A.1.7}$$

Let $H[g(t)]$ denote the Hilbert transform of $g(t)$. Because $H[\breve{g}(t)]$ results from passing $g(t)$ through two successive filters, each with transfer function $-j$ sgn(f),

$$H[\breve{g}(t)] = -g(t) \tag{A.1.8}$$

provided that $G(0) = 0$.

Equation (A.1.7) indicates that taking the Hilbert transform corresponds to introducing a phase shift of −90 degrees for all positive frequencies and +90 degrees for all negative frequencies. Consequently,

$$H[\cos 2\pi f_0 t] = \sin 2\pi f_0 t \tag{A.1.9}$$

$$H[\sin 2\pi f_0 t] = -\cos 2\pi f_0 t \tag{A.1.10}$$

These relations can be formally verified by using (A.1.7) and then taking the inverse Fourier transform of the result. If $G(f) = 0$ for $|f| > W$ and $f_0 > W$, the same method yields

$$H[g(t) \cos 2\pi f_0 t] = g(t) \sin 2\pi f_0 t \tag{A.1.11}$$

$$H[g(t) \sin 2\pi f_0 t] = -g(t) \cos 2\pi f_0 t \tag{A.1.12}$$

Other useful properties of the Hilbert transform can be found in the references.

A.2 ANALYTIC SIGNAL AND COMPLEX ENVELOPE

The *analytic signal* associated with the signal $g(t)$ is defined as the complex-valued function

$$g_a(t) = g(t) + j\breve{g}(t) \tag{A.2.1}$$

which is sometimes called the *pre-envelope* of $g(t)$. Let $G_a(f)$ denote the Fourier transform of $g_a(t)$. Using (A.1.6), (A.1.7), and (A.2.1), we obtain

$$G_a(f) = \begin{cases} 2G(f) \ , & f > 0 \\ G(0) \ , & f = 0 \\ 0 \ , & f < 0 \end{cases} \tag{A.2.2}$$

This equation indicates that the spectrum of an analytic signal is zero at all negative frequencies.

A *bandpass signal* is one with a Fourier transform that is negligible except for $f_c - W/2 \leqslant |f| \leqslant f_c + W/2$, where $0 \leqslant W < 2f_c$ and f_c is the center frequency. If $W \ll f_c$, the bandpass signal is often called a *narrowband signal.*

The *complex envelope* of $g(t)$ is defined by

$$\tilde{g}(t) = g_a(t) \exp[-j2\pi f_c t] \tag{A.2.3}$$

where f_c is the center frequency if $g(t)$ is a bandpass signal. Equations (A.2.1) and (A.2.3) imply that $g(t)$ may be expressed in terms of the complex envelope by using

$$g(t) = \text{Re}[\tilde{g}(t) \exp(j2\pi f_c t)] \tag{A.2.4}$$

The complex envelope can be expressed as

$$\tilde{g}(t) = g_c(t) + jg_s(t) \tag{A.2.5}$$

where $g_c(t)$ and $g_s(t)$ are real-valued functions, which are often called the *in-phase* and *quadrature components* of the signal, respectively. Equations (A.2.4) and (A.2.5) imply that a signal can be represented by

$$g(t) = g_c(t) \cos(2\pi f_c t) - g_s(t) \sin(2\pi f_c t) \tag{A.2.6}$$

In this representation, f_c is not necessarily either the center frequency or the carrier frequency.

If the spectrum of $g_a(t)$ is confined to the band $|f - f_c| \leqslant W/2$, (A.2.3) indicates that the spectrum of $\tilde{g}(t)$ is $G_a(f + f_c)$, which occupies the band $|f| \leqslant W/2$. Thus, the complex envelope $\tilde{g}(t)$ is a lowpass or baseband signal and is sometimes called the *lowpass equivalent signal* of $g(t)$. It then follows from (A.2.5) that $g_c(t)$ and $g_s(t)$ are lowpass signals confined to $|f| \leqslant W/2$.

A.3 STATIONARY STOCHASTIC PROCESSES

The Hilbert transform of a stochastic process $n(t)$ is the process defined by

$$\breve{n}(t) = \frac{1}{\pi} \int_{-\infty}^{\infty} \frac{n(u)}{t - u} \, du \tag{A.3.1}$$

where it is assumed that the Cauchy principal value of the integral exists for almost every sample function of $n(t)$. Suppose that $n(t)$ is a zero-mean, wide-sense stationary process with autocorrelation

$$R_n(\tau) = E[n(t)n(t + \tau)] \tag{A.3.2}$$

where $E[x]$ denotes the expected value of x. Using (A.3.1), interchanging operations, and simplifying, we obtain

$$R_{n\breve{n}}(\tau) = E[n(t)\breve{n}(t+\tau)] = \breve{R}_n(\tau) \tag{A.3.3}$$

Similarly,

$$R_{\breve{n}n}(\tau) = E[\breve{n}(t)n(t+\tau)] = -\breve{R}_n(\tau) \tag{A.3.4}$$

With the help of (A.1.8), a similar derivation yields

$$R_{\breve{n}}(\tau) = E[\breve{n}(t)\breve{n}(t+\tau)] = R_n(\tau) \tag{A.3.5}$$

Equation (A.3.1) indicates that because $n(t)$ is a zero-mean process, so is $\breve{n}(t)$. Equation (A.3.5) then implies that $\breve{n}(t)$ is a wide-sense stationary process.

The analytic signal associated with $n(t)$ is the process defined by

$$n_a(t) = n(t) + j\breve{n}(t) \tag{A.3.6}$$

The autocorrelation of the analytic signal is defined as

$$R_a(\tau) = E[n_a^*(t)n_a(t+\tau)] \tag{A.3.7}$$

where the asterisk denotes the complex conjugate. Using (A.3.2) to (A.3.7), we obtain

$$R_a(\tau) = 2R_n(\tau) + 2j\breve{R}_n(\tau) \tag{A.3.8}$$

Because $R_n(\tau)$ is an even function, a change of variables in the defining integral indicates that $\breve{R}_n(\tau)$ is an odd function and

$$\breve{R}_n(0) = 0 \tag{A.3.9}$$

It then follows from (A.3.3) and (A.3.4) that

$$R_{n\breve{n}}(0) = R_{\breve{n}n}(0) = 0 \tag{A.3.10}$$

which indicates that $n(t)$ and $\breve{n}(t)$ are uncorrelated. Equations (A.3.5), (A3.7), (A.3.8), and (A.3.9) yield

$$E[\breve{n}^2(t)] = R_n(0) = \frac{1}{2} R_a(0) = \frac{1}{2} E[|n_a(t)|^2] \tag{A.3.11}$$

The complex envelope of $n(t)$ is the stochastic process defined by

$$\tilde{n}(t) = n_a(t)\exp(-j2\pi f_c t) \tag{A.3.12}$$

where f_c is an arbitrary frequency usually chosen near the center or carrier frequency of $n(t)$. The complex envelope can be decomposed as

$$\tilde{n}(t) = n_c(t) + jn_s(t) \qquad \text{(A.3.13)}$$

where $n_c(t)$ and $n_s(t)$ are real-valued stochastic processes. Substituting (A.3.6) and (A.3.13) into (A.3.12) we find that

$$n_c(t) = n(t)\cos(2\pi f_c t) + \breve{n}(t)\sin(2\pi f_c t) \qquad \text{(A.3.14)}$$

$$n_s(t) = \breve{n}(t)\cos(2\pi f_c t) - n(t)\sin(2\pi f_c t) \qquad \text{(A.3.15)}$$

Thus, $n_c(t)$ and $n_s(t)$ are zero-mean processes. Combining these equations yields the representation

$$n(t) = n_c(t)\cos(2\pi f_c t) - n_s(t)\sin(2\pi f_c t) \qquad \text{(A.3.16)}$$

The autocorrelations of $n_c(t)$ and $n_s(t)$ are defined by

$$R_c(\tau) = E\,[n_c(t)\,n_c(t+\tau)] \qquad \text{(A.3.17)}$$

and

$$R_s(\tau) = E\,[n_s(t)\,n_s(t+\tau)] \qquad \text{(A.3.18)}$$

Using (A.3.14), (A.3.15), and (A.3.2) to (A.3.5) and trigonometric identities, we obtain

$$R_c(\tau) = R_s(\tau) = R_n(\tau)\cos(2\pi f_c \tau) + \breve{R}_n(\tau)\sin(2\pi f_c \tau) \qquad \text{(A.3.19)}$$

This equation shows explicitly that if $n(t)$ is wide-sense stationary, then $n_c(t)$ and $n_s(t)$ are wide-sense stationary with the same autocorrelation function. The variances of $n(t)$, $n_c(t)$, and $n_s(t)$ are all equal because

$$R_c(0) = R_s(0) = R_n(0) \qquad \text{(A.3.20)}$$

We define the cross correlations

$$R_{cs}(\tau) = E\,[n_c(t)\,n_s(t+\tau)] \qquad \text{(A.3.21)}$$

and

$$R_{sc}(\tau) = E\,[n_s(t)\,n_c(t+\tau)] \qquad \text{(A.3.22)}$$

A derivation similar to that leading to (A.3.19) yields

$$R_{sc}(\tau) = -R_{cs}(\tau) = R_n(\tau)\sin(2\pi f_c \tau) - \breve{R}_n(\tau)\cos(2\pi f_c \pi) \qquad \text{(A.3.23)}$$

Equations (A.3.9) and (A.3.23) give

$$R_{cs}(0) = R_{sc}(0) = 0 \tag{A.3.24}$$

which implies that $n_c(t)$ and $n_s(t)$ are uncorrelated. Equations (A.3.19) and (A.3.23) yield

$$R_n(\tau) = R_c(\tau)\cos(2\pi f_c\tau) - R_{cs}(\tau)\sin(2\pi f_c\tau) \tag{A.3.25}$$

Let $S(f)$, $S_c(f)$, and $S_s(f)$ denote the power spectral densities of $n(t)$, $n_c(t)$, and $n_s(t)$, respectively. We assume that $S_n(f)$ occupies the band $f_c - W/2 \leqslant |f| \leqslant f_c + W/2$ and that $f_c > W/2 \geqslant 0$. Taking the Fourier transform of (A.3.19), using (A.1.7), and simplifying, we obtain

$$S_c(f) = S_s(f) = \begin{cases} S_n(f - f_c) + S_n(f + f_c), & |f| \leqslant W/2 \\ 0\,, & |f| > W/2 \end{cases} \tag{A.3.26}$$

Similarly, the cross-spectral density of $n_c(t)$ and $n_s(t)$ can be derived by taking the Fourier transform of (A.3.23) and using (A.1.7). After simplification, the result is

$$S_{sc}(f) = -S_{cs}(f) = \begin{cases} j[S_n(f + f_c) - S_n(f - f_c)]\,, & |f| \leqslant W/2 \\ 0\,, & |f| > W/2 \end{cases} \tag{A.3.27}$$

If $S_n(f)$ is locally symmetric about f_c, then

$$S_n(f_c + f) = S_n(f_c - f)\,, \quad |f| \leqslant W/2 \tag{A.3.28}$$

Since a power spectral density is an even function, $S_n(f_c - f) = S_n(f - f_c)$. Equation (A.3.28) then yields $S_n(f + f_c) = S_n(f - f_c)$ for $|f| \leqslant W/2$. Therefore, (A.3.27) gives $S_{cs}(f) = 0$, which implies that

$$R_{cs}(\tau) = 0 \tag{A.3.29}$$

for all τ. Thus, $n_c(t)$ and $n_s(t + \tau)$ are uncorrelated for all τ. If $n(t)$ is a zero-mean Gaussian process, (A.3.1) implies that $\breve{n}(t)$ is Gaussian. Equations (A.3.14) and (A.3.15) then imply that $n_c(t)$ and $n_s(t)$ are Gaussian processes. Therefore, it follows from (A.3.29) that $n_c(t)$ and $n_s(t + \tau)$ are statistically independent for all τ.

The autocorrelation of the complex envelope is defined by

$$R_{\tilde{n}}(\tau) = \frac{1}{2} E[\tilde{n}^*(t)\tilde{n}(t + \tau)] \tag{A.3.30}$$

where the 1/2 is inserted so that

$$R_{\tilde{n}}(0) = R_n(0) \tag{A.3.31}$$

which follows from (A.3.11) and (A.3.12). Substituting (A.3.13) into (A.3.30) and using (A.3.19) and (A.3.23), we obtain

$$R_{\tilde{n}}(\tau) = R_c(\tau) + jR_{cs}(\tau) \tag{A.3.32}$$

The power spectral density of $\tilde{n}(t)$, which we denote by $S_{\tilde{n}}(f)$, can be derived from (A.3.32), (A.3.27), and (A.3.26). If $S_n(f)$ occupies the band $f_c - W/2 \leqslant |f| \leqslant f_c + W/2$ and $f_c > W/2 \geqslant 0$, then

$$S_{\tilde{n}}(f) = \begin{cases} 2\,S_n(f + f_c)\,, & |f| \leqslant W/2 \\ 0\,, & |f| > W/2 \end{cases} \tag{A.3.33}$$

Let P_T denote the total power in the band $f_c - B/2 \leqslant |f| \leqslant f_c + B/2$, where $B \leqslant W$. Using the fact that $S_n(-f) = S_n(f)$ and (A.3.33), we find that

$$P_T = \int_{f_c-B/2}^{f_c+B/2} S_n(f)\,df + \int_{-f_c-B/2}^{-f_c+B/2} S_n(f)\,df$$

$$= 2\int_{f_c-B/2}^{f_c+B/2} S_n(f)\,df = 2\int_{-B/2}^{B/2} S_n(f + f_c)\,df$$

$$= \int_{-B/2}^{B/2} S_{\tilde{n}}(f)\,df \tag{A.3.34}$$

Thus, P_T can be conveniently expressed as a single integral over $S_{\tilde{n}}(f)$.

Equations (A.3.32) and (A.3.25) imply that

$$R_n(\tau) = \mathrm{Re}\,[R_{\tilde{n}}(\tau)\exp(j2\pi f_c\tau)] \tag{A.3.35}$$

We expand the right-hand side of this equation by using the fact that $\mathrm{Re}\,[z] = (z + z^*)/2$. Taking the Fourier transform and observing that (A.3.33) implies that $S_{\tilde{n}}(f)$ is a real-valued function, we obtain

$$S_n(f) = \frac{1}{2}\,S_{\tilde{n}}(f - f_c) + \frac{1}{2}\,S_{\tilde{n}}(-f - f_c) \tag{A.3.36}$$

A.4 SAMPLING THEOREMS

Consider the Fourier transform $G(f)$ of an absolutely integrable function $g(t)$. It is assumed that $G(f)$ is absolutely integrable so that the inverse Fourier transform of $G(f)$ exists and yields $g(t)$. The periodic extension of $G(f)$ is defined as

$$\bar{G}(f) = \frac{1}{W}\sum_{i=-\infty}^{\infty} G(f + iW) \tag{A.4.1}$$

where W is the period of $\bar{G}(f)$ and it is assumed that the series converges uniformly. Suppose that $\bar{G}(f)$ has a piecewise continuous derivative so that it can be represented as a uniformly convergent complex Fourier series:

$$\bar{G}(f) = \sum_{k=-\infty}^{\infty} c_k \exp\left(-j2\pi k \frac{f}{W}\right) \tag{A.4.2}$$

where the Fourier coefficient c_k is given by

$$c_k = \frac{1}{W} \int_{-W/2}^{W/2} \bar{G}(f) \exp\left(j2\pi k \frac{f}{W}\right) df \tag{A.4.3}$$

Substituting (A.4.1) into (A.4.3) and interchanging the order of the summation and the integration, which is justified because of the uniform convergence, we obtain

$$c_k = \frac{1}{W} \sum_{i=-\infty}^{\infty} \int_{-W/2}^{W/2} G(f + iW) \exp\left(j2\pi k \frac{f}{W}\right) df \tag{A.4.4}$$

We change variables and observe that $\exp(j2\pi ki) = 1$ to obtain

$$c_k = \frac{1}{W} \sum_{i=-\infty}^{\infty} \int_{-W/2+iW}^{W/2+iW} G(f) \exp\left(j2\pi k \frac{f}{W} - j2\pi ki\right) df$$

$$= \frac{1}{W} \int_{-\infty}^{\infty} G(f) \exp\left(j2\pi k \frac{f}{W}\right) df \tag{A.4.5}$$

Because the last integral is recognized as the inverse Fourier transform,

$$c_k = \frac{1}{W} g\left(\frac{k}{W}\right) \tag{A.4.6}$$

Substituting (A.4.6) into (A.4.2) yields

$$\bar{G}(f) = \frac{1}{W} \sum_{k=-\infty}^{\infty} g\left(\frac{k}{W}\right) \exp\left(-\frac{j2\pi kf}{W}\right) \tag{A.4.7}$$

This equation is one version of the *Poisson sum formula.* The convergence of the series is uniform.

Suppose that the Fourier transform vanishes outside a frequency band:

$$G(f) = 0\,, \quad |f| > W/2 \tag{A.4.8}$$

It follows that

$$g(t) = \int_{-W/2}^{W/2} G(f) \exp(j2\pi ft)\, df \tag{A.4.9}$$

Because $G(f) = \bar{G}(f)$ for $|f| < W/2$, (A.4.9) and (A.4.7) and the interchange of a summation and integration yield

$$g(t) = \sum_{k=-\infty}^{\infty} g\left(\frac{k}{W}\right) \frac{1}{W} \int_{-W/2}^{W/2} \exp\left[j2\pi f\left(t - \frac{k}{W}\right)\right] df \tag{A.4.10}$$

Evaluating this integral and defining

$$\operatorname{sinc} x = \frac{\sin \pi x}{\pi x} \tag{A.4.11}$$

we obtain the *sampling theorem* for deterministic signals:

$$g(t) = \sum_{k=-\infty}^{\infty} g\left(\frac{k}{W}\right) \operatorname{sinc}(Wt - k) \cdot \tag{A.4.12}$$

Consider a wide-sense stationary stochastic process $n(t)$ with autocorrelation $R_n(\tau)$ and power spectral density $S_n(f)$. Because $S_n(f)$ is the Fourier transform of $R_n(\tau)$, if

$$S_n(f) = 0\,, \qquad |f| > W/2 \tag{A.4.13}$$

then it follows from the sampling theorem that

$$R_n(\tau) = \sum_{k=-\infty}^{\infty} R_n\left(\frac{k}{W}\right) \operatorname{sinc}(W\tau - k) \tag{A.4.14}$$

For an arbitrary constant α, the Fourier transform of $R(\tau - \alpha)$ is $S_n(f)\exp(-j2\pi f\alpha)$, which is zero for $|f| > W/2$. Therefore, (A.4.14) can be applied to $R_n'(\tau) = R_n(\tau - \alpha)$, which gives

$$R_n(\tau - \alpha) = \sum_{k=-\infty}^{\infty} R_n\left(\frac{k}{W} - \alpha\right) \operatorname{sinc}(W\tau - k) \tag{A.4.15}$$

We define the stochastic process

$$n_\nu(t) = \sum_{k=-\nu}^{\nu} n\left(\frac{k}{W}\right) \operatorname{sinc}(Wt - k) \tag{A.4.16}$$

Interchanging the orders of expectations and summations, we obtain

$$E[(n(t) - n_\nu(t))^2] = R_n(0) - 2 \sum_{k=-\nu}^{\nu} R_n\left(t - \frac{k}{W}\right) \operatorname{sinc}(Wt - k)$$

$$+ \sum_{i=-\nu}^{\nu} \operatorname{sinc}(Wt - i) \sum_{k=-\nu}^{\nu} R_n\left(\frac{i-k}{W}\right) \operatorname{sinc}(Wt - k) \qquad \text{(A.4.17)}$$

Since $R_n(\tau) = R_n(-\tau)$, the repeated use of (A.4.15) yields

$$\lim_{\nu \to \infty} E[(n(t) - n_\nu(t))^2] = 0 \qquad \text{(A.4.18)}$$

which states that the mean square difference between $n(t)$ and $n_\nu(t)$ approaches zero. Thus, the sampling theorem for stationary stochastic processes is

$$n(t) = \sum_{k=-\infty}^{\infty} n\left(\frac{k}{W}\right) \operatorname{sinc}(Wt - k) \qquad \text{(A.4.19)}$$

where the equality holds in the sense of (A.4.18).

REFERENCES

1. S. Haykin, *Communication Systems,* 2nd ed. New York: John Wiley and Sons, 1983.
2. A. Whalen, *Detection of Signals in Noise.* New York: Academic Press, 1971.

Appendix B
Cramer-Rao Inequality and its Applications

B.1 SCHWARZ INEQUALITY

Consider functions $g_i(t)$, $i = 1, 2, \ldots$, of a real variable t such that $a \leqslant t \leqslant b$ and

$$\int_a^b |g_i(t)|^2 dt < \infty\ , \quad i = 1, 2, \ldots \tag{B.1.1}$$

Let the inner product be defined by

$$(g_i, g_j) = \int_a^b g_i(t) g_j^*(t) dt \tag{B.1.2}$$

where the asterisk denotes the complex conjugate, and let $\|g_i\|^2 = (g_i, g_i)$. For any complex variable λ, and any two functions $g_1(t)$ and $g_2(t)$, an expansion and rearrangement yield

$$\|g_1 - \lambda g_2\|^2 = \|g_1\|^2 - \lambda (g_1, g_2)^* - \lambda^* (g_1, g_2) + |\lambda|^2 \|g_2\|^2 \tag{B.1.3}$$

If $\|g_2\| \neq 0$ and

$$\lambda = \frac{(g_1, g_2)}{\|g_2\|^2} \tag{B.1.4}$$

in (B.1.3), then

$$\|g_1 - \lambda g_2\|^2 = \|g_1\|^2 - \frac{|(g_1, g_2)|^2}{\|g_2\|^2} \tag{B.1.5}$$

Because $\|g_1 - \lambda g_2\|$ is nonnegative, we obtain the *Schwarz inequality:*

$$|(g_1, g_2)|^2 \leqslant \|g_1\|^2 \|g_2\|^2 \tag{B.1.6}$$

If $\|g_2\| = 0$, then $(g_1, g_2) = 0$ and the Schwarz inequality remains valid. In terms of the integrals, the Schwarz inequality is

$$\left| \int_a^b g_1(t) g_2^*(t) dt \right|^2 \leqslant \int_a^b |g_1(t)|^2 dt \int_a^b |g_2(t)|^2 dt \tag{B.1.7}$$

Equality holds if $g_1(t)$ or $g_2(t)$ is identically zero, or if

$$g_1(t) = \lambda g_2(t) \tag{B.1.8}$$

for some constant λ.

B.2 CRAMER-RAO INEQUALITY

In this section, we derive the *Cramer-Rao inequality* for the estimate of a function $g(\alpha)$ of a real nonrandom variable α. Let $\mathbf{r}$ denote a vector of observed quantites and $p(\mathbf{r}|\alpha)$ denote the conditional probability density function of $\mathbf{r}$ given the value of α. We assume the regularity conditions that

$$\frac{\partial p(\mathbf{r}|\alpha)}{\partial \alpha} \quad \text{and} \quad \frac{\partial^2 p(\mathbf{r}|\alpha)}{\partial \alpha^2}$$

exist and are absolutely integrable. These regularity conditions allow the interchange of the order of integration and differentiation in various steps in the derivation of the Cramer-Rao inequality.

Let $\hat{g}(\mathbf{r})$ denote any estimator of $g(\alpha)$ based upon observing $\mathbf{r}$. The expected value of $\hat{g}(\mathbf{r})$ is

$$\beta(\alpha) = E[\hat{g}(\mathbf{r})] = \int \hat{g}(\mathbf{r}) p(\mathbf{r}|\alpha) d\mathbf{r} \tag{B.2.1}$$

where the region of integration includes those values of $\mathbf{r}$ for which $p(\mathbf{r}|\alpha) > 0$. Because $p(\mathbf{r}|\alpha)$ is a density,

$$\int p(\mathbf{r}|\alpha) d\mathbf{r} = 1 \tag{B.2.2}$$

From (B.2.1) and (B.2.2), it follows that

$$\int [\hat{g}(\mathbf{r}) - \beta(\alpha)] \, p(\mathbf{r}|\alpha) d\mathbf{r} = 0 \tag{B.2.3}$$

Differentiating with respect to α and using a regularity condition and (B.2.2), we obtain

$$\int [\hat{g}(\mathbf{r}) - \beta(\alpha)] \, \frac{\partial p(\mathbf{r}|\alpha)}{\partial \alpha} \, d\mathbf{r} = \frac{\partial \beta(\alpha)}{\partial \alpha} \tag{B.2.4}$$

Since

$$\frac{\partial p(\mathbf{r}|\alpha)}{\partial \alpha} = \frac{\partial \ln p(\mathbf{r}|\alpha)}{\partial \alpha} \, p(\mathbf{r}|\alpha) \tag{B.2.5}$$

when $p(\mathbf{r}|\alpha) > 0$, (B.2.4) becomes

$$\int [\hat{g}(\mathbf{r}) - \beta(\alpha)] \, \frac{\partial \ln p(\mathbf{r}|\alpha)}{\partial \alpha} \, p(\mathbf{r}|\alpha)\, d\mathbf{r} = \frac{\partial \beta(\alpha)}{\partial \alpha} \tag{B.2.6}$$

Applying the Schwarz inequality, we find that

$$\int [\hat{g}(\mathbf{r}) - \beta(\alpha)]^2 p(\mathbf{r}|\alpha)\, d\mathbf{r} \int \left[\frac{\partial \ln p(\mathbf{r}|\alpha)}{\partial \alpha}\right]^2 p(\mathbf{r}|\alpha)\, d\mathbf{r} \geqslant \left[\frac{\partial \beta(\alpha)}{\partial \alpha}\right]^2 \tag{B.2.7}$$

The first integral on the left-hand side is equal to the variance of $\hat{g}(\mathbf{r})$, which we denote by $\mathrm{VAR}(\hat{g})$. Thus,

$$\mathrm{VAR}(\hat{g}) \geqslant \left[\frac{\partial \beta(\alpha)}{\partial \alpha}\right]^2 \left\{ E\left[\frac{\partial \ln p(\mathbf{r}|\alpha)}{\partial \alpha}\right]^2 \right\}^{-1} \tag{B.2.8}$$

which is a general form of the Cramer-Rao inequality. The right-hand side of this inequality is called the *Cramer-Rao bound.* From (B.1.8), it follows that equality holds if and only if

$$\frac{\partial \ln p(\mathbf{r}|\alpha)}{\partial \alpha} = h(\alpha)[\hat{g}(\mathbf{r}) - \beta(\alpha)] \tag{B.2.9}$$

for all $\mathbf{r}$ and α, where $h(\alpha)$ is an arbitrary function of α but does not depend upon $\mathbf{r}$.

Differentiating both sides of (B.2.2) with respect to α, using a regularity condition and (B.2.5), differentiating again with respect to α, and using the other regularity condition and (B.2.5), we obtain

$$\int \frac{\partial^2 \ln p(\mathbf{r}|\alpha)}{\partial \alpha^2} \, p(\mathbf{r}|\alpha)\, d\mathbf{r} + \int \left[\frac{\partial \ln p(\mathbf{r}|\alpha)}{\partial \alpha}\right]^2 p(\mathbf{r}|\alpha)\, d\mathbf{r} = 0 \tag{B.2.10}$$

Therefore,

$$E\left[\frac{\partial^2 \ln p(\mathbf{r}|\alpha)}{\partial \alpha^2}\right] = -E\left[\frac{\partial \ln p(\mathbf{r}|\alpha)}{\partial \alpha}\right]^2 \tag{B.2.11}$$

Substituting into (B.2.8), we obtain another version of the Cramer-Rao inequality:

$$\mathrm{VAR}(\hat{g}) \geqslant - \left[\frac{\partial \beta(\alpha)}{\partial \alpha}\right]^2 \left\{ E\left[\frac{\partial^2 \ln p(\mathbf{r}|\alpha)}{\partial \alpha^2}\right]\right\}^{-1} \tag{B.2.12}$$

A number of special cases are of particular interest.

1. If $g(\alpha) = \alpha$, then $\hat{g}(\mathbf{r}) = \hat{\alpha}(\mathbf{r})$ is an estimator of α.
2. If $\hat{g}(\mathbf{r})$ is an unbiased estimator of $g(\alpha)$, then $\beta(\alpha) = g(\alpha)$.
3. Suppose that $g(\alpha) = \alpha$. If $\hat{\alpha}(\mathbf{r})$ is an unbiased estimator of α or if $\beta(\alpha) = \alpha + c$ for some constant c, then

$$\mathrm{VAR}(\hat{\alpha}) \geqslant \left\{ E\left[\frac{\partial \ln p(\mathbf{r}|\alpha)}{\partial \alpha}\right]^2\right\}^{-1} = -\left\{ E\left[\frac{\partial^2 \ln p(\mathbf{r}|\alpha)}{\partial \alpha^2}\right]\right\}^{-1} \tag{B.2.13}$$

   This form of the Cramer-Rao inequality is the most commonly used one.

4. If the two estimators $\hat{g}(\mathbf{r})$ and $\hat{\alpha}(\mathbf{r})$ depend upon the same observation vector $\mathbf{r}$, then (B.2.12) and (B.2.13) imply that $\hat{g}(\mathbf{r})$, an estimator of $g(\alpha)$, satisfies

$$\mathrm{VAR}(\hat{g}) \geqslant \left[\frac{\partial \beta(\alpha)}{\partial \alpha}\right]^2 \sigma_c^2(\hat{\alpha}) \tag{B.2.14}$$

where $\sigma_c^2(\hat{\alpha})$ is the Cramer-Rao bound for an unbiased estimator of α.

B.3 FISHER INFORMATION MATRIX

When more than one unknown parameter is of concern, the Cramer-Rao inequality can be generalized [1]. Consider a vector $\boldsymbol{\alpha}$ of K non-random parameters. Let $p(\mathbf{r}|\boldsymbol{\alpha})$ denote the conditional probability density function given the vector $\boldsymbol{\alpha}$. Let $\alpha_1, \alpha_2, \ldots, \alpha_K$ denote the components of $\boldsymbol{\alpha}$. Let $\hat{\alpha}_i(\mathbf{r})$ denote an estimator of α_i. The expected value of this estimator is

$$\beta(\alpha_i) = \int \hat{\alpha}_i(\mathbf{r}) p(\mathbf{r}|\boldsymbol{\alpha})\, d\mathbf{r} \tag{B.3.1}$$

Differentiating with respect to α_j and using a regularity condition yields

$$\int \hat{\alpha}_i(\mathbf{r}) \frac{\partial \ln p(\mathbf{r}|\boldsymbol{\alpha})}{\partial \alpha_j} p(\mathbf{r}|\boldsymbol{\alpha})\, d\mathbf{r} = \delta_{ij} \frac{\partial \beta(\alpha_i)}{\partial \alpha_i} \tag{B.3.2}$$

where $\delta_{ij} = 0, i \neq j$, and $\delta_{ii} = 1$. An inequality for the variance of one of the component estimators, $\hat{\alpha}_1(\mathbf{r})$, can be determined by first defining the $(K + 1)$-dimensional vector

$$
\mathbf{x} = \begin{bmatrix} \hat{\alpha}_1(\mathbf{r}) - \beta(\alpha_1) \\ \partial \ln p(\mathbf{r}|\boldsymbol{\alpha})/\partial\alpha_1 \\ \partial \ln p(\mathbf{r}|\boldsymbol{\alpha})/\partial\alpha_2 \\ \cdot \\ \cdot \\ \cdot \\ \partial \ln p(\mathbf{r}|\boldsymbol{\alpha})/\partial\alpha_K \end{bmatrix} \qquad \text{(B.3.3)}
$$

Using (B.3.2), we derive the matrix

$$
E[\mathbf{x}\mathbf{x}^T] = \left[\begin{array}{c|ccc} \mathrm{VAR}(\hat{\alpha}_1) & \dfrac{\partial\beta(\alpha_1)}{\partial\alpha_1} & 0 \; . \; . \; . & 0 \\ \hline \dfrac{\partial\beta(\alpha_1)}{\partial\alpha_1} & & & \\ 0 & & \boldsymbol{\Upsilon} & \\ \cdot & & & \\ \cdot & & & \\ \cdot & & & \\ 0 & & & \end{array}\right] \qquad \text{(B.3.4)}
$$

where the $K \times K$ matrix $\boldsymbol{\Upsilon}$ is called the *Fisher information matrix.* Its elements are

$$
\gamma_{ij} = E\left[\frac{\partial \ln p(\mathbf{r}|\boldsymbol{\alpha})}{\partial\alpha_i} \frac{\partial \ln p(\mathbf{r}|\boldsymbol{\alpha})}{\partial\alpha_j}\right] \qquad \text{(B.3.5)}
$$

A derivation analogous to that leading to (B.2.11) gives

$$
\gamma_{ij} = -E\left[\frac{\partial^2 \ln p(\mathbf{r}|\boldsymbol{\alpha})}{\partial\alpha_i\,\partial\alpha_j}\right] \qquad \text{(B.3.6)}
$$

As shown in Appendix C, $E[\mathbf{x}\mathbf{x}^T]$ is positive semidefinite. Therefore, its determinant is nonnegative. An expansion of the determinant gives

$$
\mathrm{VAR}(\hat{\alpha}_1)\,\det(\boldsymbol{\Upsilon}) - \left[\frac{\partial\beta(\alpha_1)}{\partial\alpha_1}\right]^2 \mathrm{cof}(\gamma_{11}) \geqslant 0 \qquad \text{(B.3.7)}
$$

where $\det(\boldsymbol{\Upsilon})$ denotes the determinant of $\boldsymbol{\Upsilon}$ and $\text{cof}(\gamma_{11})$ denotes the cofactor of element γ_{11}. If $\boldsymbol{\Upsilon}$ is nonsingular, then

$$\text{VAR}(\hat{\alpha}_1) \geqslant \frac{\text{cof}(\gamma_{11})}{\det(\boldsymbol{\Upsilon})} \left[\frac{\partial \beta(\alpha_1)}{\partial \alpha_1}\right]^2 \tag{B.3.8}$$

We define

$$\boldsymbol{\psi} = \boldsymbol{\Upsilon}^{-1} \tag{B.3.9}$$

Equation (B.3.8) implies that

$$\text{VAR}(\hat{\alpha}_1) \geqslant \psi_{11} \left[\frac{\partial \beta(\alpha_1)}{\partial \alpha_1}\right]^2 \tag{B.3.10}$$

A similar derivation yields

$$\text{VAR}(\hat{\alpha}_i) \geqslant \psi_{ii} \left[\frac{\partial \beta(\alpha_i)}{\partial \alpha_i}\right]^2, \quad i = 1, 2, \ldots, K \tag{B.3.11}$$

For the determinant of $E[\mathbf{x}\mathbf{x}^T]$ to equal zero, it is necessary and sufficient that the random variable $\hat{\alpha}_1(\mathbf{r}) - \beta(\alpha_1)$ be a linear combination of the other components of $\mathbf{x}$. Therefore, equality holds in (B.3.11) if and only if

$$\hat{\alpha}_i(\mathbf{r}) - \beta(\alpha_i) = \sum_{j=1}^{K} h_{ij}(\mathbf{a}) \frac{\partial \ln p(\mathbf{r}|\mathbf{a})}{\partial \mathbf{a}_j}, \quad i = 1, 2, \ldots, K \tag{B.3.12}$$

where $h_{ij}(\mathbf{a})$ is an arbitrary function of $\mathbf{a}$ but is independent of $\mathbf{r}$.

Inequality (B.3.11) bounds the variances for the simultaneous estimation of multiple non-random parameters. It also bounds the variances of ν estimators, $1 \leqslant \nu < K$, when $K - \nu$ other non-random parameters potentially degrade the estimation of the ν parameters of interest.

In addition to the vector $\mathbf{a}$ to be estimated, $\mathbf{r}$ sometimes depends upon other parameters that can be modeled as random variables. Thus, even without the noise, there would be uncertainty in the $\hat{\alpha}_i(\mathbf{r})$. Let $\mathbf{n}$ denote the vector of these random variables, $p_2(\mathbf{n}|\mathbf{a})$ denote the conditional density of $\mathbf{n}$ given the value of $\mathbf{a}$, and $p_1(\mathbf{r}|\mathbf{a},\mathbf{n})$ denote the conditional density of $\mathbf{r}$ given both $\mathbf{a}$ and $\mathbf{n}$. To evaluate the Cramer-Rao bound or the Fisher information matrix, we first determine $p(\mathbf{r}|\mathbf{a})$ by using

$$p(\mathbf{r}|\mathbf{a}) = \int p_1(\mathbf{r}|\mathbf{a},\mathbf{n}) p_2(\mathbf{n}|\mathbf{a}) d\mathbf{n} \tag{B.3.13}$$

where the region of integration is over the possible values of $\mathbf{n}$.

B.4 WHITE GAUSSIAN NOISE

Suppose that an observed signal has the form

$$r(t) = s(t,\boldsymbol{\alpha}) + n(t)\,, \qquad 0 \leqslant t \leqslant T \tag{B.4.1}$$

where $s(t,\boldsymbol{\alpha})$ is a signal of known form, $n(t)$ is white Gaussian noise, T is the duration of the observation interval, and $\boldsymbol{\alpha}$ represents the unknown parameters. It can be shown [1, 2] that $p(\mathbf{r}|\boldsymbol{\alpha})$, which is called the *likelihood function* of $r(t)$, can be represented by

$$p(\mathbf{r}|\boldsymbol{\alpha}) = C \exp\left\{ -\frac{1}{N_0}\int_0^T [r(t) - s(t,\boldsymbol{\alpha})]^2 dt \right\} \tag{B.4.2}$$

where C is a constant and $N_0/2$ is the two-sided power spectral density of $n(t)$. We assume that the unknown parameters are nonrandom. If some of them were random, then in principle we could eliminate them by integration, as in (B.3.13).

Direct calculation and a regularity condition yield

$$\frac{\partial \ln p(\mathbf{r}|\boldsymbol{\alpha})}{\partial \alpha_i} = \frac{2}{N_0}\int_0^T [r(t) - s(t,\boldsymbol{\alpha})] \frac{\partial s(t,\boldsymbol{\alpha})}{\partial \alpha_i} dt \tag{B.4.3}$$

By definition, white Gaussian noise has the property that

$$E[n(t)n(u)] = \frac{N_0}{2}\delta(t-u) \tag{B.4.4}$$

where $\delta(t)$ is the Dirac delta function. Substituting (B.4.1), (B.4.3), and (B.4.4) into (B.3.5) and interchanging expectation and integration, we obtain

$$\gamma_{ij} = \frac{2}{N_0}\int_0^T \frac{\partial s(t,\boldsymbol{\alpha})}{\partial \alpha_i} \frac{\partial s(t,\boldsymbol{\alpha})}{\partial \alpha_j} dt \tag{B.4.5}$$

For the estimator of a single unknown parameter α, (B.3.9), (B.3.10), and (B.4.5) imply that

$$\mathrm{VAR}(\hat{\alpha}) \geqslant \left[\frac{\partial \beta(\alpha)}{\partial \alpha}\right]^2 \left\{ \frac{2}{N_0}\int_0^T \left[\frac{\partial s(t,\alpha)}{\partial \alpha}\right]^2 dt \right\}^{-1} \tag{B.4.6}$$

B.5 APPLICATIONS

In this section, we apply the preceding theory to the estimation of the frequency of a signal, the phase difference between two signals, and the arrival-time difference between two signals.

B.5.1 Frequency Estimation

For the estimation of the frequency, we assume that

$$s(t,\mathbf{a}) = s(t,f,\theta) = A\cos(2\pi ft + \theta), \quad 0 \leqslant t \leqslant T \tag{B.5.1}$$

where the frequency $f = \alpha_1$ and the phase $\theta = \alpha_2$ are assumed to be unknown nonrandom parameters and the amplitude A is assumed to be known. Applying (B.4.5) and a trigonometric identity, we obtain

$$\gamma_{11} = \frac{(2\pi A)^2}{N_0}\int_0^T t^2\,dt - \frac{(2\pi A)^2}{N_0}\int_0^T t^2\cos(4\pi ft + 2\theta)\,dt \tag{B.5.2}$$

If f and T are sufficiently large, then the second term on the right-hand side of this equation is negligible compared to the first term; thus,

$$\gamma_{11} = \frac{8\pi^2 T^2 E}{3N_0} \tag{B.5.3}$$

where $E = A^2 T/2$ is the signal energy. Similar calculations produce the other elements of the 2×2 matrix $\boldsymbol{\Upsilon}$. Equations (B.3.9) and (B.3.10) then give the Cramer-Rao inequality for an unbiased frequency estimator:

$$\mathrm{VAR}(\hat{f}) \geqslant \left(\frac{2\pi^2 T^2 E}{3N_0}\right)^{-1} \tag{B.5.4}$$

If A is assumed to be an unknown nonrandom parameter, a similar calculation of the 3×3 matrix $\boldsymbol{\Upsilon}$ leads to the same result provided that f and T are sufficiently large that γ_{13} and γ_{23} are negligible compared to the other matrix elements. At the other extreme, if both A and θ are assumed to be known, then

$$\mathrm{VAR}(\hat{f}) \geqslant \left(\frac{8\pi^2 T^2 E}{3N_0}\right)^{-1} \tag{B.5.5}$$

Thus, the Cramer-Rao bound for known θ is one-fourth of the bound for unknown θ.

B.5.2 Estimation of Phase Difference

Suppose that two signals, $r_1(t)$ and $r_2(t)$, are observed and have the form

$$r_i(t) = s_i(t,\mathbf{a}) + n_i(t), \quad 0 \leqslant t \leqslant T, \; i = 1, 2 \tag{B.5.6}$$

where the $s_i(t,\mathbf{a})$ are signals of known form, and the $n_i(t)$ are white Gaussian noises that are statistically independent of each other. The likelihood function is

$$p(\mathbf{r}|\boldsymbol{\alpha}) = C \exp\left\{ -\sum_{i=1}^{2} \frac{1}{N_{0i}} \int_0^T [r_i(t) - s_i(t,\boldsymbol{\alpha})]^2 dt \right\} \tag{B.5.7}$$

where C is a constant and $N_{0i}/2$ is the two-sided power spectral density of $n_i(t)$. Because of the statistical independence,

$$E[n_1(t)n_2(u)] = 0 \tag{B.5.8}$$

Using a regularity condition and (B.3.5), (B.4.4), and (B.5.6) to (B.5.8), we obtain

$$\gamma_{ij} = \sum_{k=1}^{2} \frac{2}{N_{0i}} \int_0^T \frac{\partial s_k(t,\boldsymbol{\alpha})}{\partial \alpha_i} \frac{\partial s_k(t,\boldsymbol{\alpha})}{\partial \alpha_j} dt \tag{B.5.9}$$

We determine the Cramer-Rao bound for the phase difference between the two signals of the form

$$s_1(t,\boldsymbol{\alpha}) = A_1 \cos[2\pi f t + \theta_0 + \theta], \quad 0 \leqslant t \leqslant T \tag{B.5.10}$$

$$s_2(t,\boldsymbol{\alpha}) = A_2 \cos[2\pi f t + \theta_0], \quad 0 \leqslant t \leqslant T \tag{B.5.11}$$

where $\alpha_1 = \theta$ is the phase difference, $\alpha_2 = \theta_0$ is a common phase term, $\alpha_3 = A_1$, $\alpha_4 = A_2$, and the frequency f is known. The four components of $\boldsymbol{\alpha}$ are assumed to be unknown but nonrandom. Equation (B.5.9) provides the elements of the 4×4 Fisher information matrix. If f and T are sufficiently large, many of the elements are negligibly small. Thus, (B.3.9) and (B.3.10) give the Cramer-Rao bound for an estimator of the phase difference:

$$\mathrm{VAR}(\hat{\theta}) \geqslant \left[\frac{\partial \beta(\theta)}{\partial \theta}\right]^2 \left[\frac{1}{2E_1/N_{01}} + \frac{1}{2E_2/N_{02}}\right] \tag{B.5.12}$$

where $E_1 = A_1^2 T/2$ and $E_2 = A_2^2 T/2$ are the signal energies. If $E_1 = E_2 = E$, $N_{01} = N_{02} = N_0$, and $\beta(\theta) = \theta$, then

$$\mathrm{VAR}(\hat{\theta}) \geqslant \left(\frac{E}{N_0}\right)^{-1} \tag{B.5.13}$$

The same results are obtained if we assume that the signal amplitudes are known functions of time, or if we assume that f is a fifth unknown component of $\boldsymbol{\alpha}$.

B.5.3 Estimation of Relative Arrival Time

Consider two signals of the form

$$s_1(t,\boldsymbol{\alpha}) = Am(t - T_0 - T_r), \quad 0 \leqslant t \leqslant T \tag{B.5.14}$$

$$s_2(t,\boldsymbol{\alpha}) = m(t - T_0), \quad 0 \leqslant t \leqslant T \tag{B.5.15}$$

where T_0 is the arrival time of $s_2(t,\boldsymbol{\alpha})$, T_r is the arrival time of $s_1(t,\boldsymbol{\alpha})$ relative to that of $s_2(t,\boldsymbol{\alpha})$, A is the amplitude of $s_1(t,\boldsymbol{\alpha})$ relative to that of $s_2(t,\boldsymbol{\alpha})$, $\alpha_1 = T_r$, and $\alpha_2 = T_0$. It is assumed that $m(t)$ is a pulse of known form and that the observation interval always includes all of $m(t - T_0)$ and $m(t - T_0 - T_r)$. From (B.5.9), we obtain

$$\gamma_{11} = \frac{2A^2}{N_{01}} \int_0^T \left[\frac{\partial m(t - T_0 - T_r)}{\partial T_r} \right]^2 dt = \frac{2A^2}{N_{01}} \int_0^T \left[\frac{\partial m(u)}{\partial u} \right]^2 \Bigg|_{u=t-T_0-T_r} dt \tag{B.5.16}$$

Because $m(t - T_0 - T_r)$ is contained within the observation interval, this equation can be simplified to

$$\gamma_{11} = \frac{2A^2}{N_{01}} \int_0^T \left[\frac{\partial m(t)}{\partial t} \right]^2 dt \tag{B.5.17}$$

Similar calculations verify that

$$\gamma_{22} = \left(1 + \frac{N_{01}}{A^2 N_{02}} \right) \gamma_{11} \tag{B.5.18}$$

$$\gamma_{12} = \gamma_{21} = \gamma_{11} \tag{B.5.19}$$

The signal energies are

$$E_i = \int_0^T s_i^2(t, \mathbf{a})\, dt\,, \quad i = 1, 2 \tag{B.5.20}$$

We define

$$\beta_r^2 = \frac{\displaystyle\int_0^T \left[\frac{\partial m(t)}{\partial t} \right]^2 dt}{\displaystyle\int_0^T m^2(t)\, dt} \tag{B.5.21}$$

Using (B.5.16) to (B.5.21), (B.3.9), and (B.3.10), we find that the Cramer-Rao inequality for an estimator of the relative arrival time is

$$\mathrm{VAR}(\hat{T}_r) \geqslant \left[\frac{\partial \beta(T_r)}{\partial T_r} \right]^2 \left(\frac{1}{2\beta_r^2} \right) \left(\frac{1}{E_1/N_{01}} + \frac{1}{E_2/N_{02}} \right) \tag{B.5.22}$$

Let $M(f)$ denote the Fourier transform of $m(t)$. It follows that $\partial m(t)/\partial t$ has a Fourier transform equal to $j 2\pi f M(f)$. By the Parseval relations,

$$\int_0^T \left[\frac{\partial m(t)}{\partial t}\right]^2 dt = (2\pi)^2 \int_{-\infty}^{\infty} f^2 |M(f)|^2 df \tag{B.5.23}$$

$$\int_0^T m^2(t)\, dt = \int_{-\infty}^{\infty} |M(f)|^2 df \tag{B.5.24}$$

Therefore, β_r can be expressed in the form

$$\beta_r^2 = \frac{(2\pi)^2 \int_0^T f^2 |M(f)|^2 df}{\int_0^T |M(f)|^2 df} \tag{B.5.25}$$

If $E_1 = E_2 = E, N_{01} = N_{02} = N_0$, and $\beta(T_r) = T_r$, then (B.5.22) simplifies to

$$\mathrm{VAR}(\hat{T}_r) \geqslant \left(\frac{E}{N_0}\, \beta_r^2\right)^{-1} \tag{B.5.26}$$

The Cramer-Rao bound for parameter estimation in the presence of white Gaussian noise is a function of E/N_0. The tightness of the bound usually increases with E/N_0. However, when E/N_0 is small, the Cramer-Rao bound may be very loose. Other bounds that are theoretically tighter are known, but they do not often reduce to simple expressions.

REFERENCES

1. H.L. Van Trees, *Detection, Estimation, and Modulation Theory,* Vol. I. New York: John Wiley and Sons, 1968.
2. A. Whalen, *Detection of Signals in Noise.* New York: Academic Press, 1971.

Appendix C
Matrix Analysis

This appendix summarizes the results from matrix analysis that are needed in Chapter 5.

C.1 ELEMENTARY RESULTS

An $m \times n$ (m by n) *matrix* is a rectangular array of mn scalar elements arranged in m rows and n columns. If $m = n$, the matrix is called a *square matrix.* A *column vector* is a set of n elements arranged in a column; thus, it is an $n \times 1$ matrix. A *row vector* is a $1 \times n$ matrix. A *diagonal matrix* is a square matrix in which only the elements along the main diagonal are nonzero. If these diagonal elements are all equal to unity, the matrix is called the *identity matrix* and is denoted by **I.** If a_{ij} is an element of an $m \times n$ matrix **A** and b_{ij} is an element of an $n \times p$ matrix **B**, then the *product* of **A** and **B** is defined to be the $m \times p$ matrix $\mathbf{C} = \mathbf{AB}$ having elements

$$c_{ij} = \sum_{k=1}^{n} a_{ik} b_{kj}\,, \qquad i = 1, 2, \ldots, m,\ j = 1, 2, \ldots, p \tag{C.1.1}$$

The *transpose* of a matrix **A** is formed by interchanging the rows and columns. It is denoted by $\mathbf{A}^T$. The transpose of a column vector is a row vector; the transpose of a row vector is a column vector. If **A** is an $m \times n$ matrix and **B** is an $n \times p$ matrix, the transpose of **AB** is a $p \times m$ matrix such that

$$(\mathbf{AB})^T = \mathbf{B}^T\mathbf{A}^T \tag{C.1.2}$$

This result follows from the definition of matrix multiplication, We denote the complex conjugate of a quantity by an asterisk. The complex conjugate of $\mathbf{A}^T$ is denoted by $\mathbf{A}^H$; that is,

$$\mathbf{A}^H = (\mathbf{A}^T)^* \tag{C.1.3}$$

A square matrix **A** is called *singular* if the determinant of its elements, which is denoted by det [**A**], is zero. If **A** is a nonsingular matrix, there exists a unique inverse matrix, denoted by $\mathbf{A}^{-1}$, such that

$$\mathbf{A}\mathbf{A}^{-1} = \mathbf{A}^{-1}\mathbf{A} = \mathbf{I} \qquad \text{(C.1.4)}$$

If **A** and **B** are nonsingular square matrices, a direct application of (C.1.4) proves that the product **AB** has an inverse that can be expressed as

$$(\mathbf{AB})^{-1} = \mathbf{B}^{-1}\mathbf{A}^{-1} \qquad \text{(C.1.5)}$$

For a nonsingular matrix **A**, it follows from (C.1.2) that

$$\mathbf{I} = \mathbf{I}^T = (\mathbf{A}\mathbf{A}^{-1})^T = (\mathbf{A}^{-1})^T\mathbf{A}^T \qquad \text{(C.1.6)}$$

Similarly,

$$\mathbf{I} = \mathbf{A}^T(\mathbf{A}^{-1})^T \qquad \text{(C.1.7)}$$

Equations (C.1.6) and (C.1.7) and the definition of an inverse yield

$$(\mathbf{A}^{-1})^T = (\mathbf{A}^T)^{-1} \qquad \text{(C.1.8)}$$

The *Euclidian norm* of a column vector **x** is a non-negative scalar defined as

$$\|\mathbf{x}\| = (\mathbf{x}^H\mathbf{x})^{1/2} \qquad \text{(C.1.9)}$$

The *Cauchy-Schwarz inequality* states that column vectors **x** and **y** satisfy

$$|\mathbf{x}^H\mathbf{y}|^2 \leqslant \|\mathbf{x}\|^2\,\|\mathbf{y}\|^2 \qquad \text{(C.1.10)}$$

Equality exists if $\mathbf{x} = \alpha\mathbf{y}$ for some scalar α.

An *eigenvector* of a square matrix **A** is any nontrivial column vector **x** that satisfies

$$\mathbf{A}\mathbf{x} = \lambda\mathbf{x} \qquad \text{(C.1.11)}$$

for some scalar λ. Each value of λ associated with an eigenvector is called an *eigenvalue.* The set of eigenvalues is determined by solving the *characteristic equation,* which is

$$\det[\mathbf{A} - \lambda\mathbf{I}] = 0 \qquad \text{(C.1.12)}$$

If **A** and **I** are $n \times n$ matrices, we expand the determinant of $\mathbf{A} - \lambda\mathbf{I}$ to obtain

$$\det \begin{bmatrix} a_{11}-\lambda & a_{12} & \cdots & a_{1n} \\ a_{21} & a_{22}-\lambda & \cdots & a_{2n} \\ \vdots & \vdots & & \vdots \\ a_{n1} & a_{n2} & \cdots & a_{nn}-\lambda \end{bmatrix}$$

$$= (-1)^n\lambda^n + c_1\lambda^{n-1} + c_2\lambda^{n-2} + \ldots + c_n \quad \text{(C.1.13)}$$

where the c_k are functions of the a_{ij}. The polynomial in (C.1.13) is called the *characteristic polynomial* of the matrix **A**. It follows from the fundamental theorem of algebra that the characteristic polynomial can be written in factored form as

$$(-1)^n\lambda^n + c_1\lambda^{n-1} + \ldots + c_n = (-1)^n \prod_{i=1}^{n} (\lambda - \lambda_i) \quad \text{(C.1.14)}$$

where the λ_i are the roots of the polynomial. Because the roots correspond to the eigenvalues, we conclude that an $n \times n$ matrix has n eigenvalues, $\lambda_1, \lambda_2, \ldots, \lambda_n$, which are not necessarily all distinct.

The expansion of the determinant in (C.1.13) gives

$$c_1 = (-1)^{n-1} \sum_{i=1}^{n} a_{ii} \quad \text{(C.1.15)}$$

Setting $\lambda = 0$ in (C.1.13), we obtain

$$c_n = \det[\mathbf{A}] \quad \text{(C.1.16)}$$

Equating terms with the same power of λ on both sides of (C.1.14), we obtain

$$c_1 = (-1)^{n+1} \sum_{i=1}^{n} \lambda_i \quad \text{(C.1.17)}$$

$$c_n = \prod_{i=1}^{n} \lambda_i \quad \text{(C.1.18)}$$

The *trace* of a square matrix is defined to be the sum of its diagonal elements. Comparing (C.1.15) and (C.1.17), we conclude that the trace of a matrix is equal to the sum of its eigenvalues; that is,

$$\text{trace}[\mathbf{A}] = \sum_{i=1}^{n} \lambda_i \quad \text{(C.1.19)}$$

Equations (C.1.16) and (C.1.18) yield

$$\det[\mathbf{A}] = \prod_{i=1}^{n} \lambda_i \tag{C.1.20}$$

which states that the determinant of a matrix is equal to the product of its eigenvalues.

Matrices may be *partitioned* into smaller submatrices that are treated as single elements when two matrices are multiplied. For example, if we define the partitioned matrices

$$\mathbf{A} = \left[\begin{array}{c|c} \mathbf{A}_{11} & \mathbf{A}_{12} \\ \hline \mathbf{A}_{21} & \mathbf{A}_{22} \end{array}\right] \qquad \mathbf{B} = \left[\begin{array}{c|c} \mathbf{B}_{11} & \mathbf{B}_{12} \\ \hline \mathbf{B}_{21} & \mathbf{B}_{22} \end{array}\right]$$

and if $\mathbf{A}_{ij}\mathbf{B}_{jk}$ is defined for $i, j, k = 1, 2$, then

$$\mathbf{AB} = \left[\begin{array}{c|c} \mathbf{A}_{11}\mathbf{B}_{11} + \mathbf{A}_{12}\mathbf{B}_{21} & \mathbf{A}_{11}\mathbf{B}_{12} + \mathbf{A}_{12}\mathbf{B}_{22} \\ \hline \mathbf{A}_{21}\mathbf{B}_{11} + \mathbf{A}_{22}\mathbf{B}_{21} & \mathbf{A}_{21}\mathbf{B}_{12} + \mathbf{A}_{22}\mathbf{B}_{22} \end{array}\right] \tag{C.1.21}$$

This result follows directly from the definition of matrix multiplication. If we partition the matrix **B** into columns, then the product **AB** yields a matrix **C**, each column of which can be obtained by premultiplying the corresponding column of **B** by the matrix **A**.

A *Hermitian matrix* is a square matrix that satisfies

$$\mathbf{A}^H = \mathbf{A} \tag{C.1.22}$$

Let λ_i and $\mathbf{e}_i$ represent an eigenvalue and its corresponding eigenvector. From (C.1.11), we have

$$\mathbf{A}\mathbf{e}_i = \lambda_i \mathbf{e}_i \tag{C.1.23}$$

Multiplying this equation by the conjugate transpose of an eigenvector $\mathbf{e}_j$ yields

$$\mathbf{e}_j^H \mathbf{A}\mathbf{e}_i = \lambda_i \mathbf{e}_j^H \mathbf{e}_i \tag{C.1.24}$$

If λ_j is the eigenvalue corresponding to $\mathbf{e}_j$, we also have

$$\mathbf{e}_i^H \mathbf{A}\mathbf{e}_j = \lambda_j \mathbf{e}_i^H \mathbf{e}_j \tag{C.1.25}$$

Taking the conjugate transpose of (C.1.25) and using (C.1.2) and (C.1.22) yield

$$\mathbf{e}_j^H \mathbf{A}\mathbf{e}_i = \lambda_j^* \mathbf{e}_j^H \mathbf{e}_i \tag{C.1.26}$$

Subtracting (C.1.26) from (C.1.24) gives

$$(\lambda_i - \lambda_j^*)\mathbf{e}_j^H \mathbf{e}_i = 0 \tag{C.1.27}$$

If $j = i$, this equation becomes

$$(\lambda_i - \lambda_i^*)\mathbf{e}_i^H \mathbf{e}_i = 0 \qquad \text{(C.1.28)}$$

Because $\mathbf{e}_i$ is not the zero vector, this equation can be satisfied only if $\lambda_i = \lambda_i^*$. Thus, we conclude that the eigenvalues of a Hermitian matrix are real. If $j \neq i$ and $\lambda_i \neq \lambda_j$, (C.1.27) implies that $\mathbf{e}_j$ and $\mathbf{e}_i$ are orthogonal; that is, the two vectors satisfy $\mathbf{e}_j^H \mathbf{e}_i = \mathbf{e}_i^H \mathbf{e}_j = 0$. Thus, the eigenvectors corresponding to distinct eigenvalues are orthogonal. In general, it can be shown that an $n \times n$ Hermitian matrix has n eigenvectors such that

$$\mathbf{e}_i^H \mathbf{e}_j = 0, \quad i \neq j, \quad i, j = 1, 2, \ldots, n \qquad \text{(C.1.29)}$$

$$\mathbf{e}_i^H \mathbf{e}_i = 1, \quad i = 1, 2, \ldots, n \qquad \text{(C.1.30)}$$

Eigenvectors satisfying these equations are called *orthonormal eigenvectors.*

A *unitary matrix* is a square matrix that satisfies

$$\mathbf{A}^H = \mathbf{A}^{-1} \qquad \text{(C.1.31)}$$

The *modal matrix* **U** of a matrix **A** is a matrix with columns that are the orthonormal eigenvectors of **A**:

$$\mathbf{U} = [\mathbf{e}_1 \ \mathbf{e}_2 \ \ldots \ \mathbf{e}_n] \qquad \text{(C.1.32)}$$

Straightforward calculations using partitioned matrices and the orthonormality yield

$$\mathbf{U}^H \mathbf{U} = \mathbf{U}\mathbf{U}^H = \mathbf{I} \qquad \text{(C.1.33)}$$

Therefore, the modal matrix of a Hermitian matrix is unitary. Using (C.1.23), (C.1.29), (C.1.30), and (C.1.32), we obtain

$$\mathbf{U}^H \mathbf{A} \mathbf{U} = \boldsymbol{\lambda} \qquad \text{(C.1.34)}$$

where $\boldsymbol{\lambda}$ is the diagonal matrix of eigenvalues:

$$\boldsymbol{\lambda} = \begin{bmatrix} \lambda_1 & 0 & 0 & \ldots & 0 \\ 0 & \lambda_2 & 0 & \ldots & 0 \\ 0 & 0 & \lambda_3 & \ldots & 0 \\ \cdot & \cdot & \cdot & & \cdot \\ \cdot & \cdot & \cdot & & \cdot \\ \cdot & \cdot & \cdot & & \cdot \\ 0 & 0 & 0 & \ldots & \lambda_n \end{bmatrix} \qquad \text{(C.1.35)}$$

The matrix $\mathbf{A}$ is said to be *diagonalized* to $\boldsymbol{\lambda}$ by the operation on the left-hand side of (C.1.34). Thus, because a Hermitian matrix has orthonormal eigenvectors, it can be diagonalized.

A *symmetric matrix* is a square matrix with real elements that satisfies

$$\mathbf{A}^T = \mathbf{A} \tag{C.1.36}$$

A symmetric matrix can be considered a special case of a Hermitian matrix. Thus, the properties of Hermitian matrices apply to symmetric matrices.

The eigenvalues of a symmetric matrix are real. There exist n real orthonormal eigenvectors such that

$$\mathbf{e}_i^T \mathbf{e}_j = 0, \quad i \neq j, \quad i, j = 1, 2, \ldots, n \tag{C.1.37}$$

$$\mathbf{e}_i^T \mathbf{e}_i = 1, \quad i = 1, 2, \ldots, n \tag{C.1.38}$$

An *orthogonal matrix* is a square matrix that satisfies

$$\mathbf{A}^T = \mathbf{A}^{-1} \tag{C.1.39}$$

The modal matrix $\mathbf{U}$ of a symmetric matrix is orthogonal. A symmetric matrix $\mathbf{A}$ can be diagonalized since

$$\mathbf{U}^T\mathbf{A}\mathbf{U} = \boldsymbol{\lambda} \tag{C.1.40}$$

where $\boldsymbol{\lambda}$ is defined by (C.1.35).

C.2 HERMITIAN AND QUADRATIC FORMS

A *Hermitian form* is a function that takes scalar values and has the form

$$H = \mathbf{x}^H\mathbf{A}\mathbf{x} \tag{C.2.1}$$

where $\mathbf{A}$ is an $n \times n$ Hermitian matrix and $\mathbf{x}$ is a column vector with n elements. It can easily be verified that $H = H^*$. Thus, a Hermitian form always assumes real values.

A Hermitian matrix is said to be *positive definite* if its corresponding Hermitian form is positive for every $\mathbf{x} \neq \mathbf{0}$. We define the column vector

$$\mathbf{y} = \mathbf{U}^H\mathbf{x} \tag{C.2.2}$$

where $\mathbf{U}$ is the unitary modal matrix defined in (C.1.32). From (C.1.33), it follows that

$$\mathbf{x} = \mathbf{U}\mathbf{y} \tag{C.2.3}$$

Using (C.1.33) to (C.1.35), (C.2.1), and (C.2.2), we obtain

$$\begin{aligned} H &= \mathbf{x}^H \mathbf{U}\mathbf{U}^{-1}\mathbf{A}\mathbf{U}\mathbf{U}^{-1}\mathbf{x} \\ &= (\mathbf{U}^H\mathbf{x})^H(\mathbf{U}^H\mathbf{A}\mathbf{U})(\mathbf{U}^H\mathbf{x}) \\ &= \mathbf{y}^H \boldsymbol{\lambda} \mathbf{y} \\ &= \sum_{i=1}^{n} \lambda_i |y_i|^2 \end{aligned} \tag{C.2.4}$$

Thus, if all the eigenvalues are positive, then $H > 0$ unless $\mathbf{y} = \mathbf{0}$, which implies $\mathbf{x} = \mathbf{0}$ by (C.2.3). Therefore, $H > 0$ for every $\mathbf{x} \neq \mathbf{0}$ if all the eigenvalues are positive. Conversely, suppose that some eigenvalue, λ_j, is negative or zero. If we set $y_i = 0, i \neq j$, and $y_j = 1$, we obtain $H \leqslant 0$; the corresponding value of $\mathbf{x}$ is $\mathbf{x} = \mathbf{U}\mathbf{y} = \mathbf{e}_j \neq \mathbf{0}$. Thus, $H > 0$ for every $\mathbf{x} \neq \mathbf{0}$ only if the eigenvalues are all positive. We conclude that a Hermitian matrix is positive definite if and only if all its eigenvalues are positive.

Because the eigenvalues are positive, (C.1.20) indicates that the determinant of a positive-definite Hermitian matrix is positive. Therefore, the matrix has an inverse.

Equations (C.2.1) and (C.2.4) imply that

$$\lambda_{min} \sum_{i=1}^{n} |y_i|^2 \leqslant \mathbf{x}^H \mathbf{A} \mathbf{x} \leqslant \lambda_{max} \sum_{i=1}^{n} |y_i|^2 \tag{C.2.5}$$

where λ_{min} and λ_{max} are the smallest and largest eigenvalues of the Hermitian matrix $\mathbf{A}$, and $\mathbf{x}$ is related to $\mathbf{y}$ by (C.2.2) and (C.2.3). Because $\mathbf{U}$ is unitary, (C.1.9) gives

$$\|\mathbf{x}\|^2 = \|\mathbf{y}\|^2 = \sum_{i=1}^{n} |y_i|^2 \tag{C.2.6}$$

Therefore, for a Hermitian matrix $\mathbf{A}$ and an arbitrary nonzero vector $\mathbf{x}$,

$$\lambda_{min} \leqslant \frac{\mathbf{x}^H \mathbf{A} \mathbf{x}}{\|\mathbf{x}\|^2} \leqslant \lambda_{max} \tag{C.2.7}$$

This relation is sometimes useful for bounding eigenvalues.

If $\mathbf{x}$ is a real n-dimensional column vector and $\mathbf{A}$ is a real $n \times n$ matrix, then

$$Q = \mathbf{x}^T \mathbf{A} \mathbf{x} \tag{C.2.8}$$

is called a *quadratic form.* A symmetric matrix is said to be positive definite if its corresponding quadratic form is positive for every $\mathbf{x} \neq \mathbf{0}$. If a symmetric matrix is positive definite, its inverse exists. A symmetric matrix is positive definite if and only if all its eigenvalues are positive.

A Hermitian or symmetric matrix is said to be *positive semidefinite* if its corresponding Hermitian or quadratic form is nonnegative for every $\mathbf{x} \neq \mathbf{0}$. It follows from (C.2.4) that a Hermitian or symmetric matrix is positive semidefinite if and only if all its eigenvalues are nonnegative.

Consider the product $\mathbf{a}^T\mathbf{x} = \mathbf{x}^T\mathbf{a}$, where $\mathbf{x}$ is an n-dimensional column vector of random variables and $\mathbf{a}$ is an n-dimensional column vector of arbitrary constants. We have

$$0 \leqslant E[\,|\mathbf{a}^T\mathbf{x}|^2\,] = E[(\mathbf{a}^T\mathbf{x})^*(\mathbf{x}^T\mathbf{a})] = E[\mathbf{a}^H\mathbf{x}^*\mathbf{x}^T\mathbf{a}] \tag{C.2.9}$$

where $E[\;]$ denotes the expected value. Therefore,

$$\mathbf{a}^H E[\mathbf{x}^*\mathbf{x}^T]\,\mathbf{a} \geqslant 0 \tag{C.2.10}$$

which indicates that $E[\mathbf{x}^*\mathbf{x}^T]$ is positive semidefinite. If

$$E[\,|\mathbf{a}^T\mathbf{x}|^2\,] > 0\,, \quad \mathbf{x} \neq 0 \tag{C.2.11}$$

then $E[\mathbf{x}^*\mathbf{x}^T]$ is positive definite. For $\mathbf{x}$ with real components, $E[\mathbf{x}\mathbf{x}^T]$ is positive definite if (C.2.11) holds; otherwise, it is positive semidefinite.

C.3 THE GRADIENT

Differentiation of a scalar function f with respect to a real vector $\mathbf{x}$ is called the *gradient* of f with respect to $\mathbf{x}$, and is defined to be the column vector

$$\nabla_x f = \begin{bmatrix} \dfrac{\partial f}{\partial x_1} \\ \dfrac{\partial f}{\partial x_2} \\ \cdot \\ \cdot \\ \cdot \\ \dfrac{\partial f}{\partial x_n} \end{bmatrix} \tag{C.3.1}$$

From this definition, it follows that for column vectors $\mathbf{x}$ and $\mathbf{y}$, we have

$$\nabla_x(\mathbf{x}^T\mathbf{y}) = \nabla_x(\mathbf{y}^T\mathbf{x}) = \mathbf{y} \tag{C.3.2}$$

Let **A** represent an $n \times n$ matrix with elements a_{ij}. Then

$$\frac{\partial}{\partial x_i}(\mathbf{x}^T\mathbf{A}\mathbf{x}) = \sum_{j=1}^{n} x_j A_{ji} + \sum_{j=1}^{n} A_{ij}x_j\,,\ i = 1, 2, \ldots, n \tag{C.3.3}$$

which implies that

$$\nabla_x(\mathbf{x}^T\mathbf{A}\mathbf{x}) = \mathbf{A}^T\mathbf{x} + \mathbf{A}\mathbf{x} \tag{C.3.4}$$

When **A** is symmetric, then

$$\nabla_x(\mathbf{x}^T\mathbf{A}\mathbf{x}) = 2\mathbf{A}\mathbf{x} \tag{C.3.5}$$

C.4 LINEAR VECTOR EQUATIONS

Consider a discrete-time equation for the vector $\mathbf{x}(k)$ that can be written in the form

$$\mathbf{x}(k+1) = \mathbf{A}\mathbf{x}(k) + \mathbf{b} \tag{C.4.1}$$

where **A** is a square matrix of constants, **b** is a column vector of constants, and k is the discrete-time variable. This equation describes a *linear time-invariant dynamic system.* A vector $\mathbf{x}_0$ is an *equilibrium point* of the dynamic system if $\mathbf{x}(k) = \mathbf{x}_0$ implies $\mathbf{x}(k+1) = \mathbf{x}_0$. Therefore, an equilibrium point must satisfy the equation

$$\mathbf{x}_0 = \mathbf{A}\mathbf{x}_0 + \mathbf{b} \tag{C.4.2}$$

If the matrix $(\mathbf{I} - \mathbf{A})$ is nonsingular, (C.4.2) has a unique solution:

$$\mathbf{x}_0 = (\mathbf{I} - \mathbf{A})^{-1}\mathbf{b} \tag{C.4.3}$$

An equilibrium point is said to be *asymptotically stable* if $\mathbf{x}(k)$ converges to $\mathbf{x}_0$ as k increases, regardless of the initial condition for $\mathbf{x}(k)$. A necessary and sufficient condition for an equilibrium point to be asymptotically stable is that the eigenvalues of **A** all have magnitudes less than unity.

Consider the linear continuous-time vector equation

$$\dot{\mathbf{x}}(t) = \mathbf{A}\mathbf{x}(t) + \mathbf{b} \tag{C.4.4}$$

where $\mathbf{x}(t)$ denotes the derivative of $\mathbf{x}(t)$ with respect to time. A vector $\mathbf{x}_0$ is an equilibrium point of the dynamic system if $\mathbf{x}(t_0) = \mathbf{x}_0$ for some t_0 implies that $\mathbf{x}(t) = \mathbf{x}_0$ for $t > t_0$. Therefore, $\dot{\mathbf{x}}(t_0) = \mathbf{0}$, and an equilibrium point must satisfy the equation

$$\mathbf{0} = \mathbf{A}\mathbf{x}_0 + \mathbf{b} \tag{C.4.5}$$

If **A** is nonsingular, there is a unique equilibrium point:

$$\mathbf{x}_0 = \mathbf{A}^{-1}\mathbf{b} \tag{C.4.6}$$

An equilibrium point is said to be asymptotically stable if $\mathbf{x}(t)$ converges to $\mathbf{x}_0$ as t increases, regardless of the initial condition for $\mathbf{x}(t)$. A necessary and sufficient condition for an equilibrium point of (C.4.4) to be asymptotically stable is that the eigenvalues of **A** all have negative real parts.

REFERENCES

1. G. Strang, *Linear Algebra and Its Applications,* 2nd ed. New York: Academic Press, 1980.
2. D.G. Luenberger, *Introduction to Dynamic Systems.* New York: John Wiley and Sons, 1979.

Index